Finite Element Software for Microwave Engineering

WILEY SERIES IN MICROWAVE AND OPTICAL ENGINEERING

KAI CHANG, Editor
Texas A&M University

FIBER-OPTIC COMMUNICATION SYSTEMS • *Govind P. Agrawal*

COHERENT OPTICAL COMMUNICATIONS SYSTEMS • *Silvello Betti, Giancarlo De Marchis and Eugenio Iannone*

HIGH-FREQUENCY ELECTROMAGNETIC TECHNIQUES: RECENT ADVANCES AND APPLICATIONS • *Asoke K. Bhattacharyya*

COMPUTATIONAL METHODS FOR ELECTROMAGNETICS AND MICROWAVES • *Richard C. Booton, Jr.*

MICROWAVE RING CIRCUITS AND ANTENNAS • *Kai Chang*

MICROWAVE SOLID-STATE CIRCUITS AND APPLICATIONS • *Kai Chang*

DIODE LASERS AND PHOTONIC INTEGRATED CIRCUITS • *Larry Coldren and Scott Corzine*

MULTICONDUCTOR TRANSMISSION-LINE STRUCTURES: MODAL ANALYSIS TECHNIQUES • *J. A. Brandão Faria*

PHASED ARRAY-BASED SYSTEMS AND APPLICATIONS • *Nick Fourikis*

FUNDAMENTALS OF MICROWAVE TRANSMISSION LINES • *Jon C. Freeman*

MICROSTRIP CIRCUITS • *Fred Gardiol*

HIGH-SPEED VLSI INTERCONNECTIONS: MODELING, ANALYSIS, AND SIMULATION • *A. K. Goel*

HIGH-FREQUENCY ANALOG INTEGRATED CIRCUIT DESIGN • *Ravender Goyal (ed.)*

FINITE ELEMENT SOFTWARE FOR MICROWAVE ENGINEERING •*Tatsuo Itoh, Giuseppe Pelosi and Peter P. Silvester (eds.)*

OPTICAL COMPUTING: AN INTRODUCTION • *M. A. Karim and A. S. S. Awwal*

MICROWAVE DEVICES, CIRCUITS AND THEIR INTERACTION • *Charles A. Lee and G. Conrad Dalman*

ANTENNAS FOR RADAR AND COMMUNICATIONS: A POLARIMETRIC APPROACH • *Harold Mott*

INTEGRATED ACTIVE ANTENNAS AND SPATIAL POWER COMBINING • *Julio A. Navarro and Kai Chang*

FREQUENCY CONTROL OF SEMICONDUCTOR LASERS • *Motoichi Ohtsu (ed.)*

SOLAR CELLS AND THEIR APPLICATIONS • *Larry D. Partain (ed.)*

ANALYSIS OF MULTICONDUCTOR TRANSMISSION LINES • *Clayton R. Paul*

INTRODUCTION TO ELECTROMAGNETIC COMPATIBILITY • *Clayton R. Paul*

INTRODUCTION TO HIGH-SPEED ELECTRONICS AND OPTOELECTRONICS • *Leonard M. Riaziat*

NEW FRONTIERS IN MEDICAL DEVICE TECHNOLOGY • *Arye Rosen and Harel Rosen (eds.)*

FREQUENCY SELECTIVE SURFACE AND GRID ARRAY • *T. K. Wu (ed.)*

OPTICAL SIGNAL PROCESSING, COMPUTING AND NEURAL NETWORKS • *Francis T. S. Yu and Suganda Jutamulia*

Finite Element Software for Microwave Engineering

Edited by

TATSUO ITOH
University of California, Los Angeles

GIUSEPPE PELOSI
University of Florence

PETER P. SILVESTER
University of British Columbia

A WILEY-INTERSCIENCE PUBLICATION

JOHN WILEY & SONS, INC.

NEW YORK / CHICHESTER / BRISBANE / TORONTO / SINGAPORE

This text is printed on acid-free paper.

Library of Congress Cataloging in Publication Data:

Finite element software for microwave engineering / [edited by] Tatsuo Itoh, Giuseppe Pelosi, Peter P. Silvester.
p. cm. -- (Wiley series in microwave and optical engineering)
"A Wiley-Interscience publication."
Includes index.
ISBN 0-471-12636-5 (cloth : alk. paper)
1. Microwaves--Mathematical models--Data processing. 2. Finite element method--Computer programs. 3. Electromagnetic devices--Mathematical models--Data processing. I. Itoh, Tatsuo. II. Pelosi, Giuseppe. III. Silvester, P. P. (Peter Peet) IV. Series.
TK7876.F54 1996
621.381'3'028553--dc20 96-2003

Printed in the United States of America

10 9 8 7 6 5 4 3 2

Contents

Contributors

M. Auborg, IRCOM–URA CNRS 356, Université de Limoges, 87060 Limoges Cedex, France

Andreas C. Cangellaris, Department of Electrical and Computer Engineering, University of Arizona, Tucson, AZ 85721, USA

Tom Cwik, Jet Propulsion Laboratory, California Institute of Technology, Pasadena, CA 91109-8099, USA

John D'Angelo, Skyblue Systems, 1223 People's Avenue, Troy, NY 12180, USA

J. B. Davies, Department of Electrical Engineering, University College London, London WC1E 7JE, England

Robert L. Eisenhart, 5982 Ellenview Avenue, Woodland Hills, CA 91367, USA

Ronald Ferrari, Department of Engineering, Cambridge University, Cambridge CB2 1PQ, England

Robert D. Ferraro, Jet Propulsion Laboratory, California Institute of Technology, Pasadena, CA 91109-8099, USA

Luis-Emilio Garciá-Castillo, Departamento de Señales, Sistemas y Radiocomunicaciones, E.T.S.I. de Telecomunicacion, Universidad Politécnica de Madrid, Madrid 28040, Spain

Jian Gong, Department of Electrical Engineering and Computer Science, University of Michigan, Ann Arbor, MI 48109-2122, USA

P. Guillon, IRCOM–URA CNRS 356, Université de Limoges, 87060 Limoges Cedex, France

Vahraz Jamnejad, Jet Propulsion Laboratory, California Institute of Technology, Pasadena, CA 91109-8099, USA

Masanori Koshiba, Division of Electronics and Information Engineering, Hokkaido University, Sapporo 060, Japan

Isaac Mayergoyz, Department of Electrical Engineering, University of Maryland, College Park, MD 20742-3285, USA

Daniel Maystre, Laboratoire d'Optique Electromagnétique, URA au CNRS no 843, Marseille Cedex 20 13397, France

R. Mittra, Department of Electrical and Computer Engineering, University of Illinois at Urbana-Champaign, Urbana, IL 61801, USA

Tayfun Ozdemir, Department of Electrical Engineering and Computer Science, University of Michigan, Ann Arbor, MI 48109-2122, USA

Ü. Pekel, Department of Electrical and Computer Engineering, University of Illinois at Urbana-Champaign, Urbana, IL 61801, USA

Andrew F. Peterson, School of Electrical and Computer Engineering, Georgia Institute of Technology, Atlanta, GA 30332-0250, USA

Gordon P. Riblet, Microwave Development Laboratories, 11 Michigan Drive, Natick, MA 01760, USA

Marc Saillard, Laboratoire d'Optique Electromagnétique, URA au CNRS no 843, Marseille Cedex 20 13397, France

Magdalena Salazar-Palma, Departamento de Señales, Sistemas y Radiocomunicaciones, E.T.S.I. de Telecomunicacion, Universidad Politécnica de Madrid, Madrid 28040, Spain

Peter P. Silvester, 1050 St. Louis Street, Victoria, BC V8S 5B6, Canada

Bill Trowbridge, D'Arcy's Field, Ford Lane, Frilford, Oxon, OX5 1JE, England

S. Verdeyme, IRCOM–URA CNRS 356, Université de Limoges, 87060 Limoges Cedex, France

Patrick Vincent, Laboratoire d'Optique Electromagnétique, URA au CNRS no 843, Marseille Cedex 20 13397, France

John L. Volakis, Department of Electrical Engineering and Computer Science, University of Michigan, Ann Arbor, MI 48109-2122, USA

Donald R. Wilton, Department of Electrical Engineering, University of Houston, Houston, TX 77204-4793, USA

Cinzia Zuffada, Jet Propulsion Laboratory, California Institute of Technology, Pasadena, CA 91109-8099, USA

Preface

Computational electromagnetics has become an essential tool in modern microwave technology, and many successful numerical methods have been proposed since its birth. Numerical procedures in general, and finite elements in particular, have attracted a great deal of attention in the microwave engineering scientific and industrial communities, as evidenced both by the great amount of published material and by the increasing number of conferences, workshops, and short courses specifically devoted to this subject. An inspection of bibliographic databases reveals that the preference of researchers and engineers is tending toward finite element methods, most likely because their flexibility allows coping with the complex structures and devices needed in microwave technology.

Finite element methods have enjoyed wide popularity in recent years, and they have come to occupy a major role as design tools for electromagnetic devices throughout the frequency spectrum. They first appeared in microwave technology in 1968, but it took more than a decade of hardware and software improvements for these techniques to migrate from the research laboratory into routine industrial use. In the past ten or twelve years, various good commercial programs have become available, so it is time to take stock of the present and attempt to assess what the future may bring.

This book is directed mainly toward the industrial microwave designer or product development engineer, for whom it brings together several key threads. It describes user experience with the programs that now exist, attempts to predict which methods might appear in the commercial products of tomorrow, and shows how the imaginative user may make an existing program achieve purposes its designers never envisaged. Its opening chapters deal with general issues of program construction and mathematical method, including the currently interesting topic of error estimation. Later chapters address the classical "interior problem," which covers most microwave device design, and the "exterior problem," which deals principally with radiated fields. These contributions describe newer methods for mesh truncation as well as ways to persuade present-day commercial software to deal with open problems. The final chapter lists a small selection of programs and their suppliers, so as to provide the reader a rough guide of where to look next.

Contrary to popular opinion, commercial engineering software is stable and evolves slowly. For example, ASKA and NASTRAN, two venerable structural finite element programs, are now in their third decade; the popular SPICE circuit analysis package is almost as old; and the MAGNET finite element magnetic field solver has reached voting age. Of course it is true that releases, versions, and updates do succeed each other in rapid sequence. But the essential paradigms change only over decades, perhaps because large-scale software creation requires substantial investment, or perhaps because users are reluctant to discard skills once learned. Whatever the cause, there is good reason to believe that most of today's finite element program suppliers will still be making and maintaining finite element software many years hence. Much the same holds for program users: design engineers accustomed to good finite element software support will continue to hone their skills, not arbitrarily replace them. The problem approaches described in this book can therefore be expected to outlast not only current software releases but their successors too.

TATSUO ITOH
GIUSEPPE PELOSI
PETER P. SILVESTER

Los Angeles, California
Florence, Italy
Victoria, British Columbia

CHAPTER ONE

Complete Modes in Uniform Waveguide

J. B. DAVIES

1. INTRODUCTION

This chapter considers the application of finite elements to a uniform waveguide, and, being the first detailed chapter, one objective is to introduce features that are common throughout this text — namely, finite elements applied to microwave engineering. The other objective is to deal with the uniform waveguide as both the fundamental component of microwave interconnection and the basis of many components. Always, concern is for the resulting software package(s). Because pre- and postprocessing and global matrix assembly are common requirements throughout finite element applications, they will not be detailed in this chapter. This chapter will concern itself with (1) the strategy, (2) the formulations, from which standard *element matrices* can be evaluated, and (3) the rather specialized matrix solvers necessary for this chapter. Some readers may find the menu of "program requirements" in Section 3.1 a focal point.

By contrast to most treatments of waveguides, consideration is given here to the *complete* spectrum of modes — of propagating modes and evanescent modes together with the less well analyzed *complex modes.* For the open waveguide a brief reference is made to *leaky modes.*

From the aspect of computer program capability, one may argue that it is sufficient to study three-dimensional structures, whether at a given frequency

Finite Element Software for Microwave Engineering, Edited by Tatsuo Itoh, Giuseppe Pelosi, and Peter P. Silvester.
ISBN 0-471-12636-5

or looking for an impulse response in the time domain. However, before examining any components, it is generally necessary to understand the simpler aspects of any interconnecting waveguides. Only when there is an analysis and understanding of the waveguide's useful modal spectrum (such as the operating bandwidth between the dominant and second mode) can one safely design and

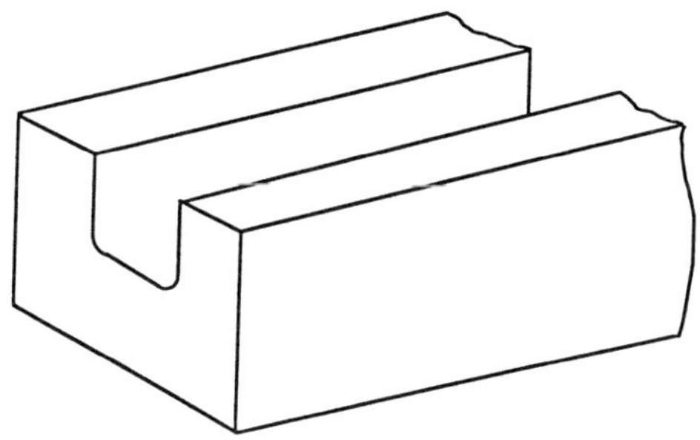

(a)

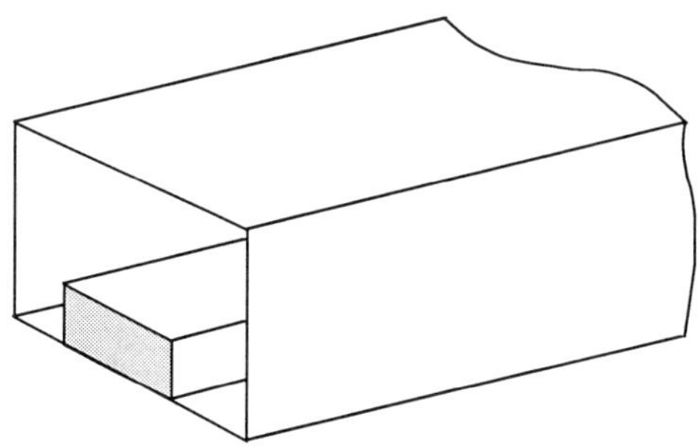

(b)

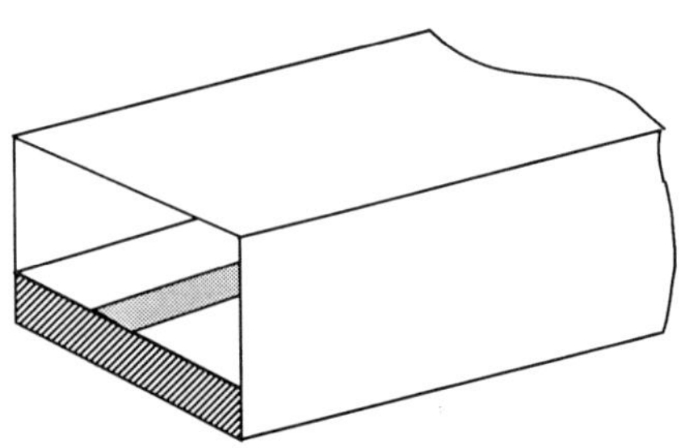

(c)

FIGURE 1. Examples of (a) hollow conducting waveguide; (b) conducting waveguide inhomogeneously loaded with dielectric or gyrotropic material; (c) same as (b) but with additional conductors, such as here with microstrip; (d) open guides such as "image guide"; and (e) waveguide with possible leakage due to limited "opening."

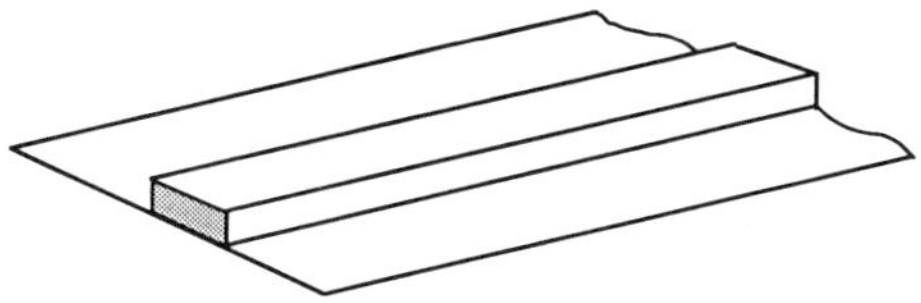

(*d*)

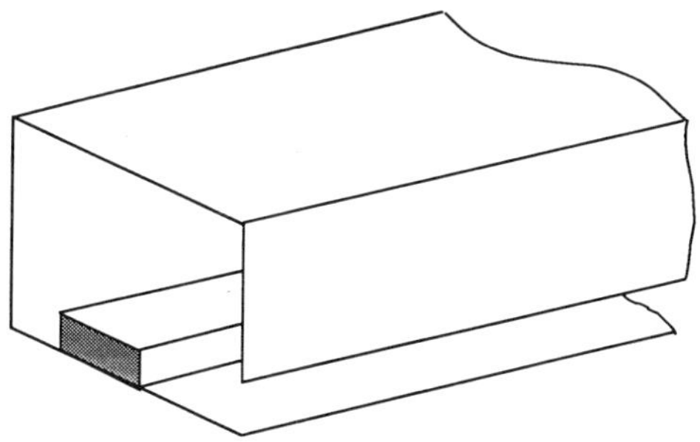

(*e*)

FIGURE 1. (*Continued*)

assemble the components. An understanding of waveguides is also useful and adequate for some component analyses. This arises firstly and directly for certain devices like sections of filters, phase shifters, and modulators, where a knowledge of the phase velocity and characteristic impedance is all that is needed. Some three-dimensional components are most efficiently computed by first doing a modal analysis of waveguide cross sections and then applying a modal expansion method, effectively matching fields across some discontinuity. A final reason for studying two-dimensional analysis before three-dimensional is the obvious rule of walking before running, since most of the methodology of two-dimensional analysis applies equally to three-dimensions and computing resources are less stretched in two-dimensional analysis.

Fully within the spirit of finite elements, consideration is now given to as wide a class of problems as seems possible and convenient, to give computer programs with maximum versatility. The class of problem is "uniform waveguide," meaning any structure that is strictly uniform in a longitudinal direction z. A range of possibilities is illustrated in Figure 1.

The main emphasis is on the waveguide that is totally contained within conducting walls (as in Figure 1a–c) and contains inhomogeneously loaded dielectric or gyrotropic media or conductors, though again all material must be uniform along z. It may be an "open" guide (such as the image guide of Figure 1d), where waves are supported with energy effectively confined to a finite region. Also considered are open waveguides (as in Figure 1e) that "leak"—that is, those having modes with finite transverse radiation away from the main guides.

For maximum versatility, the waveguide considered contains an arbitrary distribution of material defined by $\varepsilon(x, y)$ or $\mu(x, y)$, which may be lossy and may also be tensor, as intended for Figure 1b–e. Indeed, the only serious assumption is that of linearity. In principle, there is no absolute need for *a priori* approximations such as "good conductor," "low loss," or modes being "TEM-like." Subject to the assumption of material linearity, the method is taken to be "exact-in-the-limit," meaning that only computer storage and computer time limit the accuracy that can be obtained (due to errors introduced by finite word length and the finite number of elements).

Reference should be made to two recent reviews by Dillon and Webb (1994) and Davies (1993) of finite elements applied to uniform waveguides, to various chapters in the book by Jin (1993), and to a recent book by Fernandez and Lu (1996).

The chapter continues by clarifying in Section 2 the matter of completeness of modes of a uniform waveguide. Section 3 addresses the paramount matter of the important features of an appropriate computer program before developing the necessary formulation and finite element implementation. Section 4 discusses in some detail matrix algorithms, which for actual software implementation require the importing of two matrix programs (which are available, some in the public domain). Four versions (real and complex; Hermitian and non-Hermitian) are generally useful for eigenvalue/resonance problems. Finally, Section 5 gives the briefest account of the solution of *leaky modes* in open waveguides.

2. MODE CLASSIFICATION

One of the most important requirements of a microwave engineer concerned with single-mode operation of any waveguide is the answer to the question: What are the ω/β dispersion characteristics of the dominant two modes? For this and for many other reasons, it is necessary to be clear about exactly what modes can be present.

In analyzing a waveguide with a uniform cross section, the standard approach (Collin, 1991; Harrington, 1968) is employed that entitles one to look for the "modes" of the waveguide. By assuming excitation at a known frequency ω, this means looking for modes defined by $\mathbf{E}$ and $\mathbf{H}$ fields satisfying:

$$\mathbf{E}(x, y, z, t) = \mathbf{e}(x, y) \exp(j\omega t - \gamma z) \tag{1}$$

$$\mathbf{H}(x, y, z, t) = \mathbf{h}(x, y) \exp(j\omega t - \gamma z) \tag{2}$$

For any given ω and given *enclosed* loss-free waveguide, it is well known (Collin, 1991; Harrington, 1968) that there is a *discrete* spectrum of modes. This means that an ordered list can be made of the modes starting with a finite (possibly zero) number of *propagating modes* with propagation constants β_1, $\beta_2, \ldots, \beta_p$ plus an infinite number of *evanescent modes* with decay constants α_1, $\alpha_2, \ldots$.

Assuming for the time being there is no loss in the waveguide, these α's and β's are related to the complex propagation constants by $\gamma_i = j\beta_i$ if γ_i is pure imaginary and $\gamma_i = \alpha_i$ if γ_i is pure real, as illustrated in Figure 2.

So far, the common practice of most textbooks has been followed, and readers are assumed to be familiar with the practical aspects of these modes. However, there was discovered in the 1960s (Chorney, 1961; Clarricoats and Taylor, 1964) a class of *complex modes.*

Important factors in recognizing these complex modes are at least twofold. (1) In practice, failure to contemplate their existence risks using a waveguide with three modes present when a single propagating mode is naively presumed. There is plenty of evidence that these complex modes are indeed common. (2) In numerical analysis and design of waveguide filters, transitions, and so on, ignoring the complex modes yields results similar to those that result from ignoring any significant modes of the complete spectrum: inevitably errors will appear due to the incompleteness of the modal expansion. This has been demonstrated explicitly for image guides by Strube and Arndt (1985) and for the coplanar waveguide by Kooi et al. (1991).

The predictions of Chorney (1961) and Clarricoats and Taylor (1964) were followed by other works including Kalmyk et al. (1978), which included experimental checks (Schünemann and Omar, 1985; Huang and Itoh, 1988; Railton and Rozzi, 1988). A recent excellent account by Mrozowski and Mazur (1992) includes important details about the existence or otherwise of complex

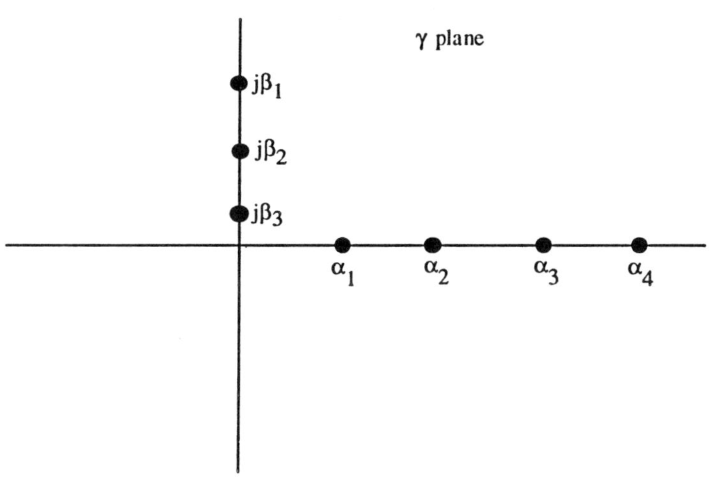

FIGURE 2. Propagation constants γ_i in loss-free waveguide.

modes. The common message is that, in addition to propagating modes and evanescent modes, one should also address the possible existence of additional modes known as *complex modes.* Complex modes can exist in loss-free waveguides and so are not to be confused with propagating or evanescent modes that, if a little loss is introduced, become modes with necessarily complex propagation constants. Introduction of a little dissipation would, for instance, make any propagating mode with complex $\gamma = j\beta$ shift slightly off the imaginary axis to $\gamma = j\beta + \alpha$; one would have a conventional propagating mode with guide wavelength $\lambda_g = 2\pi/\beta$ and an associated small attenuation constant α.

By contrast, when complex modes exist in a loss-free reciprocal guide, they occur in *conjugate pairs,* like γ_1 and $\gamma_2 = \gamma_1^*$ (and possibly $\gamma_3 = \gamma_4^* \ldots$) as shown in Figure 3. Like propagating modes, they may or may not exist (in a given loss-free waveguide) depending on the frequency, but propagating modes (as many as you like) certainly exist if you increase the frequency sufficiently. Again by contrast, there are waveguides where no complex modes have been found at any frequency, while others yield two, four, or more depending on the frequency. Complex modes have only been discovered when the waveguide contains material that is inhomogeneous or anisotropic. Of special importance to microwave engineers is that they can occur in planar guides such as finline (Schünemann and Omar, 1985), microstrip (Huang and Itoh, 1988; Railton and Rozzi, 1988), and coplanar waveguides (Kooi et al., 1991).

Having established that complex modes can exist, some explanation is needed of their behavior and importance. Consider a loss-free reciprocal waveguide with a particular pair of complex modes with constants $\gamma_1 = \alpha + j\beta$ and $\gamma_2 = \alpha - j\beta$. From Eqs. (1) and (2) the $\exp(j\omega t - \gamma z)$ dependence of fields

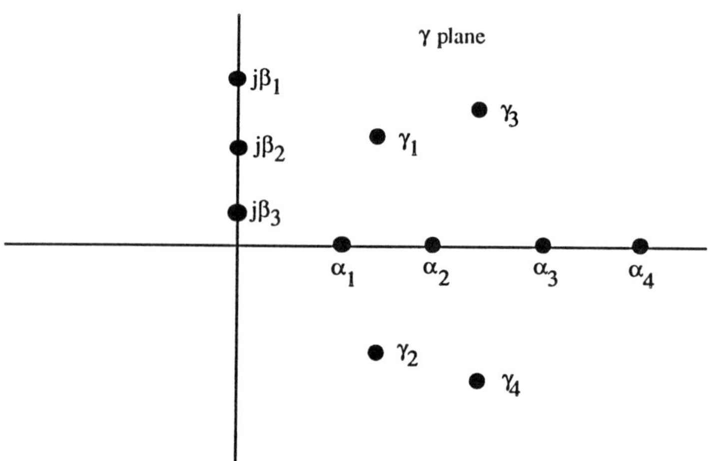

FIGURE 3. Possible range of propagation constant types γ_i in enclosed loss-free waveguide.

dictates that the first mode with γ_1 propagates in the positive z direction and attenuates in the same direction (taking α and β to be positive). The other mode attenuates in the same positive direction (having the same real part α) but propagates in the negative z direction. Expressed differently—in terms of direction of propagation one mode appears to incur an energy loss and the other an energy gain, which is possible providing the modes are inherently coupled with a common amplitude. The reason that the propagation constants of complex modes must exist in complex conjugate pairs is that the modes can only be physically present in the waveguide as inherently coupled pairs that together transmit zero power along the guide. The waveguide modal picture in the complex γ plane, as in Figure 3, is analogous to the poles and zeros of passive components in the complex s plane of frequency and the following diagrams will be familiar to workers in circuit analysis and synthesis. Figure 4 illustrates the field dependence on z for the three types of modes; for the propagating and complex modes (which "travel") they must be interpreted as "snapshots" at a given moment of time.

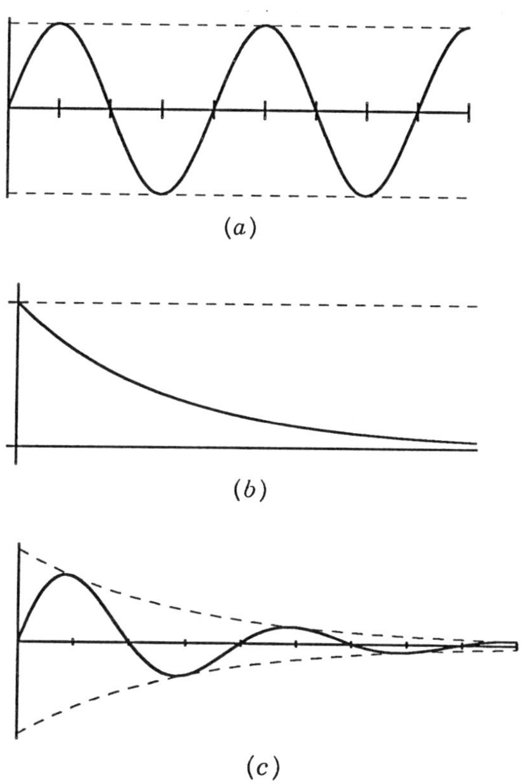

FIGURE 4. Field dependence on z for modes with (a) imaginary, (b) real, or (c) complex propagation constants γ.

Whatever the spectrum of modes for a waveguide, there arises the question: Do complex modes exist in practice? The matter of mode *excitation* (and therefore physical existence) is exactly the same for propagating, evanescent, and complex modes. Any waveguide mode incident on a discontinuity or transition to another waveguide or component generally excites all modes (besides special cases of physical symmetry such as "even" or "odd"). It follows that any propagating, evanescent, or complex modes present in the waveguide's γ-spectrum will generally be excited.

The entire discussion so far has presumed a *reciprocal* waveguide. Complex modes exist equally in nonreciprocal waveguides, but to avoid the detailed ramifications reference is made to Marqués et al. (1992), although their study does not use the finite element method.

3. REQUIREMENTS OF PROGRAM: FORMULATION AND IMPLEMENTATION

Consideration is now given to what formulation to use for a finite element software package to solve for modes in a uniform waveguide. There is no "best"—the choice depends on the user's requirements. Certainly, the more modest the requirements, the faster and the more storage-efficient the program can be; but within any finite element package, there should be scope for the user to trade-off accuracy against computer time and storage. It therefore seems wise to look for one or two programs with the most versatile capabilities, rather than risk a number of programs each with *a priori* approximations, constraints, or even potential errors. Ideally, a package should allow the user to make choices, for example, selecting the permittivity to be isotropic or tensor, lossy or loss-free, the package appropriately and invisibly taking maximum advantage of each "special case."

One very special case concerns the conducting waveguide that is hollow (or homogeneously loaded with isotropic material), like that illustrated in Figure 1a. It is such a simple, but important, case that it will be considered separately in Section 3.2 for the TE, TM, and possibly TEM modes that make up the spectrum.

3.1. Requirements of a Waveguide Solving Program

A carefully prioritized list of requirements in a program will now be presented. With high priorities first, it spirals out from high reliability and high efficiency to other "desirable features." Of course, priorities are subjective, but one might argue for the first five requirements being mandatory. Naturally, the list is of what should be available—not necessarily forced on the user!

For convenience, the 12 features are first listed and then elaborated on.

1. The program does not yield spurious modes.
2. Applied to finite elements, it gives sparse matrices.
3. The resulting matrices are in "canonical form" such as $\mathbf{Av} = \gamma^2\mathbf{Bv}$ or $\mathbf{Av} = \omega^2\mathbf{Bv}$.
4. The resulting matrices (such as $\mathbf{Av} = \gamma^2\mathbf{Bv}$) give γ as a solution for specified frequency ω.
5. It gives a *complete* spectrum, including propagating, evanescent, and complex modes.
6. It gives ultimate economy in using a 2-vector rather than the common 3-vector.

The remaining features 7 to 12 consider the waveguide shape and contents, the possibility of "sharp corners" plus the range of material type of the inhomogeneously loading given by $\varepsilon(x, y)$ or $\mu(x, y)$. The list can be extended almost without limit!

7. The program deals well with sharp corners of (a) conductor or (b) material.
8. It includes the possibility of arbitrary losses.
9. It caters for material that is magnetic *and* dielectric.
10. It caters for anisotropic material; ideally ε and μ can be arbitrary 3 by 3 tensors.
11. For rather obscure needs, one could seek to deal with bi-anisotropic materials (including the chiral family) with **E**, **D**, **H**, and **B** related by four 3 by 3 tensors, each with nine arbitrary complex numbers!
12. The waveguide can be "open" — *not* enclosed within a conducting wall.

Now we elaborate on the rationale of the above features.

1. Spurious modes are results from the program that do not correspond to reality. For about a decade they were often dealt with by the penalty method (Rahman and Davies, 1984) but that was not very satisfactory. Two classes of methods are now used to avoid these spurious modes. The most radical method is the introduction of vector elements, especially edge elements — so called because they use values of fields along the element *edges* rather than at the traditional *nodes*. In addition, there is a continuing range of re-formulations that allow traditional nodal elements to be used. The latter will be considered in this chapter.

2. Some formulations have been introduced to avoid spurious modes (feature 1), some formulations to deal with the more esoteric material tensor forms (features 9, 10, and 11), and some to deal with sharp conductor corners

(feature 7), where in so doing the resulting schemes have destroyed the matrix sparsity of, say, **A** and **B** in $\mathbf{Av} = \gamma^2\mathbf{Bv}$. For serious computation, this sacrifice of computer storage and CPU time must surely be regarded as unacceptable. As computer power increases, our appetite for large-order matrices will increase and the relative inefficiency of dense matrix work will also increase!

3. Wanting matrices in canonical form appears to be the only approach that allows robust as well as efficient matrix algorithms (Parlett, 1980). With rare exceptions, any problem that results in the solution of a general form of the equation $\det(\mathbf{A}(\gamma)) = 0$ is not fundamentally robust in solving for γ. Unlimited difficulty can arise with noncanonical equations, especially due to mode degeneracies and near degeneracies.

4. As ω is normally specified, it follows that solving for γ with given frequency ω avoids the iterative loop otherwise required. Moreover, with the right choice of matrix solving algorithm (see Section 4) a selection of modes can emerge in a single run (i.e., from solution of the same global matrices) for the one given frequency. The arguments for a formulation in terms of γ at a known ω rather than the other way around is even stronger if the decision is taken in favor of the next criterion 5, namely, to solve for all types of modes. A requirement for complex γ at given ω when the formulation is "the wrong way round" would involve iterative searches in two dimensions: first a search in the γ plane to make ω real, then a search to get the wanted ω.

5. A program package that simply does not see any complex modes seems seriously incomplete. As commercial codes are currently thin on the ground for solving the resulting matrices (sparse, complex, and non-Hermitian), it might be that this matter is still commercially premature. (The matrix **A** is Hermitian if $a_{ij} = a_{ji}^*$, where the superscript * denotes complex conjugate.) The matter of matrix software is therefore addressed in Section 4.

6. Everything else being equal, use of a 2-vector rather than a 3-vector field will give a reduction between 3:2 and 9:4 in the relevant storage and CPU time—a worthwhile but not overwhelming consideration.

7. Sharp corners of a conductor or material lead to certain field components being unbounded. Unfortunately, the measure of "dealing well with sharp corners" is imprecise; certainly sharp metal corners are more error-prone than nonsharp in any approach! One effective approach uses a formulation that avoids these particular components (e.g., using a magnetic field formulation when dielectric corners are involved). Edge elements are reported to avoid these difficulties, particularly the difficult case with metal corners.

8. Inclusion of (not necessarily small) loss again seems desirable and should be offered as an option to the user.

9–11. These features give an awkward range of possible ε and μ, ranging from depending on 2 real numbers to 18 complex numbers, ignoring the bi-anisotropy dream with 32 complex numbers! To be realistic, perhaps 90% of the microwave community would be satisfied with isotropic ε and μ, while the inclusion of ferrite capability would keep 99% happy! Again, the aim should be for fairly versatile approaches, with provision of options to the user.

12. "Open" waveguides were illustrated in Figure 1d,e. Happily, most microwave waveguides can be considered as enclosed. Any practical use of image guide or "open planar guide" can be well modeled by the guide being inside a box—albeit a large box. This allows legitimate bypassing of the difficult questions of "leaky modes" and "continuous spectra" (Collin, 1991; Harrington, 1968). A very brief reference will nevertheless be made at the end of this chapter to an approach solving for leaky modes—modes that indeed only become possible in open guides.

3.2. Solution of TE, TM, and TEM Mode Waveguides

As mentioned earlier before Section 3.1, one special case of uniform waveguide is the well-studied situation when the conducting waveguide either is hollow or contains homogeneous isotropic material. Considering first guides with no inner conductors (as, say, in stripline or coaxial cable) the modal situation is described in many microwave texts, including Collin (1991). All modes are either transverse magnetic or transverse electric, and all associated fields can be derived from $e_z(x, y)$ or $h_z(x, y)$, respectively. The dominant mode is always a TE mode and complex modes do not occur. Once the cutoff frequency (giving $\gamma = 0$) and associated $e_z(x, y)$ or $h_z(x, y)$ are known for any mode, then the mode's fields and dispersion characteristics are known for all frequencies. One other simplification is that $e_z(x, y)$ and $h_z(x, y)$ each satisfy the numerically easier problem of the scalar Helmholtz equation:

$$\nabla^2 e_z + k_c^2 e_z = 0 \quad \text{over } S \text{ subject to } e_z = 0 \text{ on } C \tag{3}$$

$$\nabla^2 h_z + k_c^2 h_z = 0 \quad \text{over } S \text{ subject to } \partial h_z/\partial n = 0 \text{ on } C \tag{4}$$

where S denotes the waveguide cross section bounded by the conductor C.

Probably the earliest public domain program using finite elements for microwave applications was for Eqs. (3) and (4), in Silvester (1969), and this can be updated to use sparse matrix solvers, such as considered in Section 4.

The remaining waveguide type of this subsection has one, two, or more inner conductors, like stripline or coaxial cable, when it is well known that there will be, in addition to the TE and TM modes described above, as many TEM modes as inner conductors. All fields at any frequency can be obtained from the single calculation of either the corresponding electrostatic or the magnetostatic problem (again see Collin, 1991). The field problem is now (for the electrostatic version) the simpler Laplace equation:

$$\nabla^2 \varphi = 0 \quad \text{over } S \text{ subject to (for 1 conductor) } \varphi = 0 \text{ on } C_0 \text{ and } \varphi = 1 \text{ on } C_1 \tag{5}$$

This static problem has been studied extensively and documented, and programs are described in regular finite element conferences and proceedings, to which readers can refer.

Formulations and finite element details for the waveguide problems of this subsection can be found in Silvester and Ferrari (1990) and Jin (1993).

3.3. Choice of Formulation

Having set down in Section 3.1 desirable requirements for a waveguide solving program, the mathematical formulation to be used in this chapter will now be considered. It is that of Hernandez-Figueroa et al. (1995), which is a variant of an earlier variational form of Fernandez and Lu (1991) expanded in Lu and Fernandez (1993). The latter allowed for dielectric materials of tensor form:

$$\|\varepsilon\| = \begin{bmatrix} \varepsilon_{xx} & \varepsilon_{xy} & 0 \\ \varepsilon_{yx} & \varepsilon_{yy} & 0 \\ 0 & 0 & \varepsilon_{zz} \end{bmatrix} \tag{6}$$

constrained to $\varepsilon_{xy} = \varepsilon_{xy}$, but this more recent formulation allows for $\varepsilon_{xy} \neq \varepsilon_{xy}$ with the five tensor elements being quite arbitrary complex functions of x and y, so that nonreciprocal and lossy waveguides can be considered. To apply to ferrites, the dual version in terms of **E** rather than **H** would be necessary.

With assumed time dependence $\exp(j\omega t)$, elimination of **E** from the Maxwell equations leads to the following vector form of the Helmholtz equation:

$$\nabla \times (\varepsilon^{-1} \cdot \nabla \times \mathbf{H}) - \omega^2 \mu_0 \mathbf{H} = 0 \tag{7}$$

Now separating the transverse and longitudinal components of **H** as

$$\mathbf{H}(x, y) = (\mathbf{H}_t(x, y) + H_z(x, y)\mathbf{u}_z) \exp(j\omega t - \gamma z) \tag{8}$$

and using the Gauss law $\nabla \cdot \mathbf{H} = 0$ leads to the following expression in terms of $\mathbf{H}_t$ alone:

$$\begin{aligned} \nabla_t \times (\varepsilon_{zz} \nabla_t \times \mathbf{H}_t) - \mathbf{u}_z \times \{\varepsilon_t^{-1} \cdot \nabla_t \times (\mathbf{u}_z \nabla_t \cdot \mathbf{H}_t)\} \\ - \omega^2 \mu_0 \varepsilon_0 \mathbf{H}_t + \gamma^2 \mathbf{u}_z \times \{\varepsilon_t^{-1} \cdot (\mathbf{u}_z \times \mathbf{H}_t)\} = 0 \end{aligned} \tag{9}$$

where the transverse part of the operator ∇ is $\nabla_t = \mathbf{u}_x \partial_x + \mathbf{u}_y \partial_y$ and ε_t denotes the 2×2 part of the full tensor in Eq. (6).

Applying the Galerkin method to Eq. (9) gives the final formulation to be used:

$$\int_S \{\varepsilon_{zz}^{-1}(\nabla_t \times \mathbf{H}_t)\cdot(\nabla_t \times \mathbf{w}_t) + (\nabla_t \cdot \mathbf{H}_t \mathbf{u}_z)\cdot(\nabla_t \times (\varepsilon_t^{-1})^T \cdot (\mathbf{u}_z \times \mathbf{w}_t))$$
$$- \omega^2 \varepsilon_0 \mu_0 \mathbf{H}_t \cdot \mathbf{w}_t - \gamma^2 \varepsilon_t^{-1}(\mathbf{H}_t \times \mathbf{u}_z)\cdot(\mathbf{u}_z \times \mathbf{w}_t)\}\, dS$$
$$+ \int_{\partial\Omega} \varepsilon_{zz}^{-1}(\nabla_t \times \mathbf{H}_t)\cdot(\mathbf{w}_t \times \mathbf{u}_n)\, \mathrm{dl}$$
$$+ \int_{\partial\Omega} (\nabla_t \cdot \mathbf{H}_t)\{\mathbf{u}_z \times (\varepsilon_t^{-1})^T \cdot (\mathbf{u}_z \times \mathbf{w}_t)\}\cdot \mathbf{u}_n\, dl = 0 \tag{10}$$

where $\mathbf{w}_t(x, y)$ refers to the continuous "testing function" of the Galerkin method (see Jin, 1993; Silvester and Ferrari, 1990).

The surface integral of Eq. (10) is over the complete cross section S of the waveguide. The line integrals over $\partial\Omega$ are over any material interfaces, namely, anywhere there is any discontinuity in $\varepsilon(x, y)$, although a variant of this is described in the leaky mode work covered in Section 5.

Checking the formulation against the list of requirements in Section 3.1, Eq. (10) indeed satisfies 1 to 6, for it yields no spurious modes, gives sparse matrices in canonical form $\mathbf{A}\mathbf{v} = \gamma^2 \mathbf{B}\mathbf{v}$ in propagation constant γ for known frequency ω, gives a complete spectrum, and uses a 2-vector field. It also satisfies items 8 and 12 (arbitrary losses and a capability for open waveguide including leaky modes) and to a fair extent item 10 in allowing the material tensor of Eq. (6). It does not consider the rare bi-anisotropy of item 12; it has not been tested for material that is magnetic *and* dielectric or for sharp conducting corners. Dielectric corners do not present a difficulty because of the boundedness of $\mathbf{H}_t$.

A brief mention can be made of alternative forms. Many have been discussed in a review by Davies (1993) and at various point in the texts by Jin (1993) and Fernandez and Yu (1996). Dillon and Webb (1994) address the choice of H-field formulations in some detail and make useful tests and comparisons of six such versions, though with only passing attention to complex modes. Of the six formulations studied, only two satisfy the perhaps mandatory items 1 to 5, and only one satisfies all of items 1 to 6. Their "best" candidate is the "modified transverse formulation" (using their terminology) in terms of the transverse field $\mathbf{H}_t$:

$$\int_\Omega \{(\nabla_t \times \mathbf{w}_t)\varepsilon_r^{-1}(\nabla_t \times \mathbf{H}_t) - (\mu_r k_0^2 + \gamma^2 \varepsilon_r^{-1})\mathbf{w}_t \cdot \mathbf{H}_t + (\nabla_t \cdot \varepsilon_r^{-1}\mathbf{w}_t)(\nabla_t \cdot \mathbf{H}_t)\}\, d\Omega$$
$$+ \int_C \{(\mathbf{n} \cdot \mathbf{w}_t)\cdot \varepsilon_r^{-1}(\nabla_t \cdot \mathbf{H}_t) - (\mathbf{n} \times \mathbf{w}_t)\cdot(\varepsilon_r^{-1}\nabla_t \times \mathbf{H}_t)\}\, dC = 0 \tag{11}$$

Equation (11) has much in common with the chosen Eq. (10) but lacks the scope for the versatile anisotropic tensor of Eq. (6).

The earlier work of Lee et al. (1991) made great progress in avoiding spurious modes by using vector elements for $\mathbf{H}_t$ together with nodal elements for H_z. It has been advanced by Lee (1994) to lossy dielectrics but the approach does not satisfy criterion 5 in apparently ignoring complex modes—or criterion 6 in using both $\mathbf{H}_t$ and H_z. The more recent work of Anderson and Cendes (1995) extends the same approach in tackling the problem of magnetized ferrites, but complex modes and lossy material are not considered. Their more ambitious capability of gyromagnetic material results in separate formulations for the two cases of longitudinal and transverse directions of the magnetic bias fields. The former needs a 3-vector and the latter a transformation leading effectively to a 5-vector at the matrix solving stage.

The approach has been established further by Valor and Zapata (1996) to achieve all items 1 to 11 from our list, except number 6. For this higher versatility, the price paid is a doubling of matrix order in addition to use of a 3-vector.

3.4. Choice of Finite Element Implementation

The first decision in the finite element implementation concerns the use of "nodal elements" or "vector elements" (Jin, 1993). As will be seen, decisions of implementation are linked to the formulation. For the chosen Eq. (10), $\mathbf{H}_t$ needs to be continuous (assuming uniform μ) and so orthodox nodal elements are appropriate. They are also appropriate for Eq. (11). On the other hand, Lee et al. (1991) and the succeeding approaches use edge elements for $\mathbf{H}_t$ with nodal elements for H_z.

Continuing to implement Eq. (10), the waveguide cross section is divided into a triangular mesh, arranging any discontinuities in $\varepsilon(x, y)$ to lie on triangle edges. Over every triangular element $\mathbf{H}_t$ can be expanded in terms of orthodox second-order nodal elements (Jin, 1993) and becomes

$$\mathbf{H}_t = \sum_{n=1}^{6} (\mathbf{u}_x a_n + \mathbf{u}_y b_n) N_n(x, y) \tag{12}$$

where $N_n(x, y)$ denotes the six interpolation polynomials with a_n and b_n as the nodal values of H_x and H_y, respectively. The testing functions $\mathbf{w}_t$ of Eq. (10) can take each component w_x and w_y as each of the same six functions $N_n(x, y)$ over every triangle. (Earlier tests using first-order nodal elements for $\mathbf{H}_t$ and $\mathbf{w}_t$ were less satisfactory.) By the Galerkin method, insisting on Eq. (10) being satisfied (to give zero "error residual") for every testing function w_x and w_y leads Eq. (10) on to the following matrix equation:

$$\mathbf{A}\mathbf{h}_t = \gamma^2 \cdot \mathbf{B}\mathbf{h}_t \tag{13}$$

where $\mathbf{h}_t$ denotes the usual column vector of $2N$ nodal values—h_x and h_y at N relevant nodes. Expressions for the elements of the matrices **A** and **B** are almost exactly those described in Lu and Fernandez (1993) even though the latter is based on their variational form.

Of prime importance is that Eq. (13) is indeed in canonical form with sparse matrices **A** and **B**. For lossy materials **A** and **B** are neither symmetric nor Hermitian, with matrix elements that are complex, as are the resulting eigenvalue and eigenvector solutions to Eq. (13). For loss-free materials **A** and **B** are real; but if, furthermore, the material is either inhomogeneous or anisotropic, **A** and **B** are nonsymmetric and so allow the possibility of complex eigenvalues. Any complex eigenvalues that do occur will be in complex conjugate pairs, exactly as found for complex modes above in Section 2. Any reality of **A** and **B** allows a 2:1 economy in computer storage and perhaps CPU time, and some economy in the evaluation of **A** and **B**, but it does not particularly ease the solution of Eq. (13).

In this context of the essential non-Hermitian form of matrices in Eq. (13) (assuming only that the waveguide loading is either inhomogeneous or anisotropic), one important comment seems appropriate. Mrozowski and Mazur (1992) point out a serious misconception in the literature: that **A** and **B** of Eq. (13) being Hermitian is regarded as a sufficient condition for all eigenvalues to be real. As they point out, sufficiency follows if additionally one knows that at least one of **A** or **B** is nonsingular. Many formulations, which give singular or indefinite matrices, therefore appear capable of yielding complex modes, even though so few finite element workers have reported them.

4. CHOICE OF SPARSE MATRIX SOLVING ALGORITHM

Equation (13), which emerges as necessary for complete modal analysis of inhomogeneously loaded waveguides, is barely in the range of commercial matrix solvers. Even if one settles for the loss-free situation, matrices **A** and **B** are real but nonsymmetric and still not well catered for. Methods are therefore described in this section that for actual software implementation require the importing of two available matrix programs, some in the public domain. The two programs needed for this chapter are to solve for **x**, with complex non-Hermitian **A** and **B**: (1) dense small order $\mathbf{Ax} = \lambda\mathbf{Bx}$ and (2) sparse large order $\mathbf{Ax} = \mathbf{y}$. However, a useful range of four sparse matrix eigenvalue solvers can be assembled by switching both programs between complex and real, and between non-Hermitian and Hermitian. The real/complex choice occurs when losses are absent or present; the non-Hermitian/Hermitian choice decides whether or not the modal analysis is complete with complex modes. Being iterative, the resulting solver can take full benefit of any available "starting values," an advantage when solving a problem over any range of physical parameter(s).

The method chosen to solve Eq. (13) is the non-Hermitian version of "subspace iteration." The Hermitian version (when **A** and **B** are real symmetric or complex Hermitian) is useful for any finite element solution of waveguide or cavity problems and so will be described in a little detail before moving on to the non-Hermitian version. Another popular method for Hermitian systems is based on the Lanczos algorithm (see Parlett, 1980).

4.1. Subspace Iteration for Hermitian Matrices

The very simplest algorithm for finding an eigenvalue and eigenvector of the basic matrix eigenvalue problem

$$\mathbf{A}\mathbf{x} = \lambda\mathbf{x} \tag{14}$$

is the "power method," which starts with some vector $\mathbf{y}_0$ and then successively premultiplies it by the matrix **A**:

$$\mathbf{y}_i = \mathbf{A}\mathbf{y}_{i-1} \quad \text{for } i = 1, 2, 3, \ldots \tag{15}$$

By considering $\mathbf{y}_0$ expanded in terms of the eigenvectors of **A**, it can be seen that each step of the iteration in Eq. (15) increases the component of the "dominant" eigenvector in the current vector $\mathbf{y}_i$ and so $\mathbf{y}_i/\mathbf{y}_{i-1}$ will approach the largest in magnitude eigenvalue of Eq. (14) and $\mathbf{y}_i$ will approach the associated dominant eigenvector.

The method of "inverse iteration" reverses the roles of $\mathbf{y}_i$ and $\mathbf{y}_{i-1}$ to give

$$\mathbf{A}\mathbf{y}_i = \mathbf{y}_{i-1} \quad \text{for } i = 1, 2, 3, \ldots \tag{16}$$

so that $\mathbf{y}_i/\mathbf{y}_{i-1}$ now approaches the eigenvalue with smallest magnitude.

"Shifted inverse iteration" modifies Eq. (16) to

$$(\mathbf{A} - \mu\mathbf{I})\mathbf{y}_i = \mathbf{y}_{i-1} \quad \text{for } i = 1, 2, 3, \ldots \tag{17}$$

to give the eigenvalue nearest the chosen μ.

Finally, applying the shifted inverse iteration scheme to the "generalized eigenvalue matrix equation,"

$$\mathbf{A}\mathbf{x} = \lambda\mathbf{B}\mathbf{x} \tag{18}$$

the iteration

$$(\mathbf{A} - \mu\mathbf{B})\mathbf{y}_i = \mathbf{B}\mathbf{y}_{i-1} \quad \text{for } i = 1, 2, 3, \ldots \tag{19}$$

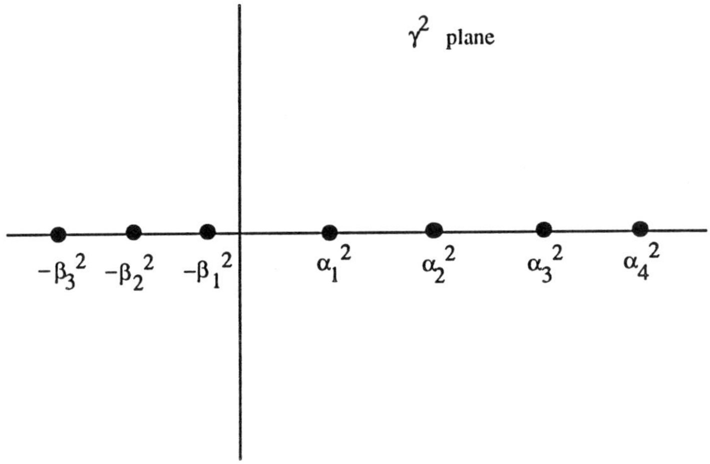

FIGURE 5. Propagation constants of Figure 2 but with γ squared.

gives a "Rayleigh quotient" at each step:

$$\rho_i = \frac{\mathbf{y}_i^*(\mathbf{A} - \mu\mathbf{B})\mathbf{y}_i}{\mathbf{y}_i^*\mathbf{B}\mathbf{y}_i} \tag{20}$$

which converges to the eigenvalue nearest the chosen μ.

Consider now application of this algorithm to the waveguide that gives a spectrum of propagation constants as in Figure 2. These are replotted in the complex γ^2 plane in Figure 5. One could evaluate each of the seven modes illustrated by choosing a starting real number for μ of Eq. (19), some sort of random vector $\mathbf{y}_0$, and applying Eqs. (19) and (20). Evaluating these seven values of γ^2 by the shifted inverse iteration scheme would need seven separate computations, each with its starting value, say, μ_1 to μ_7. Using the property of convergence always being to the eigenvalue nearest to μ would allow some systematic evaluation of the seven values, but one can see possible hazards. Degeneracies (when two or more eigenvalues are coincident) or near degeneracies can create difficulties, though not unconquerable.

Equation (20) above gives the most recent value of eigenvalue at each iteration. If it is used to provide a new shift value μ for the next iteration, it generally replaces linear convergence by quadratic convergence (Parlett, 1980), which can speed up the solution considerably. However, this has to be weighed against possible additional work (such as recalculating the triangular decomposition if Gauss is being used) and the fact that leaving μ unchanged will, if it converges, ensure convergence to the eigenvalue nearest to μ, useful in a bisection search.

The method of subspace iteration (Bathe and Wilson, 1976; Parlett, 1980) is a generalization of shifted inverse iteration; it has the very clear asset that shifted inverse iteration can be switched on as a special case. Instead of starting with seven different values of μ (to evaluate the seven values of γ^2 for Figure 5) and for each value developing separately a series of vectors $\mathbf{y}_0, \mathbf{y}_1, \mathbf{y}_2, \ldots,$ subspace iteration starts with seven vectors and *simultaneously* applies a sequence of iterations, each step involving seven vectors and giving seven current approximations to eigenvalues.

Applying the method to Eq. (18), suppose $\mathbf{A}$ and $\mathbf{B}$ are of order N and q is the order of the subspace (such as 7 above). Starting with a single shift μ and an array of q column vectors of order N (which can be assembled as an N by q matrix $\mathbf{X}_0$), the first iterate is defined by

$$(\mathbf{A} - \mu\mathbf{B})\mathbf{X}_1 = \mathbf{B}\mathbf{X}_0 \tag{21}$$

Now for $i = 1, 2, 3, \ldots$:

(i) Define the q by q matrices by

$$\begin{aligned} \mathbf{A}_i &= \mathbf{X}_i^* \mathbf{B}\mathbf{X}_{i-1} \\ \mathbf{B}_i &= \mathbf{X}_i^* \mathbf{B}\mathbf{X}_i \end{aligned} \tag{22}$$

where the superscript * denotes the complex conjugate transpose.

(ii) Solve completely the dense matrix eigenvalue problem:

$$\mathbf{A}_i\mathbf{\Phi}_i = \mathbf{B}_i\mathbf{\Phi}_i\mathbf{\Gamma}_i \tag{23}$$

to give $\mathbf{\Phi}_i$ containing q column eigenvectors and $\mathbf{\Gamma}_i$ a diagonal matrix containing the associated q eigenvalues γ_j.

(iii) Define the next iterate $\mathbf{X}_{i+1}$:

$$(\mathbf{A} - \mu\mathbf{B})\mathbf{X}_{i+1} = \mathbf{B}\mathbf{X}_i\mathbf{\Phi}_i \tag{24}$$

At each step, $\lambda_j = \mu + \gamma_j$ will be the current estimate of the q eigenvalues of Eq. (18) nearest to μ, and $\mathbf{X}_i\mathbf{\Phi}_i$ will contain the associated q eigenvectors. Rather than relying on all q eigenvalues, it is found to be better to make q larger than p, the number actually required, so that iteration is stopped when the relative changes in all p values of λ_j nearest to μ are within some chosen tolerance.

The two key parts of the above calculations are solving the q by q dense eigenvalue equation—Eq. (23)—and solving the system of linear equations—Eq. (24) These are the two programs referred to previously (before Section 4.1). The former is commonly solved by Householder with QR, or the Jacobi method. Equation (24) can be solved by any reliable solver of a complex Hermitian system. Since 1981 (Davies et al., 1982; Rahman and Davies, 1984) the subspace iteration has been applied successfully to many microwave and optical waveguide problems using Jacobi for Eq. (23) and a sparse version of Gaussian elimination from the Harwell library for Eq. (24). Various algorithms are available commercially, and in the public domain such as LAPACK.

What, then, are the virtues of subspace iteration? For a given waveguide problem at a chosen frequency, the finite element method would evaluate the global matrices **A** and **B**; their solution by subspace iteration would have the advantage of a single run categorically evaluating a range of propagation constants. For example, if Figure 5 illustrates the spectrum of γ^2, then use of a small negative shift μ with, say, $q = 7$ and $p = 4$ would reliably give $-\beta_3^2$, $-\beta_2^2$, $-\beta_1^2$, and α_1^2. There would be no fear that any eigenvalues have been "missed" between these four numbers. If $-\beta_2^2$ were near to or equal to $-\beta_1^2$, there would be no problem. The reason is that subspace iteration always works with a subspace spanning q vectors rather than q separate vectors. Shifted inverse iteration would have an impossible task in trying to approach any degeneracy, where, say, $-\beta_2^2 = -\beta_1^2$ requires two distinct eigenvectors. An additional advantage is that the method is less sensitive to choice of starting vectors. It is easier to find a starting subspace that approximately spans the set of eigenvectors than to find a set of initial vectors that indiviually approximates each of the required eigenvectors.

Returning to the choice of algorithm for solving Eq. (24), two comments can be made. First, there is the important observation that any prior knowledge of approximate values of eigenvalues and/or eigenvectors can clearly give subspace iteration a headstart (as with any iterative method). The commonest situation where this arises is when one is studying a waveguide over a range of some parameter—possibly frequency, structural dimension, or permittivity value. Solutions from one evaluation can then be used directly as the starting values for the next calculation. Second, if using a Gaussian elimination method (as mentioned above), one has the advantage that the triangular decomposition and pivoting strategy can be performed once in solving Eq. (21) and then bypassed in subsequent solutions of Eq. (24). As with dense matrices, effectively premultiplication by $\mathbf{A}^{-1}$ is then as fast as by **A**!

The shift in μ can be updated by taking advantage of new estimates of eigenvalues, in the Γ_i and γ_j of Eq. (23). Such updating is less powerful for subspace iteration than with shifted inverse iteration.

4.2. Subspace Iteration for Non-Hermitian Matrices

Having described subspace iteration in some detail, the non-Hermitian version will now be outlined. The overall strategy is the same, and many features carry over, including the eventual substitution of programs for solving the dense $\mathbf{Ax} = \lambda\mathbf{Bx}$ and sparse $\mathbf{Ax} = \mathbf{y}$. A definition is now given of the algorithm, describing the features that differ for this non-Hermitian case. It was earlier described by Fernandez et al. (1991) and used in subsequent work on complex modes by Fernandez et al. (1993) and Hernandez-Figueroa et al. (1995).

The generalized matrix equation for solution, which emerges as necessary for the solution of complex modes, is

$$\mathbf{Ax} = \lambda\mathbf{Bx} \tag{25}$$

where now $\mathbf{A}$ and $\mathbf{B}$ are not Hermitian. They are complex if the waveguide is lossy but become real (though nonsymmetric) if loss-free. Strictly speaking, Eq. (25) defines the eigenvalues and associated "right-hand" vectors. Normally the term "right-hand" is ignored because taking the complex conjugate transpose (denoted by asterisk) of Eq. (25) gives

$$\mathbf{x}^*\mathbf{A}^* = \lambda^*\mathbf{x}^*\mathbf{B}^* \tag{26}$$

and with Hermitian $\mathbf{A}$ and $\mathbf{B}$ Eq. (26) gives the same (necessarily real) eigenvalues as Eq. (25) and the left-hand vectors of Eq. (26) are simply conjugate transposes of the right-hand vectors of Eq. (22). However, with non-Hermitian $\mathbf{A}$ and $\mathbf{B}$ the N left-hand vectors are distinct from the N wanted eigenvectors of Eq. (25). This double system pervades the non-Hermitian subspace iteration.

Assume now that there are these sets of p right and left initial vectors $\mathbf{X}_0$ and $\mathbf{Y}_0$, respectively ($p \ll N$) to Eq. (25), and μ is the chosen shift. Starting as in Eq. (21), two sets of new vectors are simultaneously iterated for $i = 1, 2, 3, \ldots$ using

$$(\mathbf{A} - \mu\mathbf{B})\mathbf{X}_i = \mathbf{B}\mathbf{X}_{i-1} \tag{27}$$

$$\mathbf{Y}_i(\mathbf{A} - \mu\mathbf{B}) = \mathbf{Y}_{i-1}\mathbf{B} \tag{28}$$

p by p matrices are defined by

$$\mathbf{A}_i = \mathbf{Y}_i(\mathbf{A} - \mu\mathbf{B})\mathbf{X}_i \tag{29}$$

$$\mathbf{B}_i = \mathbf{Y}_i\mathbf{B}\mathbf{X}_i \tag{30}$$

The resulting dense systems are solved completely as

$$\mathbf{A}_i\mathbf{\Phi}_i = \mathbf{B}_i\mathbf{\Phi}_i\mathbf{\Gamma}_i \tag{31}$$

$$\mathbf{\Psi}_i\mathbf{A}_i = \mathbf{\Gamma}_i\mathbf{\Psi}_i\mathbf{B}_i \tag{32}$$

and the eventual right and left eigenvectors of Eq. (25) are given by

$$\mathbf{X} = \mathbf{X}_i\mathbf{\Phi}_i \tag{33}$$

$$\mathbf{Y} = \mathbf{\Psi}_i\mathbf{Y}_i \tag{34}$$

The non-Hermitian complex solver of the Harwell library MA28 (now succeeded by ME48) has been used successfully by Fernandez et al. (1991) in the key equations (27) and (28). Once the LU decomposition has been executed, the triangulation can be used for both systems—Eqs. (27) and (28)—as well as throughout succeeding iterations. As with the Hermitian matrices, setting $p = 1$ reduces the subspace method to shifted inverse iteration, and successive updating of the shift μ will accelerate convergence.

A compact version was studied using only the right-hand set of eigenvectors, with the conjugate transpose of $\mathbf{X}_i$ substituting for $\mathbf{Y}_i$. This is analytically valid and requires a few more (even though faster) iterations. Experience is that the full set of iterations of Eqs. (27) to (34) provides the more robust algorithm. Computation time has been found to be close to $O(N^{1.6})$ compared with the usual $O(N^3)$ of any dense algorithm.

5. MODAL SOLUTION OF "LEAKY MODES" IN OPEN WAVEGUIDE

Any uniform waveguide that is at all "open," and so able to have fields existing to infinity, may permit leaky modes as well as the already mentioned evanescent and possibly propagating and complex modes. For the sake of definiteness, the example of Figure 1e is chosen, with its cross section shown in Figure 6, of rectangular conducting waveguide with a gap in one of its walls.

A leaky mode will indeed exist here, satisfying the definition of Eqs. (1) and (2) but including a finite radiation of energy away from the main guiding structure. Such finite radiation implies a complex propagation constant—just as in a lossy waveguide—to conserve power. Although such modes and the "absorbing boundary condition" will be properly addressed in a later chapter, a brief mention will be made here regarding the problem. This is because of the special context here—when a computer program has the capability of dealing properly with complex propagation constants it has the potential to be extended to deal with these leaky modes.

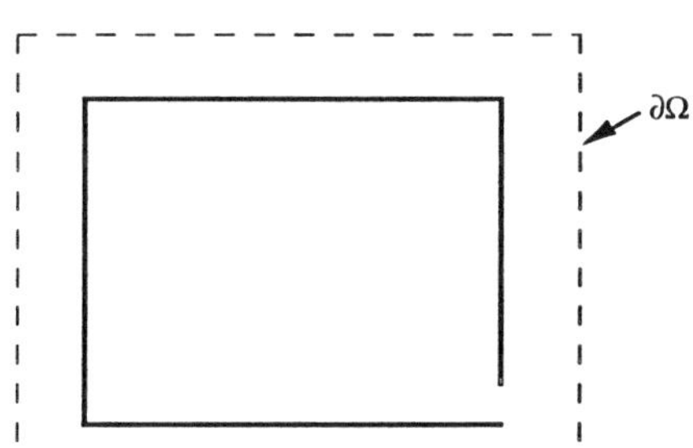

FIGURE 6. Slotted rectangular guide with surrounding artificial boundary $\partial\Omega$.

Details have been given by Hernandez-Figueroa et al. (1995), but a very brief account will be given now.

As in Figure 6, the physical waveguide structure is surrounded by an artificial boundary $\partial\Omega$. The essence of the approach is to make the single *a priori* approximation—that the fields along and outside $\partial\Omega$ associated with the leaky mode can be approximated by a single cylindrically radiating wave, centered on a suitable axis within $\partial\Omega$. For Figure 6, the axis would naturally be chosen centered on the gap. Equation (10) is now applied to the waveguide cross section, where the artificial boundary contributes to $\partial\Omega$, the line integral of Eq. (10). For a longitudinal complex propagation constant γ, the cylindrical wave has presumed Hankel function dependence with asymptotic form:

$$H_z(x, y, z) = C(kr)^{0.5} \exp(-jkr) \exp(-\gamma z) \tag{35}$$

where k is given by $(\omega^2 \mu_0 \varepsilon - \gamma^2)^{0.5}$ and ε is the value along $\partial\Omega$.

Finite element analysis is applied as described in this chapter, using Eq. (10) and with second order nodal triangular elements within $\partial\Omega$. The presumption of Eq. (35) acts like an absorbing boundary condition and allows evaluation of the line integral of Eq. (10) without spoiling the sparsity of $\mathbf{A}$ and $\mathbf{B}$, and results in a nonlinear version of the earlier described matrix equation:

$$\mathbf{A}(\gamma)\mathbf{x} = \gamma^2 \mathbf{B}\mathbf{x} \tag{36}$$

Using subspace iteration with $p = q = 1$, so that only one mode is being sought at a time, Eq. (36) is solved for some guessed value of γ, which is then used to reevaluate $\mathbf{A}(\gamma)$, which is re-solved in Eq. (36), and so on. Successful results are reported by Hernandez-Figueroa et al. (1995) for the leaky modes of a four-layer open waveguide on a high permittivity substrate; results have also been obtained for the structure of Figure 6. Details of the method of this section can be found in the above reference.

REFERENCES

Anderson, B. C., and Cendes, Z. J. (1995). Solution of ferrite loaded waveguide using vector finite elements. *IEEE Trans. Magn.* **MAG-31**, 1578–1581.

Bathe, K. J., and Wilson, E. L. (1976). *Numerical Methods in Finite Element Analysis.* Prentice-Hall, Englewood Cliffs, NJ.

Chorney, P. (1961). *Power and Energy Relations in Bi-directional Waveguide*, MIT Res. Lab. Tech. Rep. No. 396. Massachusetts Institute of Technology, Cambridge, MA.

Clarricoats, P. J. B., and Taylor, B. C. (1964). Evanescent and propagating modes of dielectric-loaded circular waveguide. *Proc. Inst. Electr. Eng.* **111**, 1951–1956.

Collin, R. E. (1991). *Field Theory of Guided Waves*, 2nd ed. IEEE Press, NJ.

Davies, J. B. (1993). Finite element analysis of waveguides and cavities—a review. *IEEE Trans. Magn.* **MAG-29**, 1578–1583.

Davies, J. B., Fernandez, F. A., and Philippou, G. Y. (1982). Finite element analysis of all modes in cavities with circular symmetry. *IEEE Trans. Microwave Theory Tech.* **MTT-30**, 1975–1980.

Dillon, B. M., and Webb, J. P. (1994). A comparison of formulations for the vector finite element analysis of waveguides. *IEEE Trans. Microwave Theory Tech.* **MTT-42**, 308–316.

Fernandez, F. A., and Lu, Y. (1991). A variational finite element formulation for dielectric waveguides in terms of transverse magnetic fields. *IEEE Trans. Magn.* **MAG-27**, 3864–3867.

Fernandez, F. A., and Lu, Y. (1996). *Microwave and Optical Waveguide Analysis by the Finite Element Method.* Research Studies Press, Wiley, Chichester and New York.

Fernandez, F. A., Davies, J. B., Zhu, S., and Lu, Y. (1991). Sparse matrix eigenvalue solver for finite element solution of dielectric waveguides. *Electron. Lett.* **27**, 1824–1826.

Fernandez, F. A., Lu, Y., Davies, J. B., and Zhou, S. (1993). Finite element analysis of complex waves in inhomogeneous waveguides. *IEEE Trans. Magn.* **MAG-29**, 1601–1604.

Harrington, R. F. (1968). *Time Harmonic Electromagnetic Fields.* McGraw-Hill, New York.

Hernandez-Figueroa, H. E., Fernandez, F. A., Lu, Y., and Davies, J. B. (1995). Vectorial finite element modelling of 2-D leaky waveguides. *IEEE Trans. Magn.* **MAG-31**, 1710–1713.

Huang, W., and Itoh, T. (1988). Complex modes in lossless shielded microstrip lines. *IEEE Trans. Microwave Theory Tech.* **MTT-36**, 163–165.

Jin, J. (1993). *The Finite Element Method in Electromagnetics.* Wiley, New York.

Kalmyk, V. A., Rayevskiy, S. B., and Vgryumov, V. P. (1978). An experimental verification of existence of complex waves in a two-layer circular shielded waveguide. *Radio Eng. Electron. Phys. (Engl. Transl.)* **23**, 16–19.

Kooi, P. S., Leong, M. S., Widjojo, B., and Yeo, T. S. (1991). Role of complex modes in coplanar-waveguide-to-coplanar-waveguide discontinuity calculations. *Microwave Opt. Technol. Lett.* **4**, 378–381.

Lee, J. F. (1994). Finite element analysis of lossy dielectric waveguides. *IEEE Trans. Microwave Theory Tech.* **MTT-42**, 1025–1031.

Lee, J. F., Sun, D. K., and Cendes, Z. J. (1991). Full-wave analysis of dielectric waveguides using tangential vector finite elements. *IEEE Trans. Microwave Theory Tech.* **MTT-39**, 1262–1271.

Lu, Y., and Fernandez, F. A. (1993). An efficient finite element solution of inhomogeneous anisotropic and lossy dielectric waveguides. *IEEE Trans. Microwave Theory Tech.* **MTT-41**, 1215–1223.

Marqués, R., Mesa, F., and Horno, M. (1992). On the complex nature of higher order modes in lossless nonreciprocal transversely magnetized waveguides. *IEEE Microwave Guided Wave Lett.* **2**, 278–280.

Mrozowski, M., and Mazur, J. (1992). Matrix theory approach to complex waves. *IEEE Trans. Microwave Theory Techn.* **MTT-40**, 781–785.

Parlett, B. N. (1980). *The Symmetric Eigenvalue Problem.* Prentice-Hall, Englewood Cliffs, NJ.

Rahman, B. M., and Davies, J. B. (1984). Penalty function improvement of waveguide solution by finite elements. *IEEE Trans. Microwave Theory Tech.* **MTT-32**, 922–928.

Railton, C. J., and Rozzi, T. (1988). Complex modes in boxed microstrip. *IEEE Trans. Microwave Theory Tech.* **MTT-36**, 865–873.

Schünemann, K., and Omar, A. S. (1985). Formulation of the singular integral equation technique for planar transmission lines. *IEEE Trans. Microwave Theory Tech.* **MTT-33**, 1313–1322.

Silvester, P. (1969). A general high-order finite element waveguide analysis program. *IEEE Trans. Microwave Theory Tech.* **MTT-17**, 204–210.

Silvester, P. P., and Ferrari, F. L. (1990). *Finite Elements for Electrical Engineers.* Cambridge Univ. Press, Cambridge, UK.

Strube, J., and Arndt, F. (1985). Rigorous hybrid-mode analysis of the transition from rectangular waveguide to shielded dielectric image guide. *IEEE Trans. Microwave Theory Techn.* **MTT-33**, 391–401.

Valor, L., and Zapata, J. (1996). An efficient finite element formulation to analyze waveguides with lossy inomogeneous bi-anistropic materials. *IEEE Trans. Microwave Theory Tech.* **MTT-44**, 291–296.

CHAPTER TWO

Spatially Periodic Structures

RONALD FERRARI

1. INTRODUCTION

Quite a few practical microwave and optical devices utilize the phenomenon of electromagnetic wave guidance by a spatially periodic structure. The simplest form of such a device is a closed corrugated waveguide; variants of this general configuration include traveling wave tube slow-wave distributed circuits and open periodic dielectric waveguides. The physical behavior of these arrangements hinges on the fact that as a wave propagates through a periodic structure, at any given periodic perturbation of the plain waveguide there may exist reflected waves from perturbations ahead, which tend to cancel the current oscillation. Thus frequency bands can arise over which propagation through the waveguide is severely attenuated, while the tortuous path represented by such wholesale reflections, when not canceling, can result in a useful slowing down of the effective wave velocity. There is a direct analogy between such electromagnetic phenomena and the wave behavior of electrons in a periodic crystal lattice; indeed, three-dimensional periodic electromagnetic wave structures having optical "stop" and "pass" bands corresponding to solid-state energy bands have been proposed. Other devices involve external radiation interacting with an open periodic structure. The classical diffraction grating is a prime example here, constituting a directional filter/polarizer, examples of which are important over the whole electromagnetic spectrum. Among the other important applications, open periodic structures are used in holography and employed in the fabrication of microwave lenses and nonref-

Finite Element Software for Microwave Engineering, Edited by Tatsuo Itoh, Giuseppe Pelosi, and Peter P. Silvester.
ISBN 0-471-12636-5

lective surfaces. The phased array antenna, considered as an infinite array of radiators having a constant phase difference between each element, is also amenable to treatment by the methods detailed here.

The finite element method is particularly suitable for tackling periodic waveguide problems because the constraint required from Floquet's theorem at periodic boundaries turns out to be rather simple, while the computation reduces to one carried out over a unit cell. Moaveni (1979) was the first to apply the finite element method to solve for plane wave scattering from a two-dimensional periodic grating; however, only the restricted case of unit cells isolated from one another was considered. Davies et al. (1982) introduced the critical application of Floquet's theorem in a finite element analysis to work out the modes of propagation in a circularly symmetric periodic waveguide. Nakata and Koshiba (1988) first applied the periodicity condition to the finite element analysis of plane wave scattering.

1.1. Floquet's Theorem

Originally postulated in purely mathematical terms relating to the solution of a differential equation, Floquet's theorem is capable of very simple formulation and justification (Slater, 1950). Consider a quasi-uniform wave propagating through an infinitely extending medium having spatial x-periodicity D_x. Making observations at some particular location and then at a periodic neighbor, the wave could be found to be attenuated, amplified, or just changed in phase, *but its functional form must otherwise be precisely unchanged.* If this were not so, the observer's disorientating experience of never being able to determine in which particular periodic cell he or she was positioned would not occur. In complex phasor notation, omitting the implied harmonic time factor $\exp(j\omega t)$, the supposed change in amplitude and/or phase over a spatial period, may be represented as $\exp(-\gamma D_x)$, where γ is a complex constant that in general depends on ω. In electromagnetic terms, say, using the electric field vector as a working variable, put

$$\mathbf{E} = \mathbf{E}_p(x, y, z)\exp(-\gamma x) \tag{1}$$

where

$$\mathbf{E}_p(x + D_x, y, z) = \mathbf{E}_p(x, y, z) \tag{2}$$

so that the electric field does indeed change by the same complex constant on traversing any single spatial period. Equations (1) and (2) represent a statement of Floquet's theorem. If the medium is lossless then $\gamma = j\beta$, a purely imaginary constant unless relating to a stop band, when γ is purely real. The computational problem associated with propagation through a periodic medium may be, in the first instance, to find $\mathbf{E}_p$ and β given ω. However, it is often more practical instead to compute ω and $\mathbf{E}_p$, given β. The addition of $2\pi/D_x$ to β, any number of times, has no effect on the corresponding values of ω and $\mathbf{E}_p$, so it is not generally necessary to computate β outside the range, say, $-\pi/D_x$ to $+\pi/D_x$.† As in ordinary waveguide problems, a multiplicity of different modal solutions may arise.

Floquet's theorem can be generalized to cover media with periodicity in two or three dimensions. In that case define a vector

$$\mathbf{R} = \mathrm{n}_1\mathbf{a} + n_2\mathbf{b} + n_3\mathbf{c} \tag{3}$$

where n_1, n_2, and n_3 are integers while **a**, **b**, and **c** are noncoplanar primitive vectors representing the lattice periodicity. Then the proposition becomes

$$\mathbf{E} = \mathbf{E}_p(\mathbf{r})\exp(-\boldsymbol{\gamma}\cdot\mathbf{r}) \tag{4}$$

where $\boldsymbol{\gamma}$ is a complex vector constant depending on ω while

$$\mathbf{E}_p(\mathbf{r} + \mathbf{R}) = \mathbf{E}_p(\mathbf{r}) \tag{5}$$

for all possible **R**. When the theorem is applied in quantum mechanics to the electronic wavefunction associated with a periodic crystal lattice, it is known as "Bloch's theorem."

2. FINITE ELEMENT FORMULATION FOR THE CLOSED PERIODIC SYSTEM

A singly periodic arrangement such as illustrated in Figure 1 is considered. Choosing **E** as the working vector variable, Maxwell's equations give

$$\nabla \times \mu_r^{-1}\nabla \times \mathbf{E} - k^2\varepsilon_r\mathbf{E} = 0 \tag{6}$$

where $k^2 = \omega^2\mu_0\varepsilon_0$, while the relative constitutive constants μ_r and ε_r may be complex tensors representing a lossy, anisotropic material. The corresponding dual in complex phasor **H** may also be used as a starting equation for the analysis. The periodicity of the arrangement is expressed by the fact that

$$\mu_r(x, y, z) = \mu_r(x + D_x, y, z) \tag{7}$$

while ε_r and the external boundary conditions are similarly periodic. A single "unit" cell, comprising the smallest region Ω of the waveguide that repeats itself in x, is considered. The boundary S of the cell comprises an external surface

†The "space" represented by β can be regarded as reciprocal to the Euclidean space of x and is similarly periodic. This concept is much used in the quantum mechanics of crystal lattices. The range of β described here is termed a "Brillouin zone."

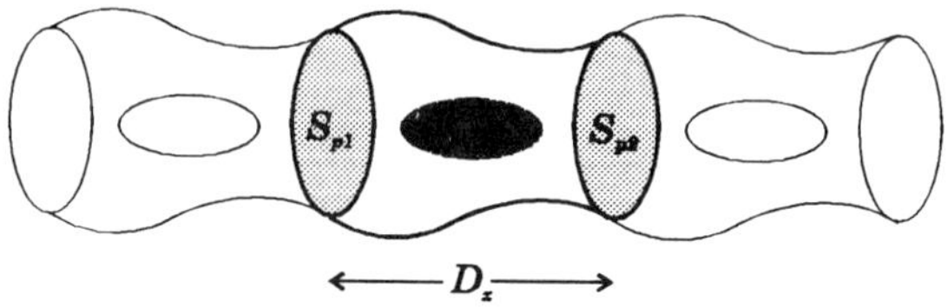

FIGURE 1. A spatially periodic waveguide.

S_X, an arbitrarily placed transverse surface S_{p1} (conveniently but not necessarily plane), and a surface S_{p2}, identical with S_{p1} but displaced from it by one spatial period D_x. The external surface S_X is made up of portions S_{Xe} and S_{Xm}, electric and magnetic walls, respectively, while the transverse boundary planes are subject to the Floquet condition expressed by Eqs. (1) and (2).

The Galerkin weighted residual route to the finite element formulation is followed. Thus Eq. (6) is multiplied throughout by an arbitrary vector weighting function $\mathbf{W}(x, y, z)$, integrated over Ω, and a vector integration by parts performed to yield the weak-form weighted residual

$$R = \int_{\Omega} (\nabla \times \mathbf{W} \cdot \mu_r^{-1} \nabla \times \mathbf{E} - k^2 \varepsilon_r \mathbf{W} \cdot \mathbf{E})\, d\Omega - \oint_S (\mathbf{W} \times \mu_r^{-1} \nabla \times \mathbf{E}) \cdot \mathbf{1}_n \, dS = 0 \tag{8}$$

where $\mathbf{1}_n$ is the unit outward normal to the surface S. In the usual fashion, $\mathbf{E} \times \mathbf{1}_n$ and $\mathbf{W} \times \mathbf{1}_n$ are set to zero on S_{Xe}, corresponding to an essential boundary constraint, while $\mathbf{E}$ is left free on the natural boundary S_{Xm}, in which case the surface integral over S_X vanishes. There remains in Eq. (8) the sum of surface integrals

$$\int_{S_{p1}} (\mathbf{W} \times \mu_r^{-1} \nabla \times \mathbf{E}) \cdot \mathbf{1}_n \, dS + \int_{S_{p2}} (\mathbf{W} \times \mu_r^{-1} \nabla \times \mathbf{E}) \cdot \mathbf{1}_n \, dS \tag{9}$$

Choose, without loss of generality, a reciprocal weighting function

$$\mathbf{W} = \mathbf{W}_p \exp(\gamma x) \tag{10}$$

where the complex function of position $\mathbf{W}_p$ is periodic in x,

$$\mathbf{W}_p(x + D_x, y, z) = \mathbf{W}_p(x, y, z) \tag{11}$$

Because of the Floquet constraint on $\mathbf{E}$, Eqs. (1) and (2), and the fact that, by definition, surfaces S_{p1} and S_{p2} are separated by a spatial period D_x, whereas their unit normal vectors are equal and opposite, the last remnant of the weak-form weighted residual surface integral, Eq. (9), now vanishes. Thus the weighted residual formulation may be rewritten as

$$R = \int_{\Omega} [(\nabla + \gamma \mathbf{1}_x) \times \mathbf{W}_p \cdot \mu_r^{-1}(\nabla - \gamma \mathbf{1}_x) \times \mathbf{E}_p - k^2 \varepsilon_r \mathbf{W}_p \cdot \mathbf{E}_p] \, d\Omega = 0 \qquad (12)$$

The Galerkin form of the weighted residual expression now follows by selecting the $\mathbf{W}_p$'s from the expansion functions chosen for $\mathbf{E}_p$; the result then is the same as that obtained variationally. Given identical discretization of the planes $S_{p1,2}$, the periodicity required for $\mathbf{E}_p$ simply needs the coefficients of the expansion of $\mathbf{E}_p$ to match between these two planes; the expansion functions there will automatically be the same, hence guaranteeing the appropriate periodicity for $\mathbf{W}_p$. The analysis here is valid for the general case of anisotropic, lossy materials.

2.1. Two-Dimensional Problems

Davies et al. (1982) took the essentially two-dimensional problem of a periodic, isotropic circular waveguide with azimuthal symmetry and introduced the coupling of the finite element method with the Floquet periodicity condition. They employed a simple variational approach, which limited them to lossless arrangements, with a vector variable using nodal finite elements, liable to spuriously corrupted modes. Ferrari (1991) introduced the alternative reciprocal weighted residual formulation above, valid for the general anisotropic case with loss, and applied it to the simple two-dimensional example given here.

An empty, perfectly conducting corrugated waveguide, having two-dimensional translational symmetry (Figure 2), is considered. The wave supported separates into transverse electric and transverse magnetic modes. Considering the latter, with Cartesian axes chosen such that there is no z-variation but periodicity D_x in the x-direction, the TM wave may be represented by the fields

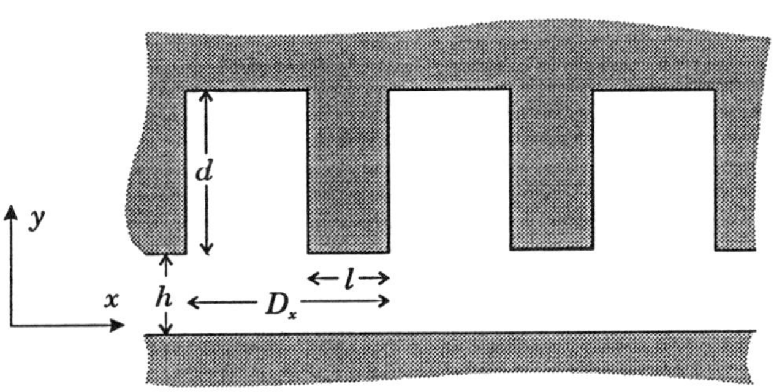

FIGURE 2. A corrugated waveguide with two-dimensional symmetry.

either $(0, 0, H_{pz})\exp(-\gamma x)$ or $(E_{px}, E_{py}, 0)\exp(-\gamma x)$, functions of x and y only. In such a case the simpler option is to choose the scalar representation in H_{pz} requiring the **H**-form of the weighted residual, dual to Eq. (12). The volume Ω may now be treated as an area A and the weighted residual is

$$R = \int_A \frac{1}{\varepsilon_r}\left[\nabla W_{pz}\cdot\nabla H_{pz} - \gamma\left(W_{pz}\frac{\partial H_{pz}}{\partial x} - \frac{\partial W_{pz}}{\partial x}H_{pz}\right)\right]dA - \int_A \left(\frac{\gamma^2}{\varepsilon_r} + k^2\mu_r\right)W_{pz}H_{pz}\,dA \qquad (13)$$

Since here **H** is guaranteed to be solenoidal, a nodal finite element expansion of $H_{pz}(x, y)$, using scalar interpolation functions defined, say, over triangular simplexes or isoparametric quadrilaterals (Silvester and Ferrari, 1990),

$$H_{pz} = \sum_j H_{pj}\alpha_j(x, y) \qquad (14)$$

may be used. Then, considering the lossless case $(\gamma = j\beta)$ of an empty $(\mu_r = \varepsilon_r = 1)$ corrugated waveguide, Figure 2, and putting

$$W_{pz} = \alpha_i(x, y) \qquad (15)$$

a sparse matrix equation

$$\mathbf{Sh} + j\beta\mathbf{Xh} + (\beta^2 - k^2)\mathbf{Th} = 0 \qquad (16)$$

where **h** is the global vector of unknown nodal coefficients H_{pj}, ensues from the Galerkin treatment of the weighted residual, Eq. (13). In Eq. (16) arrays **S** and **T** are the well-known ones, given in their local form for a single finite element A_e by

$$S_{ij} = \int_{A_e} \nabla\alpha_i\cdot\nabla\alpha_j\,dx\,dy \qquad (17)$$

$$T_{ij} = \int_{A_e} \alpha_i\alpha_j\,dx\,dy \qquad (18)$$

An unfamiliar matrix **X** appears (Stone, 1972), given locally by

$$X_{ij} = \int_{A_e}\left(\alpha_i\frac{\partial\alpha_j}{\partial x} - \frac{\partial\alpha_i}{\partial x}\alpha_j\right)dx\,dy \qquad (19)$$

As in evaluating the local matrices S_{ij} and T_{ij}, the integrations to obtain the coefficients X_{ij} can be performed analytically for triangular elements, while a simple Gaussian quadrature is needed for isoparametric curvilinear elements. The assembly of **S**, **T**, and **X** to form the corresponding global matrices, incorporating the essential boundary constraints and leaving natural boundaries unconstrained, follows the standard finite element pattern. In addition, it must be ensured that the H_{pj} corresponding to equivalent points on S_{p1} and S_{p2} are treated as one and the same unknown, thus guaranteeing the spatial periodicity of $\mathbf{H}_p$. Now $(0, 0, H_{pz})$ is a continuous variable even when either or both of ε_r and μ_r change discontinuously with x and y. Thus if the relative constitutive parameters vary with sufficiently small gradient, they can be modeled as constants within each finite element and the assembly of the global matrix proceeds with only minor modification. If the TE wave is to be modeled, then the electric field formulation similarly involves a single scalar variable, now E_z, subject to an essential boundary constraint $E_z = 0$ on the waveguide walls.

Equation (16) represents a complex matrix "eigenproblem." Let **h** be separated into its real and imaginary parts,

$$\mathbf{h} = \mathbf{h}_{\mathrm{Re}} + j\mathbf{h}_{\mathrm{Im}} \tag{20}$$

Noting that **S**, **T**, and **X** are purely real matrices, Eq. (16) may be rewritten in partitioned form as

$$\begin{bmatrix} \mathbf{S} & 0 \\ 0 & \mathbf{S} \end{bmatrix} \begin{bmatrix} \mathbf{h}_{\mathrm{Re}} \\ \mathbf{h}_{\mathrm{Im}} \end{bmatrix} + \begin{bmatrix} 0 & \beta\mathbf{X} \\ -\beta\mathbf{X} & 0 \end{bmatrix} \begin{bmatrix} \mathbf{h}_{\mathrm{Re}} \\ \mathbf{h}_{\mathrm{Im}} \end{bmatrix} + (\beta^2 - k^2) \begin{bmatrix} \mathbf{T} & 0 \\ 0 & \mathbf{T} \end{bmatrix} \begin{bmatrix} \mathbf{h}_{\mathrm{Re}} \\ \mathbf{h}_{\mathrm{Im}} \end{bmatrix} = 0 \tag{21}$$

While **S** and **T** are sparse and symmetric, **X** is sparse and *antisymmetric*, so that if the problem is formulated as "given β, find normalized frequency k," then one of the standard matrix solvers for dealing with

$$\mathbf{Mx} + \lambda\mathbf{Nx} = 0 \tag{22}$$

when **M** and **N** are real, sparse and symmetric, can be employed to compute the eigenvalues k and eigenvectors **h**. Figure 3 shows the results of the TM finite element computation for the waveguide of Figure 2 (Ferrari, 1991). In this instance the boundary conditions for the unit cell, $\partial H_{pz}/\partial n = 0$ on the waveguide walls, together with the Floquet condition at the periodic closure required no constraint other than ensuring that periodic nodes are matched on the closure planes. The computations were carried out over an unnecessarily large range of βD to demonstrate the expected reciprocal periodicity in β-space; taking symmetry into account, the range $0 \leqslant \beta D \leqslant \pi$ would have been sufficient.

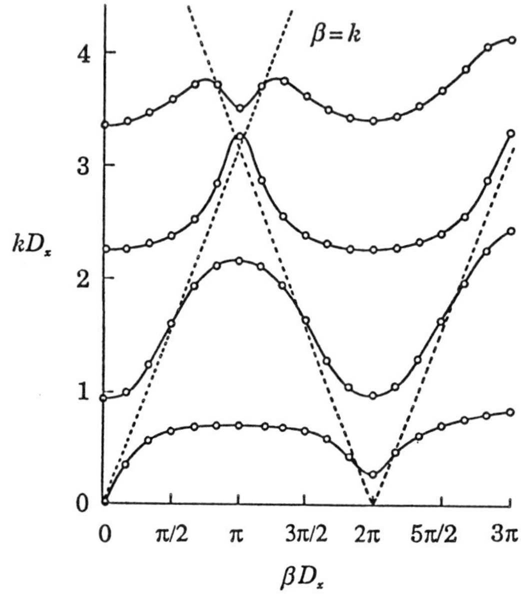

FIGURE 3. Computed k–β curves for the corrugated waveguide of Figure 2, $d/D_x = 2$, $h/D_x = 1$, $l/D_x = 0.5$, showing four modes. (After Ferrari, 1991.)

2.2. Three-Dimensional Problems

Following the introductory work by Davies et al. (1982), the topic of periodic waveguides was reopened by Inoue et al. (1989) in a fully three-dimensional treatment but still taking the earlier variational route, which implicitly assumes the lossless case. Spatially periodic configurations, which are irreducibly three-dimensional, require the use of a vector variable for finite element modeling, so that the residual expression, Eq. (12), or its dual, must be used in full vector form. The simplest example of this situation is that of propagation through a closed waveguide with single periodicity, such as illustrated in Figure 1. Arrangements with twofold and threefold periodicity may also be classed in the general three-dimensional category; such configurations are not waveguides in the ordinary sense because the direction of propagation relative to the periodic structure can be chosen at will. The Floquet periodicity condition Eq. (1) is replaced by Eq. (4) expressed in terms of a general vector propagation constant $\boldsymbol{\gamma}$; thus Eq. (12) must be modified, replacing $\gamma \mathbf{1}_x$ by $\boldsymbol{\gamma}$ (or, in view of the complexity now represented, more likely by $j\boldsymbol{\beta}$ corresponding to the lossless case). The treatment, however, is no different in principle except that curl-conforming vector expansion functions (see Chapter 5) must be used to express the vector variables in the finite element procedure so as to avoid contamination by spurious modes. Thus set

$$\mathbf{E}_p = \sum_j E_j \boldsymbol{\tau}_j \tag{23}$$

where E_j are unknown scalar coefficients to be determined and $\boldsymbol{\tau}_j$ are tangentially continuous vector interpolation functions, such as first-order tetrahedral edge elements or mixed-order covariant projection elements. Choosing, appropriate to Galerkin's method, the $\mathbf{W}_p$'s from the set of $\boldsymbol{\tau}$'s and substituting into the modified Eq. (12) gives, for $\boldsymbol{\gamma} = j\boldsymbol{\beta}$, element matrices

$$A_{ij} = \int_{\Omega_e} (\nabla + j\boldsymbol{\beta}) \times \boldsymbol{\tau}_i \cdot \mu_r^{-1}(\nabla - \mathrm{j}\boldsymbol{\beta}) \times \boldsymbol{\tau}_j \, d\Omega \tag{24}$$

$$B_{ij} = \int_{\Omega_e} \varepsilon_r \boldsymbol{\tau}_i \cdot \boldsymbol{\tau}_j \, d\Omega \tag{25}$$

which may be evaluated and assembled in standard fashion, as in the scalar case, to yield a global sparse matrix equation

$$\mathbf{A}(j\boldsymbol{\beta})\mathbf{e} + k^2\mathbf{B}\mathbf{e} = 0 \tag{26}$$

Now in the general case, in order to exploit standard matrix eigensolvers, a *vector* $\boldsymbol{\beta}$ must be chosen, equivalent to specifying the direction of propagation as well as the magnitude of the wave constant, in order to work out at what (normalized) frequency k such a wave could occur. It may be noted that, provided μ_r and ε_r are real, $\mathbf{A}$ is a complex Hermitian matrix while $\mathbf{B}$ is symmetric; thus, as would be expected for lossless situations, real frequencies k ensue when $\boldsymbol{\beta}$ itself is real.

Mias and Ferrari (1994) give an example of the methodology described here for a closed three-dimensional singly periodic waveguide; Ferrari (1993) describes the analogous finite element procedure applying to electronic wavefunctions in a two-dimensional, doubly periodic lattice, while Mias (1995) has modeled triply periodic configurations representing optical bandgap materials.

3. PLANE-WAVE SCATTERING FROM SINGLY PERIODIC STRUCTURES WITH TWO-DIMENSIONAL SYMMETRY

A two-dimensional finite element analysis of plane-wave scattering from a singly periodic structure was originally worked out by Nakata and Koshiba (1988). The method described here, solving for the periodic part of the Floquet wavefield using reciprocal weight-function Galerkin analysis in the finite element region and Rayleigh harmonics externally, is essentially that given by Delort and Maystre (1993). Pelosi et al. (1993) a little earlier had given an equivalent analysis in terms of the full field; based on the latter work, Cecchini et al. (1995) give a decription of a program that computes the surface impedance and reflection coefficient for a plane wave incident on the singly periodic structure. A further alternative is the combined finite element/method of moments approach given by Gedney et al. (1992).

3.1. Singly Periodic Structures with Two-Dimensional Symmetry

A singly periodic "grating" with two-dimensional symmetry and no variation in the z-direction is illustrated in Figure 4. The arrangement is illuminated by a plane wave, also independent of z, propagating in a direction parallel to the xy-plane. Such a wave can be resolved into transverse electric (TE) and transverse magnetic (TM) constituents. The TE case only will be considered in detail, the analysis for a TM-incident wave following with only trivial modification. An incident TE wave with direction of propagation making angle ϕ with the x-axis is considered; it has a z-directed complex phasor electric field E_z only. Clearly, the resultant scattered field in this case will be expressible also with no z-variation and entirely in terms of E_z. The problem space, depicted in Figure 4, is divided into three regions separated by planes parallel to the xz-plane. Region 1 contains the grating, comprised of material with relative constitive constants ε_r, μ_r varying with spatial periodicity D_x, perhaps also with periodically distributed perfect conductors. It will be supposed that ε_r, μ_r are complex scalars, representing an isotropic medium, which, however, may be lossy or active according to the sign of the imaginary parts. Here E_z has to

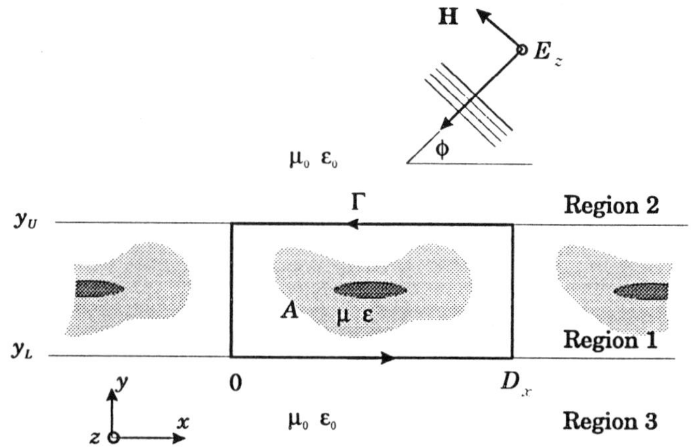

FIGURE 4. A transverse electric wave incident on a two-dimensional singly periodic structure.

satisfy the Helmholtz equation

$$\nabla \cdot \frac{1}{\mu_r} \nabla E_z + \varepsilon_r k^2 E_z = 0 \tag{27}$$

where $k^2 = \omega^2 \mu_0 \varepsilon_0$.† The boundaries of region 1 are positioned arbitrarily but stand clear of (or just touch) the grating. Regions 2 and 3 are free space (a simple modification of the analysis takes care of cases where region 3, say, is a semi-infinite substrate of homogeneous or even layered material), with E_z subject to

$$\nabla^2 E_z + k^2 E_z = 0 \tag{28}$$

In region 2 the incident field may be expressed as

$$E_z^I = \mathscr{E}^I \exp jk(x \cos\phi + y \sin\phi) \tag{29}$$

where $\mathscr{E}^I$ is a given constant. Note that in order to preserve the conventional cylindrical coordinates angle ϕ, the incident radiation is chosen so as to proceed in the $-x$ and $-y$ directions. The reflected waves in region 2 and those transmitted through the grating into region 3 form plane-wave "Rayleigh" expansions

$$E_z^R = \sum_{n=-\infty}^{\infty} \mathscr{E}_n^R \exp -j(k_{xn}x + k_{yn}y) \tag{30}$$

$$E_z^T = \sum_{n=-\infty}^{\infty} \mathscr{E}_n^T \exp -j(k_{xn}x - k_{yn}y) \tag{31}$$

where $\mathscr{E}_n^R$ and $\mathscr{E}_n^T$ are unknown reflection and transmission amplitudes whose determination will be the ultimate goal here. Here the sign convention adopted is such that for positive k_{xn}, k_{yn} a reflected plane-wave component travels in the $+x$ and $+y$ directions and transmitted wave in the $+x$ and $-y$ directions. Corresponding to the incident wave, these scattered waves are TE, while their phase dependence on x can only differ from the incident x-dependence by a factor due to the traversing of an arbitrary whole number of spatial periods D_x. Thus

$$k_{xn} = \frac{2n\pi}{D_x} - k\cos\phi, \quad n = 0, \pm 1, \pm 2, \ldots \tag{32}$$

†To analyze the TM case it is necessary to substitute H_z for E_z and to interchange μ_r and ε_r everywhere.

In order to satisfy Eq. (28) the y-spatial dependence $\exp(-jk_{yn}y)$ must be such that

$$k_{yn} = \sqrt{k^2 - k_{xn}^2}, \quad \text{for } k^2 > k_{xn}^2 \tag{33}$$

There will be instances where the inequality in Eq. (33) above does not hold, in which case

$$k_{yn} = -j\sqrt{k_{xn}^2 - k^2}, \quad \text{for } k^2 < k_{xn}^2 \tag{34}$$

For the latter case the Rayleigh harmonics decay exponentially in y away from the grating and represent waves guided along the structure. The interdependence of k_{xn} and k_{yn}, Eqs. (33) and (34), sets known directions on the individual Rayleigh waves while the dependence of the amplitudes $\mathscr{E}_n^{R,T}$ on n then fixes the qualitatively familiar physical form of the diffraction pattern of a grating. The classical treatment of the diffraction grating proceeds by finding a general solution to Eq. (27) inside the spatially periodic region 1 and applying the constraint that both tangential **E** and tangential **H** must be continuous across the interfaces between the regions. Examination of Maxwell's equations reveals that, for the TE case here, this amounts just to the continuity of E_z and $(1/\mu_r)(\partial E_z/\partial y)$.

From Eqs. (29) and (32), the Floquet wavenumber appropriate for use with the periodic system under discussion is

$$\gamma = -jk\cos\phi \tag{35}$$

Thus any Floquet periodic quantity in this analysis may be obtained from the nonperiodic equivalent through multiplication by $\exp(-jk\cos\phi)$. Substituting Eq. (32) into Eqs. (30) and (31), there results for the sum of incident and reflected waves in region 2, using Eq. (29),

$$E_{pz}(x, y) = \mathscr{E}^I \exp(jky\sin\phi) + \sum_{n=-\infty}^{\infty} \mathscr{E}_n^R \exp -j\left(\frac{2n\pi x}{D_x} + k_{yn}y\right), \quad y \geqslant y_U \tag{36}$$

while in region 3 for the transmitted wave,

$$E_{pz}(x, y) = \sum_{n=-\infty}^{\infty} \mathscr{E}_n^T \exp -j\left(\frac{2n\pi x}{D_x} - k_{yn}y\right), \quad y \leqslant y_L \tag{37}$$

The Rayleigh harmonics are orthogonal functions; thus once the Floquet periodic fields at $y = y_U$ and $y = y_L$ are known, the reflected and transmitted amplitudes $\mathscr{E}^{R,T}$ become determined:

$$\mathscr{E}_n^R = \frac{1}{D_x}\int_0^{D_x} [E_{pz}(x, y_U) - \mathscr{E}^I \exp j(ky_U \sin\phi)] \exp j\left(\frac{2n\pi x}{D_x} + k_{yn}y_U\right) dx$$

$$= \frac{1}{D_x}\int_0^{D_x} E_{pz}(x, y_U) \exp j\left(\frac{2n\pi x}{D_x} + k_{yn}y_U\right) dx, \quad n \neq 0 \tag{38}$$

$$\mathscr{E}_0^R = \frac{1}{D_x}\int_0^{D_x} E_{pz}(x, y_U) \exp j(k_{y0}y_U)\, dx - \mathscr{E}^I \exp j(ky_U \sin\phi + k_{y0}y_U) \tag{39}$$

$$\mathscr{E}_n^T = \frac{1}{D_x}\int_0^{D_x} E_{pz}(x, y_L) \exp j\left(\frac{2n\pi x}{D_x} - k_{yn}y_L\right) dx \tag{40}$$

From Eqs. (32) and (33), note that $k_{y0} = k \sin\phi$.

3.2. Finite Element Treatment of the Periodic Unit Cell

In order to set up a finite element formulation taking into account the Rayleigh expansions established at the $y = y_{U,L}$ interfaces, the weighted residual Eq. (8) integrated over a periodic unit cell is again the starting point. Putting in the reciprocal periodic trial and weight functions, Eqs. (1) and (10), respectively, the surface integrations over the periodic faces of the unit cell cancel as already described but the rest of that integral remains. The residual, to be set to zero in establishing the finite element matrices, now becomes (say)

$$R = R_A + R_\Gamma \tag{41}$$

where R_A is the dual of Eq. (13), however, using the Floquet wavefactor $\gamma = -jk\cos\phi$. Thus

$$R_A = \int_A \frac{1}{\mu_r}\left[\nabla W_{pz}\cdot\nabla E_{pz} + jk\cos\phi\left(W_{pz}\frac{\partial E_{pz}}{\partial x} - \frac{\partial W_{pz}}{\partial x}E_{pz}\right)\right] dA$$
$$+ k^2 \int_A \left(\frac{\cos^2\phi}{\mu_r} - \varepsilon_r\right) W_{pz}E_{pz}\, dA \tag{42}$$

On the other hand, R_Γ has to be worked out from the surface integral

$$R_\Gamma = -\int_{\Gamma_{U,L}} \left[\mathbf{W}_p \times \frac{1}{\mu_r}(\nabla - \gamma \mathbf{1}_x) \times \mathbf{E}_p\right]\cdot \mathbf{1}_n\, d\Gamma$$
$$= -\int_{\Gamma_{U,L}} \frac{1}{\mu_r}\left[W_{pz}\left(\frac{\partial E_{pz}}{\partial x} - \gamma E_{pz}\right), W_{pz}\frac{\partial E_{pz}}{\partial y}, 0\right]\cdot \mathbf{1}_n\, d\Gamma \tag{43}$$

see Eq. (8), evaluated in two dimensions over $\Gamma_{U,L}$, the combined upper and lower parts of the periodic unit cell; for simplicity assume that the bounding surface $\Gamma_{U,L}$ lies in nonmagnetic material, so that μ_r may be set to unity in Eq. (43). Noting the sense of the line integration illustrated in Figure 4, so that the minus sign in front of Eq. (43) is subsumed either because $d\Gamma = -dx$ on Γ_U or because $\mathbf{1}_n = -\mathbf{1}_y$ on Γ_L, this equation is found to reduce to

$$R_\Gamma = \int_0^{D_x} \left[W_{pz}(x, y_U) \frac{\partial E_{pz}(x, y_U)}{\partial y} + W_{pz}(x, y_L) \frac{\partial E_{pz}(x, y_L)}{\partial y} \right] dx \tag{44}$$

Equation (44) is the key to the finite element process here since, through Eqs. (36)–(40), the derivatives of E_{pz} with respect to y in Eq. (43) can explicitly be linked with the external excitation due to the incident plane wave. Equation (44) becomes

$$\begin{aligned} R_\Gamma = &\int_0^{D_x} [W_{pz}(x, y_U) jk \sin\phi\, \mathscr{E}^I \exp(jky_U \sin\phi)] dx \\ &- \int_0^{D_x} \left\{ \sum_{n=-\infty}^{\infty} W_{pz}(x, y_U) jk_{yn} \mathscr{E}_n^R \exp\left[-j\left(\frac{2n\pi x}{D_x} + k_{yn} y_U \right) \right] \right\} dx \\ &+ \int_0^{D_x} \left\{ \sum_{n=-\infty}^{\infty} W_{pz}(x, y_L) jk_{yn} \mathscr{E}_n^T \exp\left[-j\left(\frac{2n\pi x}{D_x} - k_{yn} y_L \right) \right] \right\} dx \end{aligned} \tag{45}$$

Then substituting for $\mathscr{E}_n^R$ and $\mathscr{E}_n^T$ from Eqs. (38)–(40) finally gives

$$\begin{aligned} R_\Gamma = 2 &\int_0^{D_x} [W_{pz}(x, y_U) jk \sin\phi\, \mathscr{E}^I \exp(jky_U \sin\phi)]\, dx \\ &- \frac{1}{D_x} \int_0^{D_x} \int_0^{D_x} \sum_{n=-\infty}^{\infty} W_{pz}(x, y_U) E_{pz}(x', y_U) jk_{yn} \exp\left[j\, \frac{2n\pi(x' - x)}{D_x} \right] dx\, dx' \\ &+ \frac{1}{D_x} \int_0^{D_x} \int_0^{D_x} \sum_{n=-\infty}^{\infty} W_{pz}(x, y_L) E_{pz}(x', y_L) jk_{yn} \exp\left[j\, \frac{2n\pi(x' - x)}{D_x} \right] dx\, dx' \end{aligned} \tag{46}$$

The factor 2 multiplying the first integral in Eq. (46) arises from the extra term in the expression for $\mathscr{E}_n^R$ when $n = 0$, Eq. (39). Now representing E_{pz} by a nodal finite element expansion

$$E_{pz} = \sum_j E_j \alpha_j(x, y) \tag{47}$$

and choosing the weight functions W_p from the α's gives a finite element matrix equation

$$\mathbf{S}\mathbf{e}_p - jk\cos\phi\,\mathbf{X}\mathbf{e}_p + k^2\mathbf{T}\mathbf{e}_p + \mathbf{M}_U\mathbf{e}_p + \mathbf{M}_L\mathbf{e}_p = -2jk\sin\phi\,\mathscr{E}^I\exp(jky_U\sin\phi)\mathbf{b} \tag{48}$$

where $\mathbf{e}_p$ is the vector of all unknown nodal values of E_{pz} while the local versions of the rest of the arrays in Eq. (48), relating to a single finite element comprising the area A_e, are given below:

$$S_{ij} = \int_{A_e} \frac{1}{\mu_r} \nabla\alpha_i \cdot \nabla\alpha_j\, dA \tag{49}$$

$$X_{ij} = \int_{A_e} \frac{1}{\mu_r} \left(\alpha_i \frac{\partial\alpha_j}{\partial x} - \alpha_j \frac{\partial\alpha_i}{\partial x}\right) dA \tag{50}$$

$$T_{ij} = \int_{A_e} \left(\frac{\cos^2\phi}{\mu_r} - \varepsilon_r\right) \alpha_i\alpha_j\, dx\, dy \tag{51}$$

expressed here in a more general form than in Eqs. (17)–(19). The coefficients M_{Uij} and M_{Uij} all vanish for any finite element e that does not include one or other of the boundaries $y = y_{U,L}$. For an element that does have boundary edges b extending from x_1 to x_2 lying, say, on $y = y_U$, then, for nodes i in b and j in b' (extending from x'_1 to x'_2) and truncating the infinite sum to that between integers $-N_1$ and N_2,

$$M_{Uij} = -\frac{1}{D_x}\int_{x_1}^{x_2}\int_{x'_1}^{x'_2} \sum_{n=-N_1}^{N_2} \alpha_i(x, y_U)\alpha_j(x', y_U)\, jk_{yn} \exp\left[j\,\frac{2n\pi(x'-x)}{D_x}\right] dx\, dx' \tag{52}$$

similarly (with a change of sign and integration along $y = y_L$) for M_{Lij}. Finally, the vector $\mathbf{b}$ shown on the right-hand side of Eq. (48) is, for a node i in b,

$$b_i = \int_{x_1}^{x_2} \alpha_i(x, y_U)\, dx \tag{53}$$

otherwise zero. The evaluation of the element array coefficients Eqs. (49)–(53) for triangular or isoparametric quadrilateral elements is a routine matter, as is their assembly to form the global matrix equation, Eq. (48) (see Silvester and Ferrari, 1990).

There are a number of practical considerations involved in the setting up and solution of Eq. (48) in any given instance. In choosing the finite element mesh, triangular elements offer element matrices that have precalculated area integrations, whereas the isoparametric quadrilaterals, requiring a Gaussian area integration, allow curved boundaries to be modeled with facility. The size of the mesh needs to be chosen capable of faithfully reproducing harmonic variations on the local wavelength scale $\lambda_l = 2\pi/\omega\sqrt{\mu\varepsilon}$, say, a mesh-point separation of at most $\lambda_l/4$. Adequate modeling of the modes propagating along and bound to the structure, corresponding to Eq. (34), has been found to require the employment of an excessive number of Rayleigh harmonics as set by the choice of the integers $N_{1,2}$. Thus the boundaries of the finite element region at $y = y_{U,L}$ are best defined standing clear of the periodic structure so that any bound modes have decayed away sufficiently with $\pm y$. On the other hand, at these boundaries the finite element process is required to model the quantity

$$\exp\left[-j\left(\frac{2n\pi x}{D_x} \pm k_{yn} y_{U,L}\right)\right], \quad -N_1 \leqslant n \leqslant N_2 \tag{54}$$

If the integrations required by Eq. (52) are carried out numerically, the finite element process may fail to meet this obligation if $|n|$ is too large because the spatial oscillations represented by Eq. (54) are too fine. Indeed, it has been found that there exist optimum values of $N_{1,2}$, choosing values beyond which the overall accuracy of the finite element simulation actually worsens. Mias (1995) finds a "rule of thumb" that this optimum lies around the choice $N_1 = N_2 = N$, where N is the number of quadratic, 3-node, line elements contained within a spatial period D_x. On the other hand, if the integrations are performed analytically, as is certainly possible using triangular simplex elements, then this difficulty does not arise and the infinite sum may be truncated at any sufficiently large values $N_{1,2}$ that are convenient.

The matrices **S** and **T** are sparse and symmetric, whereas **X** is sparse and antisymmetric, all three having dimensions representing the total number of nodes in a region 1 unit cell, while $\mathbf{M}_U$, $\mathbf{M}_L$, and the vector **b** have dimensions corresponding to the much smaller number of nodes lying on the upper and lower boundaries of the unit cell. However, the two latter matrices are dense, complex, and unsymmetric; thus while to some extent the sparse, Hermitian character of the **S**, **T**, and **X** combination can be exploited in the solution of Eq. (48) for $\mathbf{e}_p$ given **b**, the solver employed has to be able to cope with unsymmetrical matrices. Once the vector $\mathbf{e}_p$ has been determined, simple postprocessing recovers both the total field within the grating finite element unit cell and that, for any order $n \in [-N_1, N_2]$, outside the grating; other physical attributes of the system are then obtainable without difficulty.

3.3. Alternative Solution Using a Green's Function for the Exterior Problem

Gedney et al. (1992) developed a different approach for the exterior part of the problem, using a Green's function technique. Thus an exterior problem solution by Galerkin's method is described as the "method of moments." In two dimensions, the kernel of the Green operation is a Hankel function; as a result of the periodicity an infinite sum of Hankel functions turns up, giving, after some manipulation, the same results as obtained from the Rayleigh plane-wave expansion above.

4. SCATTERING FROM DOUBLY PERIODIC STRUCTURES

The topic of scattering of plane waves and radiation from doubly periodic structures follows from the singly periodic analysis; it has applications including frequency-selective surface absorbers and phased array antennas. Jorgenson and Mittra (1991) describe solution of the scattering problem by the method of moments, a technique closely allied to finite element analysis but in itself not capable of tackling a general inhomogeneous periodic cell. Jin and Volakis (1993) treat an array of arbitrary cavities set in a perfectly conducting infinite half-space. They use the equivalence principle to replace the aperture fields of the (otherwise uncoupled) cavities by spatially periodic surface electric and magnetic currents residing on an infinite ground plane. Thence a Green's function approach taking the Floquet principle into account results in the requirement to solve a dense matrix system. Because the matrices refer to integrations entirely over a flat plane, it is possible to use a fast Fourier transform technique in conjunction with iterative matrix solvers employing the conjugate gradient method. Such a convolutional approach retains an $O(N)$ memory requirement for the solution procedure notwithstanding the dense matrices. The formulation derived here is based on that given by Mias et al. (1994), using the reciprocal-weight Galerkin method with a Rayleigh plane-wave expansion and is a generalization of that given for the singly periodic two-dimensional problem; Oelhoffen et al. (1994) independently also employed this approach. Lucas and Fontana (1995) use an equivalent "partial variational principle" for the doubly periodic problem, which overcomes the difficulties encountered with standard variational approaches when it is required to combine complex μ_r, ε_r, associated with lossy materials, with the complex Floquet phase factor $\exp(-\gamma x)$. The latter workers also include consideration of radiating sources within the periodic cell. There is no difficulty in including lossy materials and/or a periodic array of sources in the present Galerkin treatment.

4.1. The Doubly Periodic Configuration

Figure 5 shows the basic geometry considered; this is of some complexity. The periodicity is embodied in an array of hexahedral periodic cells lying parallel to the xy-plane. However, the primitive vectors **a** and **b** of the double periodicity (see Section 1.1) will not be restricted here to an orthogonal pair. Thus

$$\mathbf{a} = D_u \mathbf{1}_u, \quad \mathbf{b} = D_v \mathbf{1}_v \tag{55}$$

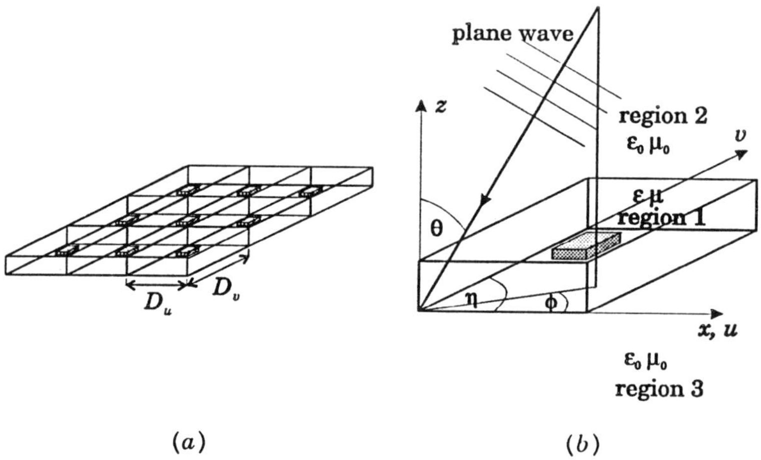

FIGURE 5. (a) Doubly periodic planar structure. (b) Plane wave incident on the structure at left.

where $\mathbf{1}_u$ and $\mathbf{1}_v$ are a pair of unit vectors lying in the xy-plane at an angle η to each other, so that the cross section of the unit cell viewed from above the xy-plane is a parallelogram having internal angle η.

Inside the unit cell the relative constitutive parameters, ε_r and μ_r, are arbitrary functions of position (x, y, z) with doubly periodic repetition outside. Electric and magnetic currents, $\mathbf{J}^S(x, y, z)$ and $\mathbf{M}^S(x, y, z)$, respectively, also disposed in a doubly periodic fashion, may exist as the excitation source for the system; a constant phase change between adjacent sources is precribed separately along each axis of periodicity. Alternatively, the arrangement is excited by an incident plane wave of arbitrary polarization; see Figure 5. In order to keep down the analytic complexity, as in the singly periodic case, regions 2 and 3 above and below the structure have been assumed to be free space.

4.2. Galerkin Formulation for Region 1

To facilitate the inclusion of periodic array of sources in the problem, Maxwell's equations, omitting the factor of $\exp(j\omega t)$, become (Kong, 1986)

$$\nabla \times \mathbf{E} = -j\omega_0\mu_r\mathbf{H} - \mathbf{M}^S \tag{56}$$

$$\nabla \times \mathbf{H} = j\omega\varepsilon_0\varepsilon_r\mathbf{E} + \mathbf{J}^S \tag{57}$$

Now suppose that the $\mathbf{E}$ variable has been chosen as the working variable while assuming that the sources $\mathbf{J}^S$ and $\mathbf{M}^S$ do not intersect the boundary S. An extended weighted residual expression, the sum of volume and surface integrals relating to region 1,

$$
\begin{aligned}
R = R_\Omega + R_S = & \int_\Omega \left[\frac{1}{\mu_r} (\nabla \times \mathbf{W}) \cdot (\nabla \times \mathbf{E}) - k^2 \varepsilon_r \mathbf{W} \cdot \mathbf{E} \right] d\Omega \\
& + \int_\Omega \left[jkZ\mathbf{W} \cdot \mathbf{J}^S + \mathbf{W} \cdot \nabla \times \left(\frac{1}{\mu_r} \mathbf{M}^S \right) \right] d\Omega \\
& - \oint_S \left(\mathbf{W} \times \frac{1}{\mu_r} \nabla \times \mathbf{E} \right) \cdot \mathbf{1}_n \, dS = 0
\end{aligned}
\tag{58}
$$

where $Z = \sqrt{\mu_0/\varepsilon_0}$, results from elimination of $\mathbf{H}$ between Eqs. (56) and (57) followed by multiplication by a vector weight $\mathbf{W}$ and integration by parts. Using the Maxwell equation (56), the surface integral part of the residual may be rewritten as

$$
R_S = jkZ \oint_S (\mathbf{W} \times \mathbf{H}) \cdot \mathbf{1}_n \, dS \tag{59}
$$

In the case of a doubly periodic structure, there is no simplification allowing the use of a scalar variable within the periodic cell while a vector Floquet factor γ, see Eq. (4), must be employed. Here this vector will be parallel to the xy-plane (indicated by the use of a subscript t) and known either from parameters relating to the incident wave or specified from the given phase between radiators in adjacent cells. Thus put

$$
\mathbf{E} = \mathbf{E}_p(\mathbf{r}) \exp(-\gamma_t \cdot \mathbf{r}), \quad \mathbf{H} = \mathbf{H}_p(\mathbf{r}) \exp(-\gamma_t \cdot \mathbf{r}) \tag{60}
$$

$$
\mathbf{J}^S = \mathbf{J}_p^S(\mathbf{r}) \exp(-\gamma_t \cdot \mathbf{r}), \quad \mathbf{M}^S = \mathbf{M}_p^S(\mathbf{r}) \exp(-\gamma_t \cdot \mathbf{r}) \tag{61}
$$

$$
\mathbf{W} = \mathbf{W}_p(\mathbf{r}) \exp(\gamma_t \cdot \mathbf{r}) \tag{62}
$$

where the subscript p denotes the double periodicity expressed by Eq. (55), and substitute into the residual Eq. (58), which becomes

$$
\begin{aligned}
R = R_\Omega + R_S = & \int_\Omega \left[\frac{1}{\mu_r} (\nabla + \gamma_t) + \mathbf{W}_p \cdot (\nabla - \gamma_t) \times \mathbf{E}_p - k^2 \varepsilon_r \mathbf{W}_p \cdot \mathbf{E}_p \right] d\Omega \\
& + \int_\Omega \left[jkZ\mathbf{W}_p \cdot \mathbf{J}_p^S + \mathbf{W}_p \cdot (\nabla - \gamma_t) \times \frac{1}{\mu_r} \mathbf{M}_p^S \right] d\Omega \\
& + jkZ \oint_{S_{U,L}} [\mathbf{W}_p \times \mathbf{H}_p] \cdot \mathbf{1}_n \, dS = 0
\end{aligned}
\tag{63}
$$

where $S_{U,L}$ denotes the upper and lower surfaces of the unit cell, all integrals over the periodic faces having cancelled out.

4.3. Rayleigh Expansions for Regions 2 and 3

The case $\mathbf{J}^S = \mathbf{M}^S = 0$, with excitation arising solely from an external incident plane wave, is analyzed first. Consider a general free-space plane wave, such as here the incident wave or one of the reflected or transmitted Rayleigh component waves, defined in terms of its electric field $\mathbf{E}_g$ and propagation vector $\mathbf{k}_g$ as

$$\mathbf{E}_g = \mathscr{E}_g \exp(-j\mathbf{k}_g \cdot \mathbf{r}), \quad |\mathbf{k}_f| = k \tag{64}$$

$$\mathbf{E}_g \cdot \mathbf{k}_g = 0 \tag{65}$$

while the magnetic field vector is recovered as

$$\mathbf{H}_g = \frac{\mathbf{k}_g \times \mathbf{E}_g}{kZ} \tag{66}$$

In its most general form a plane wave is elliptically polarized and thus expressible in terms of just two independent complex scalar amplitudes. Here it is convenient to write

$$\mathbf{E}_g = \mathbf{E}_{gt} + E_{gz}\mathbf{1}_z \tag{67}$$

$$\mathbf{k}_g = \mathbf{k}_{gt} + k_{gz}\mathbf{1}_z, \quad |\mathbf{k}_{gt}|^2 + k_{gz}^2 = k^2 \tag{68}$$

where the subscript t continues to be used as indicating a vector parallel to the xy-plane. From Eqs. (64), (67), and (68),

$$E_{gz} = \pm \frac{\mathbf{E}_{gt} \cdot \mathbf{k}_{gt}}{\sqrt{k^2 - \mathbf{k}_{gt}^2}} \tag{69}$$

the + sign for incident and transmitted waves, the − sign for reflected waves. It is seen that the wave amplitude expressed by Eq. (67) does indeed have just two degrees of freedom. Thus assuming k is given, $\mathbf{k}_{gt}$ and $\mathbf{E}_{gt}$ together are sufficient to define any elliptically polarized plane wave. Expanding Eq. (66) reveals further that for the respective types of wave

$$kZ\mathbf{H}_{gt} = \pm \frac{[k_{gx}k_{gy}E_{gx} + (k_{gy}^2 + k_{gz}^2)E_{gy}]\mathbf{1}_x - [k_{gx}k_{gy}E_{gy} + (k_{gx}^2 + k_{gz}^2)E_{gx}]\mathbf{1}_y}{\sqrt{k^2 - k_{gx}^2 - k_{gy}^2}} \tag{70}$$

Unlike in the singly periodic case, the transverse electric and magnetic reflected and transmitted waves are coupled together regardless as to the incident wave type, both polarizations occurring in the scattered waves. Thus there is little to be gained by separating the plane waves into component polarizations; instead, two-component vectors of the form $\mathbf{E}_{gt}$ representing both polarizations will be used, deriving $\mathbf{H}_{gt}$ from Eq. (70) where required. Such a plane-wave definition fits in very nicely with the fact that continuous tangential field vectors will figure prominently in the analysis at the interfaces of region 1 with regions 2 and 3.

The general elliptically polarized incident plane wave in region 2 traveling toward the periodic structure in a direction defined by the spherical angles θ, ϕ (see Figure 5) will be represented by

$$\mathbf{E}_t^I = \mathscr{E}_t^I \exp j(\mathbf{k}_t^I \cdot \mathbf{r}_t + k_z^I z) \tag{71}$$

where

$$k_x^I = k \sin\theta \cos\phi, \quad k_y^I = k \sin\theta \sin\phi, \quad k_z^I = k \cos\theta \tag{72}$$

For the time being assume that the primitive vectors **a** and **b** of the double periodicity are in fact orthogonal,

$$\mathbf{a} = D_x \mathbf{1}_x, \quad \mathbf{b} = D_y \mathbf{1}_y \tag{73}$$

The Rayleigh expansions for reflected and transmitted waves, introduced as scalars in Section 3.1, must be written in a more general vector form, the total field in region 2 being

$$\begin{aligned} \mathbf{E}_t^{(2)} = \mathbf{E}_t^I + \mathbf{E}_t^R &= \mathscr{E}_t^I \exp j(\mathbf{k}_t^I \cdot \mathbf{r}_t + k_z^I z) \\ &+ \sum_{m=-\infty}^{m=\infty} \sum_{n=-\infty}^{n=\infty} \mathscr{E}_{mnt}^R \exp -j(k_{xm}x + k_{yn}y + k_{zmn}z) \end{aligned} \tag{74}$$

while in region 3 there is just a transmitted field,

$$\mathbf{E}_t^{(3)} = \mathbf{E}_t^T = \sum_{m=-\infty}^{m=\infty} \sum_{n=-\infty}^{n=\infty} \mathscr{E}_{mnt}^T \exp -j(k_{xm}x + k_{yn}y - k_{zmn}z) \tag{75}$$

Using the same reasoning as in Section 3.1,

$$k_{xm} = \frac{2m\pi}{D_x} - k_x^I, \quad m = 0, \pm 1, \pm 2, \ldots \tag{76}$$

$$k_{yn} = \frac{2n\pi}{D_y} - k_y^I, \quad n = 0, \pm 1, \pm 2, \ldots \tag{77}$$

while

$$k_{zmn} = \sqrt{k^2 - k_{xm}^2 - k_{yn}^2} \quad \text{or} \quad -j\sqrt{k_{xm}^2 + k_{yn}^2 - k^2} \tag{78}$$

depending on whether $(k^2 - k_{xm}^2 - k_{yn}^2)$ is >0 or <0, again allowing the possibility of waves that are guided over the surface of the periodic structure. Determination of the unknown coefficients in the Rayleigh expansions may again be regarded as the ultimate goal here; this end can be achieved exploiting the orthogonality between individual Rayleigh components and from the matching of tangential Rayleigh field-components with those in the finite element domain, region 1, at appropriate interfaces. Equation (60) shows that the Floquet wavenumber is

$$\gamma_t = -j\mathbf{k}_t^I \tag{79}$$

while the periodic vectors here may be obtained from the corresponding "laboratory" vectors through multiplication by the factor $\exp -j(\mathbf{k}_t^I \cdot \mathbf{r}_t)$. Thus the total Floquet periodic fields in regions 2 and 3, respectively, are

$$\mathbf{E}_{pt}^{(2)} = \mathscr{E}_t^I \exp jk_z^I z + \sum_{m=-\infty}^{m=\infty} \sum_{n=-\infty}^{n=\infty} \mathscr{E}_{mnt}^R \exp -j\left(\frac{2m\pi}{D_x} x + \frac{2n\pi}{D_y} y + k_{zmn} z\right) \tag{80}$$

$$\mathbf{E}_{pt}^{(3)} = \sum_{m=-\infty}^{m=\infty} \sum_{n=-\infty}^{n=\infty} \mathscr{E}_{mnt}^T \exp -j\left(\frac{2m\pi}{D_x} x + \frac{2n\pi}{D_y} y - k_{zmn} z\right) \tag{81}$$

From the orthogonality of the Rayleigh components there results, in a similar fashion to the singly periodic expansion and with the case when m and n vanish together having to be considered separately,

$$\mathscr{E}_{mnt}^R = \frac{1}{D_x D_y} \int_0^{D_x} \int_0^{D_y} \mathbf{E}_{pt}^{(2)}(x', y', z_U) \times \exp j\left(\frac{2m\pi}{D_x} x' + \frac{2n\pi}{D_y} y' + k_{zmn} z_U\right) dx'\,dy', \quad (m, n) \neq (0, 0) \tag{82}$$

$$\mathscr{E}_{00t}^R = \frac{1}{D_x D_y} \int_0^{D_x} \int_0^{D_y} \left[\mathbf{E}_{pt}^{(2)}(x', y', z_U) \exp jk_{z00} z_U - \mathscr{E}_t^I \exp j(k_z^I + k_{z00}) z_U\right] dx'\,dy' \tag{83}$$

$$\mathscr{E}_{mnt}^T = \frac{1}{D_x D_y} \int_0^{D_x} \int_0^{D_y} \mathbf{E}_{pt}^{(3)}(x', y', z_L) \exp j\left(\frac{2m\pi}{D_x} x' + \frac{2n\pi}{D_y} y' - k_{zmn} z_L\right) dx'\,dy' \tag{84}$$

noting that, from Eqs. (76)–(78), in fact $k_{z00} = k_z^I$. The z-coordinates for the double Fourier (x', y') inversions Eqs. (82)–(84), recovering $\mathscr{E}_{mnt}^R$ and $\mathscr{E}_{mnt}^T$, have been chosen, respectively, as those of the unit cell upper and lower boundaries, $z_{U,L}$, since it is there that $\mathbf{E}_{pt}^{(2,3)}$ must, by elementary rules of electromagnetism, be made continuous with the finite element fields of region 1. Now substitute the complex Rayleigh coefficients, Eqs. (82)–(84), back into the Rayleigh expansions, Eqs. (80) and (81):

$$\mathbf{E}_{pt}^{(2)}(x, y, z) = \mathscr{E}_t^I \exp jk_z^I z - \mathscr{E}_t^I \exp jk_z^I(2z_U - z)$$

$$+ \sum_{m=-\infty}^{m=\infty} \sum_{n=-\infty}^{n=\infty} \frac{1}{D_x D_y} \int_0^{D_x} \int_0^{D_y} \mathbf{E}_{pt}^{(2)}(x', y', z_U)$$

$$\times \exp j\left[\frac{2m\pi}{D_x}(x' - x) + \frac{2n\pi}{D_y}(y' - y) + k_{zmn}(z_U - z)\right] dx'\,dy' \qquad (85)$$

$$\mathbf{E}_{pt}^{(3)}(x, y, z) = \sum_{m=-\infty}^{m=\infty} \sum_{n=-\infty}^{n=\infty} \frac{1}{D_x D_y} \int_0^{D_x} \int_0^{D_y} \mathbf{E}_{pt}^{(3)}(x', y', z_L)$$

$$\times \exp j\left[\frac{2m\pi}{D_x}(x' - x) + \frac{2n\pi}{D_y}(y' - y) - k_{zmn}(z_U - z)\right] dx'\,dy' \qquad (86)$$

say

$$\mathbf{E}_{pt}^{(2)}(x, y, z) = \mathscr{E}_t^I \exp jk_z^I z - \mathscr{E}_t^I \exp jk_z^I(2z_U - z)$$

$$+ \sum_{m=-\infty}^{m=\infty} \sum_{n=-\infty}^{n=\infty} \mathbf{E}_{ptmn}^U(x, y, z_U) \exp(-jk_{zmn} z) \qquad (87)$$

$$\mathbf{E}_{pt}^{(3)}(x, y, z) = + \sum_{m=-\infty}^{m=\infty} \sum_{n=-\infty}^{n=\infty} \mathbf{E}_{ptmn}^L(x, y, z_L) \exp(jk_{zmn} z) \qquad (88)$$

The first term in Eq. (87) represents the given incident wave, the second a direct reflected wave, while the third expresses the reflected Rayleigh harmonics with amplitude coefficients written in terms of the transverse electric field distributions $\mathbf{E}_{pt}^{(2)}$ at $z = z_U$, all given in Floquet periodic form. Equation (88) similarly represents the transmitted wave, now just Rayleigh harmonics, with amplitudes written in terms of $\mathbf{E}_{pt}^{(3)}$ at $z = z_L$.

Equation (70), relating general plane-wave fields $\mathbf{H}_{gt}$ to $\mathbf{E}_{gt}$, applies equally well to the Floquet periodic form since the Floquet components are all obtained from the laboratory form through multiplication by $\exp -j(\mathbf{k}_t^I \cdot \mathbf{r}_t)$. Thus from Eq. (70), the transverse $kZ\mathbf{H}_{pt}$ corresponding to each of the separate parts of Eqs. (87) and (88), that is incident wave, direct reflected wave and individual Rayleigh harmonics, can be determined. Each of these separate parts has its own propagation constant, $\mathbf{k} = (k_x^I, k_y^I, \mp k_z^I)$, or $\mathbf{k} = (k_{xm}, k_{yn}, \mp k_{zmn})$, while Eq. (70) requires this information. The $\mp$ sign of the k_z-component corresponds respectively to the $\pm$ overall sign of Eq. (70); thus, although the first two terms in Eq. (87) cancel when $z = z_U$ is substituted, because waves propagating in opposite z-directions are being represented, the corresponding magnetic fields from Eq. (70) *add*. It is clear, without entering into further algebraic detail, that the total $\mathbf{H}_{pt}(x, y, z)$ in regions 2 and 3 can be formed, and the substitution $z = z_{U,L}$ made. This yields the transverse Floquet magnetic field at the fintie element boundaries, say as

$$\mathbf{H}_{pt}^{U} = \mathscr{H}_t^I \exp jk_z^I z + \sum_{m=-\infty}^{m=\infty} \sum_{n=-\infty}^{n=\infty} \mathbf{H}_{pmnt}^{U} \tag{89}$$

$$\mathbf{H}_{pt}^{L} = \sum_{m=-\infty}^{m=\infty} \sum_{n=-\infty}^{n=\infty} \mathbf{H}_{pmnt}^{L} \tag{90}$$

written entirely in terms of the given parameters of the incident wave and the transverse electric fields, $\mathbf{E}_{pt}^{(2,3)}(x, y, z_{U,L})$, the latter, because of the continuity laws, being immediately accessible from the finite element process.

4.4. Rayleigh-Wave and Finite Element Region Coupling

The analysis above refers to plane waves in the free-space regions 2, 3 up to and including the boundaries $y_{U,L}$, respectively. However, invoking the rule of electromagnetics that tangential **E** and **H** shall be continuous across any interface, it is evident that the residual Eq. (63), significantly including the surface integral part R_S, Eq. (59), but for the moment ignoring internal sources, can now be written entirely in terms of an electric field vector $\mathbf{E}_p$ valid everywhere in Ω *up to and on* $S_{U,L}$. Note that further integrations over S_U and S_L, say, with respect to unprimed x and y are required, reproducing the equivalent of Eq. (46). The vector $\mathbf{E}_p$ will now be consistent with the elliptically polarized plane wave incident on S_U and the consequent Rayleigh plane-wave harmonics (also elliptically, but differently, polarized) reflected from S_U and transmitted through S_L. Thus a finite element discretization of Eq. (63) may be made using trial and weight functions

$$\mathbf{E}_p = \sum_j e_{pj}\boldsymbol{\tau}_j, \quad \mathbf{W}_p = \boldsymbol{\tau}_i \tag{91}$$

respectively, where the $\boldsymbol{\tau}$'s are tangentially continuous vector interpolation functions. This enables a matrix equation analogous to Eq. (48) for the scalar, singly periodic case,

$$\mathbf{A}\mathbf{e}_p + \mathbf{M}_U\mathbf{e}_p + \mathbf{M}_L\mathbf{e}_p = \mathbf{b}^I \tag{92}$$

with a right-hand side vector $\mathbf{b}^I$ dependent on the incident-wave amplitude $\mathscr{E}_t^I$, to be set up and whose solution for the unknown coefficients e_{pj} essentially completes the problem posed here. The discretization of the unit cell needs to be carried out in some standard fashion, employing, say, tetrahedral or hexahedral vector elements, so defining the interpolation functions $\boldsymbol{\tau}$. The lengthy expressions for the elements of the matrices **A** and $\mathbf{M}_{U,L}$ and vector $\mathbf{b}^I$ will not be written down here explicitly. The doubly infinite summations arising from the Rayleigh expansions, Eqs. (74) and (75), must be appropriately truncated, also following the procedure adopted for singly periodic gratings. The ultimate goal here—that of determining the amplitudes and polarizations of the reflected and transmitted Rayleigh harmonics, or perhaps the overall power reflection and transmission coefficients—can be achieved by a straightforward if analytically tedious postprocessing procedure applied to the finite element solution vector $\mathbf{e}_p$. Once again the final matrix loses some of the usual symmetry, expected of finite element arrays, in regions of its structure associated with the relatively small submatrices $\mathbf{M}_{U,L}$ for the boundaries $S_{U,L}$.

4.5. Internal Sources

The problem represented by a doubly infinite periodic planar array of radiators, without external excitation, requires just a simple modification of the procedure already given. It may be assumed that the radiators are modeled by electric and magnetic current sources,

$$\mathbf{J}^S = \mathbf{J}_p^S(\mathbf{r})\exp(-j\boldsymbol{\beta}_t\cdot\mathbf{r}), \quad \mathbf{M}^S = \mathbf{M}_p^S(\mathbf{r})\exp(-j\boldsymbol{\beta}_t\cdot\mathbf{r}_t), \quad \boldsymbol{\beta}_t = (\beta_x, \beta_y) \tag{93}$$

displaying constant phase shifts $\theta_x = \beta_x D_x$ and $\theta_y = \beta_y D_y$ on traversing between adjacent periodic cells in the x- and y-directions, respectively.† The "reflected" and "transmitted" wave terms in the Rayleigh expansion are retained in Eqs. (74) and (75) but interpreted as "radiated" into regions 2 and 3, respectively. The residual expression is still Eq. (63), but putting $\boldsymbol{\gamma}_t = j\boldsymbol{\beta}_t$ and having chosen $\mathbf{W} = \mathbf{W}_p(\mathbf{r})\exp(\boldsymbol{\beta}_t\cdot\mathbf{r}_t)$, with the source terms now included, while Eqs. (76) and (77) are replaced by

$$k_{xm} = \frac{2m\pi}{D_x} - \beta_x, \quad m = 0, \pm 1, \pm 2, \ldots \tag{94}$$

$$k_{yn} = \frac{2n\pi}{D_y} - \beta_y, \quad n = 0, \pm 1, \pm 2, \ldots \tag{95}$$

The procedure outlined in Sections 4.2–4.4, but with the new parameters here, remains valid so that a new finite element matrix equation,

$$\mathbf{A}\mathbf{e}_p + \mathbf{M}_U\mathbf{e}_p + \mathbf{M}_L\mathbf{e}_p = \mathbf{b}^S \tag{96}$$

arises now with a right-hand side $\mathbf{b}^S$ resulting from the phase-delayed periodic sources, Eq. (93).

†The vector $\boldsymbol{\beta}_t$ will be assumed purely real, since the general case of a complex $\boldsymbol{\gamma}_t$, to be associated with an exponential increase in source amplitude in some direction or other, does not seem to be a practical one.

4.6. Nonorthogonal Axes of Periodicity

The modifications required to take into account the general geometry with nonorthogonal axes of periodicity $\mathbf{1}_u$ and $\mathbf{1}_v$ aligned forming an angle η, Figure 5, are now outlined. It is convenient to take the x-axis to lie along the $\mathbf{1}_u$ direction while the z-axis coincides with $\mathbf{1}_u \times \mathbf{1}_v$. From Figure 5 it is seen that the transformations from (x, y) to (u, v) are

$$x = u + v\cos\eta, \quad y = v\sin\eta \tag{97}$$

The parameters of the general plane wave, Eq. (64), separated into transverse and normal components in Eqs. (67) and (68), were implicitly assumed to be represented in the Cartesian system, whereas the notation there could in fact refer to the (u, v, z) system. These remarks certainly apply to the incident wave (if any), as expressed by Eq. (71), so that the Floquet wavenumber remains as given in Eq. (79), $\gamma_t = -j\mathbf{k}_t^I$.

Hence it may be shown using the transformation Eq. (97) that, with skew axes of periodicity, the finite element procedure for solving the scattering problem may proceed using the (x, y) coordinates exactly as already described for orthogonal axes of periodicity, provided the pair of periodic lengths $(D_u, D_v\sin\eta)$ replaces (D_x, D_y) everywhere. In the case where there are sources $\mathbf{J}^S$ and $\mathbf{M}^S$ but no incident wave, the Floquet wavenumber $\gamma_t = j\boldsymbol{\beta}_t$ in its Cartesian representation is given by

$$\beta_x = \frac{\theta_u}{D_u}, \quad \beta_y = \frac{\theta_v}{D_v\sin\eta} \tag{98}$$

where θ_u, θ_v are the phase changes between adjacent radiating elements along the u- and v-axes. As in the orthogonal axes case, the driving-term right-hand side in the finite element matrix equations now results either from the incident wave or from the radiating source terms in the volume integral part of the residual, Eq. (63).

ACKNOWLEDGMENTS

The author is grateful for helpful discussions with Christos Mias.

REFERENCES

Cecchini, R., Coccioli, R., and Pelosi, G. (1995). PERIODIC[3]: A software package for the analysis of artificially anisotropic surfaces. *IEEE Antennas Propag. Mag.*, **APM-37**, 83–86.

Davies, J. B., Fernandez, F. A., and Philippou, G. Y. (1982). Finite element analysis of all modes in cavities with circular symmetry. *IEEE Trans. Microwave Theory Tech.* **MTT-30**, 1975–1982.

Delort, T. and Maystre, D. (1993). Finite-element method for gratings. *J. Opt. Soc. Am. A*, **10**, 2592–2601.

Ferrari, R. L. (1991). Finite element solution of time-harmonic modal fields in periodic structures. *Electron. Lett.* **27**, 33–34.

Ferrari, R. L. (1993). Electronic band structure for two-dimensional periodic lattice quantum configurations by the finite element method. *Int. J. Numer. Modell.: Electron. Networks, Devices Fields* **6**, 283–297.

Gedney, S. D., Lee, J. F., and Mittra, R. (1992). A combined FEM/MOM approach to analyze the plane-wave diffraction by arbitrary gratings. *IEEE Trans. Microwave Theory Tech.* **MTT-40**, 363–370.

Inoue, K., Hayata, K., and Koshiba, M. (1989). Finite-element solution of three-dimensional periodic waveguide problems. *Electron. Commun. Jpn.* **38**, Part 2, 68–77; translated from *Denshi Joho Gakkai Ronbunshi* **71-C**, 1404–1411 (1988).

Jin, J.-M., and Volakis, J. L. (1993). Scattering and radiation analysis of three-dimensional cavity arrays via a hybrid finite-element method. *IEEE Trans. Antennas Propag.* **AP-41**, 1580–1586.

Jorgenson, R. E., and Mittra, R. (1991). Scattering from structural slabs having two-dimensional periodicity. *IEEE Trans. Antennas Propag.* **AP-39**, 151–157.

Kong, J. A. (1986). *Electromagnetic Wave Theory.* Wiley, New York.

Lucas, E. W., and Fontana, T. P. (1995). A 3-D hybrid finite element/boundary element method for the unified radiation and scattering analysis of general infinite periodic arrays. *IEEE Trans. Antennas Propag.* **AP-43**, 145–153.

Mias, C. (1995). Finite element modelling of the electromagnetic behaviour of spatially periodic structures. Ph.D. Dissertation, University of Cambridge, Cambridge, UK.

Mias, C., and Ferrari, R. L. (1994). Closed singly periodic three dimensional waveguide analysis using vector finite elements. *Electron. Lett.* **30**, 1863–1865.

Mias, C., Webb, J. P., and Ferrari, R. L. (1994). Finite element analysis of electromagnetic plane wave scattering from an infinite doubly periodic structure. *COMPEL—Int. J. Comput. Math. Electr. Electron. Eng.* **13** Suppl. A, 393–398.

Moaveni, M. K. (1979). Analysis of transmission gratings by method of finite elements. *Proc. Inst. Electr. Eng.* **126**, 35–40.

Nakata, Y., and Koshiba, M. (1988). Finite-element analysis of plane-wave diffraction from metallic gratings with arbitrary complex permittivity. *Electron. Commun. Jpn.* **71**, Part 2, 80–91.

Oelhoffen, F., Lubrano, F., and Farail, F. (1994). Scattering by doubly periodic gratings. *COMPEL—Int. J. Comput. Math. Electr. Electron. Eng.* **13**, Suppl. A, 365–370.

Pelosi, G., Freni, A., and Coccioli, R. (1993). Hybrid technique for analyzing scattering from periodic structures. *IEE Proc., Part H: Microwaves Antennas Propag.* **140**, 65–70.

Silvester, P. P., and Ferrari, R. L. (1990). *Finite Elements for Electrical Engineers.* Cambridge Univ. Press, Cambridge, UK.

Slater, J. C. (1950). *Microwave Electronics.* Van Nostrand, New York.

Stone, G. O. (1972). Coupling matrices for high-order finite-element analysis of acoustic-wave propagation. *Electron. Lett.* **8**, 466–468.

CHAPTER THREE

Microwave Passive Devices

M. AUBOURG, S. VERDEYME, and P. GUILLON

1. INTRODUCTION

Reducing the cost of microwave circuits goes hand in hand with reducing their size. In this respect, microstrip and strip lines have been essential in eliminating bulky waveguides and coaxial lines in a great majority of microwave systems. A more recent advance in miniaturization of microwave circuits has been the appearance of the low-loss temperature-stable dielectric resonators (DRs). They are used to replace waveguide filters in such demanding applications as satellite communications, where planar metallic resonators cannot be used because of their inherently high losses. Furthermore, carefully designed microwave oscillators and power combiners, which utilize dielectric resonators, can equal the temperature stability of conventional microwave resonant cavities machined from Invar.

Although the geometrical form of a dielectric resonator is extremely simple, an exact solution of Maxwell equations is considerably more difficult than for hollow metal cavities. This difficulty holds true for an isolated dielectric resonator mounted in a microstrip or placed within a shielding metal cavity. For this reason the exact electromagnetic parameters of certain resonant modes (TE, TM, or HEM) can only be computed by rather complicated procedures. For an estimation of these parameters approximate models have been developed. But in the major practical applications these approximate models are not significant; that is the reason for which high-precision procedures for evaluating the electromagnetic parameters of DRs have been developed (Kajfez and Guillon, 1986).

Finite Element Software for Microwave Engineering, Edited by Tatsuo Itoh, Giuseppe Pelosi, and Peter P. Silvester.
ISBN 0-471-12636-5

These methods are the mode matching (both radial and axial) method, the Green's function method, the perturbations methods, and others. While the mathematical formulation in any of these methods is rather involved, once they are programmed on a computer, they provide rather accurate results.

Numerical solutions are another way to compute rigorously electromagnetic parameters of resonant modes of dielectric resonators. In this chapter we present one of these methods: the finite element one.

2. FINITE ELEMENT METHOD

We use now the finite element method (Zienkiewicz, 1979) to compute electromagnetic and electrical parameters of dielectric resonators by taking into account the influence of the environment on the characteristics of shielded dielectric resonators.

2.1. Maxwell's Equations

We consider here the general problem of the electromagnetic analysis of shielded microwave circuits in the frequency domain. Then these theoretical considerations are applied to dielectric resonator microwave devices.

The system (Figure 1) contains an arbitrary number of homogeneous materials, not necessarily isotropic, defined by tensor permittivity ε_r and tensor permeability μ_r. σ is the conductivity.

The domain Ω has a boundary $\partial\Omega$, which is made up of at most three parts: electric walls S_e, magnetic walls S_m, and ports $S_{\text{exc}}^{(i)}$ carrying electromagnetic

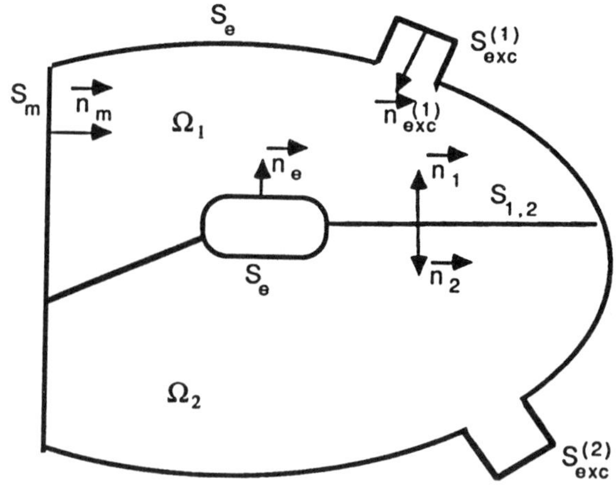

FIGURE 1. General structure to be tested.

power. These ports are plane sections of microwave guides and/or microwave lines.

We could consider two working modes:

- The system has no access and its mode of operation is the free oscillation one.
- The system has at least one access and its mode of operation is called the forced oscillation mode.

Note here that the boundary of Ω is the support of surface currents and surface charges. These currents are necessary for the existence of the electromagnetic field. So electric (magnetic) walls are defined as surfaces where magnetic (electric) current vanishes.

The currents at the ports are the sum of currents due to input and output propagation waves. We assume that at each port one propagation mode can exist, and we call $\mathbf{J}_m^{(i)}\delta_{S_{\text{exc}}}^{(i)}$ and $\mathbf{J}_e^{(i)}\delta_{S_{\text{exc}}}^{(i)}$ the currents due to the input propagation wave, which carries a unit electromagnetic power.

The currents at the ith access may be written

$$\begin{aligned} \mathbf{J}_m^{(i)}\partial_{S_{\text{exc}}}^{(i)} &= (a_i + b_i)\mathbf{J}_m^{(i)}\delta_{S_{\text{exc}}}^{(i)} \\ \mathbf{J}_e^{(i)}\delta_{S_{\text{exc}}}^{(i)} &= (a_i - b_i)\mathbf{J}_e^{(i)}\delta_{S_{\text{exc}}}^{(i)} \end{aligned} \tag{1}$$

a_i and b_i are, respectively, the magnitude coefficients of input and output waves. If a and b are column vectors whose components are a_i and b_i, the scattering matrix S is

$$b = Sa \tag{2}$$

Using normalized fields, in Eqs. (3) and (4) currents and charges are expressed as indicated in Eqs. (5)–(8):

$$\mathbf{E} \rightarrow \sqrt{\varepsilon_0}\,\mathbf{E} \tag{3}$$

$$\mathbf{H} \rightarrow \sqrt{\mu_0}\,\mathbf{H} \tag{4}$$

$$\mathbf{J}_e\delta_s \rightarrow \sqrt{\mu_0}\,\mathbf{J}_e\delta_s \tag{5}$$

$$\mathbf{J}_m\delta_s \rightarrow \sqrt{\varepsilon_0}\,\mathbf{J}_m\delta_s \tag{6}$$

$$\rho_e\delta_s \rightarrow \frac{\rho_e\delta_s}{\sqrt{\varepsilon_0}} \tag{7}$$

$$\rho_m\delta_s \rightarrow \frac{\rho_m\delta_s}{\sqrt{\mu_0}} \tag{8}$$

With these assumptions, Maxwell's equations satisfy Eqs. (9)–(12):

$$\nabla \times \mathbf{E} = -jk_0\mu_r\mathbf{H} - \mathbf{J}_m\delta_{S_m} - \sum_i \mathbf{J}_m^{(i)}\delta_{S_{\text{exc}}}^{(i)} \tag{9}$$

$$\nabla \times \mathbf{H} = jk_0\varepsilon_r\mathbf{E} + \mathbf{J}_e\delta_{S_e} + \sum_i \mathbf{J}_e^{(i)}\delta_{S_{\text{exc}}}^{(i)} \tag{10}$$

$$\nabla \cdot (\mu_r\mathbf{H}) = \rho_m\delta_{S_m} + \sum_i \rho_m^{(i)}\delta_{S_{\text{exc}}}^{(i)} \tag{11}$$

$$\nabla \cdot (\varepsilon_r\mathbf{E}) = \rho_e\delta_{S_e} + \sum_i \rho_e^{(i)}\delta_{S_{\text{exc}}}^{(i)} \tag{12}$$

with $k_0 = \omega/c$, c being the velocity of light in vacuum.

As usual in the frequency domain, the conductivity is taken into account by substituting in these equations the permittivity ε_r by the factor $\varepsilon_r - j\sigma/\omega\varepsilon_0$.

2.2. Electrical and Magnetic Field Formulations

It is well known that, in the frequency domain, both equations that contain divergence operators may be induced from ones that contain curl operators.

Note that the validity of this comment is extremely dependent on the choice of test functions. A bad choice of these test functions is responsible for the existence of spurious nonphysical modes, which appear in solving Maxwell's equations by the finite element method (Nedelec, 1986; Lee et al., 1991).

The electromagnetic field is modeled by both distributions **E** and **H** defined by functions that are as smooth as we want inside each volume element.

So the initial problem is slightly altered since faces of the volume elements have to be considered as interfaces between materials. This alteration has no physical effect if the mesh represents correct geometric features of the system.

As an example, we derive here the electrical field **E** formulation. We start by using the curl equation (9) to write **H** in terms of **E** and then we solve Eq. (10.)

We must do that carefully because of the presence of singular distributions. Noting that supports of surface distributions are negligible in R^3, Eq. (9) is equivalent to the following ones:

$$\nabla \times \mathbf{E} = -jk_0\mu_r\mathbf{H} \tag{13}$$

$$(\mathbf{n}_i \times \mathbf{E}_i + \mathbf{n}_j \times \mathbf{E}_j)\delta_{S_{ij}} = 0, \quad \forall i, j\ (i < j) \tag{14}$$

$$(\mathbf{n}_e \times \mathbf{E})\delta_{S_e} = 0 \tag{15}$$

$$(\mathbf{n}_m \times \mathbf{E})\delta_{S_m} = -\mathbf{J}_m\delta_{S_m} \tag{16}$$

$$(\mathbf{n}_{S_{\text{exc}}^{(p)}} \times \mathbf{E})\delta_{S_{\text{exc}}^{(p)}} = -\mathbf{J}_m^{(p)}\delta_{S_{\text{exc}}^{(p)}}, \quad \forall p \tag{17}$$

Equation (13) gives the expression of **H** in terms of **E** since the equality of distributions implies the unicity of the functions almost everywhere:

$$\mathbf{H} = -\frac{1}{jk_0}\mu_r^{-1}\nabla \times \mathbf{E} \tag{18}$$

Equation (14) means that there is no magnetic current on the interfaces S_{ij}, and to satisfy this equation, we have to force the continuity of the tangential component of $\mathbf{E}$ at the interface.

In the same way, we have to impose the nullity of the tangential component of $\mathbf{E}$ on electric walls, so that Eq. (15) is satisfied.

Since, on the magnetic walls, magnetic current is induced, Eq. (16) defines the current in terms of field, and so becomes a relation.

Using Eq. (10), we obtain $\mathbf{E}$ in such a way that its tangential component is continuous on S_{ij} and vanishes on S_e:

$$\nabla \times (\mu_r^{-1}\nabla \times \mathbf{E}) - k_0^2\varepsilon_r\mathbf{E} = -jk_0\mathbf{J}_e\delta_{S_e} - jk_0\sum_p \mathbf{J}_e^{(p)}\delta_{S_{\mathrm{exc}}^{(p)}} \tag{19}$$

$$(\mathbf{n}_{S_{\mathrm{exc}}^{(p)}} \times \mathbf{E})\delta_{S_{\mathrm{exc}}^{(p)}} = -\mathbf{J}_m^{(p)}\delta_{S_{\mathrm{exc}}^{(p)}}, \quad \forall p \tag{20}$$

From Eq. (19) we note that

$$[\mathbf{n}_e \times (\mu_r^{-1}\nabla \times \mathbf{E})]\delta_{S_e} = -jk_0\mathbf{J}_e\delta_{S_e} \tag{21}$$

When we write physical equations, we do not measure an existing field but we impose some quantities that will define the field. Since the electric current on electric walls is induced, we cannot impose any value. One way to do that is to use test functions that cannot interact with this electric current; so test functions will be normal to electric walls.

Since the only differential operator present in equations is the curl one, we can use test functions for which only the tangential component to S_{ij} is continuous. As a conclusion, we deduce that test functions and solutions are in the same space.

Let us call U the vectorial space of functions defined over $\mathbf{\Omega}$, valued in R^3, and for which tangential components are continuous on S_{ij}. In this space we define the following norm:

$$\forall \mathbf{\Phi} \in U, \quad \|\mathbf{\Phi}\|_u = \left(\int_\Omega \|\mathbf{\Phi}\|^2\,dx + \int_\Omega \|\nabla \times \mathbf{\Phi}\|^2\,dx\right)^{1/2} \tag{22}$$

Let us call U_0 the subspace of U for which functions are normal to electric walls:

$$U_0 = \{\mathbf{\Phi} \in U | \mathbf{n}_e \times \mathbf{\Phi} = 0 \text{ on } S_e\} \tag{23}$$

The problem now is to find $\mathbf{E} \in U_0$ such that $\forall \boldsymbol{\phi} \in U_0$:

$$(\mu_r^{-1} \nabla \times \mathbf{E}, \nabla \times \mathbf{\Phi}) - k_0^2 \langle \varepsilon_r \mathbf{E}, \mathbf{\Phi} \rangle = -jk_0 \sum_p \mathbf{J}_e^{(p)} \delta_{S_{\text{exc}}^{(p)}}, \mathbf{\Phi} \tag{24}$$

$$\langle (\mathbf{n} \delta_{S_{\text{exc}}^{(p)}} \times \mathbf{E}) \delta_{S_{\text{exc}}^{(p)}} \mathbf{\Phi} \rangle = \langle -\mathbf{J}_m^{(p)} \delta_{S_{\text{exc}}^{(p)}}, \mathbf{\Phi} \rangle \tag{25}$$

From Eq. (12) and considering the conservation of charge law, we obtain Eq. (26):

$$\nabla \cdot \mathbf{J}_e^{(p)} \delta_{S_{\text{exc}}^{(p)}} + jk_0 \rho_e^{(p)} d_{S_{\text{exc}}^{(p)}} = 0 \rightarrow \rho_e^{(p)} \delta_{S_{\text{exc}}^{(p)}} = -\frac{1}{jk_0} \nabla \cdot (\mathbf{J}_e^{(p)} \delta_{S_{\text{exc}}^{(p)}}) \tag{26}$$

An analogous way for solving Eq. (12) leads one to the choice of continuous test functions that vanish on electric walls. Let V be the space of functions defined over $\mathbf{\Omega}$, valued in R and continuous.

On V we define the following norm:

$$\forall \Phi \in V, \quad \|\mathbf{\Phi}\|_V = \left(\int_\Omega |\Phi|^2 \, dx + \int_\Omega \|\nabla \Phi\|^2 \, dx \right)^{1/2} \tag{27}$$

V_0 is the subspace of V for which functions vanish on electric walls:

$$V_0 = \{\Phi \in V \,|\, \Phi = 0 \text{ on } S_e\} \tag{28}$$

Comparing now Eqs. (29) and (30):

$$(\mu_r^{-1} \nabla \times \mathbf{E}, \nabla \times \mathbf{\Phi}) - k_0^2 \langle \varepsilon_r \mathbf{E}, \mathbf{\Phi} \rangle = -jk_0 \sum_p \langle \mathbf{J}_e^{(p)} \delta_{S_{\text{exc}}^{(p)}}, \mathbf{\Phi} \rangle, \quad \forall \mathbf{\Phi} \in U_0 \tag{29}$$

$$-k_0^2 \langle \varepsilon_r \mathbf{E}, \nabla \Phi \rangle = -jk_0 \sum_p \langle \mathbf{J}_e^{(p)} \delta_{S_{\text{exc}}^{(p)}}, \nabla \mathbf{\Phi} \rangle, \quad \Phi \in V_0 \tag{30}$$

It is clear that to obtain Eq. (30) we must have

$$\{\mathbf{\Phi} \in U_0 \,|\, \nabla \times \mathbf{\Phi} = 0\} = \{\nabla \Phi \,|\, \Phi \in V_0\} \tag{31}$$

Note that when discretizing the curl–curl problem, we will verify that this relation is always satisfied.

It is a remarkable fact that the space V_0 of the continuous scalar test functions, which does not appear explicitly in the formulation, imposes the discontinuity of normal components on vectorial test functions of space U_0.

When using nodal elements we have only

$$\{\mathbf{\Phi} \in U_0 \,|\, \nabla \times \mathbf{\Phi} = 0\} C \{\nabla \Phi \,|\, \Phi \in V_0\} \tag{32}$$

The explanation of the generation of spurious nonphysical modes is contained in this consideration, since the space of test functions is too small.

From the strict point of view of physics, all the interesting information is in the space of test functions since they define the forced values of the electromagnetic field.

As previously stated, we can consider two working modes named, respectively, free oscillation mode and forced oscillation mode.

Free Oscillation Working Mode. In the case of the free oscillation working mode, the access of the microwave device is short circuited. Equation (24) becomes

$$\{\langle \mu_r^{-1} \nabla \times \mathbf{E}, \nabla \times \mathbf{\Phi} \rangle - k_0^2 \langle \varepsilon_r \mathbf{E}, \mathbf{\Phi} \rangle\} = 0, \quad \forall \mathbf{\Phi} \in U_0 \tag{33}$$

This problem is more mathematical than physical since there are no physical sources. Its discretization leads to a numerical eigenvalue problem usually solved by projection methods.

Forced Oscillation Working Mode. In the case of forced oscillations we consider, for each propagative mode in the ports, a distribution called the elementary solution $\mathbf{\Psi}_p$, which is solution of the first modified Eq. (34). So for $\forall \mathbf{\Phi} \in U_0$,

$$\{(\mu_r^{-1} \nabla \times \Psi_p, \nabla \times \mathbf{\Phi}) - k_0^2 \langle \varepsilon_r \mathbf{\Psi}_p, \mathbf{\Phi} \rangle\} = \langle \tilde{\mathbf{J}}_e^{(p)} \delta_{S_{\text{exc}}^{(p)}}, \mathbf{\Phi} \rangle \tag{34}$$

$\tilde{\mathbf{J}}_e^{(p)}$ is a function defined on $S_{\text{exc}}^{(p)}$ due to the inward propagation wave of the pth port, which carries unitary power.

This function may be expressed in terms of the transverse electromagnetic field $(\tilde{\mathbf{E}}_t^{(p)}, \tilde{\mathbf{H}}_t^{(p)})$, which can be obtained from analytical or numerical two-dimensional analysis. This current on $S_{\text{exc}}^{(p)}$ is given by Eq. (35):

$$\tilde{\mathbf{J}}_e^{(p)} = \mathbf{n}_{\text{exc}}^{(p)} \times \tilde{\mathbf{H}}_t^{(p)} \tag{35}$$

Elementary solutions being known, the electric field may be expressed as

$$\mathbf{E} = -jk_0 \sum_p (a_p - b_p) \mathbf{\Psi}_p \tag{36}$$

With this assumption, Eq. (25) leads to

$$-jk_0 \sum_p (a_q - b_q) \langle \mathbf{n}_{S_{\text{exc}}^{(p)}} \times \mathbf{\Psi}_q \delta_{S_{\text{exc}}^{(p)}}, \mathbf{\Phi} \rangle = -(a_p + b_p) \langle \tilde{\mathbf{J}}_m^{(p)} \delta_{S_{\text{exc}}^{(p)}}, \mathbf{\Phi} \rangle, \quad \forall p, \forall \Phi \in U_0 \tag{37}$$

This expression could be rewritten:

$$
\begin{aligned}
&(\langle \tilde{\mathbf{J}}_m^{(p)} \delta_{S_{\text{exc}}^{(p)}}, \mathbf{\Phi} \rangle + jk_0 \langle (\mathbf{n}_{S_{\text{exc}}^{(p)}} \times \mathbf{\Psi}_p) \delta_{S_{\text{exc}}^{(p)}}, \mathbf{\Phi} \rangle) b_p + \sum_{q \neq p} jk_0 \langle (\mathbf{n}_{S_{\text{exc}}^{(p)}} \times \mathbf{\Psi}_q) \delta_{S_{\text{exc}}^{(p)}}, \mathbf{\Phi} \rangle b_q \\
&\quad = -(\langle \tilde{\mathbf{J}}_m^{(p)} \delta_{S_{\text{exc}}^{(p)}}, \mathbf{\Phi} \rangle - jk_0 \langle (\mathbf{n}_{S_{\text{exc}}^{(p)}} \times \mathbf{\Psi}_p) \delta_{S_{\text{exc}}^{(p)}}, \mathbf{\Phi} \rangle) a_p \\
&\qquad + \sum_{q \neq p} jk_0 \langle (\mathbf{n}_{S_{\text{exc}}^{(p)}} \times \mathbf{\Psi}_q) \delta_{S_{\text{exc}}^{(p)}}, \mathbf{\Phi} \rangle a_q
\end{aligned} \tag{38}
$$

This system (38) clearly defines linear relations between a_p and b_p but coefficients are distribution values rather than scalar ones.

Choosing $\mathbf{\Phi} = \mathbf{H}_t^{(p)}$ as a test function, we have

$$
\begin{aligned}
\langle \tilde{\mathbf{J}}_m^{(p)} \delta_{S_{\text{exc}}^{(p)}}, \mathbf{\Phi} \rangle &= \int_{S_{\text{exc}}^{(p)}} \tilde{\mathbf{J}}_m^{(p)} \cdot \tilde{\mathbf{H}}_t^{(p)} \, dS \\
&= -\int_{S_{\text{exc}}^{(p)}} \tilde{\mathbf{E}}_t^{(p)} \times \tilde{\mathbf{H}}_t^{(p)} \cdot \mathbf{n}_{S_{\text{exc}}^{(p)}} \, dS = -2
\end{aligned} \tag{39}
$$

Equation (38) becomes

$$
\begin{aligned}
&\left(2 - jk_0 \int_{S_{\text{exc}}^{(p)}} (\mathbf{\Psi}_p \times \tilde{\mathbf{H}}_t^{(p)}) \cdot \mathbf{n}_{S_{\text{exc}}^{(p)}} \, dS \right) b_p - \sum_{q \neq \pi} jk_0 \int_{S_{\text{exc}}^{(p)}} (\mathbf{\Psi}_q \times \tilde{\mathbf{H}}_t^{(p)}) \cdot \mathbf{n}_{S_{\text{exc}}^{(p)}} \, dS \, b_q \\
&\quad = -\left(2 + jk_0 \int_{S_{\text{exc}}^{(p)}} (\mathbf{\Psi}_p \times \tilde{\mathbf{H}}_t^{(p)}) \cdot \mathbf{n}_{S_{\text{exc}}^{(p)}} \, dS \right) a_p \\
&\qquad - \sum_{q \neq \pi} jk_0 \int_{S_{\text{exc}}^{(p)}} (\mathbf{\Psi}_q \times \tilde{\mathbf{H}}_t^{(p)}) \cdot \mathbf{n}_{S_{\text{exc}}^{(p)}} \, dS \, a_q, \quad \forall p
\end{aligned} \tag{40}
$$

This last system has the matrix form $Ub = Va$ and so the scattering matrix is

$$
S = U^{-1} V \tag{41}
$$

2.3. Discretization

The Galerkin procedure of discretization solves the problem in a subspace of the original one, this subspace having finite dimensions.

More accurately, we use test functions in a finite-dimensional subspace of U_0 and we look for the solution in its dual, which is also finite dimensional.

U_{h0} is this finite-dimensional subspace of U_0 and N, its dimension $\{\mathbf{W}_1, \ldots, \mathbf{W}_N\}$, is a basis.

It is easy to note here that, because of the linearity of operators, "for all functions in U_{h0}" is equivalent to writing "for all basis functions $\mathbf{W}_i$."

As for the original problem, we suppose that this space has Hilbertian features; so the space of solutions may be identified to U_{h0}. Thus the field may be expressed in terms of basis functions:

$$\mathbf{E} = \sum_{j=1}^{N} E_j \mathbf{W}_j \tag{42}$$

The approximative free oscillation problem may be written

$$\sum_{j=1}^{N} \left(\left(\int_\Omega (\mu_r^{-1} \nabla \times \mathbf{W}_j) \cdot \nabla \times \mathbf{W}_i \, dx \right) E_j = k_0^2 \sum_{j=1}^{N} \left(\left(\int_\Omega (\varepsilon_r \mathbf{W}_j) \cdot \mathbf{W}_i \, dx \right) E_j, \quad \forall i = 1, \ldots, N \tag{43}$$

This system may be written in an obvious matrix form:

$$Av = k_0^2 Bv \tag{44}$$

where A and B are square matrices of order N, v being a column vector whose jth component is E_j.

The forced oscillation problem is discretized in the same way.

Discretization of Space. The mesh is a powerful tool for spatial representation of numerous systems in physics, which gives the finite element method its success.

We consider here n-dimensional spaces with $1 < n < 3$.

A mesh is a list of elementary geometrical elements, which forms a partition of the system called triangulation. These elements are segments in R^1; quadrilaterals and triangles in R^2; and tetrahedrons, pentrahedrons, and hexahedrons in R^3.

Elements called simplexes are segments, triangles, and tetrahedrons. Since we can build any type of element using simplexes, we can use meshes composed only of simplexes without reducing the generality of the problem.

To define the geometrical element we can consider that the boundary of a simplex is composed of simplexes of lower dimensions. Using this assumption in a three-dimensional space we have:

- Three-dimensional element: tetrahedron.
- Two-dimensional element: triangle (face of tetrahedron).
- One-dimensional element: segment (edge of triangle).
- Zero-dimensional element: apex (edge of segment).

Some elementary rules must also be satisfied. In particular, the intersection of both elements must be

- A common face.
- A common segment.
- A common apex.
- The empty set.

The mesh must also be done in such a way that characteristic surfaces (like electric and magnetic walls), interfaces betweeen both materials, must be described by two-dimensional elements. So a volume element is entirely included in one and only one material.

Consideration of the geometry of the system may lead to the use of curvilinear elements.

Functional Discretization. The space called U is generally expressed by

$$H\nabla\times:\{\mathbf{V}\in[L^2(\Omega)]^3 \mid \nabla\times v\in[L^2(\Omega)]^3\}$$

The subspace U_h contained functions that have components in each element; these functions are polynomials of degree k. At the boundary between two elements the tangential components are continuous.

For polynomials of degree $k = 1$, a basis function $\mathbf{W}$ is defined for each edge of the mesh. In a triangle or a tetrahedron, this basis function is defined by

$$\mathbf{W} = \lambda_i \nabla\lambda_j - \lambda_j \nabla\lambda_i \tag{45}$$

This type of element is also called the Whitney form (Bossavit, 1988).

3. APPLICATION OF FEM TO DIELECTRIC RESONATOR DEVICES

3.1. Electromagnetic Analysis of DR Structures

Resonant Frequencies. In practical applications a dielectric resonator is expected to operate within a given frequency range and the electromagnetic field in the vicinity of the resonator is expected to exhibit behavior specific to the desired modes of operation. One of the disadvantages associated with the use of resonant circuits, however, is that one may find resonant frequencies of undesired modes in proximity to the resonant frequency of the desired mode. It is therefore of great practical importance to be able to determine the resonant frequencies, field patterns, energy, and dissipated power of unwanted modes of the resonator as well as those of the desired mode, so that operation of and coupling to the resonator can be obtained. Note also that the electromagnetic characteristics of the various modes are of course influenced by the surroundings of the resonator, such as metal cavity walls, metal tuning screws, or dielectric supports.

In practical applications the shape of the dielectric resonator most often used is a short solid cylinder, but one can also find tubular spherical and parallelepiped shapes. The commonly used resonant modes in cylindrical dielectrics is denoted $TE_{0,m,\delta}$, $TM_{0,m,\delta}$, and $HEM_{n,m,\delta}$, respectively, for transverse electric, transverse magnetic, and hybrid mode. The subscripts n, m, δ are, respectively, the azimuthal, radial, and axial wavenumbers.

When the dielectric resonator structure under consideration exhibits rotational symmetry with respect to the z axis only, one-half of the meridian (ϕ = constant) cross section is considered.

Now we direct our attention to the analysis of a shielded dielectric resonator, that is, a resonator placed in a cylindrical cavity on a dielectric substrate (Figure 2).

In the absence of a source, we assume here that the region of interest is characterized by permeability μ_0 and permittivity $\varepsilon_0\varepsilon_i$, where ε_i, the relative permittivity, can vary with position.

We have $\varepsilon_i = \varepsilon_s$ in the substrate, $\varepsilon_i = \varepsilon_r$ in the dielectric resonator, and $\varepsilon_r = 1$ in the remaining region bounded by the metallic walls, which are assumed to be electrically conducting. To obtain resonant frequencies and the field pattern, Eq. (33) is solved by using 2D FEM.

Figure 3 shows the dependence of the resonant frequencies of different TE, TM, and HEM modes of the same dielectric resonator as a function of the height H of the shield ($D = 10$ mm, $h = 5$ mm, $\varepsilon_r = 36$, $\varepsilon_s = 9.6$, $h_s = 1$ mm, $R = 15$ mm).

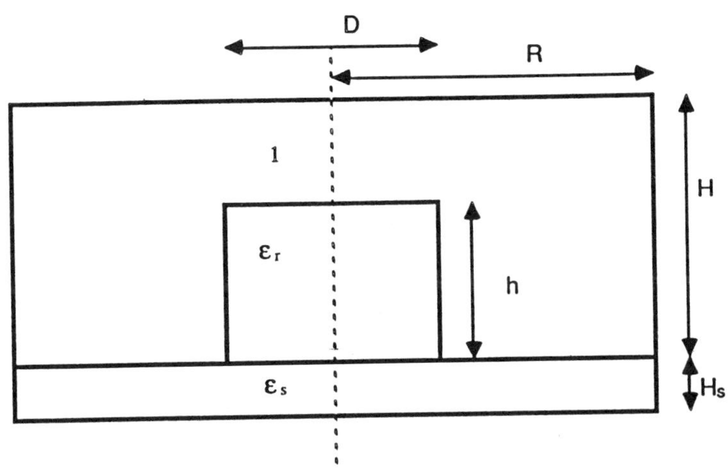

FIGURE 2. Dielectric resonator in a microstrip structure.

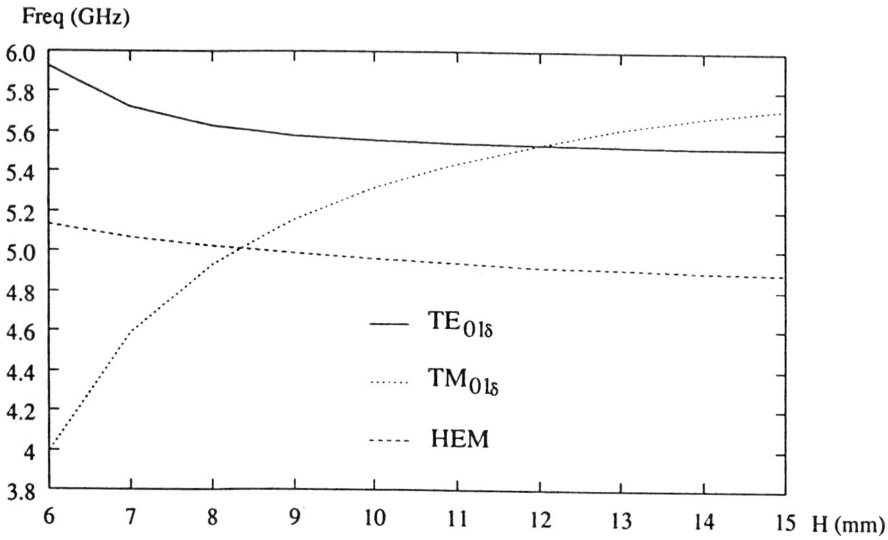

FIGURE 3. Resonant frequencies as a function of distance from the top cover.

Note that when the distance between the shield and the top of the resonator is small, the frequencies vary rapidly and the curves for various modes intersect, making the identification of modes extremely difficult. When this separation is reduced, the TE resonant frequencies increase, while the TM frequencies decrease.

Electromagnetic Field Pattern. Figures 4 through 10 show electric magnetic field plots of the modes. For the $TE_{0,1,\delta}$ mode, the electric field is shown in the equatorial plane. (The magnetic field in a meridian plane is shown in Figure 5.) Since this is an azimuthally symmetric mode, the plot of the electric field would be the same in any meridian plane. We observe that the electric field is quite strong everywhere within the equatorial plane of the resonator except near the resonator center.

The magnetic and electric fields of the $TM_{0,1,\delta}$ mode in the equatorial and meridian planes are shown in Figures 6 and 7, respectively. Observe that the magnetic field of this mode is well contained within the resonator.

The hybrid mode with lowest resonant frequency is the $HEM_{1,1,\delta}$ mode, which is depicted in Figure 8. This $HEM_{1,1,\delta}$ has two polarizations and consequently use of this mode is possible in dual mode filters.

Note that the magnetic field is very weak outside each resonator face, while the electric field is strongest there.

Finally, an example of a field plot of whispering gallery modes (WGMs), essentially used in the millimeter wavelength band, is given. WGMs are described as comprising waves running against the concave side of the

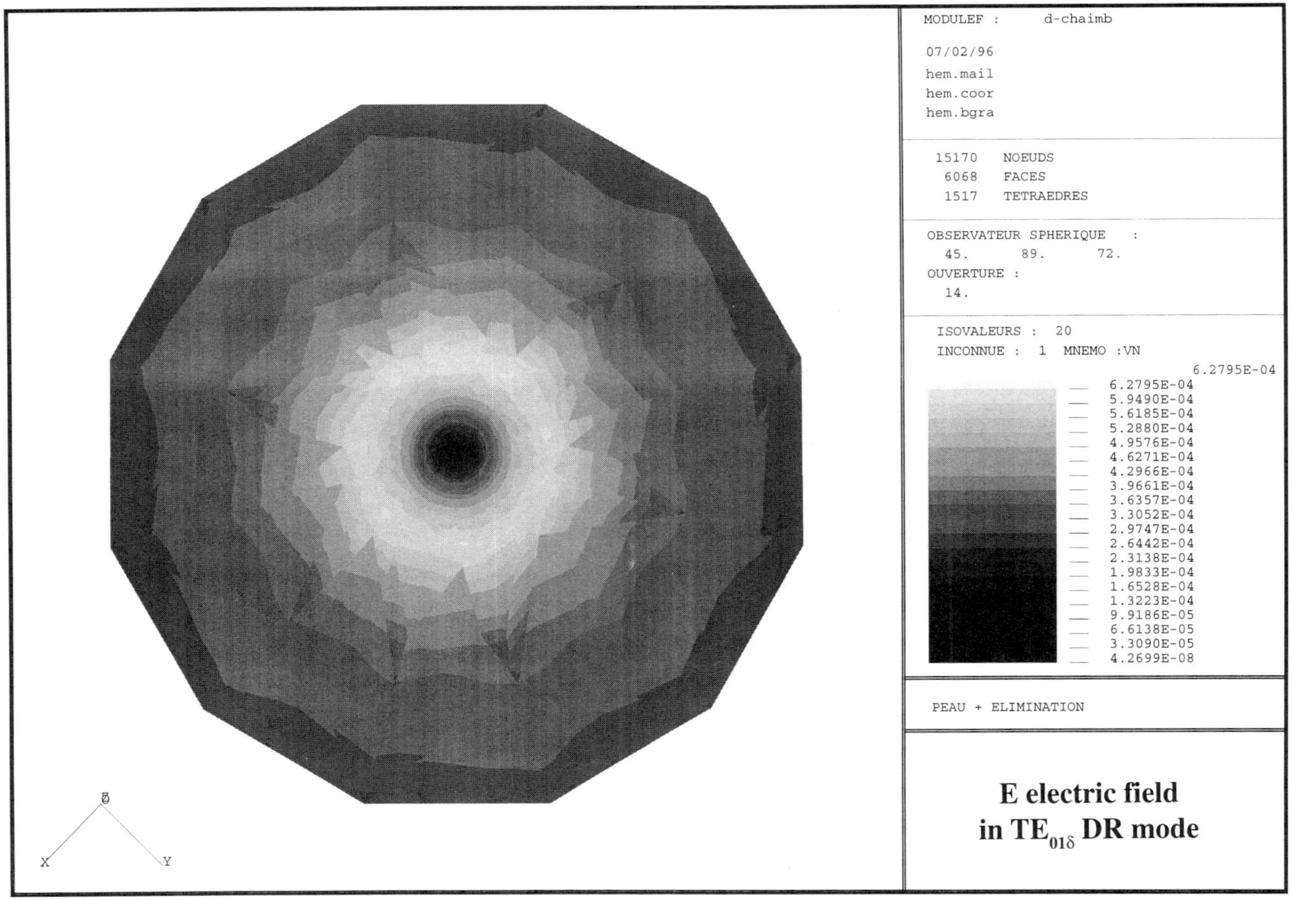

FIGURE 4. Electric field line of $TE_{0,1,\delta}$ DR mode.

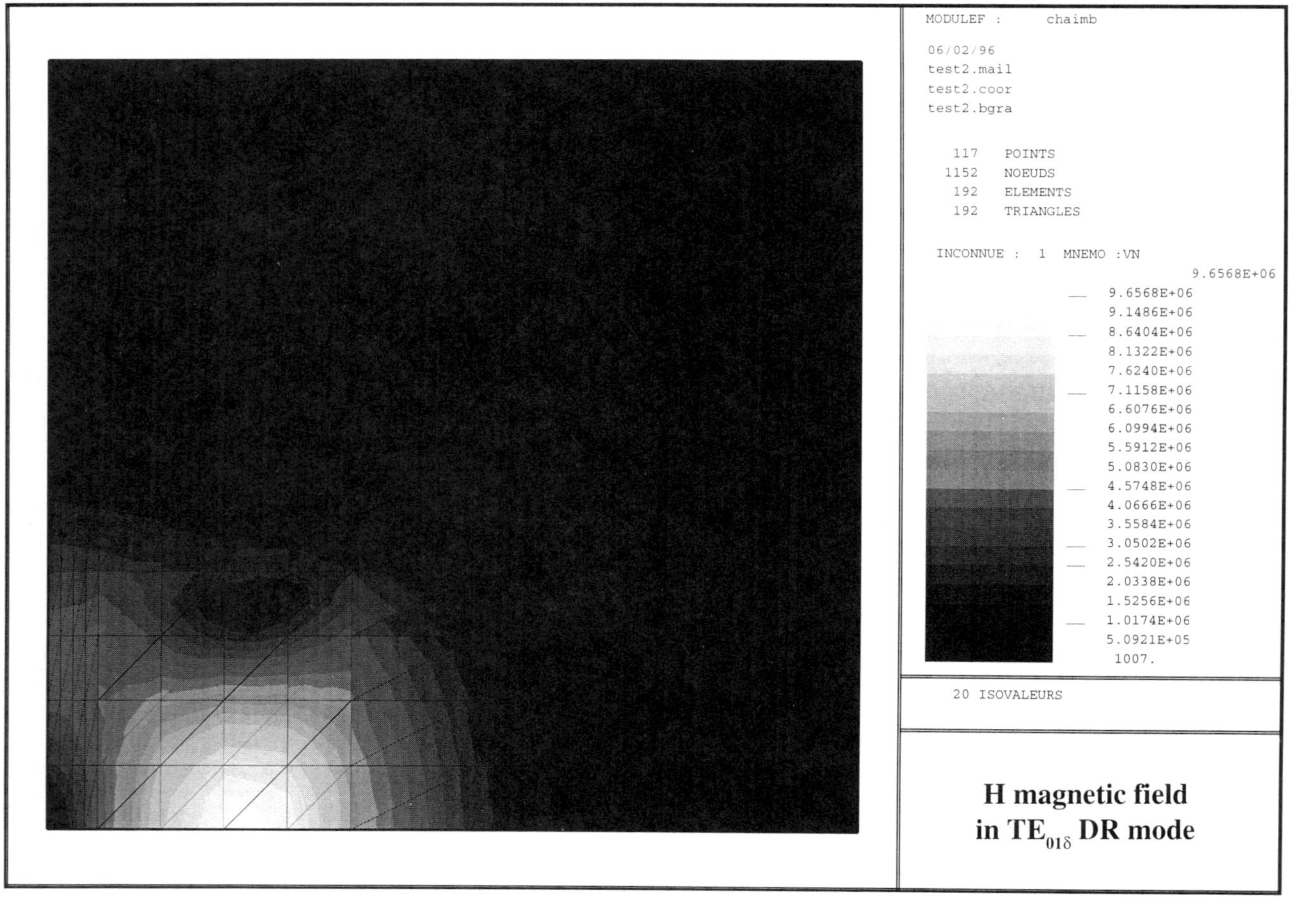

FIGURE 5. Magnetic field line of $TE_{0,1,\delta}$ DR mode.

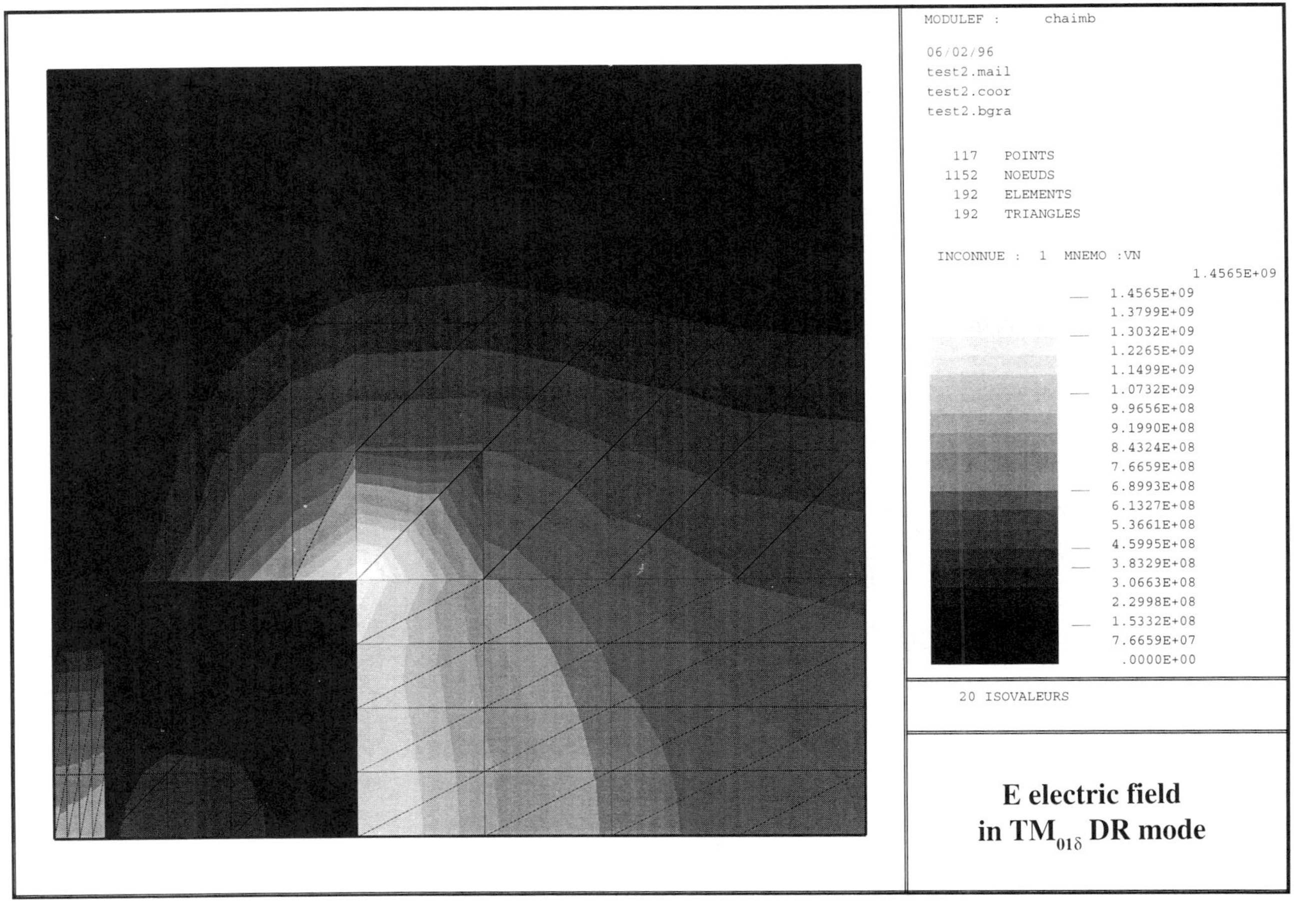

FIGURE 6. Electric field line of $TM_{0,1,\delta}$ DR mode.

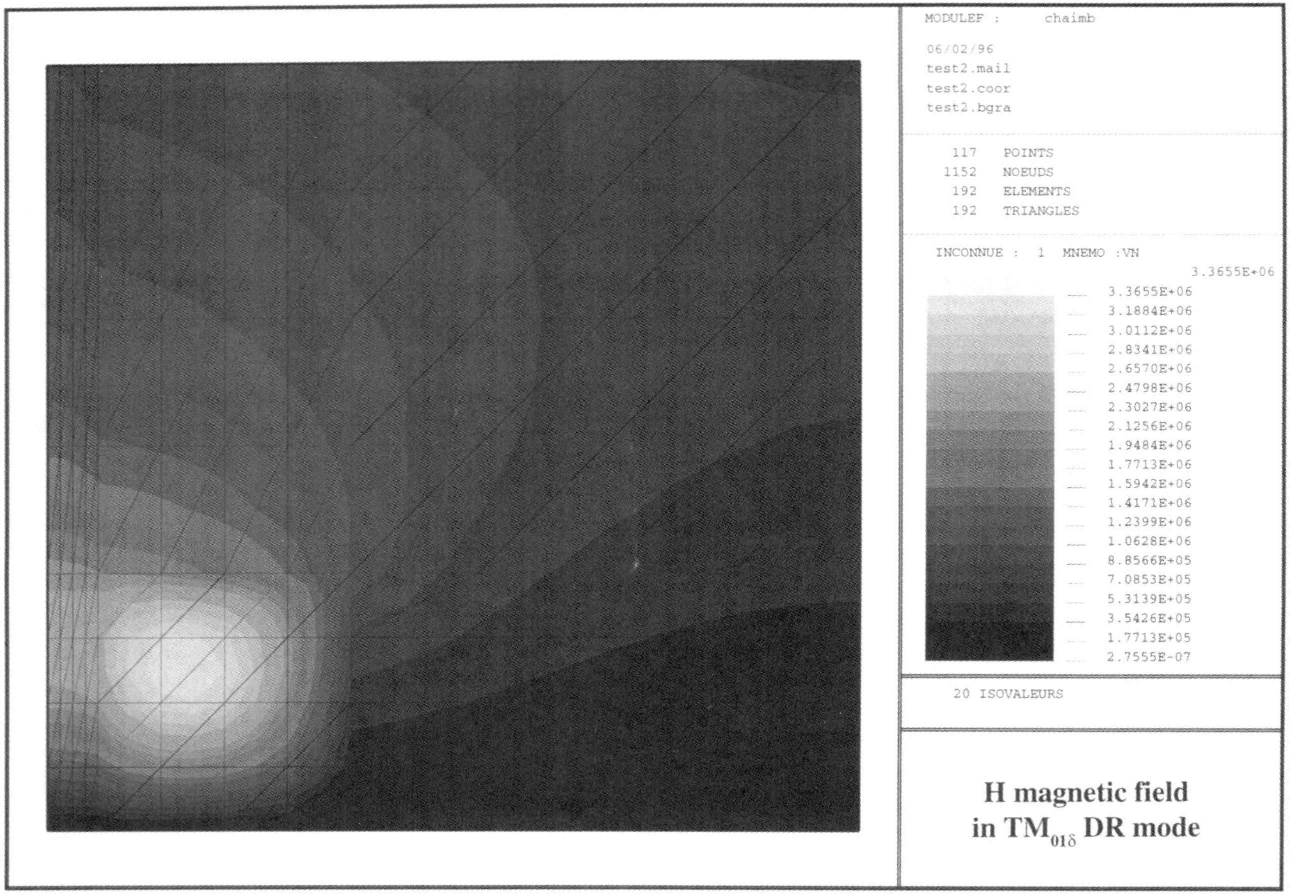

FIGURE 7. Magnetic field line of $TM_{0,1,\delta}$ DR mode.

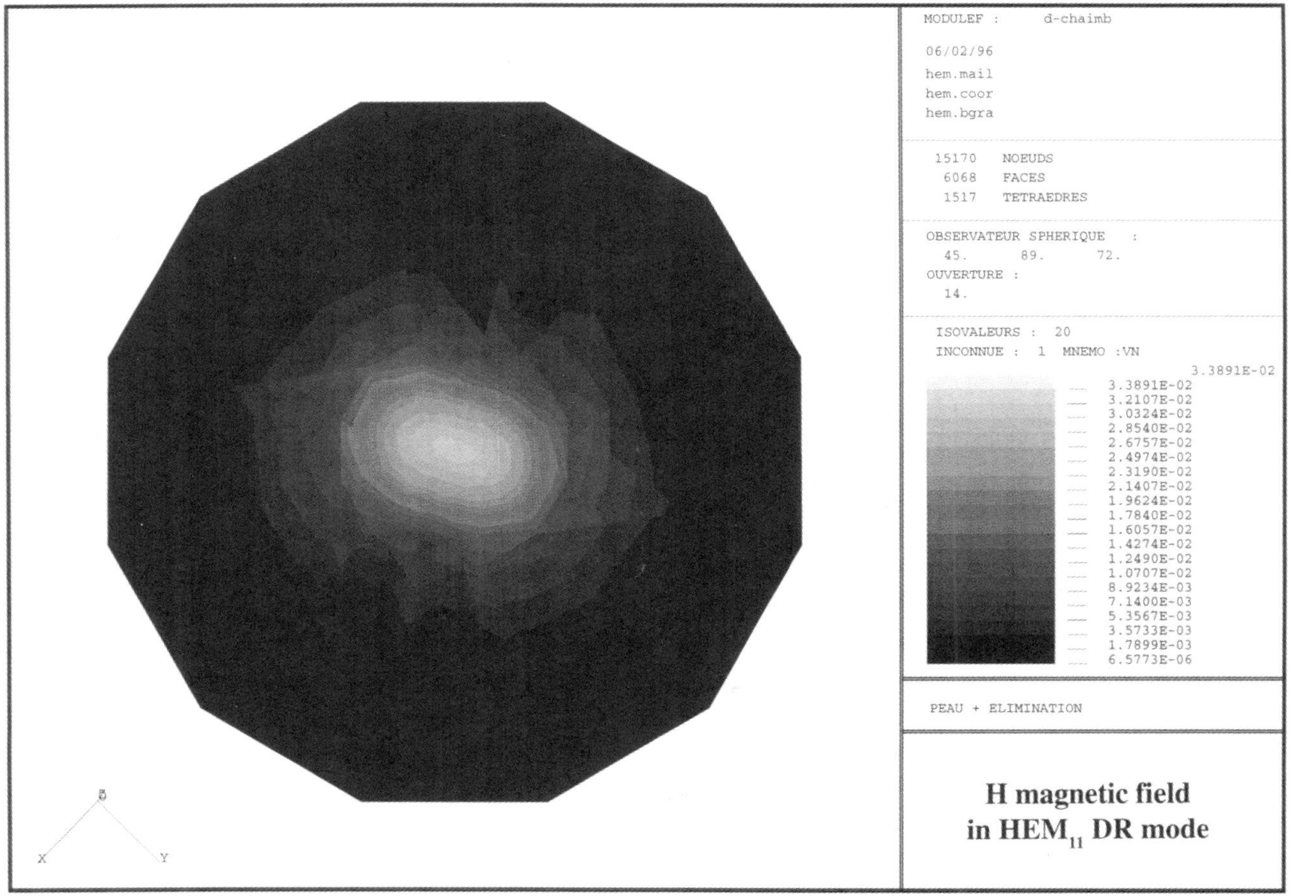

FIGURE 8. Magnetic field line of $HEM_{1,1,\delta}$ DR mode.

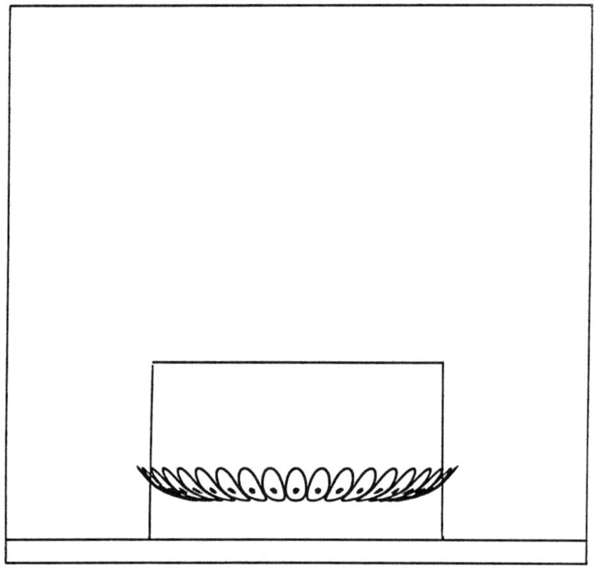

FIGURE 9. Magnetic field line of $WGH_{19,0,0}$ DR mode.

cylindrical boundary of the resonator. The waves move essentially in the plane of the circular cross section of the resonator, which has a diameter much larger than its height.

These modes, which run in the azimuthal direction, offer very good confinement of the field inside the dielectric resonator and so radiation losses are reduced. They are classified as either $WGE_{n,,m,\delta}$ modes, in which the electric field is essentially transverse, and $WGH_{n,m,\delta}$ modes, for which the electric field is essentially axial.

In Figures 9 and 10 magnetic field lines of $WGH_{19,0,0}$ and $WGE_{18,0,0}$ modes are shown.

***Q* Factor.** The unloaded Q is the Q factor due to losses in the shielded resonator itself:

$$Q_u = \omega_0 \frac{W}{P}$$

where P is the power dissipation (for resonators, power losses by conductors and dielectric materials contribute to unloaded Q) and W is the stored energy inside and outside the volume of the dielectric resonator.

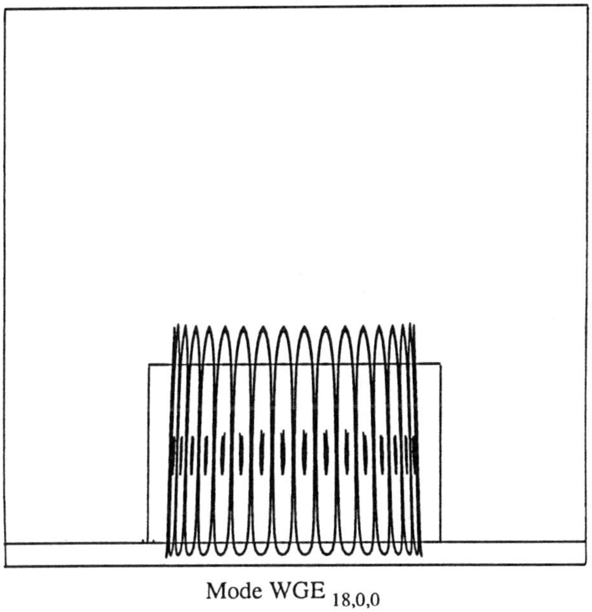

FIGURE 10. Magnetic field line of $WGE_{18,0,0}$ DR mode.

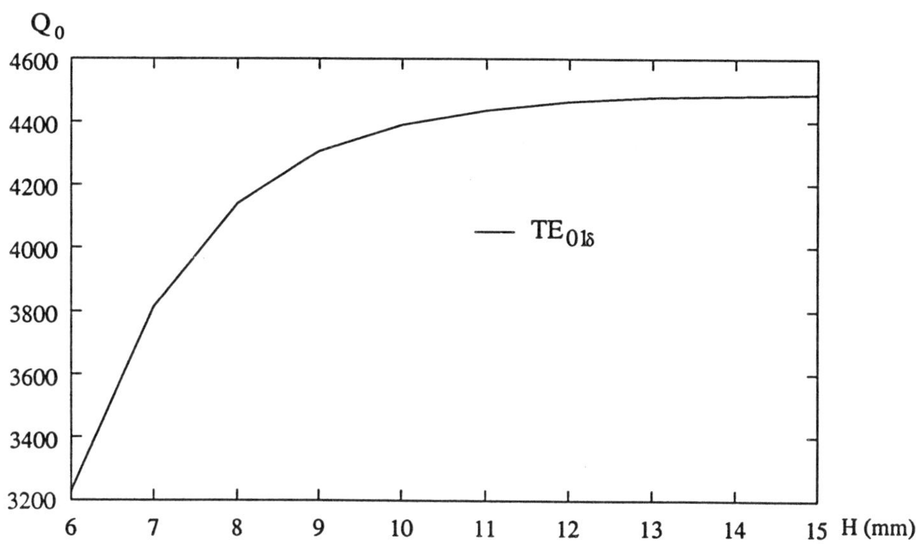

FIGURE 11. Q variation of DR in microstrip structure.

Using a complex formulation in FEM we can evaluate the Q_u factor of the resonant modes of a shielded dielectric resonator. Figure 11 shows the variation of the Q_u factor of $TE_{0,1,\delta}$ modes as a function of the top plane of the resonator ($D = 10\,\text{mm}$, $h = 5\,\text{mm}$, $\varepsilon_r = 36$, $\varepsilon_s = 9.6$, $h_s = 1\,\text{mm}$, $R = 15\,\text{mm}$).

3.2. Electrical Analysis of DR Structures

Microwave oscillators and filters form an important part of all microwave systems, such as those used in radar, communication links, navigation, and electric warfare. Dielectric resonators offer the advantages of small size, low cost, and high temperature stability. One important factor concerns coupling of the dielectric resonator to microstrip line, inductor loop, electric probe, or waveguide.

We first consider the coupling of a dielectric resonator with input and output ports.

As far as coupling of the $TM_{0,1,\delta}$ dielectric resonator mode with a probe, this application is one of the most popular. Use of the $TM_{0,1,\delta}$ mode has been proposed for resonator combiners and filters. Also, low frequency filters for cellular telephones were designed using the triple-mode TM.

Coupling of the $TM_{0,1,\delta}$ mode to a microwave structure could be achieved by using an electrical probe. A dielectric resonator centered in a metallic cavity is coupled electrically and so forms a one-pole bandpass filter (Figure 12). The coupling can easily be adjusted by moving the resonator and the probe away.

Solving Eq. (34) by means of FEM we obtain simultaneously S_{ij} parameters at the access planes of the microwave structure (Figure 13).

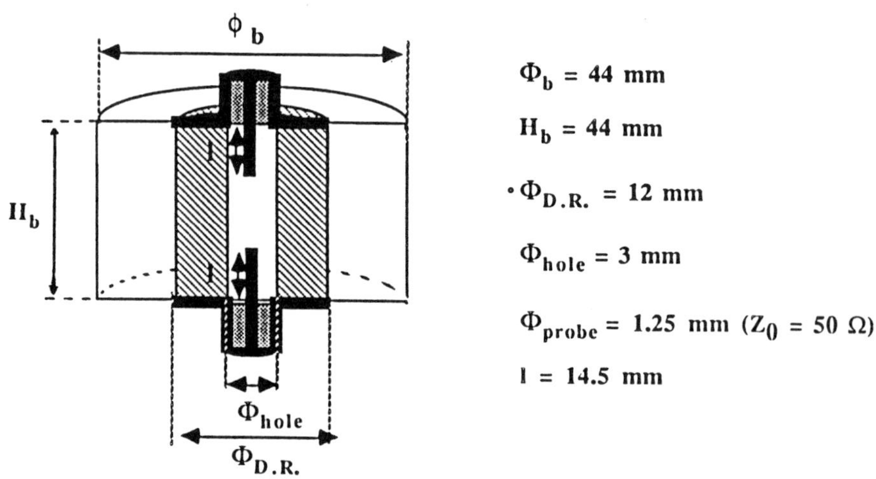

FIGURE 12. Coupling between $TM_{0,1,\delta}$ mode and a probe.

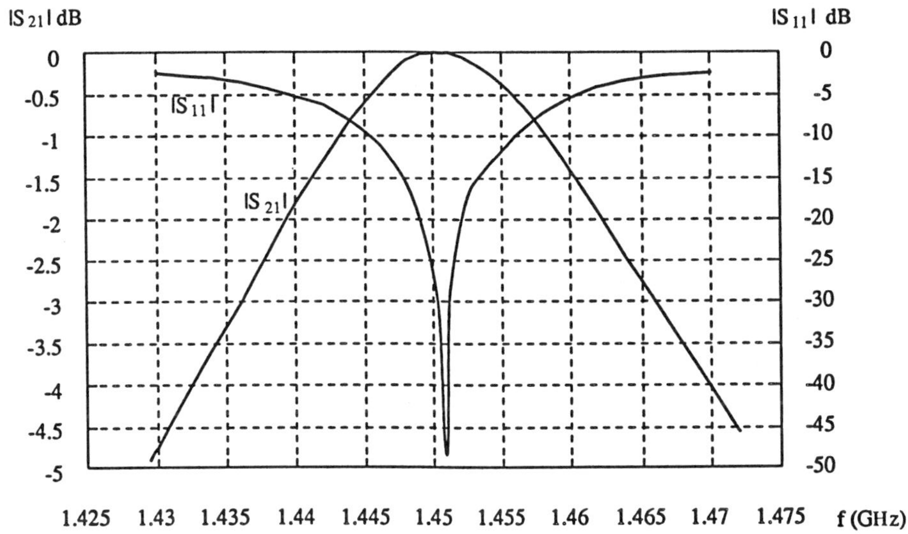

FIGURE 13. S_{ij} parameters as a function of frequencies.

Note that from the S_{12} curve we can obtain successively the loaded $Q(Q_L)$, which is the total Q for the system including power losses both internal and external to the resonator system, and the external quality factor Q_e, which takes into account the power loss due to the presence of an external load in the resonator system.

Coupling of the $TE_{0,1,\delta}$ Dielectric Resonator Mode with a Microstrip Line. The most commonly used mode in a cylindrical dielectric resonator is the $TE_{0,1,\delta}$. It is used in single-mode filters and oscillators. In this case, it is often coupled magnetically with one or two microstrip lines, forming a bandstop or a bandpass filter.

We start with a discussion of the coupling of the $TE_{0,1,\delta}$ mode in a dielectric cylindrical resonator with a microstrip line. It is well known that the $TE_{0,1,\delta}$ mode in the dielectric resonator can be approximated by a magnetic dipole. The coupling between the line and the resonator is accomplished by orienting the magnetic moment of the resonator perpendicular to the microstrip plane so that the magnetic lines of the resonator link with those of the microstrip line.

The dielectric resonator placed adjacent to the microstrip line operates like a reaction cavity, which reflects the radio frequency (RF) energy at the resonant frequency (Figure 14).

The coupling has been formulated in many papers, but these formulations use a low-frequency equivalent network approach that is only approximate.

Solving Eq. (34) we evaluate rigorously the scattering parameter of the dielectric resonator microwave structure. In particular, as shown in Figure 15,

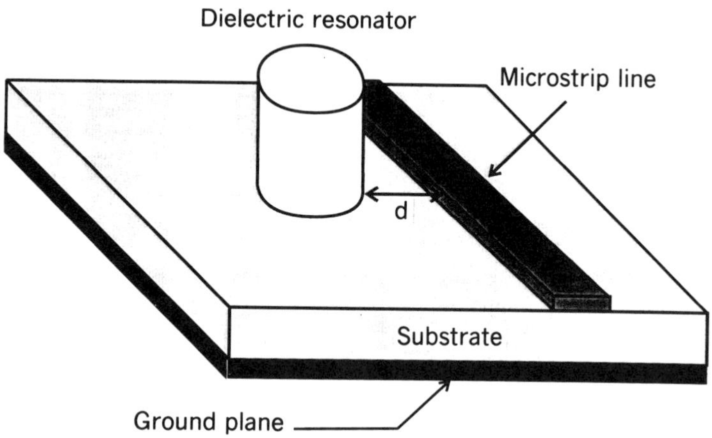

FIGURE 14. $TE_{0,1,\delta}$ DR mode coupled with a microstrip line.

the coupling can easily be adjusted by moving the resonator away or toward the center from the microstrip line.

The configuration in which a dielectric resonator is coupled with a microstrip line has essentially been used with a Gunn diode oscillator to reduce

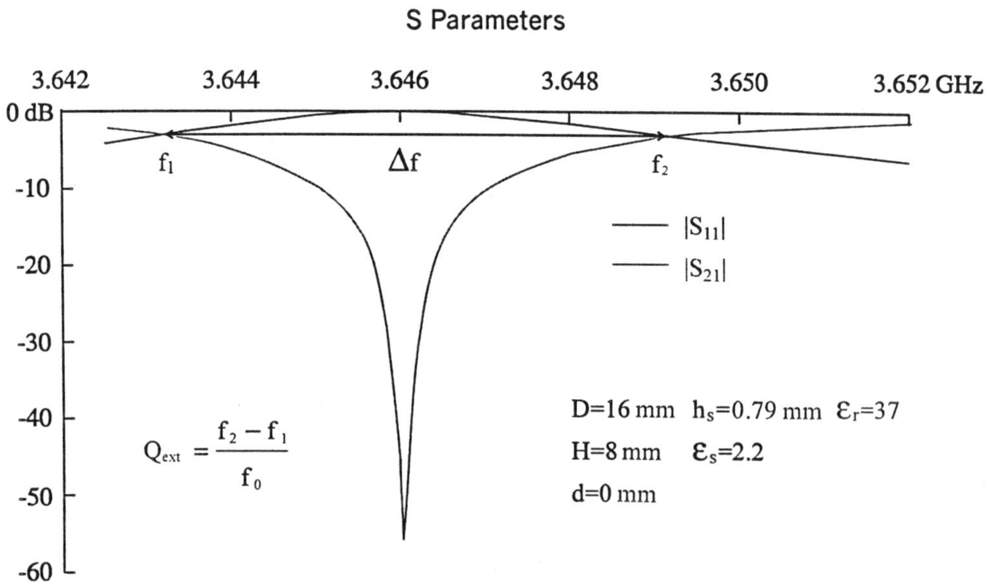

FIGURE 15. Scattering parameters of a $TE_{0,1,\delta}$ DR mode coupled with a microstrip line.

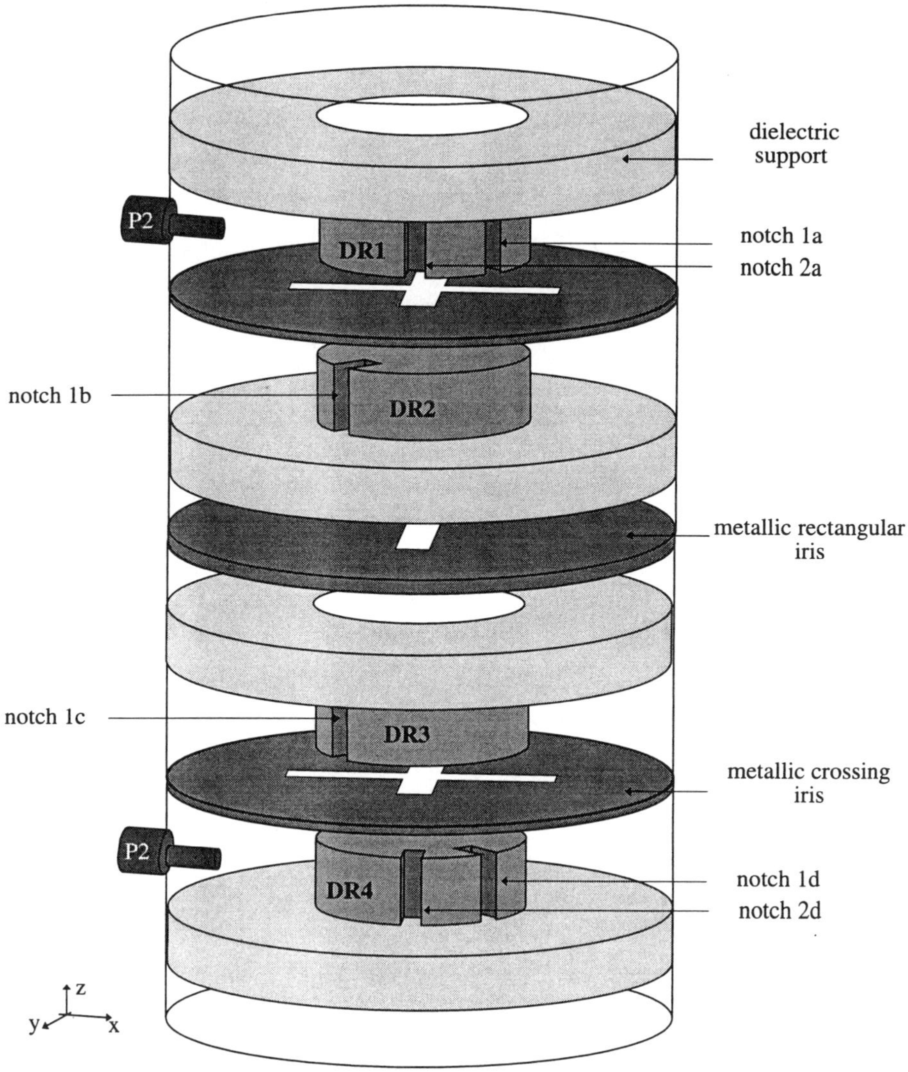

metallic cavity :

H_C = (14.1*4) mm
D_C = 31.0 mm

dielectric resonator :

H_{RD} = 3.4 mm
D_{RD} = 15.2 mm
ε_{RD} = 3 7

crossing iris :

ép = 1 mm
(1.00*3.16) mm
(1.00*6.90) mm

rectangular iris :

ép= 1 mm
(1.00 *13.80) mm

tuning notches (2a = 2d) :

l2 = 0.62 mm
L2 = 1.00 mm

coupling notches (1a = 1d) :

l1 = 0.94 mm
L1 = 1.00 mm

probes - DR distance :

SRD1 = SRD4 = 1.20 mm

FIGURE 16. Quasi-elliptic eight-pole DR dual-mode filter.

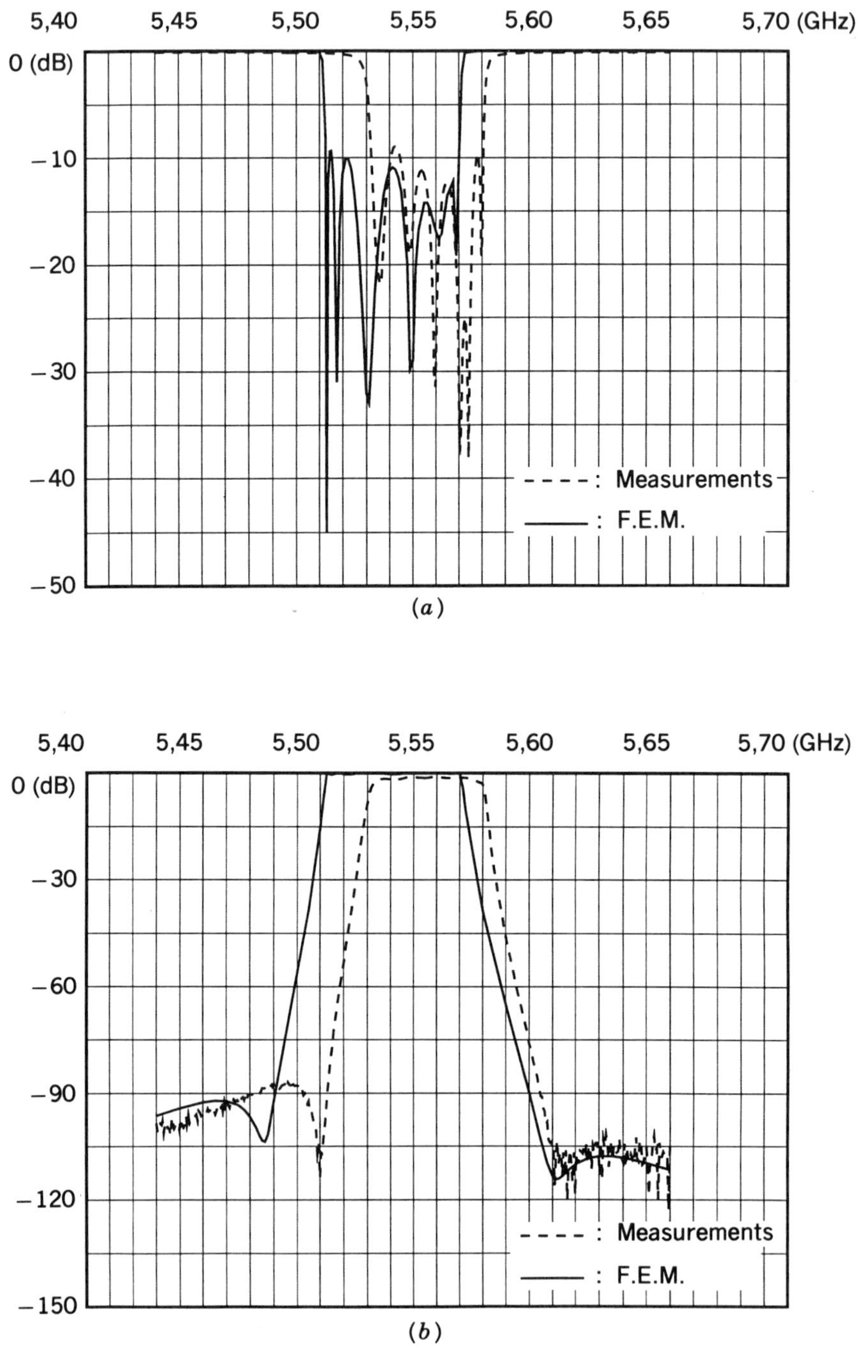

FIGURE 17. Theoretical and experimental transmission (a) and reflection (b) responses of quasi-elliptic eight-pole DR dual-mode filter.

frequency variations due to temperature or bias changes and to lower oscillator phase noise. FET or bipolar oscillators using a dielectric resonator as a stabilizing element are classified as reflection or feedback oscillators. Actually, these types of oscillators are working at up to 60 GHz. Their phase noise performance demonstrates the suitability of such oscillators for strong communications systems.

Dual-Mode Filters. Dual-mode configuration is preferred for satellite applications and allows simple realization of high performance elliptic filters. A reduction in size and weight with minimum degradation of resonator Q is achieved.

Dual-mode filters use the $HEM_{1,1,\delta}$ mode of a cylindrical dielectric resonator to design the microwave cavity filter. A determination of basic factors like resonant frequencies, Q factor, and coupling coefficient between individual modes (or input/output) is necessary.

A typical filter configuration is presented in Figure 16 in which dielectric resonators are mounted axially in the center of each evanescent cavity. Electrical probes are used as a means of input/output coupling. Coupling between modes within a single dielectric resonator is achieved via a mode coupling screw or notch at an angle of 45° with respect to the orthogonal tuning screws. Intercavity coupling is achieved by means of selective polarization discriminative coupling slots. This kind of filter can be synthesized using the FEM approach.

Electrical performance for this dielectric filter is displayed in Figure 17. This type of filter is very similar to the standard $TE_{1,1,1}$ metal wall cavity filter. However, the volume of this one is reduced by 10 or 12.

Excellent insertion loss performance of the filter indicates that degradation of the dielectric resonator Q by the metal wall of the housing is minimal. An additional advantage of this type of configuration is that tuning and coupling adjustments are very easy.

4. CONCLUSION

The finite element method is a very convenient tool for the analysis of dielectric resonator devices. It permits rigorous evaluation of both electromagnetic and electrical parameters of dielectric resonators inserted into any kind of microwave circuit. The same results could be obtained for other microwave devices, in particular, traveling wave tubes or coplanar circuits.

REFERENCES

Bossavit, A. (1988). Whitney forms: A class of finite element for three dimensional computations in electromagnetism. *IEEE Proc.* **135**, Part A, No. 8.

Lee, J. F., Sun, D. K., and Cendes, Z. J. (1991). Full wave analysis of dielectric waveguides using tangential rector finite elements. *IEEE Microwave Theory Tech.* **MTT-39**, 1262–1271.

Kajfez, D., and Guillon, P. (1986). *Dielectric Resonators.* Artech House, Dedham, MA.

Nedelec, J. C. (1986). A new family of mised finite elements in R^3. *Numer. Mathe.* **50**, 57–81.

Zienkiewicz, O. C., (1979) *La méthode des éléments finis.* McGraw-Hill, New York.

CHAPTER FOUR

Computer-Aided Design of Integrated Optical Waveguide Devices

MASANORI KOSHIBA

1. INTRODUCTION

As optical waveguide devices penetrate into many photonic systems, there is a demand for user-friendly design and modeling techniques. For this purpose, various numerical techniques have been developed. In particular, the finite element method (FEM) is a powerful and efficient tool for general optical waveguiding problems. A very good review of the progress to date on the application of FEM to modeling of microwave and optical waveguides has been described by Rahman et al. (1991) and Davies (1993). Recently, a full-vector FEM has been utilized as a waveguide solver of CAD packages, as in Young (1988) and Sewell and Cvetkovic (1989). Although the vector FEM enables one to compute accurately the mode spectrum of a waveguide with arbitrary cross section, it has been known to include nonphysical spurious solutions and to require a long computer time and/or large computer memory. In order to cure these problems, different types of the vector FEM have been developed, and, more recently, a comparison of the different FEM formulations has been reported by Dillon and Webb (1994).

In this chapter a simple scalar FEM proposed by Mabaya et al. (1981) and Koshiba et al. (1982) is introduced for the design and modeling of optical waveguide devices and is utilized as a solver of an analysis/design package for

Finite Element Software for Microwave Engineering, Edited by Tatsuo Itoh, Giuseppe Pelosi, and Peter P. Silvester.
ISBN 0-471-12636-5

step-index and/or graded-index anisotropic optical waveguides. The scalar FEM has as its main advantages: a smaller matrix dimension, less computer time, and no spurious solutions. In fact, the scalar FEM has been successfully applied to various optical waveguides, such as optical fibers of arbitrary cross section by Chiang (1986) and Eguchi and Koshiba (1994), inhomogeneous optical waveguides by Strake et al. (1988), Mustacich (1988), and Bersiner et al. (1991), optical rib waveguides by Koshiba et al. (1992), multiple-quantum-well waveguides by Osman et al. (1994), lossy optical waveguides by Hayata et al. (1986) and Themistors et al. (1994), leaky optical waveguides by Grant et al. (1994), magnetooptic waveguides by Koshiba and Zhuang (1993), and nonlinear optical waveguides by Li et al. (1992), Zoboli et al. (1993), and Niiyama et al. (1995).

In addition to the optical waveguide solver based on the scalar FEM, the present CAD package has three other solvers for the electrostatic problem, the elasticity problem, and the heat-conduction problem. The FEM is also useful because the solvers can be linked to the optical waveguide solver. For instance, an electrostatic solver, a stress solver, or a thermal solver is linked to the optical waveguide solver to produce a two-step analysis of electrooptic, acoustooptic, or thermooptic modulation of optical waveguides. These solvers are incorporated into an optical CAD system, ELM (expert lightwave modeling system), which has been developed on a workstation at Hokkaido University, Sapporo, Japan.

2. BASIC EQUATIONS

An optical waveguide as shown in Figure 1 will support the propagation of waves that have two possible field configurations, classified as the E^x and E^y modes. The main field components of the E^x_{mn} modes are E_x and H_y, while those of the E^y_{mn} modes are E_y and H_x. The subscripts m and n designate the number of maxima of the dominant field in the x and y directions, respectively.

Noting that the E^x and E^y modes are well approximated by the TE^y ($E_y \equiv 0$, the leading function is E_x) and TM^y ($H_y \equiv 0$, the leading function is H_x) modes, respectively, the following Helmholtz equations are derived from Maxwell's equations:

$$\frac{n_x^2}{n_z^2}\frac{\partial^2 E_x}{\partial x^2} + \frac{\partial^2 E_x}{\partial y^2} - \beta^2 E_x + k_0^2 n_x^2 E_x = 0 \quad \text{for } E^x \text{ modes} \tag{1}$$

$$\frac{1}{n_y^2}\frac{\partial^2 H_x}{\partial x^2} + \frac{1}{n_z^2}\frac{\partial^2 H_x}{\partial y^2} - \frac{\beta^2}{n_y^2} H_x + k_0^2 H_x = 0 \quad \text{for } E^y \text{ modes} \tag{2}$$

where k_0 is the free-space wavenumber; β is the propagation constant in the z (axial) direction; and n_x, n_y, and n_z are the refractive indices in the x, y, and z directions, respectively.

3. FINITE ELEMENT APPROACH

Application of the standard finite element technique, as in Koshiba (1992) and Jin (1993), to Eqs. (1) and (2) gives the following matrix equations:

$$[K]\{E_x\} = n_{\text{eff}}^2[M]\{E_x\} \tag{3}$$

with

$$[K] = \sum_e \iint_e \left[k_0^2 n_x^2 \{N\}\{N\}^T - \frac{n_x^2}{n_z^2} \frac{\partial\{N\}}{\partial x}\frac{\partial\{N\}^T}{\partial x} - \frac{\partial\{N\}}{\partial y}\frac{\partial\{N\}^T}{\partial y} \right] dx\, dy \tag{4}$$

$$[M] = \sum_e \iint_e k_0^2 \{N\}\{N\}^T\, dx\, dy \tag{5}$$

for the E^x modes, and

$$[K]\{H_x\} = n_{\text{eff}}^2[M]\{H_x\} \tag{6}$$

with

$$[K] = \sum_e \iint_e \left[k_0^2 \{N\}\{N\}^T - \frac{1}{n_y^2} \frac{\partial\{N\}}{\partial x}\frac{\partial\{N\}^T}{\partial x} - \frac{1}{n_z^2}\frac{\partial\{N\}}{\partial y}\frac{\partial\{N\}^T}{\partial y} \right] dx\, dy \tag{7}$$

$$[M] = \sum_e \iint_e \frac{k_0^2}{n_y^2} \{N\}\{N\}^T\, dx\, dy \tag{8}$$

for the E^y modes, where n_{eff} is the effective index and is given by $n_{\text{eff}} = \beta/k_0$, the components of the $\{E_x\}$ and $\{H_x\}$ vectors are the values of E_x and H_x at all nodes, respectively, $\{N\}$ is the shape function vector, the summation $\sum_e$ extends over all different elements, and T denotes a transpose. The form of Eqs. (3) and (6) is a generalized eigenvalue problem whose eigenvalue and eigenvector directly correspond to the effective index and modal field, respectively.

The scalar FEM approach to anisotropic optical waveguides with off-diagonal elements in a permittivity tensor has also been formulated by Koshiba et al. (1984).

4. PRINCIPAL FUNCTIONS OF THE CAD PACKAGE

The CAD package described here consists of three modules: a preprocessor, a solver, and a postprocessor.

The preprocessor is used to define the guide geometry and specify the optical wavelength in micrometers, the refractive indices, the boundary conditions (Neumann and/or Dirichlet conditions on artificial boundaries far from the core region, and conditions on the planes of symmetry), the desired mode (E^x_{mn} or E^y_{mn} modes, m, $n = 1, 2$), and the type of elements (linear or quadratic triangular elements). In this package isoparametric quadratic triangular elements that fit the curved boundaries of a waveguide are used. Furthermore, numerical integration formulas necessary to construct the finite element matrices, $[K]$ and $[M]$, are introduced, and therefore, for graded-index waveguides formed by diffusion, electrooptic effect, acoustooptic effect, thermooptic effect, nonlinear effect, and so on, the refractive index change in each element can be modeled accurately. Isoparametric quadratic triangular elements and numerical integration formulas are described in the textbook by Koshiba (1992).

The solver is constructed by using the scalar FEM. It evaluates the phase constant, attenuation constant, confinement factor, spot size (near-field pattern), and beam divergence angle (far-field pattern). Although the phase and attenuation constants can be obtained directly by solving the complex eigenvalue problem, the attenuation constant can also be evaluated with the perturbation techniques by Hayata et al. (1986).

The postprocessor is used to display the near-field pattern and the far-field pattern. Results can be displayed as contour lines, color contours, or three-dimensional views.

5. PRACTICAL WORK

5.1. Benchmark Tests

In order to check the accuray of the scalar FEM, a typical rib waveguide as shown in Figure 1 is considered, where the refractive indices of GaAs and GaAlAs are taken as 3.44 and 3.40, respectively, and the operating wavelength is 1.15 μm. The input data are entered on the pop-up window as shown in

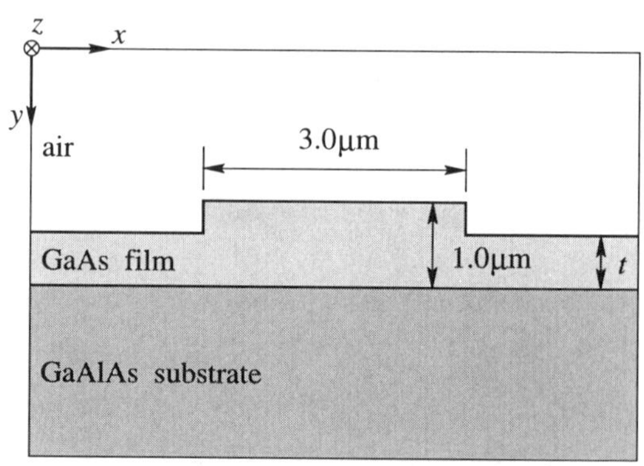

FIGURE 1. Optical rib waveguide.

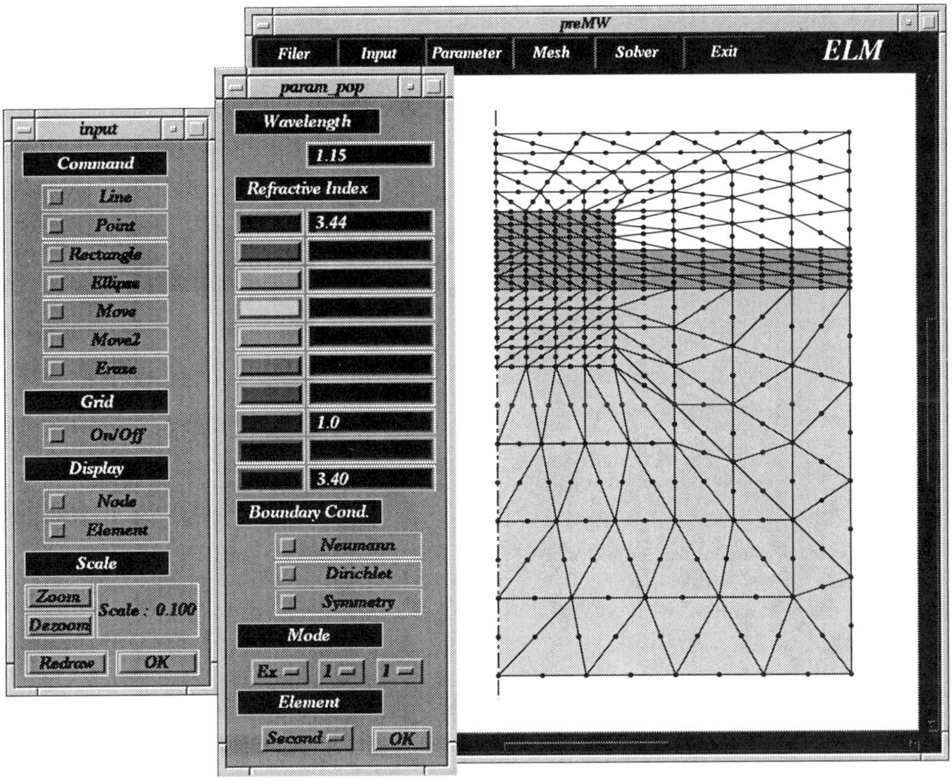

FIGURE 2. Input menu and mesh generation.

TABLE 1. Comparison of Vector and Scalar Formulations

	E^x_{11} Mode				E^y_{11} Mode		
t (μm)	Scalar FEM	Vector FEM[a]	Vector FEM[b]	Vector BPM	Scalar FEM	Vector FEM[b]	Vector BPM
0.0	3.41204	3.41200	3.41194	3.41209	3.41028	3.41043	3.41071
0.1	3.41214	3.41220	3.41209	3.41221	3.41037	3.41053	3.41081
0.2	3.41229	3.41235	3.41224	3.41241	3.41051	3.41067	3.41096
0.3	3.41249	3.41255	3.41246	3.41263	3.41070	3.41086	3.41115
0.4	3.41276	3.41285	3.41276	3.41291	3.41097	3.41114	3.41140
0.5	3.41311	3.41315	3.41312	3.41326	3.41132	3.41147	3.41171
0.6	3.41353	3.41365	3.41358	3.41371	3.41174	3.41191	3.41211
0.7	3.41404	3.41410	3.41414	3.41419	3.41227	3.41242	3.41256
0.8	3.41468	3.41475	3.41480	3.41473	3.41293	3.41305	3.41305
0.9	3.41553	3.41560	3.41567	3.41557	3.41383	3.41393	3.41386

[a]Vector FEM: vector FEM combined with the penalty function method.
[b]Vector FEM: vector FEM with the high-order hybrid edge/nodal element.

Figure 2. After inputting all the required data, the finite element mesh is generated automatically. A sample mesh is also shown in Figure 2, where symmetry conditions are used, and only one-half of the waveguide cross section is subdivided into quadratic triangular elements. The number of elements is 224, and the number of nodes is 485.

Table 1 shows the effective index for the fundamental E^x (E^x_{11}) and the fundamental E^y (E^y_{11}) modes. The results of the vector FEM by Rahman and Davies (1985) and Koshiba et al. (1994) and of the vector beam propagation method (BPM) by Huang and Xu (1993) are also given in Table 1. The results of the scalar FEM by Koshiba et al. (1992) agree excellently with the more accurate results of the full-vector analysis methods.

5.2. Electrooptic Waveguides

High-speed optical modulators are important components for high-speed and broad-band optical communication systems and signal processing systems. For broad-band operation, a Ti:$LiNbO_3$ modulator with a traveling-wave (TW) type electrode is usually employed.

Figure 3 shows the cross-sectional view of a Ti:$LiNbO_3$ Mach-Zehnder optical modulator with a coplanar waveguide (CPW) electrode structure by Noguchi et al. (1995). Since the electrode consists of two conductors (for the CPW case, both wide electrodes are connected to each other), a quasi-TEM wave transmission is assumed, and hence, the electric field calculation is reduced to a two-dimensional electrostatic field problem governed by Laplace's equation. The FEM matrix equation for Laplace's equation is given by

$$[K]\{V\} = \{Q\} \tag{9}$$

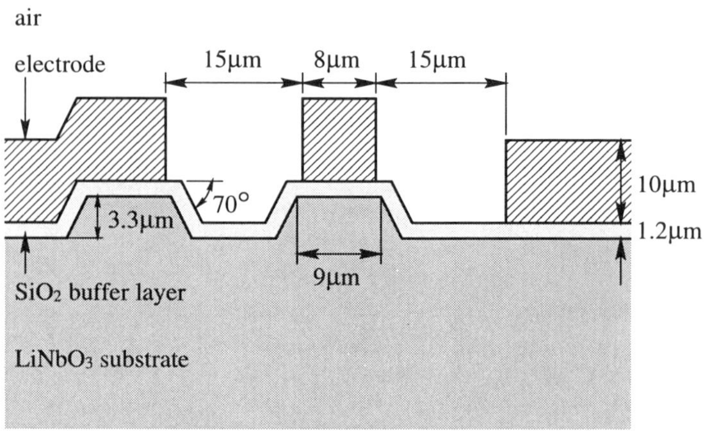

FIGURE 3. Electrooptic waveguide.

with

$$[K] = \sum_e \iint_e \left[\varepsilon_x \frac{\partial\{N\}}{\partial x} \frac{\partial\{N\}^T}{\partial x} + \varepsilon_y \frac{\partial\{N\}}{\partial y} \frac{\partial\{N\}^T}{\partial y} \right] dx\, dy \tag{10}$$

where $\{V\}$ is the nodal electrostatic potential vector, the vector $\{Q\}$ originates from imposed voltages and corresponds to the surface charge density distribution on the electrode, and ε_x and ε_y are the relative permittivities along the x and y directions, respectively.

From the nodal potential values, the capacitance per unit length of the TW electrode structure, C, is evaluated as

$$C = \{V\}^T[K]\{V\}/V_0^2 \tag{11}$$

where V_0 is the applied voltage. The microwave effective index, n_m, and the characteristic impedance, Z_c, are given by

$$n_m = \sqrt{C/C_0} \tag{12}$$

$$Z_c = \frac{1}{c\sqrt{CC_0}} \tag{13}$$

where C_0 is the free-space capacitance of the electrode, and c is the light velocity in free space. The refractive index changes due to the electrooptic effect are calculated by using the microwave electric fields and the electrooptic coefficients, as in Mustacich (1988). The modified waveguide with electrooptic-induced changes in the refractive index is then analyzed using the optical waveguide solver, and the optical response, $m(f)$, as in Uehara (1973), is given by

$$m(f) = \left| \frac{1 - S_1 S_2}{(1 + S_2)[\exp(j2u_+) - S_1 S_2 \exp(-j2u_-)]} \times \left[\exp(ju_+) \frac{\sin u_+}{u_+} + S_2 \exp(-ju_-) \frac{\sin u_-}{u_-} \right] \right| \tag{14}$$

with

$$u_\pm = \pi f l (n_m \mp n_{\text{eff}})/c - j\alpha l/2 \tag{15}$$

where S_1 and S_2 are the reflection coefficients at the input and output terminals, respectively, l is the electrode length, f is the microwave frequency, and n_{eff} is the optical effective index. The attenuation constant due to the conductor loss, α, can be calculated using the incremental inductance formula by Wheeler (1942).

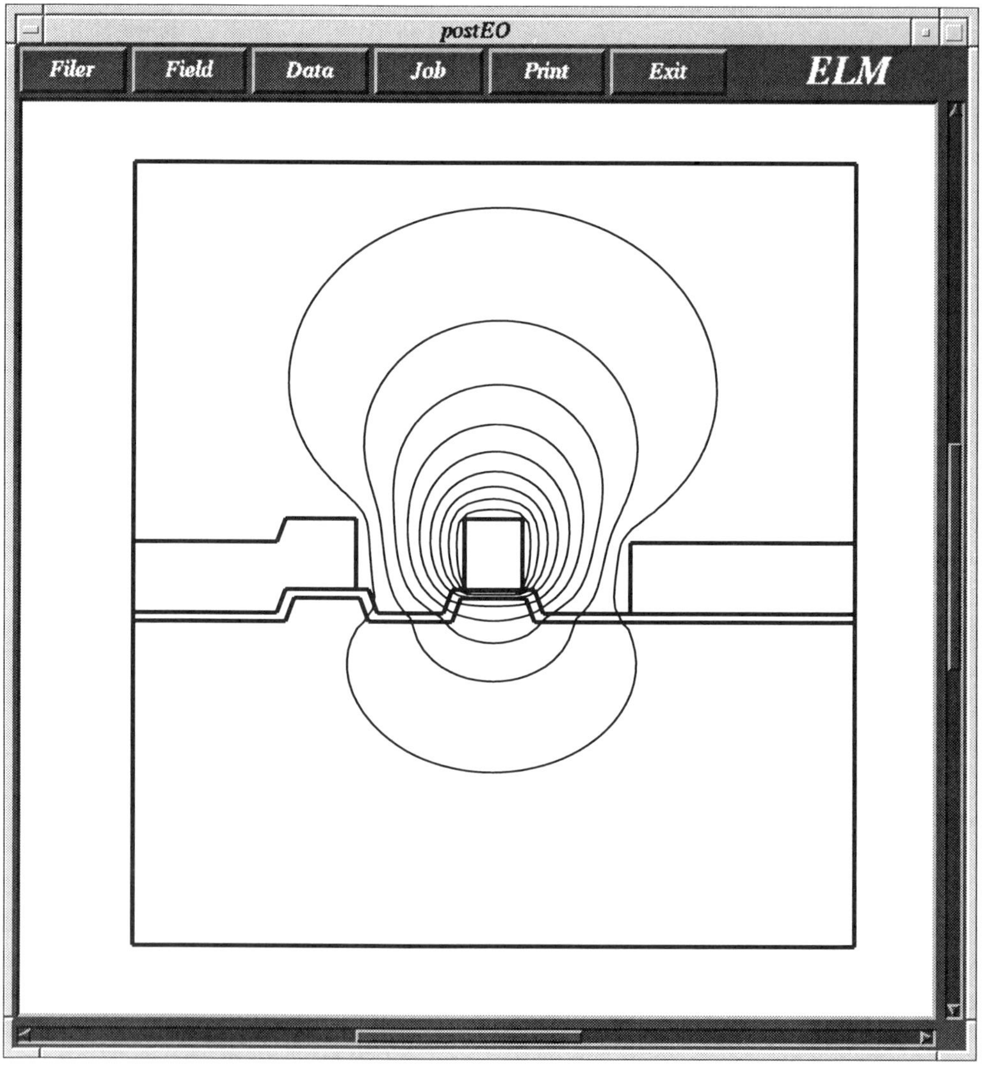

FIGURE 4. Window showing an equipotential plot in 10% intervals.

Figure 4 shows the equipotential plot in 10% intervals for the Z-cut Ti:$LiNbO_3$ Mach-Zehnder optical modulator with gold CPW electrodes as in Figure 3, where the relative permittivities of the Z-cut $LiNbO_3$ are 28 in the y direction and 43 in other directions. The relative permittivity of the SiO_2 buffer layer is 3.9. The calculated microwave effective index, characteristic impedance, and attenuation constant are, respectively, $n_m = 2.17$, $Z_c = 47.3\,\Omega$ and $\alpha = 0.44\sqrt{f}\,\text{dB}/(\sqrt{\text{GHz}}\cdot\text{cm})$. Figure 5 shows the optical response of this modulator, where the optical effective index is assumed to be 2.14, and the

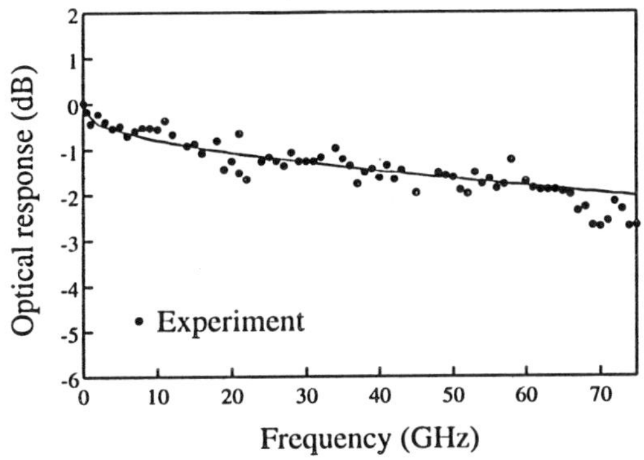

FIGURE 5. Small signal optical response of a Ti:LiNbO$_3$ optical modulator.

microwave generator internal impedance and the shunted load impedance are both taken as 50 Ω. The calculated response is in good agreement with the experimental one by Noguchi et al. (1995).

5.3. Acoustooptic Waveguides

Residual thermal stress in optical fibers and optical channel waveguides can be large enough to cause substantial changes in the refractive index and to create optical anisotropy. When the dimension of the waveguide structure in one direction (z direction) is very large in comparison with the other two directions (x and y directions), and the applied forces act in the xy plane and do not vary in the z direction, the problem is called the plane strain problem.

The FEM matrix equation for the plane strain problem is given by, as in Okamoto et al. (1981),

$$\begin{bmatrix} [K_{xx}] & [K_{xy}] \\ [K_{yx}] & [K_{yy}] \end{bmatrix} \begin{bmatrix} \{u_x\} \\ \{u_y\} \end{bmatrix} = \begin{bmatrix} \{f_x\} \\ \{f_y\} \end{bmatrix} \tag{16}$$

with

$$[K_{xx}] = \sum_e \iint_e \left[\frac{(1-\nu)E}{(1+\nu)(1-2\nu)} \frac{\partial\{N\}}{\partial x} \frac{\partial\{N\}^T}{\partial x} + \frac{E}{2(1+\nu)} \frac{\partial\{N\}}{\partial y} \frac{\partial\{N\}^T}{\partial y} \right] dx\,dy \tag{17}$$

$$[K_{xy}] = [K_{yx}]^T = \sum_e \iint_e \left[\frac{\nu E}{(1+\nu)(1-2\nu)} \frac{\partial\{N\}}{\partial x} \frac{\partial\{N\}^T}{\partial y} + \frac{E}{2(1+\nu)} \frac{\partial\{N\}}{\partial y} \frac{\partial\{N\}^T}{\partial x} \right] dx\,dy \tag{18}$$

$$[K_{yy}] = \sum_e \iint_e \left[\frac{E}{2(1+v)} \frac{\partial\{N\}}{\partial x} \frac{\partial\{N\}^T}{\partial x} + \frac{(1-v)E}{(1+v)(1-2v)} \frac{\partial\{N\}}{\partial y} \frac{\partial\{N\}^T}{\partial y} \right] dx\,dy \tag{19}$$

$$\{f_x\} = \sum_e \iint_e \frac{E\alpha_T \Delta T}{1-2v} \frac{\partial\{N\}}{\partial x} dx\,dy \tag{20}$$

$$\{f_y\} = \sum_e \iint_e \frac{E\alpha_T \Delta T}{1-2v} \frac{\partial\{N\}}{\partial y} dx\,dy \tag{21}$$

where the components of the $\{u_x\}$ and $\{u_y\}$ vectors are the values of the particle displacements, u_x and u_y, at all nodes, respectively, E is Young's modulus, v is Poisson's ratio, α_T is the thermal expansion coefficient, and ΔT is the temperature change (negative on cooling). The refractive index changes due to the acoustooptic effect are calculated by using the normal stresses and the photoelastic (stress-optic or strain-optic) coefficients, as in Okamoto et al. (1981).

Figure 6 shows the cross-sectional view of an acoustooptic channel waveguide, and the stress distributions are depicted in Figure 7, where the temperature change is assumed to be $-1000°C$. Material constants by Okamoto et al. (1981) and Kirkby et al. (1979) are, respectively, used for SiO_2 and GaAs. Horizontal and vertical stresses in kg/mm^2 are also depicted at a depth of $2\,\mu m$ beneath the SiO_2 stripe and at the center, respectively. Positive and negative values correspond to tensile and compressive stresses, respectively. The stress builds up rapidly as the edge of the oxide is approached, and the contour lines crowd very closely together. Figure 8 shows the stress-induced refractive index changes at a depth of $2\,\mu m$ beneath the SiO_2 stripe. The modified waveguide with acoustooptic-induced changes in the refractive index is then analyzed using the optical waveguide solver, and the field-amplitude contour plots in 10% intervals for the E_{11}^x and the E_{11}^y modes are shown in Figure 9, where the operating wavelength is $1.15\,\mu m$, and the refractive indices of n-GaAs, n^+-GaAs, and SiO_2 are taken as 3.440, 3.434, and 1.449, respectively. For comparison, the mode profiles of the unstressed case are also shown. In the

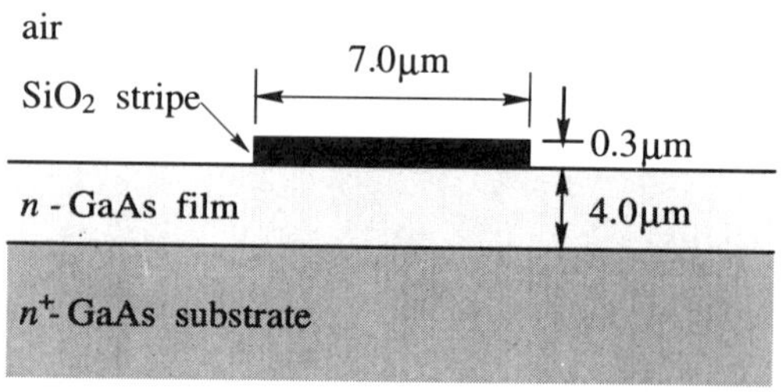

FIGURE 6. Acoustooptic waveguide.

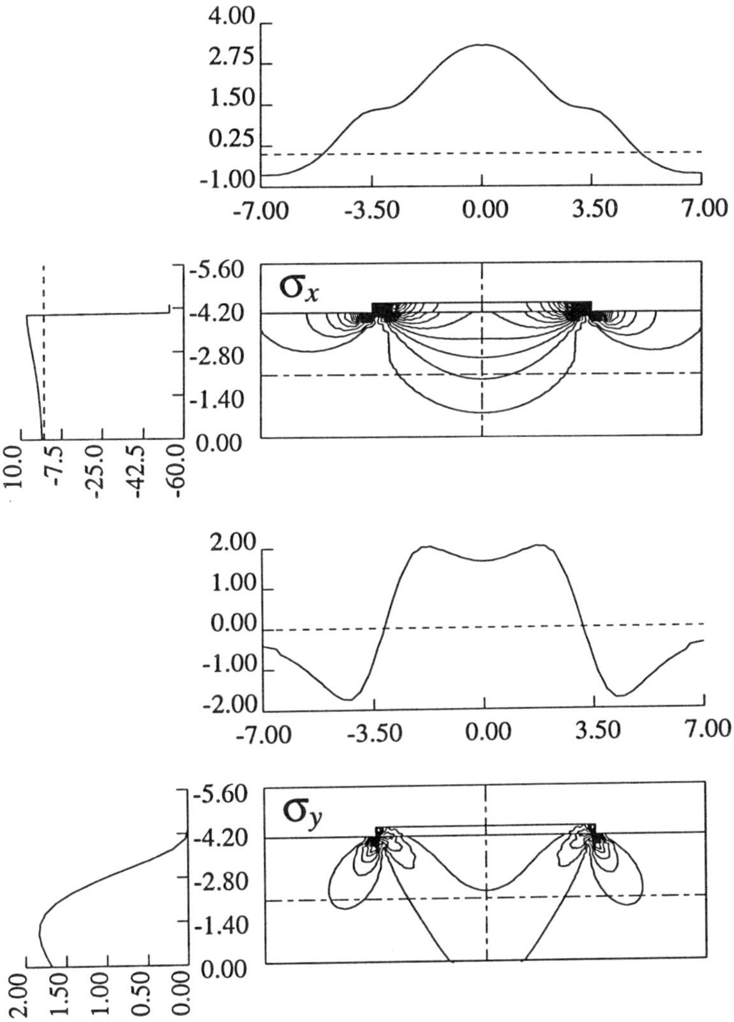

FIGURE 7. Stress distributions.

stressed case, the strong confinement in the horizontal (x) direction is produced by the large photoelastic perturbation in the refractive index of GaAs shown in Figure 8. The calculated mode profiles agree well with the experimental ones by Westbrook et al. (1980).

5.4. Thermooptic Waveguides

Thermooptic waveguide devices consisting of a thin-film heater deposited on a waveguide are useful in optical switching and/or tuning functions. Although

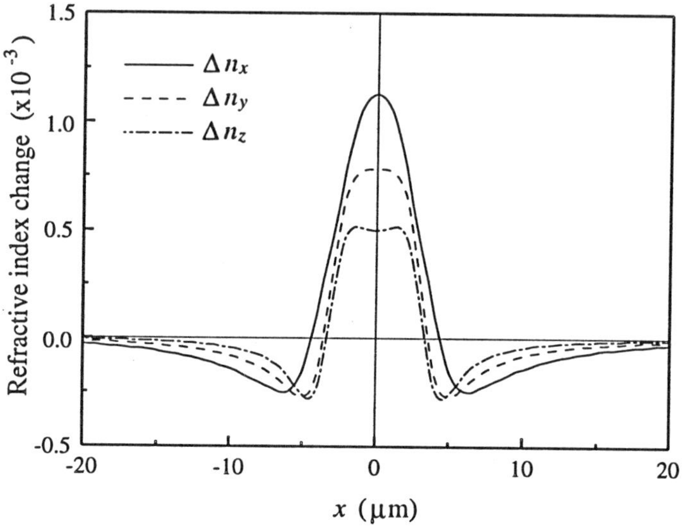

FIGURE 8. Stress-induced refractive index changes.

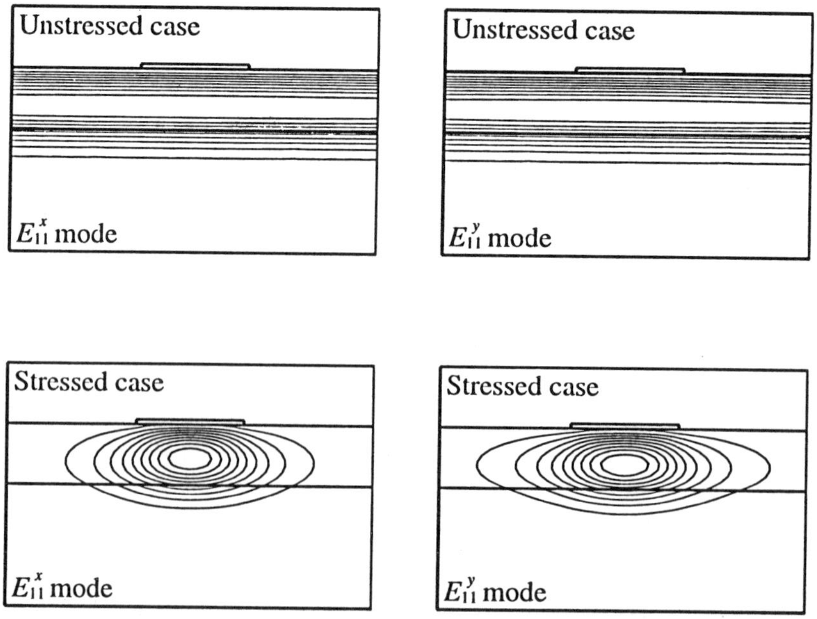

FIGURE 9. Field-amplitude contour plots in 10% intervals.

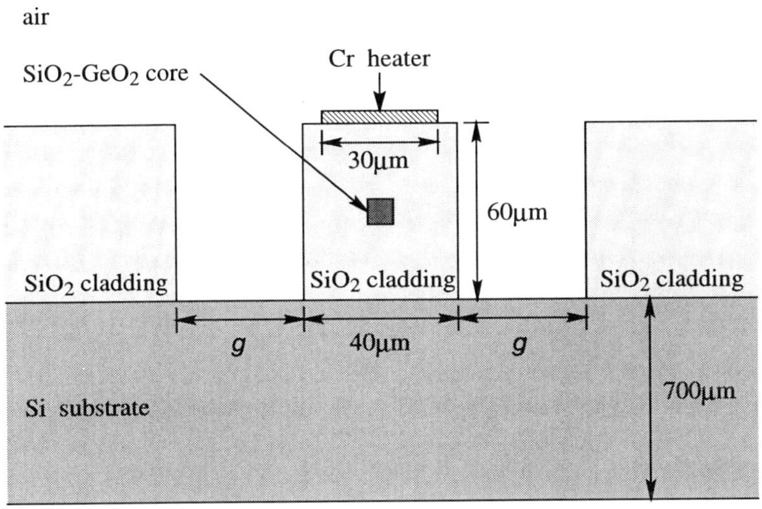

FIGURE 10. Thermooptic waveguide.

the switching speed is not so high, even passive silica-based waveguides can be utilized.

Figure 10 shows the cross-sectional view of a thermooptic phase shifter with grooves by Inoue et al. (1992). Since the heater length is much longer than the heater width, the two-dimensional analysis is used. The FEM matrix equation for the steady-state heat-conduction problem is given by

$$[K]\{T\} = \{Q_T\} \tag{22}$$

with

$$[K] = \sum_e \iint_e \left[\kappa \frac{\partial\{N\}}{\partial x} \frac{\partial\{N\}^T}{\partial x} + \kappa \frac{\partial\{N\}}{\partial y} \frac{\partial\{N\}^T}{\partial y} \right] dx\, dy \tag{23}$$

where $\{T\}$ is the nodal temperature vector, the vector $\{Q_T\}$ originates from the input power density and corresponds to the rate at which heat is generated at the heater–SiO_2 interface per unit area, and κ is the thermal conductivity.

Figure 11 shows the temperature distribution for the silica-based waveguide with an $8 \times 8\,\mu m$ SiO_2–GeO_2 core and a 5-mm-long Cr heater, where the thermal conductivities of SiO_2 and Si are taken as 0.014 W/(cm · °C) and 1.5 W/(cm · °C), respectively. It is assumed that the heat does not diffuse into the air and that the temperature of the rear surface of the Si substrate is kept

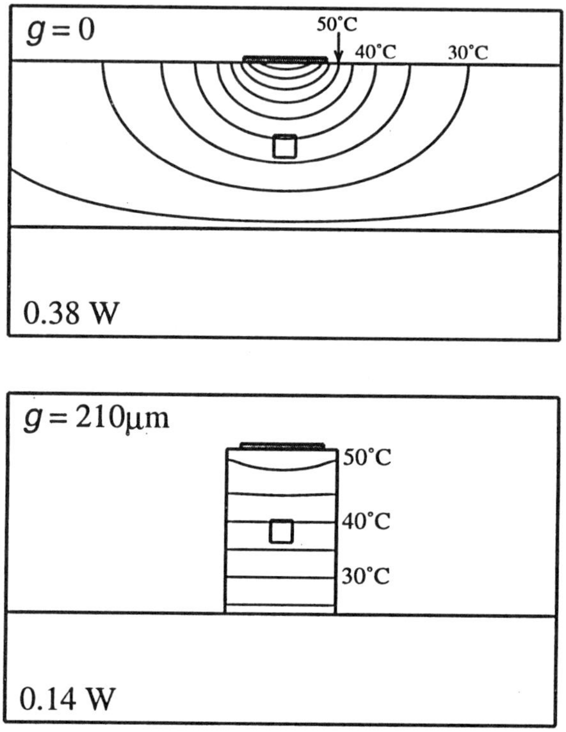

FIGURE 11. Temperature distributions.

constant at a room temperature of 23°C. As the depth increases, the temperature decreases and drops down to almost 23°C at the SiO_2–Si interface. This result indicates that the Si substrate works as a heat sink. As the thermooptic coefficient (refractive index change with temperature) is about 1×10^{-5}/°C for the silica glass, the phase shift becomes π for the operating wavelength of 1.53 μm when the temperature of the core region under the 5-mm-long heater rises by 15.3°C. The required switching power for π phase shift is 0.38 W in the thermooptic waveguide without grooves ($g = 0$), while the switching power is reduced to 0.14 W in the waveguide with grooves ($g = 210\ \mu$m), because the grooves restrict heat diffusing. The modified waveguide with thermooptic-induced changes in the refractive index is then analyzed using the optical waveguide solver. The effective index changes for the E_{11}^x mode due to the thermooptic effect are 1.5327×10^{-4} and 1.5320×10^{-4} for the switching power of 0.38 W and 0.14 W, respectively. When the operating wavelength is 1.53 μm and the device length is 5 mm, the required effective index change for π phase shift is 1.53×10^{-4}. The results of thermal analysis coincide almost with those of optical waveguide analysis.

5.5. Magnetooptic Waveguides

Nonreciprocal waveguide devices such as isolators and circulators are important integrated optical components in optical communication systems. Magnetooptic waveguides are the basic elements for realizing these nonreciprocal waveguide devices.

Figure 12 shows the cross-sectional view of a magnetooptic rib waveguide, where the DC magnetic field is applied in the horizontal (x) direction. Under the scalar-wave approximation, the FEM matrix equation is given by Eq. (3) for the E^x modes, while for the E^y modes the following matrix equation is derived:

$$[K(n_{\text{eff}})]\{H_x\} = n_{\text{eff}}^2 [M]\{H_x\} \tag{24}$$

with

$$\begin{aligned}[K(n_{\text{eff}})] = \sum_e \iint_e \Bigg[& k_0^2\{N\}\{N\}^T - \frac{n^2}{n^4-\delta^2}\left(\frac{\partial\{N\}}{\partial x}\frac{\partial\{N\}^T}{\partial x} - \frac{\partial\{N\}}{\partial y}\frac{\partial\{N\}^T}{\partial y}\right) \\ & + \frac{sn_{\text{eff}}k_0\delta}{n^4-\delta^2}\left(\{N\}\frac{\partial\{N\}^T}{\partial y} + \frac{\partial\{N\}}{\partial y}\{N\}^T\right)\Bigg] dx\,dy \end{aligned} \tag{25}$$

$$[M] = \sum_e \iint_e \frac{n^2k_0^2}{n^4-\delta^2}\{N\}\{N\}^T dx\,dy \tag{26}$$

where n is the refractive index; δ is related to the specific Faraday rotation, θ_F, by the relation $\delta = 2n\theta_F/k_0$; and s is a directional parameter defined as $s = +1$ and -1 for $+z$ and $-z$ directions, respectively. Although Eq. (3) for the E^x modes is a linear generalized eigenvalue equation, Eq. (24) for the E^y modes is a nonlinear generalized eigenvalue equation and is solved by using the iterative scheme by Koshiba and Zhuang (1993).

The magnitude of the nonreciprocal phase shift as a function of magnetooptic-film thickness is shown in Figure 13, where the refractive index and the

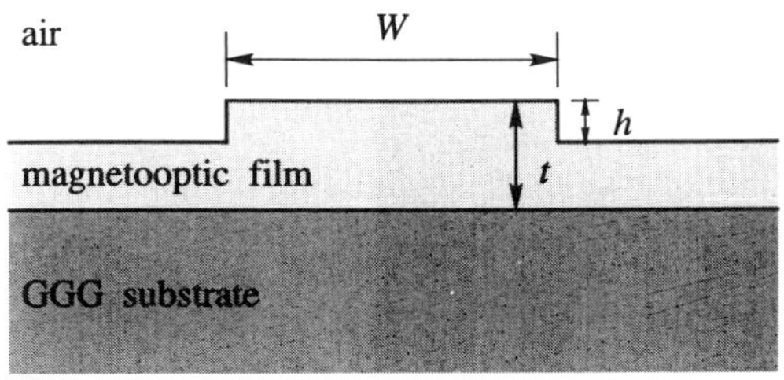

FIGURE 12. Magnetooptic waveguide.

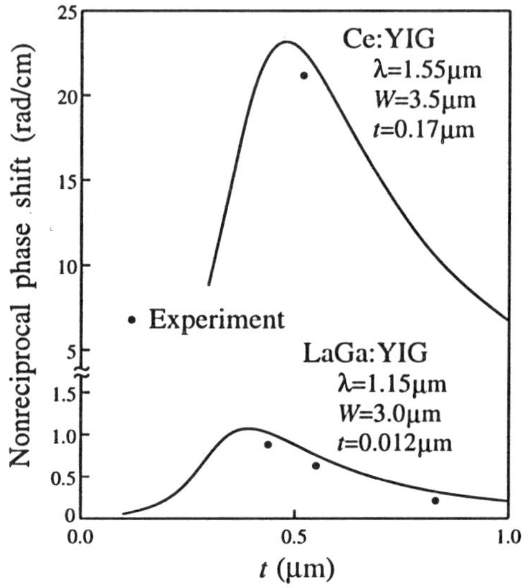

FIGURE 13. Amplitude of the normalized phase shift as a function of magnetooptic-film thickness.

Faraday rotation are, respectively, 2.18 and 230 deg/cm for LaGa:YIG and are, respectively, 2.21 and -4800 deg/cm for Ce:YIG. The refractive index of GGG is taken as 1.95 and 1.93 at $\lambda = 1.15\,\mu$m and $\lambda = 1.55\,\mu$m, respectively. The results of the present scalar FEM agree well with experimental results by Inuzuka et al. (1988) and Shintaku et al. (1993).

5.6. Nonlinear Optical Waveguides

Nonlinear optical waveguides have attracted considerable attention because of their fascinating features and their use in all-optical signal processing. In particular, intensity-dependent refractive indices caused by the optical Kerr-effect nonlinearity are applicable to various all-optical signal processing devices, such as bistability, switching, and upper and lower threshold devices and optical limiters.

Considering the intensity-dependent refractive index given by

$$n = n(x, y; |\mathbf{E}|) \tag{27}$$

with $\mathbf{E}$ being the electric field vector, the FEM matrix equation is obtained as

$$[K(E_x)]\{E_x\} = n_{\text{eff}}^2[M]\{E_x\} \tag{28}$$

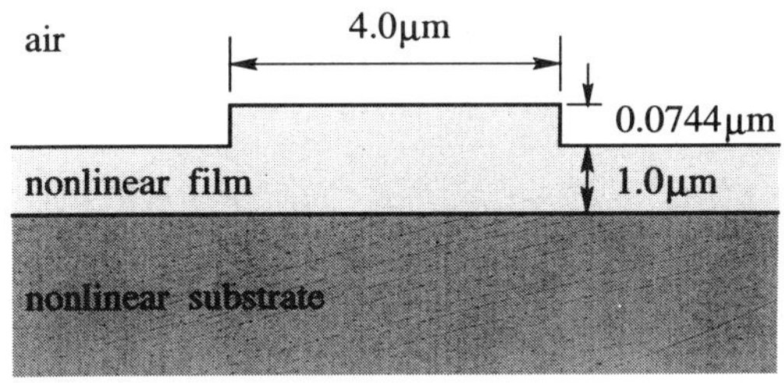

FIGURE 14. Nonlinear optical waveguide.

for the E^x modes, and

$$[K(H_x)]\{H_x\} = n_{\text{eff}}^2[M(H_x)]\{H_x\} \tag{29}$$

for the E^y modes, where the contents of the coefficient matrices, $[K]$ and $[M]$, are similar to those in Eqs. (4), (5), (7), and (8).

Equations (28) and (29) are nonlinear generalized eigenvalue equations and are solved by using the iterative scheme by Niiyama et al. (1995).

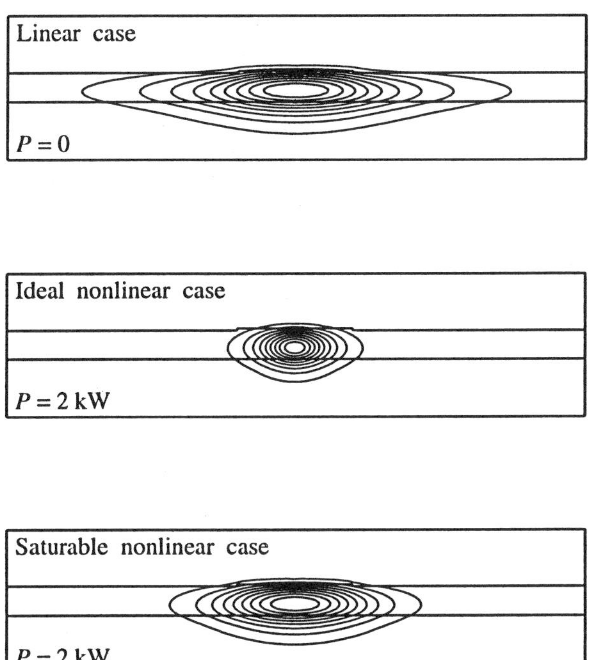

FIGURE 15. Field-amplitude contour plots in 10% intervals.

Figure 14 shows the cross-sectional view of a nonlinear rib waveguide, and the intensity-dependent refractive index is given by

$$n = n_{\mathrm{LN}} + n_{\mathrm{NL}}|\mathbf{E}|^2 \tag{30}$$

for ideal nonlinear materials, and

$$n = n_{\mathrm{LN}} + \Delta n_{\mathrm{sat}}[1 - \exp(-n_{\mathrm{NL}}|\mathbf{E}|^2/\Delta n_{\mathrm{sat}})] \tag{31}$$

for saturable nonlinear materials, where n_{LN} is the linear part, n_{NL} is the nonlinear optical coefficient, and Δn_{sat} is the maximum nonlinear index change (saturable nonlinearity). The field-amplitude contour plots in 10% intervals for the E_{11}^x mode are shown in Figure 15, where the operating wavelength is $1.32\,\mu$m, the guided wave power is $P = 2\,$kW, the refractive indices of the film and the substrate are taken as 1.6 and 1.5, respectively, and it is assumed that the film and the substrate have the same nonlinear optical coefficient of $2 \times 10^{-19}\,\mathrm{m}^2/\mathrm{V}^2$ and saturable nonlinearity of 0.01, as in Krijnen et al. (1994), which is roughly comparable to the values found for some polymers. A strong field confinement is found for the ideal nonlinearity, while for the saturable nonlinearity, the field profile is obviously more confined in the central part of the waveguide than in the linear limit ($P = 0$), but the enhancement is not as strong as for the ideal nonlinearity.

6. CONCLUDING REMARKS

The scalar FEM, which should not give rise to spurious solutions, has been applied to construct an optical waveguide solver. It has been demonstrated that the scalar FEM is a powerful and flexible approach to optical waveguide analysis, including inhomogeneities, nonreciprocities, and/or nonlinearities. In addition to the optical waveguide solver, three other solvers—an electrostatic solver, a stress solver, and a thermal solver—have also been developed. These solvers can be linked to the optical waveguide solver to produce a two-step analysis of electrooptic, acoustooptic, and thermooptic waveguides. In all the solvers, isoparametric quadratic triangular elements are used, and the sparseness and bandedness of the finite element matrices are effectively exploited. The subspace iteration method by Fernandez et al. (1991) is employed to find the eigenvalues and the corresponding eigenvectors of generalized eigenvalue equations in the optical waveguide solver. To solve linear algebraic equations in the electrostatic, stress, and thermal solvers, on the other hand, the upper and lower triangular decomposition and the forward and backward substitution processes are performed.

The present solvers have been incorporated into an optical CAD system, ELM (expert lightwave modeling system). A full-vector optical waveguide solver, as in Koshiba et al. (1994), and a longitudinally varying optical waveguide solver, as in Tsuji and Koshiba (1996), are now under consideration and will be incorporated into ELM in future.

REFERENCES

Bersiner, L., Hempelmann, U., and Strake, E. (1991). Numerical analysis of finite-element method and beam-propagation method results. *J. Opt. Soc. Am. B* **8**, 422–433.

Chiang, K. S. (1986). Finite element analysis of weakly guiding fibers with arbitrary refractive-index distribution. *J. Lightwave Technol.* **LT-8**, 980–990.

Davies, J. B. (1993). Finite element analysis of waveguides and cavities—A review. *IEEE Trans. Magn.* **MAG-29**, 1578–1583.

Dillon, B. M., and Webb, J. P. (1994). A comparison of formulations for the vector finite element analysis of waveguides. *IEEE Trans. Microwave Theory Tech.* **MTT-42**, 308–316.

Eguchi, M., and Koshiba, M. (1994). Accurate finite-element analysis of dual-mode highly elliptical-core fibers. *J. Lightwave Technol.* **LT-12**, 607–613.

Fernandez, F. A., Davies, J. B., Zhu, S., and Lu, Y. (1991). Sparse matrix eigenvalue solver for finite element solution of dielectric waveguides. *Electron. Lett.* **27**, 1824–1826.

Grant, J. C., Beal, J. C., and Frentte, N. J. P. (1994). Finite element analysis of the ARROW leaky optical waveguide. *IEEE J. Quantum Electron.* **QE-30**, 1250–1253.

Hayata, K., Koshiba, M., and Suzuki, M. (1986). Lateral mode analysis of buried heterostructure diode lasers by the finite-element method. *IEEE J. Quantum Electron.* **QE-22**, 781–788.

Huang, W. P., and Xu, C. L. (1993). Simulation of three-dimensional optical waveguides by a full-vector beam propagation method. *IEEE J. Quantum Electron.* **QE-29**, 2639–2649.

Inoue, Y., Katoh, K., and Kawachi, M. (1992). Polarization sensitivity of a silica waveguide thermooptic phase shifter for planar lightwave circuits. *IEEE Photon. Technol. Lett.* **4**, 36–38.

Inuzuka, H., Okamura, Y., and Yamamoto, S. (1988). Magneto-optic nonreciprocal phase shift in $(YLa)_3(FeGa)_5O_{12}$ single-mode channel waveguides. *J. Appl. Phys.* **63**, 1575–1577.

Jin, J. (1993). *The Finite Element Method in Electromagnetics.* Wiley, New York.

Kirkby, P. A., Selway, P. R., and Westbrook, L. D. (1979). Photoelastic waveguides and their effect on stripe-geometry $GaAs/Ga_{1-x}Al_xAs$ lasers. *J. Appl. Phys.* **50**, 4567–4579.

Koshiba, M. (1992) *Optical Waveguide Theory by the Finite Element Method.* KTK Scientific Publishers/Kluwer Academic Publishers, Tokyo/Dordrecht, The Netherlands.

Koshiba, M., and Zhuang, X.-P. (1993). An efficient finite-element analysis of magnetooptic channel waveguides. *J. Lightwave Technol.* **LT-11**, 1453–1458.

Koshiba, M., Hayata, K., and Suzuki, M. (1982). Approximate scalar finite-element analysis of anisotropic optical waveguides. *Electron. Lett.* **18**, 411–413.

Koshiba, M., Hayata, K., and Suzuki, M. (1984). Approximate scalar finite-element analysis of anisotropic optical waveguides with off-diagonal elements in a permittivity tensor. *IEEE Trans. Microwave Theory Tech.* **MTT-32**, 587–593.

Koshiba, M., Saitoh, H., Eguchi, M., and Hirayama, K. (1992). Simple scalar finite element approach to optical rib waveguides. *IEE Proc., Part J: Optoelectron.* **139**, 166–171.

Koshiba, M., Maruyama, S., and Hirayama, K. (1994). A vector finite element method with the high-order mixed-interpolation-type triangular elements for optical waveguiding problems. *J. Lightwave Technol.* **LT-12**, 495–502.

Krijnen, G. J. M., Hoekstra, J. W. M., and Lambeck, P. V. (1994). A new method for the calculation of propagation constants and field profile of guided modes of nonlinear channel waveguides based on the effective index method. *J. Lightwave Technol.* **LT-12**, 1550–1559.

Li, Q. Y., Sammut, R. A., and Pak, C. (1992). Variational and finite element analysis of nonlinear strip optical waveguides. *Opt. Commun.* **94**, 37–43.

Mabaya, N., Lagasse, P. E., and Vandenbulcke, P. (1981). Finite-element analysis of optical waveguides. *IEEE Trans. Microwave Theory Tech.* **MTT-29**, 600–605.

Mustacich, R. V. (1988). Scalar finite element analysis of electrooptic modulation in diffused channel waveguides and poled waveguides in polymer thin films. *Appl. Opt.* **27**, 3732–3737.

Niiyama, A., Koshiba, M., and Tsuji, Y. (1995). An efficient scalar finite element formulation for nonlinear optical channel waveguides. *J. Lightwave Technol.* **LT-13**, 1919–1925.

Noguchi, K., Mitomi, O., Miyazawa, H., and Seki, S. (1995). A broadband Ti:$LiNbO_3$ optical modulator with a ridge structure. *J. Lightwave Technol.* **LT-13**, 1164–1168.

Okamoto, K., Hosaka, T., and Edahiro, T. (1981). Stress analysis of optical fibers by a finite element method. *IEEE J. Quantum Electron.* **QE-17**, 2123–2129.

Osman, N., Koshiba, M., and Kaji, R. (1994). A comprehensive analysis of multilayer channel waveguides. *J. Lightwave Technol.* **LT-12**, 821–826.

Rahman, B. M. A., and Davies, J. B. (1985). Vector-H finite element solution of GaAs/GaAlAs rib waveguides. *IEE Proc., Part J: Optoelectron.* **132**, 349–353.

Rahman, B. M. A., Fernandez, F. A., and Davies, J. B. (1991). Review of finite element methods for microwave and optical waveguides. *Proc. IEEE* **79**, 1442–1448.

Sewell, G., and Cvetkovic, S. (1989). WAVEGIDE—An interactive waveguide program. *Adv. Eng. Software* **11**, 169–175.

Shintaku, T., Uno, T., and Kobayashi, M. (1993). Magneto-optic channel waveguides in Ce-substituted yttrium iron garnet. *J. Appl. Phys.* **74**, 4877–4881.

Strake, E., Bava, G. P., and Montrosset, I. (1988). Guided modes of Ti:$LiNbO_3$ channel waveguides: A novel quasi-analytical technique in comparison with the scalar finite-element method. *J. Lightwave Technol.* **LT-6**, 1126–1135.

Themistors, C., Rahman, B. M. A., and Grattan, K. T. V. (1994). Finite element analysis of lossy optical waveguides by using perturbation techniques. *IEEE Photon. Technol. Lett.* **6**, 537–539.

Tsuji, Y., and Koshiba, M. (1996). A finite element beam propagation method for strongly guiding and longitudinally varying optical waveguides. *J. Lightwave Tech.* **LT-14**, 217–222.

Uehara, S. (1973). Focusing-type optical modulator. *IEEE J. Quantum Electron.* **QE-9**, 984–986.

Westbrook, L. D., Fiddyment, P. J., and Robson, P. N. (1980). Photoelastic channel optical waveguides in epitaxial GaAs layers. *Electron. Lett.* **16**, 169–170.

Wheeler, H. A. (1942). Formulas for the skin effect. *Proc. IRE* **30**, 412–414.

Young, T. P. (1988). Design of integrated optical circuits using finite elements. *IEE Proc., Part A: Phys. Sci., Meas. Instrum., Manage. Educ. Rev.* **135**, 135–144.

Zoboli, M., Di Pasqual, F., and Selleri, S. (1993). Full-vectorial and scalar solutions of nonlinear optical fibers. *Opt. Commun.* **97**, 11–15.

CHAPTER FIVE

Curl-Conforming Mixed-Order Edge Elements for Discretizing the 2D and 3D Vector Helmholtz Equation

ANDREW F. PETERSON and DONALD R. WILTON

1. INTRODUCTION

Microwave devices generally involve three-dimensional, time-harmonic electromagnetic fields, a complete description of which can be obtained using either Maxwell's equations or the curl–curl form of the vector Helmholtz equation. Finite element discretizations of the vector Helmholtz equation using conventional scalar basis functions produce numerous "spurious" eigensolutions that can not easily be separated from the true, physical solutions. In recent years, certain types of mixed-order vector basis functions ("edge elements") have been shown to alleviate the problem. These basis functions are known as "curl-conforming," since they do not impose continuity of the normal field component across cell boundaries, and thus are differentiable only in the sense of the curl operation.

Two different families of curl-conforming basis functions are considered in this chapter. The first type, which are complete to a given polynomial order, eliminate spurious nonzero eigenvalues and permit an easier imposition of

Finite Element Software for Microwave Engineering, Edited by Tatsuo Itoh, Giuseppe Pelosi, and Peter P. Silvester.
ISBN 0-471-12636-5

boundary conditions. This type of expansion function can be separated into two subsets, one of which has zero curl and can not be used to represent fields in a source-free region. Members of this set can be discarded to improve the efficiency of solution, leaving a second basis type equivalent to the representation proposed by Nedelec (1980). Specific examples of linear, quadratic, and cubic-order basis functions are presented for the mixed-order functions. Numerical results are used for illustration.

2. TWO EIGENFUNCTION FAMILIES ASSOCIATED WITH THE CURL–CURL OPERATOR

Consider the vector Helmholtz equation

$$\nabla \times \nabla \times \mathbf{E} = k^2 \mathbf{E} \tag{1}$$

for the electric field in a homogeneous source-free region. Eigenfunctions of this equation can generally be separated into two families, one of which includes the valid electromagnetic fields of the form $\{\mathbf{E} = \nabla \times \mathbf{V}\}$, where $\mathbf{V}$ is proportional to the magnetic field $\mathbf{H}$, and the other of which has the form $\{\mathbf{E} = \nabla\Phi\}$. The gradient $\nabla\Phi$ is a mathematical solution to Eq. (1) but does not represent an electromagnetic field in a source-free region. Since $\nabla \times \nabla\Phi = 0$, these solutions only satisfy Eq. (1) for $k = 0$ and are said to form the "nullspace" of the curl–curl operator. For a given geometry, both eigenfamilies satisfy the boundary conditions as well as the Helmholtz equation.

As a simple example, consider a rectangular two-dimensional cavity occupying the region $(0 < x < a, 0 < y < b)$. The true electromagnetic modes are

$$\mathbf{E}_{mn} = -\hat{\mathbf{x}} \frac{n\pi}{b} \cos\left(\frac{m\pi x}{a}\right) \sin\left(\frac{n\pi y}{b}\right) + \hat{\mathbf{y}} \frac{m\pi}{a} \sin\left(\frac{m\pi x}{a}\right) \cos\left(\frac{n\pi y}{b}\right) \tag{2}$$

These are easily shown to be eigenfunctions of Eq. (1), with associated eigenvalues

$$k^2 = \left[\left(\frac{m\pi}{a}\right)^2 + \left(\frac{n\pi}{b}\right)^2\right] \tag{3}$$

For this example, the nullspace eigenfunctions can be obtained from

$$\nabla \left\{\sin\left(\frac{m\pi x}{a}\right) \sin\left(\frac{n\pi y}{b}\right)\right\} \tag{4}$$

and therefore have the form

$$\mathbf{E}_{mn}^{\text{nullspace}} = \hat{\mathbf{x}} \frac{m\pi}{a} \cos\left(\frac{m\pi x}{a}\right) \sin\left(\frac{n\pi y}{b}\right) + \hat{\mathbf{y}} \frac{n\pi}{b} \sin\left(\frac{m\pi x}{a}\right) \cos\left(\frac{n\pi y}{b}\right) \quad (5)$$

It is easily verified that both families of solution satisfy the boundary conditions on tangential **E** at the perfect electric walls of the cavity. The divergence of the eigenfunctions in Eq. (2) is identically zero, while the divergence of the nullspace eigenfunctions is nonzero.

Unless specific attempts are made to exclude the nullspace eigensolutions from numerical results, members of both families will appear in the discrete operator. This may appear to be merely a problem of numerical efficiency, as long as the order of approximation is sufficient to easily distinguish between the physical modes and the nullspace solutions. However, traditional discretizations of Eq. (1), using scalar Lagrangian expansion functions that impose complete continuity of the primary field components, give rise to grossly inaccurate approximations of the zero eigenvalues associated with the nullspace (Winkler and Davies, 1984; Rahman and Davies, 1984; Crowley, 1988). In fact, the nullspace eigenvalues obtained in this manner are usually of the same order of magnitude as the physical eigenvalues and not easily distinguished from them. Consequently, these solutions have been known as "spurious modes" in the finite element literature.

3. CONTINUITY CONDITIONS AND SPURIOUS MODES

During the past decade, it has been observed that the problem of spurious modes can be alleviated by employing basis functions that do not impose complete continuity between cells (Hano, 1984; Mur, 1994). Specifically, representations that impose tangential-vector continuity but not normal-vector continuity produce nullspace eigenvalues that are in principle zero (and are nonzero in practice only because of finite-precision arithmetic). Although these basis functions generally capture a large number of nullspace eigenfunctions, the true eigenvalues are easily distinguished from the nullspace solutions.

There are other advantages associated with a field representation that does not impose the complete continuity of all components. First, since the true fields exhibit jump discontinuities at material interfaces, a robust expansion should also. In addition, traditional scalar expansion functions suffer from the cumbersome nature of the boundary and continuity conditions that must be imposed on interfaces that do not coincide with constant-coordinate surfaces. In other words, one can represent the x- and y-components of the field by independent scalar representations. At a boundary of the region, however, the boundary conditions couple these representations together, destroying the purely local nature of traditional finite elements and complicating the sparse storage scheme used to organize the data.

Because of these properties, it appears that numerical solutions of Eq. (1) and the more general vector Helmholtz equation

$$\nabla \times (\mu_r^{-1} \nabla \times \mathbf{E}) = k^2 \varepsilon_r \mathbf{E} \tag{6}$$

should employ a basis set that imposes tangential continuity but not normal continuity. Vector basis functions that impose tangential continuity have finite curl and are known as "curl-conforming" representations.

4. LOCAL COORDINATES

Basis functions used in finite element applications for triangular and tetrahedral cell shapes are most conveniently described in terms of simplex coordinates (Silvester and Ferrari, 1990). Coordinates $\{\xi_1, \xi_2, \xi_3\}$ specify the position of a point within a triangle by giving the relative perpendicular distance measured from each side to the point, with the distance expressed as a fraction of the triangle altitude. For tetrahedral cells, a point (x, y, z) divides the tetrahedron into four smaller tetrahedra, each with a volume that is some fraction of the total. Simplex coordinates $\{\xi_1, \xi_2, \xi_3, \xi_4\}$ may be defined by the respective ratio of the volume of each of the smaller cells to that of the original.

For cells of quadrilateral or hexahedral shape, it is convenient to define the basis functions using a "standard" square or cubical cell. A mapping of these functions (Section 8) can be used to define vector basis functions on skewed or even curved cells. In the two-dimensional case the standard square cell is centered at the origin of the local coordinate system (ξ, η), with the cell defined to span the region $(-1 < \xi < 1, -1 < \eta < 1)$, while in the three-dimensional case the standard cubical cells spans the volume $(-1 < \xi < 1, -1 < \eta < 1, -1 < \zeta < 1)$.

5. POLYNOMIAL-COMPLETE CURL-CONFORMING EXPANSIONS FOR TRIANGLES

The first basis function to be considered provides a polynomial-complete representation (Mur and de Hoop, 1985; Nedelec, 1986). These vector basis functions eliminate two difficulties associated with a traditional scalar expansion: they do not impose normal continuity at material interfaces and they interpolate to tangential components along cell edges and therefore make it easy to implement boundary conditions.

For the purpose of illustration, consider the case of triangular cells. Within a cell, a two-dimensional vector has the linear-order basis representation

$$\mathbf{B}(x, y) = \hat{\mathbf{x}}(A + Bx + Cy) + \hat{\mathbf{y}}(D + Ex + Fy) \tag{7}$$

containing six degrees of freedom. These degrees of freedom can be distributed among six basis functions that each interpolate to the tangential component of the vector at one end of a cell edge. Such functions maintain tangential continuity between cells by sharing their coefficient with the analogous function defined in the adjacent cell. Since there are no remaining degrees of freedom to assign to normal components, the functions do not impose normal continuity from cell to cell. From a global perspective, each basis function straddles two triangular cells and interpolates to the tangential component along one end of the common edge. Consequently, the global representation requires two unknown coefficients per nonboundary edge. The six functions are sketched in Figure 1. If expressed in simplex coordinate, the functions have the form

$$w_i \xi_j \nabla \xi_k, \quad i \neq j \neq k \tag{8}$$

where w_i denotes the length of the edge opposite node i, and i, j, and k assume values between 1 and 3. Since these functions produce a linear representation of both vector components throughout the cell, we label them "linear tangential/linear normal" (LT/LN) basis functions.

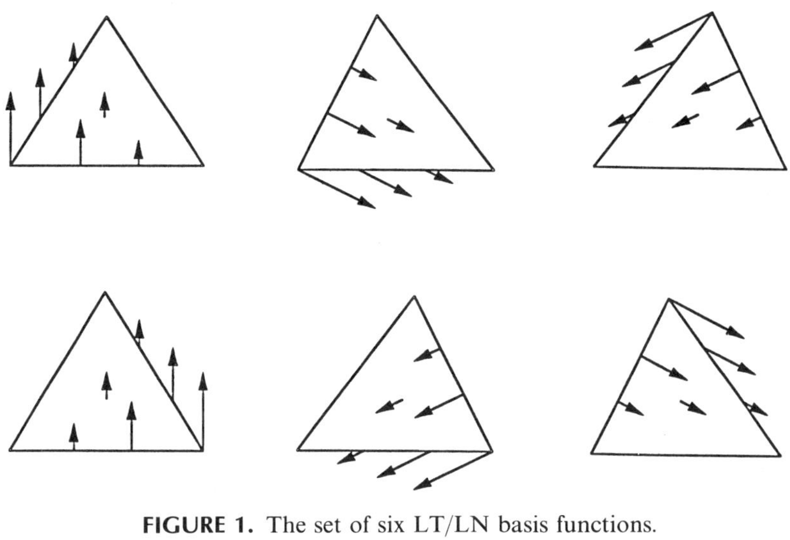

FIGURE 1. The set of six LT/LN basis functions.

Because the coefficients of the vector basis functions in Eq. (8) represent the tangential component along cell edges, boundary conditions such as $\hat{\mathbf{n}} \times \mathbf{E} = 0$ can be implemented in a purely local fashion by zeroing the appropriate basis function coefficient. Therefore these basis functions make it easier to incorporate essential boundary conditions into the solution process than a Lagrangian expansion of independent field components. Although the LT/LN functions are complete to linear polynomial order, they do not produce the same representation as a linear Lagrangian expansion since they are not completely continuous between cells. Consequently, there are more degrees of freedom (and more unknowns) associated with the vector functions than with the usual Lagrangian expansion on the same mesh. However, as demonstrated below, the vector basis functions in Eq. (8) properly separate the nullspace eigenvalues from the physical eigenvalues.

Analogous basis functions can be created that provide a complete quadratic representation. An expansion of the form

$$\begin{aligned}\mathbf{B}(x, y) = {} & \hat{\mathbf{x}}(A + Bx + Cy + Dx^2 + Exy + Fy^2) \\ & + \hat{\mathbf{y}}(G + Hx + Iy + Jx^2 + Kxy + Ly^2)\end{aligned} \tag{9}$$

contains 12 degrees of freedom, and basis functions can be defined that interpolate to the tangential components at three locations along each edge of a triangular cell and the normal component at the middle of each edge. Using simplex coordinates, one can define six functions of the form

$$w_i \xi_j (2\xi_j - 1) \nabla \xi_k, \quad i \neq j \neq k \tag{10}$$

that interpolate to a unity tangential component at the outside nodes of each edge, three functions of the form

$$4 w_i \xi_j \xi_k (\nabla \xi_j - \nabla \xi_k), \quad i \neq j \neq k \tag{11}$$

that interpolate to a unity tangential component at the midside nodes, and three functions of the form

$$\xi_i \xi_j \nabla \xi_k, \quad i \neq j \neq k \tag{12}$$

that provide a linearly independent normal component along each edge.

To maintain tangential continuity between cells, the nine quadratic basis functions in Eqs. (10) and (11) share coefficients with the analogous functions in the adjacent cells. The three basis functions per cell defined in Eq. (12) are entirely local and have coefficients that are independent from those of the neighboring cells. Consequently, the normal component can be discontinuous between cells. The global representation produced by the quadratic functions requires three unknowns per nonboundary edge and three unknowns per cell. These functions will be denoted "quadratic tangential/quadratic normal" (QT/QN) basis functions.

Consider the use of the vector basis functions for discretizing Eq. (1) for a circular, homogeneous cavity bounded by perfect electric walls, to obtain the transverse-electric (TE) cavity modes. The solution procedure is based on the weak form of the vector Helmholtz equation

$$\iint_{\Omega} \nabla \times \mathbf{T} \cdot \nabla \times \mathbf{E} - k^2 \mathbf{T} \cdot \mathbf{E} \, dx \, dy = -\int_{\Gamma} \mathbf{T} \cdot \hat{\mathbf{n}} \times (\nabla \times \mathbf{E}) \, dt \tag{13}$$

where Ω denotes the two-dimensional interior region, Γ denotes the boundary, and $\mathbf{T}$ is a suitably differentiable vector testing function. For the TE polarization, the boundary condition $\hat{\mathbf{n}} \times \mathbf{E} = 0$ on the conducting boundary is enforced by omitting any basis function with a nonzero tangential value on the boundary from the system of equations. Since there are no basis functions remaining with nonzero tangential components along the boundary Γ, there will be no need for testing functions with tangential components along Γ and the boundary integral in Eq. (13) will not contribute to the system. The discretization process yields a matrix eigenvalue equation $\mathbf{Ae} = k^2 \mathbf{Be}$ with entries

$$A_{mn} = \iint_{\Omega} \nabla \times \mathbf{T}_m \cdot \nabla \times \mathbf{B}_n \, dx \, dy \tag{14}$$

and

$$B_{mn} = \iint_{\Omega} \mathbf{T}_m \cdot \mathbf{B}_n \, dx \, dy \tag{15}$$

As is common practice, we employ the same vector functions for expansion and testing.

The region of interest is divided into triangular cells, with the mesh terminating on the cavity boundary Γ. The matrices $\mathbf{A}$ and $\mathbf{B}$ are constructed on a cell-by-cell basis, using a connectivity array that identifies the rows and columns of the matrix associated with a particular cell. The ordinary connectivity array employed with "node-based" scalar finite element analysis identifies the three nodes associated with each cell. However, the coefficients of the vector basis functions represent the tangential or normal fields along cell edges. In this "edge-based" formulation, it is much more convenient to employ a connectivity array that identifies the three edges associated with each cell. (A second pointer

TABLE 1. Smallest Eigenvalues Produced by a Discretization of a Circular Cavity with $\varepsilon_r = 1$, $\mu_r = 1$, and Unit Radius Using Polynomial-Complete Linear (LT/LN) Functions from Eq. (8) and Polynomial-Complete Quadratic (QT/QN) Basis Functions from Eqs. (10)–(12), for the TE Polarization[a]

LT/LN	QT/QN	Exact
0.0(67)	0.0(163)	
1.87(2)	1.84(2)	1.841(2)
3.20(2)	3.06(2)	3.054(2)
4.13(1)	3.84(1)	3.832(1)
4.6(2)	4.21(2)	4.201(2)
6.0(2)	5.35(2)	5.318(2)
6.1(2)	5.37(2)	5.331(2)

[a]The model consisted of 42 triangular cells and resulted in a matrix of order 108 for the LT/LN functions and 288 for the QT/QN functions.

Source: After Peterson and Wilton (1995).

array is also necessary to identify the two nodes associated with each edge.) The matrices may be constructed by scanning through the global model, computing 6×6 element matrices for the LT/LN basis functions and 12×12 element matrices for the QT/QN basis functions from the integrals in Eqs. (14) and (15). The connectivity arrays identify the appropriate locations of the element matrix entries within the global system. It is necessary to ensure that the basis functions that overlap two cells have a common orientation on either side of a given edge.

Table 1 shows numerical results for the eigenvalues k. Clearly, the nonzero eigenvalues appear to have a one-to-one correlation with analytical results for the electromagnetic cavity modes. The numerical spectrum also contains a large number of zero eigenvalues that represent the nullspace. Overall, these results support the hypothesis that spurious nonzero eigenvalues are eliminated by a polynomial-complete expansion that only imposes tangential continuity between cells.

Unfortunately, Table 1 also shows that almost two-thirds of the available degrees of freedom in the vector expansions are used to capture nullspace eigensolutions. Since eigenvalue computation is a process that can grow as fast as N^3, where N is the order of the system, more than 90% of the computational effort in these examples can be wasted on nonphysical eigensolutions. This drawback of polynomial-complete curl-conforming representations is serious, and consequently we turn our attention to a different basis family that eliminates some of the degrees of freedom wasted on nullspace eigenfunctions.

6. MIXED-ORDER EXPANSION FUNCTIONS OF NEDELEC FOR TRIANGULAR CELLS

The data in Table 1 suggest that polynomial-complete functions are sufficient for eliminating spurious nonzero eigenvalues, as long as they impose only tangential continuity from cell to cell. However, it is possible to eliminate some of the degrees of freedom in the LT/LN and QT/QN expansions that are used to capture nullspace eigensolutions. After observing that the degrees of freedom associated with the gradient of order-$(n + 1)$ polynomial terms belong to the nullspace of an order-n representation, Nedelec (1980) proposed mixed-order functions that eliminate some of the nullspace degrees of freedom.

The linear vector basis function in Eq. (7) can be divided into two functions:

$$\mathbf{B}(x, y) = \hat{\mathbf{x}}[A + [(C - E)/2]y\} + \hat{\mathbf{y}}\{D + [(E - C)/2]x\} \tag{16}$$

and

$$\mathbf{B}_{\text{grad}}(x, y) = \hat{\mathbf{x}}\{Bx + [(C + E)/2]y\} + \hat{\mathbf{y}}\{[(E + C)/2]x + Fy\} \tag{17}$$

Equations (16) and (17) each contain three degrees of freedom, embodied in the coefficients A, $(C - E)/2$, D, B, $(C + E)/2$, and F. The functions in Eq. (17) are the gradient of a polynomial

$$\Phi(x, y) = \{(B/2)x^2 + [(E + C)/2]xy + (F/2)y^2\} \tag{18}$$

and consequently have identically zero curl within the cell. If constrained to have tangential continuity from cell to cell, basis functions of the form of Eq.

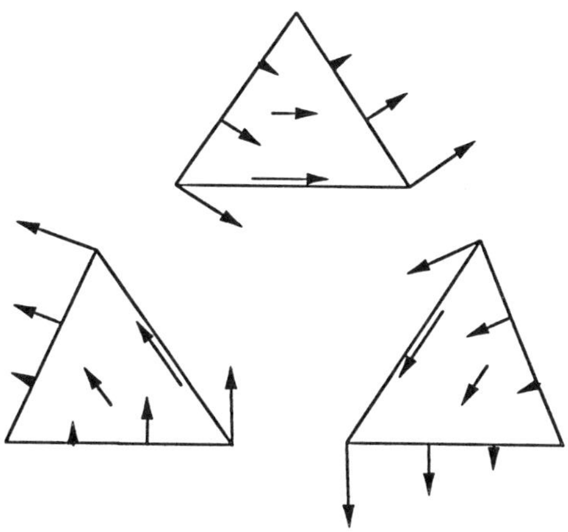

FIGURE 2. The set of three CT/LN basis functions.

(17) have zero curl over the entire problem domain and can only represent functions in the nullspace of the curl–curl operator. Thus it seems desirable to restrict the basis set to the degrees of freedom in Eq. (16).

The reduced expansion defined by Eq. (16) consists of mixed-order polynomial functions that provide a constant tangential component and a linear normal (CT/LN) component along any cut through the cell. Basis functions can be defined for triangular cells so that each function straddles two adjacent cells and interpolates to the tangential component along the common edge (Figure 2). Thus the global representation requires one unknown per nonboundary edge. In common with the polynomial-complete functions, the mixed-order functions are curl-conforming—they impose tangential continuity between cells but do not impose normal continuity and should not produce spurious nonzero eigenvalues. These basis functions have been widely used under the name "edge elements" (Bossavit and Verite, 1982; Barton and Cendes, 1987). Table 2 provides expressions in simplex coordinates for the CT/LN functions within a triangular cell.

In the quadratic case, the polynomial-complete representation in Eq. (9) can be decomposed into a function

$$\mathbf{B}(x, y) = \hat{\mathbf{x}}\{A + Bx + Cy + [(E - 2J)/3]xy + [(2F - K)/3]y^2\} + \hat{\mathbf{y}}\{G + Hx + Iy + [(2J - E)/3]x^2 + [(K - 2F)/3]xy\} \quad (19)$$

TABLE 2. Summary of the Simplex-Coordinate Representation of the First Three Types of Mixed-Order Basis Functions Within a Triangular Cell[a]

CT/LN	LT/QN	QT/CuN
$\xi_1\nabla\xi_2 - \xi_2\nabla\xi_1$	$\xi_1\nabla\xi_2$	$\xi_2(2\xi_2 - 1)\nabla\xi_1$
$\xi_1\nabla\xi_3 - \xi_3\nabla\xi_1$	$\xi_2\nabla\xi_1$	$\xi_3(2\xi_3 - 1)\nabla\xi_1$
$\xi_2\nabla\xi_3 - \xi_3\nabla\xi_2$	$\xi_1\nabla\xi_3$	$\xi_1(2\xi_1 - 1)\nabla\xi_2$
	$\xi_3\nabla\xi_1$	$\xi_3(2\xi_3 - 1)\nabla\xi_2$
	$\xi_2\nabla\xi_3$	$\xi_1(2\xi_1 - 1)\nabla\xi_3$
	$\xi_3\nabla\xi_2$	$\xi_2(2\xi_2 - 1)\nabla\xi_3$
	$\xi_2\xi_3\nabla\xi_1 - \xi_1\xi_2\nabla\xi_3$	$\xi_2\xi_3(\nabla\xi_2 - \nabla\xi_3)$
	$\xi_1\xi_3\nabla\xi_2 - \xi_1\xi_2\nabla\xi_3$	$\xi_1\xi_3(\nabla\xi_3 - \nabla\xi_1)$
		$\xi_1\xi_2(\nabla\xi_1 - \nabla\xi_2)$
		$\xi_1(2\xi_1 - 1)(\xi_2\nabla\xi_3 - \xi_3\nabla\xi_2)$
		$\xi_2(2\xi_2 - 1)(\xi_3\nabla\xi_1 - \xi_1\nabla\xi_3)$
		$\xi_3(2\xi_3 - 1)(\xi_1\nabla\xi_2 - \xi_2\nabla\xi_1)$
		$\xi_1^2(\xi_2\nabla\xi_3 - \xi_3\nabla\xi_2)$
		$\xi_2^2(\xi_3\nabla\xi_1 - \xi_1\nabla\xi_3)$
		$\xi_3^2(\xi_1\nabla\xi_2 - \xi_2\nabla\xi_1)$

[a]The last two LT/QN and last six QT/CuN functions are entirely local. The functions are not normalized.

Source: After Peterson and Wilton (1995).

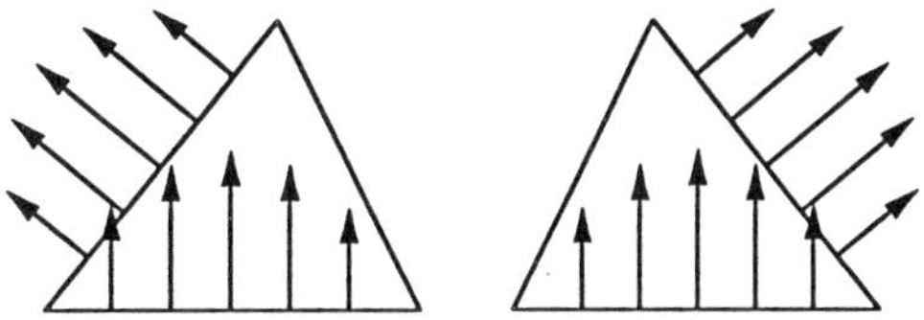

FIGURE 3. The LT/QN basis functions associated with the normal vector component at the middle of the cell edges. Combination of these functions with the six LT/LN functions shown in Figure 1 makes up the entire LT/QN set of eight basis functions.

containing eight degrees of freedom and a complementary function

$$\mathbf{B}_{\text{grad}}(x, y) = \hat{\mathbf{x}}\{Dx^2 + [2(E + J)/3]xy + [(F + K)/3]y^2\} \\ + \hat{\mathbf{y}}\{[(E + J)/3]x^2 + [2(K + F)/3]xy + Ly^2\} \tag{20}$$

containing four degrees of freedom. Equation (20) is the gradient of a cubic-order polynomial and consequently has identically zero curl. Therefore those degrees of freedom can be discarded to improve computational efficiency.

Equation (19) provides a linear-tangential, quadratic-normal (LT/QN) representation along any cut through a triangular cell. A simplex-coordinate realization of these functions (Peterson, 1994) appears in Table 2. The LT/QN basis set is a combination of the six LT/LN functions in Eq. (8) and two additional quadratic functions shown in Figure 3. The six functions in Eq. (8) interpolate to a unity tangential component at the cell edges and share a coefficient with similar functions in adjacent cells to maintain tangential continuity. The other two functions each contribute a quadratic normal component to two of the three edges and have no tangential component along any of the edges. Their coefficients are independent of those in neighboring cells, allowing them to reproduce a discontinuity in the normal component of the vector field at cell junctions.

To verify that the mixed-order basis functions provide a robust discretization of the curl–curl operator, Table 3 presents numerical eigenvalues obtained using the CT/LN and LT/QN discretizations for the identical finite element mesh used to generate Table 1. The mixed-order vector basis functions clearly separate the true electromagnetic eigenvalues from the zero eigenvalues associated with the nullspace. Furthermore, the CT/LN results in Table 3 and the LT/LN results in Table 1 contain the same number of nonzero eigenvalues, but the CT/LN functions produce far fewer zero eigenvalues. The number of zero eigenvalues is reduced by exactly the number of basis functions excluded by eliminating the degrees of freedom in Eq. (17), which in this case is half the original matrix order. The LN/QT results also contain the same number of nonzero eigenvalues, but far fewer zero eigenvalues, than produced by the QT/QN expansion in Table 1. For the LN/QT functions, the number of zero

TABLE 3. Lowest Eigenvalues Produced by a Discretization of a Circular Cavity with $\varepsilon_r = 1$, $\mu_r = 1$, and Unit Radius Using Mixed-Order Basis Functions, for the TE Polarization[a]

CT/LN	LT/QN	Exact
0.0(13)	0.0(67)	
1.86(2)	1.84(2)	1.841(2)
3.10(2)	3.05(2)	3.054(2)
3.82(1)	3.84(1)	3.832(1)
4.28(2)	4.20(2)	4.201(2)
5.27(2)	5.32(2)	5.318(2)
5.39(2)	5.35(2)	5.331(2)

[a]The 42-cell model produced a matrix with order 54 for the CT/LN functions and 192 for the LT/QN functions.

Source: After Peterson and Wilton (1995).

eigenvalues is reduced by exactly the number of basis functions excluded by discarding the degrees of freedom in Eq. (20).

It is noteworthy that the six LT/LN functions form a subset of the LT/QN functions. By comparing Tables 1 and 3 we observe that the LT/QN expansion produces exactly the same number of zero eigenvalues as the complete linear expansion. This suggests that the additional basis functions used to build up the set of eight LT/QN functions do not contribute to nullspace eigensolutions. However, their addition to the set of six polynomial-complete linear functions clearly improves the accuracy of the nonzero eigenvalues.

In summary, the triangular-cell CT/LN and LT/QN basis functions enable a reduction in the degrees of freedom used to represent nullspace eigensolutions, without reducing the number of true electromagnetic eigenfunctions. Furthermore, for the cavity application they appear to produce eigenvalues that are no less accurate than those obtained using a polynomial-complete expansion of comparable order. This is expected, since the degrees of freedom excluded in the mixed-order expansions can not be used to represent source-free electromagnetic fields. Of course, the polynomial-complete functions may offer better accuracy than the mixed-order functions when modeling regions containing sources.

7. MIXED-ORDER EXPANSION FUNCTIONS IN TWO AND THREE DIMENSIONS

The preceding section presented CT/LN and LT/QN mixed-order basis functions for triangular cells. Higher-order functions can be obtained by an extension of the process. A two-dimensional (2D) vector expansion of polynomial degree k will have $k(k + 2)$ degrees of freedom remaining after eliminating the gradient of order-$(k + 1)$ polynomial terms. For example, a general cubic polynomial representation of a 2D vector contains 20 degrees of freedom. Five of these degrees are associated with the gradient of a fourth-order polynomial and can be excluded to reduce the expansion set to 15 basis functions. Nine of these functions interpolate to the tangential vector component along cell edges, while the remaining six functions build up the normal component. Consequently, these 15 basis functions provide a quadratic tangential and cubic normal (QT/CuN) representation.

The CT/LN, LT/QN, and QT/CuN basis functions for triangular cells are sumarized in Table 2. Note that there are equivalent basis functions that differ from these but still provide a representation consistent with Nedelec's spaces. On the other hand, alternative mixed-order functions have been proposed (Lee et al., 1991) that are not equivalent to those presented here and do not eliminate the same degrees of freedom associated with the nullspace of the curl operator.

For quadrilateral cells, mixed-order basis functions were also originally proposed by Nedelec, 1980). These functions have a different number of degrees of freedom—and a different mathematical expression in Cartesian coordinates—than those used on triangles. However, they are based on the common idea of discarding degrees of freedom associated with nullspace eigensolutions.

The triangular-cell functions discard all the degrees of freedom associated with the gradient of a higher-order polynomial. In the case of rectangular cells, some of these degrees of freedom must be kept to provide a symmetric distribution of basis functions. Consider the quadratic case. As an alternative to the decomposition defined by Eqs. (19) and (20), a quadratic expansion can be separated into

$$\mathbf{B}(x, y) = \hat{\mathbf{x}}\{A + Bx + Cy + Dxy + Ey^2\} + \hat{\mathbf{y}}\{F + Gx + Hy + Ix^2 + Jxy\} \tag{21}$$

and

$$\mathbf{B}_{\text{grad}}(x, y) = \hat{\mathbf{x}}Kx^2 + \hat{\mathbf{y}}Ly^2 \tag{22}$$

This decomposition eliminates two degrees of freedom, leaving ten in Eq. (21). However, 12 degrees of freedom are required to build up a linear tangential and quadratic normal (LT/QN) component along the sides of a rectangular cell. The polynomial components in Eq. (22) do not contribute to an LT/QN behavior, and instead it is convenient to add two cubic-order degrees

$$\hat{\mathbf{x}}Kxy^2 + \hat{\mathbf{y}}Lx^2y \tag{23}$$

to Eq. (21). The resulting function provides an LT/QN representation along the cell edges.

Specific vector basis functions can be defined on the standard cell $(-1 < \xi < 1, -1 < \eta < 1)$ and mapped to other cell shapes (Section 8). The two-dimensional functions for the standard cell can be obtained from the Lagrangian polynomial

$$\phi_j^{(N)} = \frac{(x - x_1)(x - x_2)\cdots(x - x_{j-1})(x - x_{j+1})\cdots(x - x_N)(x - x_{N+1})}{(x_j - x_1)\cdots(x_j - x_{j-1})(x_j - x_{j+1})\cdots(x_j - x_{N+1})} \tag{24}$$

where $[x_1, \ldots, x_{N+1}]$ denotes a uniform subdivision of the interval $-1 < x < 1$. The set of nth order functions, where n refers to the degree of the normal-component representation, can be defined in general by the expressions

$$\mathbf{B}_{ij}^{\xi} = \hat{\boldsymbol{\xi}}\phi_i^{(n-1)}(\xi)\phi_j^{(n)}(\eta), \quad i = 1, \ldots, n; j = 1, \ldots, n + 1 \tag{25}$$

and

$$\mathbf{B}_{ij}^{\eta} = \hat{\boldsymbol{\eta}}\phi_i^{(n)}(\xi)\phi_j^{(n-1)}(\eta), \quad i = 1, \ldots, n + 1; j = 1, \ldots, n \tag{26}$$

Specific CT/LN, LT/QN, and QT/CuN functions are presented in Table 4. Some of these functions interpolate to tangential fields along the cell edges and share coefficients with analogous functions in adjacent cells to maintain tangential continuity. Other functions are associated only with normal fields

TABLE 4. Summary of the Local-Coordinate Representation of the First Three Types of Mixed-Order Basis Functions[a] Within a Standard Square Cell $(-1 < \xi < 1, -1 < \eta < 1)$

CT/LN	LT/QN	QT/CuN
(4 Functions)	(12 Functions)	(24 Functions)
All Edge-based:	8 Edge-based:	12 Edge-based:
$\hat{\boldsymbol{\xi}}(1 - \eta)$	$\hat{\boldsymbol{\xi}}(1 \pm \xi)\eta(\eta \pm 1)$	$\hat{\boldsymbol{\xi}}\xi(\xi \pm 1)(\eta \pm 1)(\eta + \frac{1}{3})(\eta - \frac{1}{3})$
$\hat{\boldsymbol{\xi}}(1 + \eta)$	$\hat{\boldsymbol{\eta}}(1 \pm \eta)\xi(\xi \pm 1)$	$\hat{\boldsymbol{\xi}}(\xi + 1)(\xi - 1)(\eta \pm 1)(\eta + \frac{1}{3})(\eta - \frac{1}{3})$
$\hat{\boldsymbol{\eta}}(1 - \xi)$		$\hat{\boldsymbol{\eta}}\eta(\eta \pm 1)(\xi \pm 1)(\xi + \frac{1}{3})(\xi - \frac{1}{3})$
$\hat{\boldsymbol{\eta}}(1 + \xi)$	4 Cell-based:	$\hat{\boldsymbol{\eta}}(\eta + 1)(\eta - 1)(\xi \pm 1)(\xi + \frac{1}{3})(\xi - \frac{1}{3})$
	$\hat{\boldsymbol{\xi}}(1 \pm \xi)(1 - \eta)(\eta + 1)$	
	$\hat{\boldsymbol{\eta}}(1 \pm \eta)(1 - \xi)(\xi + 1)$	12 Cell-based:
		$\hat{\boldsymbol{\xi}}\xi(\xi \pm 1)(\eta + 1)(\eta \pm \frac{1}{3})(\eta - 1)$
		$\hat{\boldsymbol{\xi}}(\xi + 1)(\xi - 1)(\eta + 1)(\eta \pm \frac{1}{3})(\eta - 1)$
		$\hat{\boldsymbol{\eta}}\eta(\eta \pm 1)(\xi + 1)(\xi \pm \frac{1}{3})(\xi - 1)$
		$\hat{\boldsymbol{\eta}}(\eta + 1)(\eta - 1)(\xi + 1)(\xi \pm \frac{1}{3})(\xi - 1)$

[a]These functions are not normalized.

TABLE 5. Simplex-Coordinate Definition of Mixed-Order Basis Functions[a] Within a Tetrahedral Cell

CT/LN	LT/QN	QT/CuN
(6 Functions)	(20 Functions)	(45 Functions)
All Edge-based:	Edge-based:	Edge-based:
$\xi_1\nabla\xi_2 - \xi_2\nabla\xi_1$	$\xi_1\nabla\xi_2$	12 of the form:
$\xi_1\nabla\xi_3 - \xi_3\nabla\xi_1$	$\xi_2\nabla\xi_1$	$\xi_i(2\xi_i - 1)\nabla\xi_j, \quad i \neq j$
$\xi_1\nabla\xi_4 - \xi_4\nabla\xi_1$	$\xi_1\nabla\xi_3$	
$\xi_2\nabla\xi_3 - \xi_3\nabla\xi_2$	$\xi_3\nabla\xi_1$	6 of the form:
$\xi_2\nabla\xi_4 - \xi_4\nabla\xi_2$	$\xi_1\nabla\xi_4$	$\xi_i\xi_j(\nabla\xi_i - \nabla\xi_j), \quad i \neq j$
$\xi_3\nabla\xi_4 - \xi_4\nabla\xi_3$	$\xi_4\nabla\xi_1$	
	$\xi_2\nabla\xi_3$	Face-based:
	$\xi_3\nabla\xi_2$	12 of the form:
	$\xi_2\nabla\xi_4$	$\xi_i(2\xi_i - 1)(\xi_j\nabla\xi_k - \xi_k\nabla\xi_j)$
	$\xi_4\nabla\xi_2$	$i \neq j \neq k$
	$\xi_3\nabla\xi_4$	
	$\xi_4\nabla\xi_3$	12 of the form:
		$\xi_i^2(\xi_j\nabla\xi_k - \xi_k\nabla\xi_j)$
	Face-based:	$i \neq j \neq k$
	$\xi_1\xi_2\nabla\xi_3 - \xi_1\xi_3\nabla\xi_2$	
	$\xi_2\xi_3\nabla\xi_1 - \xi_1\xi_3\nabla\xi_2$	3 Cell-based:
	$\xi_1\xi_2\nabla\xi_4 - \xi_1\xi_4\nabla\xi_2$	$\xi_1\xi_2\xi_3\nabla\xi_4 - \xi_2\xi_3\xi_4\nabla\xi_1$
	$\xi_2\xi_4\nabla\xi_1 - \xi_1\xi_4\nabla\xi_2$	$\xi_1\xi_2\xi_4\nabla\xi_3 - \xi_2\xi_3\xi_4\nabla\xi_1$
	$\xi_2\xi_3\nabla\xi_4 - \xi_2\xi_4\nabla\xi_3$	$\xi_1\xi_3\xi_4\nabla\xi_2 - \xi_2\xi_3\xi_4\nabla\xi_1$
	$\xi_3\xi_4\nabla\xi_2 - \xi_2\xi_4\nabla\xi_3$	
	$\xi_1\xi_3\nabla\xi_4 - \xi_1\xi_4\nabla\xi_3$	
	$\xi_3\xi_4\nabla\xi_1 - \xi_1\xi_4\nabla\xi_3$	

[a]These functions are not normalized.

along the cell edges and will not share coefficients with neighboring cells. At a node where four cells come together in the global model, there will be four independent coefficients, one associated with the tangential field along each incoming or outgoing edge. The global CT/LN representation involves one unknown per edge, while the LN/QT expansion involves two basis functions per edge and four basis functions per cell.

Three-dimensional counterparts of the mixed-order curl-conforming vector basis functions can be developed by extending the two-dimensional process to the three-dimensional case. Table 5 summarizes the CT/LN, LT/QN, and QT/CuN basis functions within a tetrahedral cell. Six CT/LN functions overlap each tetrahedron, with each interpolating to the constant tangential component along one edge. There are 20 LT/QN functions, 12 of which interpolate to the tangential component along edges, and 8 of which build up a tangential component on the faces that is normal to the cell edges. The 45 QT/CuN basis functions consist of edge-based, face-based, and cell-based functions. A three-dimensional (3D) expansion of polynomial degree k has

$$N = \frac{k(k+2)(k+3)}{2} \tag{27}$$

coefficients within a tetrahedral cell after eliminating the degrees of freedom that can represent the gradient of a $(k+1)$-order polynomial. Closed-form expressions for the element matrix entries associated with tetrahedral-cell CT/LN and LT/QN basis functions have been developed (Savage and Peterson, 1996).

Mixed-order vector basis functions can also be defined for general hexahedral cell shapes, by transforming standard functions defined on a cube as described in Section 8. The standard functions can be constructed according to

$$\mathbf{B}_{ijk}^{\xi} = \hat{\boldsymbol{\xi}}\phi_i^{(n-1)}(\xi)\phi_j^{(n)}(\eta)\phi_k^{(n)}(\zeta), \quad i = 1,\ldots,n;\ j = 1,\ldots,n+1;\ k = 1,\ldots,n+1 \tag{28}$$

$$\mathbf{B}_{ijk}^{\eta} = \hat{\boldsymbol{\eta}}\phi_i^{(n)}(\xi)\phi_j^{(n-1)}(\eta)\phi_k^{(n)}(\zeta), \quad i = 1,\ldots,n+1;\ j = 1,\ldots,n;\ k = 1,\ldots,n+1 \tag{29}$$

$$\mathbf{B}_{ijk}^{\zeta} = \hat{\boldsymbol{\zeta}}\phi_i^{(n)}(\xi)\phi_j^{(n)}(\eta)\phi_k^{(n-1)}(\zeta), \quad i = 1,\ldots,n+1;\ j = 1,\ldots,n+1;\ k = 1,\ldots,n \tag{30}$$

where $\phi_j^{(N)}$ is the Nth order Lagrangian polynomial defined in Eq. (24), and the cell domain spans $(-1 < \eta < 1, -1 < \xi < 1, -1 < \zeta < 1)$.

The $n = 1$ basis set consists of 12 functions exhibiting a constant-tangential, linear-normal (CT/LN) behavior along the edges of the standard cube. Each basis function interpolates to the tangential field along an edge of the global mesh and shares a coefficient with all the cells adjacent to that edge to ensure tangential continuity. The $n = 2$ basis set consists of 54 functions per cell that provide a linear-tangential, quadratic-normal (LT/QN) representation along the faces of the cube, with 24 functions associated with tangential components along edges, 24 functions associated with tangential components on cell faces, and six functions that are confined to a single cell and provide a linearly independent normal field component at the center of each face. Specific basis functions are summarized in Table 6.

Vector basis functions of this type were first proposed for hexahedral cells by Nedelec (1980). Their ability to suppress spurious modes was confirmed by Crowley (1988), who also presented a subroutine for calculating the 54×54 element matrix entries for the LT/QN functions within general cell shapes.

The mixed-order spaces of Nedelec eliminate some of the degrees of freedom from the nullspace of the curl–curl operator. One might wonder if all the nullspace eigenfunctions could be eliminated. A recent article proposes a simple

TABLE 6. Mixed-Order Basis Functions[a] Within a Standard Cubical Cell ($-1<\xi<1, -1<\eta<1, -1<\zeta<1$)

CT/LN	LT/QN
(12 Functions)	(54 Functions)
All Edge-based:	24 Edge-based:
$\hat{\xi}(1 \pm \eta)(1 \pm \zeta)$	$\hat{\xi}(1 \pm \xi)\eta(\eta \pm 1)\zeta(\zeta \pm 1)$
$\hat{\eta}(1 \pm \xi)(1 \pm \zeta)$	$\hat{\eta}(1 \pm \eta)\xi(\xi \pm 1)\zeta(\zeta \pm 1)$
$\hat{\zeta}(1 \pm \xi)(1 \pm \eta)$	$\hat{\zeta}(1 \pm \zeta)\xi(\xi \pm 1)\eta(\eta \pm 1)$
	24 Face-based:
	$\hat{\xi}(1 \pm \xi)(\eta + 1)(\eta - 1)\zeta(\zeta \pm 1)$
	$\hat{\xi}(1 \pm \xi)\eta(\eta \pm 1)(\zeta + 1)(\zeta - 1)$
	$\hat{\eta}(1 \pm \eta)(\xi + 1)(\xi - 1)\zeta(\zeta \pm 1)$
	$\hat{\eta}(1 \pm \eta)\xi(\xi \pm 1)(\zeta + 1)(\zeta - 1)$
	$\hat{\zeta}(1 \pm \zeta)(\xi + 1)(\xi - 1)\eta(\eta \pm 1)$
	$\hat{\zeta}(1 \pm \zeta)\xi(\xi \pm 1)(\eta + 1)(\eta - 1)$
	6 Cell-based:
	$\hat{\xi}(1 \pm \xi)(\eta + 1)(\eta - 1)(\zeta + 1)(\zeta - 1)$
	$\hat{\eta}(1 \pm \eta)(\xi + 1)(\xi - 1)(\zeta + 1)(\zeta - 1)$
	$\hat{\zeta}(1 \pm \zeta)(\xi + 1)(\xi - 1)(\eta + 1)(\eta - 1)$

[a]These functions are not normalized.

procedure for discarding the remaining nullspace degrees of freedom, based on the tree and cotree of the finite element mesh (Manges and Cendes, 1995). The approach is built on a foundation provided by mixed-order basis functions and was demonstrated using the CT/LN functions defined on triangular cells. However, the reduction process appears to destroy the sparsity of the matrix equation (Manges and Cendes, 1995).

8. COVARIANT PROJECTION MAPPING TO DEFINE CURL-CONFORMING BASIS FUNCTIONS FOR GENERAL CELL SHAPES

The vector basis functions introduced in the preceding sections can be mapped to curved cells in two dimensions or three dimensions. However, it is important to define the vector projection in a way that preserves the curl-conforming nature of the functions. A detailed discussion of the mapping procedure was presented by Crowley (1988) and is summarized below. A hexahedral or quadrilateral cell shape is considered for illustration; the extension to tetrahedral or triangular cells follows by modifying the limits of integration. Consider a reference cell described by (ξ, η, ζ) coordinates and a specific mapping into (x, y, z) space. In order to define and manipulate vector quantities within a curvilinear cell, we introduce the base vectors

$$\boldsymbol{\xi} = \frac{\partial x}{\partial \xi}\hat{\mathbf{x}} + \frac{\partial y}{\partial \xi}\hat{\mathbf{y}} + \frac{\partial z}{\partial \xi}\hat{\mathbf{z}} \tag{31}$$

$$\boldsymbol{\eta} = \frac{\partial x}{\partial \eta}\hat{\mathbf{x}} + \frac{\partial y}{\partial \eta}\hat{\mathbf{y}} + \frac{\partial z}{\partial \eta}\hat{\mathbf{z}} \tag{32}$$

$$\boldsymbol{\zeta} = \frac{\partial x}{\partial \zeta}\hat{\mathbf{x}} + \frac{\partial y}{\partial \zeta}\hat{\mathbf{y}} + \frac{\partial z}{\partial \zeta}\hat{\mathbf{z}} \tag{33}$$

and the reciprocal base vectors $\boldsymbol{\xi}' = \nabla\xi$, $\boldsymbol{\eta}' = \nabla\eta$, $\boldsymbol{\zeta}' = \nabla\zeta$. The base vectors are tangential to the constant coordinate surfaces (cell boundaries) while the reciprocal base vectors are normal to the cell boundaries. The components of the base vectors are the entries of the Jacobian matrix

$$\begin{bmatrix} \dfrac{\partial}{\partial \xi} \\[2ex] \dfrac{\partial}{\partial \eta} \\[2ex] \dfrac{\partial}{\partial \zeta} \end{bmatrix} = \begin{bmatrix} \dfrac{\partial x}{\partial \xi} & \dfrac{\partial y}{\partial \xi} & \dfrac{\partial z}{\partial \xi} \\[2ex] \dfrac{\partial x}{\partial \eta} & \dfrac{\partial y}{\partial \eta} & \dfrac{\partial z}{\partial \eta} \\[2ex] \dfrac{\partial x}{\partial \zeta} & \dfrac{\partial y}{\partial \zeta} & \dfrac{\partial z}{\partial \zeta} \end{bmatrix} \begin{bmatrix} \dfrac{\partial}{\partial x} \\[2ex] \dfrac{\partial}{\partial y} \\[2ex] \dfrac{\partial}{\partial z} \end{bmatrix} = \mathbf{J} \begin{bmatrix} \dfrac{\partial}{\partial x} \\[2ex] \dfrac{\partial}{\partial y} \\[2ex] \dfrac{\partial}{\partial z} \end{bmatrix} \tag{34}$$

A vector can be represented by its covariant components

$$\mathbf{B} = (\mathbf{B} \cdot \boldsymbol{\xi})\boldsymbol{\xi}' + (\mathbf{B} \cdot \boldsymbol{\eta})\boldsymbol{\eta}' + (\mathbf{B} \cdot \boldsymbol{\zeta})\boldsymbol{\zeta}' \tag{35}$$

ot its contravariant components

$$\mathbf{B} = (\mathbf{B} \cdot \boldsymbol{\xi}')\boldsymbol{\xi} + (\mathbf{B} \cdot \boldsymbol{\eta}')\boldsymbol{\eta} + (\mathbf{B} \cdot \boldsymbol{\zeta}')\boldsymbol{\zeta} \tag{36}$$

When constructing a curl-conforming basis function on a curvilinear cell, the appropriate mapping is given by

$$\begin{bmatrix} B_x \\ B_y \\ B_z \end{bmatrix} = \mathbf{J}^{-1} \begin{bmatrix} B_\xi \\ B_\eta \\ B_\zeta \end{bmatrix} \tag{37}$$

where B_ξ, B_η, and B_ζ denote the covariant components of the basis function defined in the reference cell. The covariant mapping in Eq. (37) maintains the tangential-vector continuity of the basis functions in the xyz space.

When working with mapped basis functions, it is convenient to perform most of the calculations directly in the local coordinates. Consider the element matrix calculations necessary for the vector Helmholtz equation. In the 2D case, the curl operation may be evaluated, using the 2D equivalent of Eq. (37), to produce

$$\hat{\mathbf{z}} \cdot \nabla \times \mathbf{B} = \left[\frac{\partial B_y}{\partial x} - \frac{\partial B_x}{\partial y} \right] = \frac{1}{(\det \mathbf{J})} \left[\frac{\partial B_\eta}{\partial \xi} - \frac{\partial B_\xi}{\partial \eta} \right] \tag{38}$$

Therefore element matrix entries for the 2D vector Helmholtz operator can be computed directly in local coordinates via the integrals

$$\begin{aligned} A_{mn}^{(1)} &= \int_x \int_y \nabla \times \mathbf{B}_m \cdot \nabla \times \mathbf{B}_n \, dx \, dy \\ &= \int_{\eta=-1}^{1} \int_{\xi=-1}^{1} \frac{1}{(\det \mathbf{J})} \left[\frac{\partial B_{m\eta}}{\partial \xi} - \frac{\partial B_{m\xi}}{\partial \eta} \right] \left[\frac{\partial B_{n\eta}}{\partial \xi} - \frac{\partial B_{n\xi}}{\partial \eta} \right] d\xi \, d\eta \end{aligned} \tag{39}$$

and

$$\begin{aligned} A_{mn}^{(2)} &= \int_x \int_y \mathbf{B}_m \cdot \mathbf{B}_n \, dx \, dy = \int_x \int_y [B_{mx} \quad B_{my}] \begin{bmatrix} B_{nx} \\ B_{ny} \end{bmatrix} dx \, dy \\ &= \int_{\eta=-1}^{1} \int_{\xi=-1}^{1} [B_{m\xi} \quad B_{m\eta}] \mathbf{J}^{-T} \mathbf{J}^{-1} \begin{bmatrix} B_{n\xi} \\ B_{n\eta} \end{bmatrix} (\det \mathbf{J}) \, d\xi \, d\eta \end{aligned} \tag{40}$$

where $dx\,dy = (\det \mathbf{J})\, d\xi\, d\eta$.

In the 3D case, the curl of a basis function defined according to Eq. (37) can be expressed in the form

$$\nabla \times \mathbf{B} = \frac{1}{(\det \mathbf{J})} \left\{ \boldsymbol{\xi} \left[\frac{\partial B_\zeta}{\partial \eta} - \frac{\partial B_\eta}{\partial \zeta} \right] + \boldsymbol{\eta} \left[\frac{\partial B_\xi}{\partial \zeta} - \frac{\partial B_\zeta}{\partial \xi} \right] + \boldsymbol{\zeta} \left[\frac{\partial B_\eta}{\partial \xi} - \frac{\partial B_\xi}{\partial \eta} \right] \right\} \tag{41}$$

where B_ξ, B_η, and B_ζ denote the covariant components in the reference cell. Note that Eq. (41) involves the contravariant components of the curl and therefore must be transformed via the contravariant mapping

$$\begin{bmatrix} V_x \\ V_y \\ V_z \end{bmatrix} = \mathbf{J}^T \begin{bmatrix} V_\xi \\ V_\eta \\ V_\zeta \end{bmatrix} \tag{42}$$

For notational purposes, let us introduce a column matrix $\mathbf{V}_n$ to represent the contravariant components of the curl of the nth basis function, scaled by the determinant, namely,

$$\mathbf{V}_n^T = \left[\left(\frac{\partial B_{n\zeta}}{\partial \eta} - \frac{\partial B_{n\eta}}{\partial \zeta} \right) \left(\frac{\partial B_{n\xi}}{\partial \zeta} - \frac{\partial B_{n\zeta}}{\partial \xi} \right) \left(\frac{\partial B_{n\eta}}{\partial \xi} - \frac{\partial B_{n\xi}}{\partial \eta} \right) \right] \tag{43}$$

The necessary element matrix entries can be computed from the integrals

$$\begin{aligned} A_{mn}^{(1)} &= \int_x \int_y \int_z \nabla \times \mathbf{B}_m \cdot \nabla \times \mathbf{B}_n \, dx \, dy \, dz \\ &= \int_{\xi=-1}^{1} \int_{\eta=-1}^{1} \int_{\zeta=-1}^{1} \frac{1}{(\det \mathbf{J})} \mathbf{V}_m^T \mathbf{J} \mathbf{J}^T \mathbf{V}_n \, d\xi \, d\eta \, d\zeta \end{aligned} \tag{44}$$

and

$$\begin{aligned} A_{mn}^{(2)} &= \int_x \int_y \int_z \mathbf{B}_m \cdot \mathbf{B}_n \, dx \, dy \, dz = \int_x \int_y \int_z [B_{mx} \quad B_{my} \quad B_{mz}] \begin{bmatrix} B_{nx} \\ B_{ny} \\ B_{nz} \end{bmatrix} dx \, dy \, dz \\ &= \int_{\xi=-1}^{1} \int_{\eta=-1}^{1} \int_{\zeta=-1}^{1} [B_{m\xi} \quad B_{m\eta} \quad B_{m\zeta}] \mathbf{J}^{-T} \mathbf{J}^{-1} \begin{bmatrix} B_{n\xi} \\ B_{n\eta} \\ B_{n\zeta} \end{bmatrix} (\det \mathbf{J}) \, d\xi \, d\eta \, d\zeta \end{aligned} \tag{45}$$

In practice, the entries of the Jacobian matrix $\mathbf{J}$ are obtained directly from the mapping, which typically specifies (x, y, z) as functions of (ξ, η, ζ). The inverse Jacobian and the determinant vary with location within a cell and are usually computed numerically from $\mathbf{J}$. The required integrations can be performed by numerical quadrature.

9. ANALYSIS OF DIELECTRIC-LOADED WAVEGUIDES USING CURL-CONFORMING VECTOR BASIS FUNCTIONS

Consider a waveguide structure aligned with the z-axis, with an inhomogeneous cross section that may contain dielectric and magnetic materials. A number of waveguide formulations are described in the literature, and we focus our attention on an approach originally proposed by Lee et al. (1991). For illustration, we consider a waveguide bounded by perfectly conducting electric walls and base the analysis on the vector Helmholtz equation

$$\nabla \times \left(\frac{1}{\varepsilon_r} \nabla \times \mathbf{H} \right) = k^2 \mu_r \mathbf{H} \tag{46}$$

where ε_r and μ_r are functions of transverse position throughout the waveguide cross section. The magnetic field can be written in the form

$$\mathbf{H}(x, y, z) = \{\mathbf{H}_t(x, y) + \hat{\mathbf{z}}\mathbf{H}_z(x, y)\}e^{-j\beta z} \tag{47}$$

and used to separate the vector Helmholtz equation into its transverse and z-components. To facilitate the use of either k or β as the eigenvalue, Lee et al. introduced the scaling $\mathbf{h}_t = \beta\mathbf{H}_t$. They also used the substitution $h_z = -jH_z$, which produces a real-valued system of equations in the lossless case. These substitutions convert Eq. (46) into the coupled equations

$$\nabla_t \times \left\{\frac{1}{\varepsilon_r}\nabla_t \times \mathbf{h}_t\right\} + \frac{1}{\varepsilon_r}(\beta^2\nabla_t h_z + \beta^2\mathbf{h}_t) = k^2\mu_r\mathbf{h}_t \tag{48}$$

$$-\nabla_t \cdot \left\{\frac{1}{\varepsilon_r}(\nabla_t h_z + \mathbf{h}_t)\right\} = k^2\mu_r h_z \tag{49}$$

where ∇_t is the transverse part of the ∇ operator.

Equation (48) is a vector equation with the same leading-order derivatives as the 2D vector Helmholtz equation. This equation can be converted into a weak form by introducing a transverse vector testing function $\mathbf{T}(x, y)$ and integrating the equation over the domain Ω to produce (Crain and Peterson, 1995)

$$\iint_\Omega \frac{1}{\varepsilon_r}\{\nabla_t \times \mathbf{T}\cdot\nabla_t \times h_t + \beta^2\mathbf{T}\cdot\mathbf{h}_t + \beta^2\mathbf{T}\cdot\nabla_t h_z\} = k^2\iint_\Omega \mu_r\mathbf{T}\cdot\mathbf{h}_t - \int_\Gamma \frac{1}{\varepsilon_r}\hat{\mathbf{n}}\times\nabla_t\times\mathbf{h}_t \tag{50}$$

where Ω denotes the guide interior and Γ denotes the guide walls. By introducing a scalar testing function $T(x, y)$, an analogous weak equation can be written for Eq. (49) in the form

$$\iint_\Omega \frac{1}{\varepsilon_r}\{\nabla_t T\cdot\nabla_t h_z + \nabla_t T\cdot\mathbf{h}_t\} = k^2\iint_\Omega \mu_r T h_z + \int_\Gamma \frac{1}{\varepsilon_r}\left\{T\frac{\partial h_z}{\partial n} + T\hat{\mathbf{n}}\cdot\mathbf{h}_t\right\} \tag{51}$$

Since the boundary Γ consists of perfectly conducting material, appropriate boundary conditions dictate that

$$\hat{\mathbf{n}} \times \nabla_t \times \mathbf{h}_t = 0 \tag{52}$$

$$\frac{\partial h_z}{\partial n} = 0 \tag{53}$$

and

$$\hat{\mathbf{n}}\cdot\mathbf{h}_t = 0 \tag{54}$$

along Γ. These conditions eliminate the boundary integrals in Eqs. (50) and (51).

Suppose the function $\mathbf{h}_t(x, y)$ is represented by mixed-order vector basis functions, while h_z is represented by ordinary scalar Lagrangian basis functions. Given sets of linearly independent vector and scalar testing functions, Eqs. (50) and (51) can be written in the form of the matrix eigenvalue equation

$$\begin{bmatrix} \mathbf{A}^{tt} & \mathbf{A}^{tz} \\ \mathbf{A}^{zt} & \mathbf{A}^{zz} \end{bmatrix} \begin{bmatrix} \mathbf{h}_t \\ \mathbf{h}_z \end{bmatrix} = k^2 \begin{bmatrix} \mathbf{B}^{tt} & \mathbf{0} \\ \mathbf{0} & \mathbf{B}^{zz} \end{bmatrix} \begin{bmatrix} \mathbf{h}_t \\ \mathbf{h}_z \end{bmatrix} \tag{55}$$

where

$$A^{tt}_{mn} = \iint \frac{1}{\varepsilon_r} \{\nabla_t \times \mathbf{T}_m \cdot \nabla_t \times \mathbf{B}_n + \beta^2 \mathbf{T}_m \cdot \mathbf{B}_n\} \tag{56}$$

$$A^{tz}_{mn} = \beta^2 \iint \frac{1}{\varepsilon_r} \{\mathbf{T}_m \cdot \nabla_t B_n\} \tag{57}$$

$$A^{zt}_{mn} = \beta^2 \iint \frac{1}{\varepsilon_r} \{\nabla_t T_m \cdot \mathbf{B}_n\} \tag{58}$$

$$A^{zz}_{mn} = \beta^2 \iint \frac{1}{\varepsilon_r} \{\nabla_t T_m \cdot \nabla_t B_n\} \tag{59}$$

$$B^{tt}_{mn} = \iint \mu_r \mathbf{T}_m \cdot \mathbf{B}_n \tag{60}$$

and

$$B^{zz}_{mn} = \beta^2 \iint \mu_r T_m B_n \tag{61}$$

and where the integrals encompass the appropriate cells of the mesh. The scalar equation has been scaled by a factor β^2 to improve the symmetry in Eq. (55). By rearranging terms, Eq. (55) can be converted to an eigenvalue equation for β^2 instead of k^2 (Crain and Peterson, 1995).

To illustrate the approach, Figure 4 shows a boxed microstrip geometry and a portion of the k–β diagram produced by the finite element solution, compared with data from an integral equation formulation (Railton and Rozzi, 1988). The region was modeled using quadrilateral-cell LT/QN functions defined via the mapping procedure discussed in Section 8. Because of symmetry, only half the structure was modeled. The mesh contained 36 cells and produced a generalized eigenvalue equation of order 463, which was solved using a subspace iteration algorithm (Crain and Peterson, 1994). No spurious eigenvalues were observed in the results.

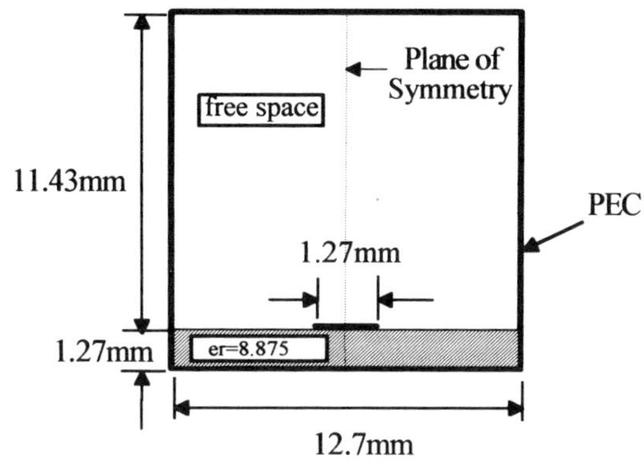

(*a*)

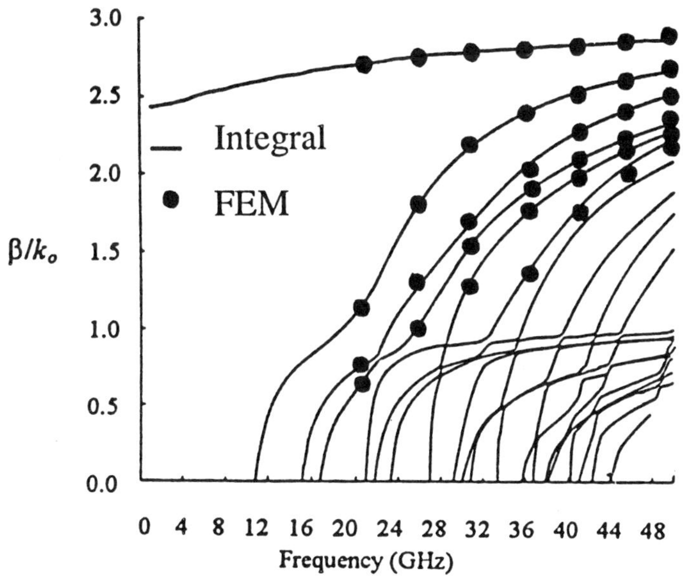

(*b*)

FIGURE 4. (a) Geometry of the boxed microstrip structure. Because of left-to-right symmetry, only half of the region is modeled with finite elements. (b) Dispersion curves for the boxed microstrip waveguide shown in (a), produced by a solution of an order-463 eigenvalue equation. (After Crain and Peterson, 1994.)

In summary, this waveguide formulation incorporates ordinary scalar basis functions to represent the z-component of the field and mixed-order vector basis functions to represent the transverse component (Lee et al., 1991; Crain and Peterson, 1995). The approach offers the possibility of treating material with simultaneous discontinuities in both ε_r and μ_r, and it eliminates spurious eigenvalues since the vector basis functions permit jump discontinuities in the normal field component. The approach can also be modified to treat the electric field as the primary unknown, in which case slightly different boundary conditions must be imposed along $\partial\Gamma$. In addition, absorbing boundary conditions can be incorporated into the formulation in order to treat open-region waveguiding structures such as microstrip (Crain and Peterson, 1995).

10. SUMMARY

Mixed-order curl-conforming basis functions have found widespread use for numerical solutions of the vector Helmholtz equation, since they eliminate spurious eigenvalues. This chapter has discussed their properties, presented higher-order basis functions of the same type, and summarized the mapping of these functions to curvilinear cells. Applications from cavity and waveguide analysis were presented to illustrate the concepts.

REFERENCES

Barton, M. L. and Cendes, Z. J. (1987). New vector finite elements for three-dimensional magnetic field computations. *J. Appl. Phys.* **61**, 3919–3921.

Bossavit, A., and Verite, J. C. (1982). A mixed FEM-BIEM method to solve 3D eddy current problems. *IEEE Trans. Magn.* **MAG-18**, 431–435.

Crain, B. R., and Peterson, A. F. (1994). Analysis of dielectric-loaded waveguides using covariant projection vector finite elements. *Proc. 10th Annu. Rev. Prog. Appl. Comput. Electromagn.*, Monterey, CA, 1994, Vol. 1, pp. 144–151.

Crain, B. R., and Peterson, A. F. (1995). Analysis of propagation on open microstrip lines using mixed-order covariant projection vector finite elements. *Int. J. Microwave Millimeter-Wave Comput.-Aided Eng.* **5**, 59–67.

Crowley, C. W. (1988). Mixed-order covariant projection finite elements for vector fields. Ph.D. Dissertation, McGill University, Montreal, Quebec.

Hano, M. (1984). Finite element analysis of dielectric-loaded waveguides. *IEEE Trans. Microwave Theory Tech.* **MTT-32**, 1275–1279.

Lee, J.-F., Sun, D.-K., and Cendes, Z. J. (1991). Full-wave analysis of dielectric waveguides using tangential vector finite elements. *IEEE Trans. Microwave Theory Tech.* **MTT-39**, 1262–1271.

Manges, J. B., and Cendes, Z. J. (1995). A generalized tree-cotree gauge for magnetic field computation. *IEEE Trans. Magn.* **MAG-31**, 1342–1347.

Mur, G. (1994). Edge elements, their advantages and their disadvantages. *IEEE Trans. Magn.* **MAG-30**, 3552–3557.

Mur, G., and de Hoop, A. T. (1985). A finite element method for computing three-dimensional electromagnetic fields in inhomogeneous media. *IEEE Trans. Magn.* **MAG-21**, 2188–2191.

Nedelec, J. C. (1980). Mixed finite elements in R3. *Numer. Math.* **35**, 315–341.

Nedelec, J. C. (1986). A new family of mixed finite elements in R3. *Numer. Math.* **50**, 57–81.

Peterson, A. F. (1994). Vector finite element formulation for scattering from two-dimensional heterogeneous bodies. *IEEE Trans. Antennas Propag.* **AP-43**, 357–365.

Peterson, A. F., and Wilton, D. R. (1995). A rationale for the use of mixed-order basis functions within finite element solutions of the vector Helmholtz equation. *Proc. 11th Annu. Rev. Prog. Appl. Comput. Electromagn.*, Monterey, CA, 1995, pp. 1077–1084.

Rahman, B. M. A., and Davies, J. B. (1984). Penalty-function improvement of waveguide solution by finite elements. *IEEE Trans. Microwave Theory Tech.* **MTT-32**, 922–928.

Railton, C. J., and Rozzi, T. (1988). Complex modes in boxed microstrip. *IEEE Trans. Microwave Theory Tech.* **MTT-36**, 865–874.

Savage, J. S., and Peterson, A. F. (1966). Higher order vector finite elements. *IEEE Trans. Microwave Theory Tech.* **MTT-44** (accepted for publication).

Silvester, P. P., and Ferrari, R. L. (1990). *Finite Elements for Electrical Engineers.* Cambridge: Cambridge University Press.

Winkler, J. R., and Davies, J. B. (1984). Elimination of spurious modes in finite element analysis. *J. Comput. Physics* **56**, 1–14.

CHAPTER SIX

Two-Dimensional Bandpass Filters

GORDON P. RIBLET

1. INTRODUCTION

High Q devices such as filters have not often been treated using the three-dimensional finite element approach due to the large number of elements needed to achieve adequate accuracy. Waveguide bandpass filters can usually be treated as two-dimensional entities, however. If the filter cross section is uniform in the y direction, then only the field components E_y, H_x, and H_z are nonzero. These can be represented by higher order scalar elements. The physical symmetry of such filters further reduces the number of two-dimensional elements needed so that the finite element approach becomes practical. The irises used in such filters typically have or can be made to have a rectangular cross section. This permits a simple triangulation. Personal computer-based programs for the accurate design and analysis of multicavity waveguide filters have been written. These have been constructed by modifying and assembling existing FORTRAN subroutines such as those given by Silvester and Ferrari (1990).

2. THE SYMMETRY OF TWO-PORT FILTERS

Filters with the best passband and stopband characteristics tend to have a high degree of physical symmetry. Schematic diagrams of the cross sections of the

Finite Element Software for Microwave Engineering, Edited by Tatsuo Itoh, Giuseppe Pelosi, and Peter P. Silvester.
ISBN 0-471-12636-5

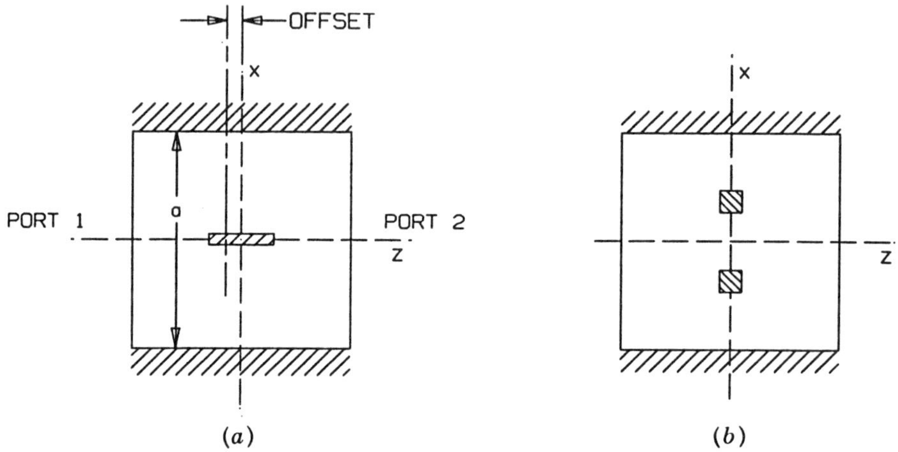

FIGURE 1. Cross-sectional views of two styles of symmetrical rectangular waveguide obstacles used in waveguide filters. The offset distance is the difference between the mechanical and electrical center.

two most common iris styles used in waveguide bandpass filters are given in Figure 1. The width of the irises is W and their thickness T. The waveguide width is a. The centered iris given in Figure 1a is the most economical to produce. The dual irises given in Figure 1b tend to give better stopband rejection characteristics above the passband. In both cases there is symmetry about the x and z axes. Consequently, only one quadrant need be analyzed by the finite element approach. This reduces the number of finite elements needed to achieve a given level of accuracy by a factor of 4. Before the advent of numerical control manufacturing, such irises were almost always made as round posts. However, the choice between a round or an equivalent square iris has a negligible effect on filter performance.

An N-section bandpass filter is formed by cascading $N + 1$ such irises together. Each iris is separated from its neighbor by about a half-guide wavelength. The irises at each end of the filter will be identical and so forth down the line. That is, the N-section filter is also symmetrical about the x and z centerline axes. A finite element analysis of such an N-section filter also requires that only one quadrant be considered.

3. EIGENVALUE TREATMENT OF SYMMETRICAL TWO PORTS

The eigenvalue treatment of symmetrical two ports is the basis for the methods to be discussed in this chapter. The S matrix of a symmetrical two port has the form

$$S = \begin{bmatrix} S_{11} & S_{12} \\ S_{12} & S_{11} \end{bmatrix} \tag{1}$$

In the case of a filter it is generally the quantity of interest. By multiplication it is readily seen that the eigenvectors of this S matrix are (1, 1) and (1, −1) independent of the actual values of S_{11} and S_{12}. These eigenvectors correspond to the even and odd mode excitations: that is, the eigenvectors are determined by symmetry alone. These vectors are also eigenvectors of the impedance matrix Z and admittance matrix Y. Let (s_1, s_2), (z_1, z_2), and (y_1, y_2) be the pairs of eigenvalues of the S, Z, and Y matrices, respectively. They are related in the following way:

$$s_1 = \frac{z_1 - 1}{z_1 + 1} = \frac{1 - y_1}{1 + y_1} \tag{2}$$

$$s_2 = \frac{z_2 - 1}{z_2 + 1} = \frac{1 - y_2}{1 + y_2} \tag{3}$$

Furthermore,

$$S_{11} = \frac{s_1 + s_2}{2} \tag{4}$$

$$S_{12} = \frac{s_1 - s_2}{2} \tag{5}$$

It follows from these equations that if the impedance matrix or admittance matrix eigenvalues can be determined, then the scattering parameters can be found. The eigenvalues s_1 and s_2 correspond to the reflection coefficients of the even and odd mode excitations. A more detailed treatment has been given by Helszajn (1978).

What is the significance of this for the finite element approach? Whereas S_{11} and S_{12} are generally complex, in the case of a lossless device z_1, z_2, y_1, and y_2 will be purely imaginary. They correspond to real reactances X_1 and X_2 and real susceptances B_1 and B_2, respectively. In fact, it will be seen that these real quantities can be derived from real finite element nodal potentials. The use of real as opposed to complex numbers reduces the storage requirements by another factor of 2. The use of real arithmetic also greatly reduces computation times.

4. FINITE ELEMENT EXPRESSIONS FOR EIGENSUSCEPTANCES (B_e) AND EIGENREACTANCES (X_e)

The approach proposed in the previous section requires that either the even and odd mode eigensusceptances or the even and odd mode eigenreactances be determined. First, however, the standing wave E-field nodal potentials E_y or the standing wave H-field potentials H_x and H_z are solved for assuming a unit amplitude sinusoidal field at the input. The emphasis in this chapter is on solving for the E-field potentials and eigensusceptances B_e. There are several reasons for this. For a given triangulation there are twice as many H-field nodal potentials as E-field nodal potentials. Since the field E_y goes to zero on conducting boundaries unlike H_x and H_z, the solution is more accurate for a given triangulation with such potentials. Finally, the component E_y satisfies the two-dimensional Helmholtz equation. Consequently, standard FORTRAN subroutines for the solution of this common equation can be used.

The two-dimensional Helmholtz equation is solved with an open circuit boundary condition along the yz symmetry plane, a short circuit boundary condition along the waveguide wall, and $E_y = \cos(\pi x/a)$ at the input reference plane. This is illustrated in Figure 2 for the iris style of Figure 1a. For the even mode eigensusceptance an open circuit boundary condition applies along the xy symmetry plane while for the odd mode eigenadmittance a short circuit boundary condition applies along this plane. In the finite element analysis

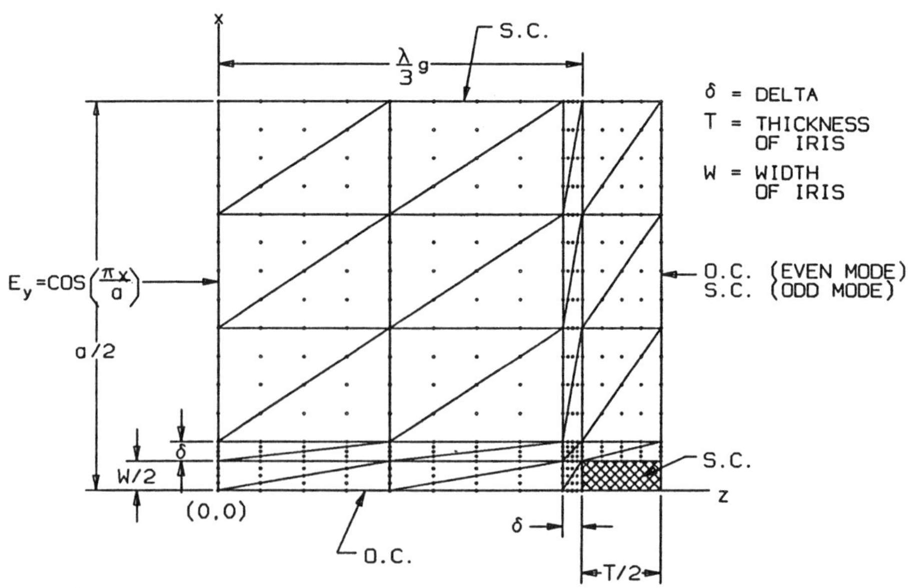

FIGURE 2. Suitable triangulation of the reactive region of the iris type given in Figure 1a.

fourth-order elements have been used to improve the accuracy as discussed by Scott (1992). By considering the case of an empty waveguide, where the exact solution is known, it was determined that an excellent approximation to the sinusoidal fields is obtained by having the sides of three such elements lie along the direction of any half-wavelength change in the fields. Including further elements produced only marginal improvement in accuracy. These conclusions suggest the triangulation of the reactive region given in Figure 2. Additional fourth-order elements of width δ have been included next to the iris walls in order to better model the exponentially decaying field modes close to the obstacle. The input reference plane is set at a distance of one-third of a guide wavelength λ_g away from the obstacle in order to allow these higher order mode fields to decay to a negligible value.

4.1. Eigensusceptances

It is known that in the lossless case the eigensusceptances and eigenreactances must be proportional to the difference of the electric field and magnetic field stored energies $W_e - W_m$. It is a question of determining the proportionality factor applicable to rectangular waveguide. This can be done using formulas that relate the maximum electric or magnetic field strengths to the incident power as given, for instance, by Montgomery et al. (1987). In the case of eigensusceptances one finds that

$$B_e = -\frac{8\pi\lambda_g}{\varepsilon a\lambda^2}\int \{\mu \mathbf{H}^* \cdot \mathbf{H} - \varepsilon \mathbf{E}^* \cdot \mathbf{E}\}\, dV \tag{6}$$

$$= -\frac{2\lambda_g}{\pi a}\int \left[\left(\frac{\partial E_y}{\partial z}\right)^2 + \left(\frac{\partial E_y}{\partial x}\right)^2 - k^2 E_y^2\right] dV \tag{7}$$

where λ is the wavelength, λ_g is the guide wavelength, and a is the waveguide width. The first expression applies to the general case and the second to the two-dimensional case. The two-dimensional finite element expression for B_e follows from the second equation and is given by

$$B_e \cong -\frac{2\lambda_g}{\pi a}\,[e^+[S]e - k^2 e^+[T]e] \tag{8}$$

where $[S]$ and $[T]$ are the global finite element matrices for the two-dimensional scalar Helmholtz equation and e is the vector of nodal E-field potentials within the reactive region. Expressions similar to Eqs. (6), (7), and (8) can also be stated for eigenreactances X_e. FORTRAN code for determining B_e from the E-field nodal potentials is given in Section 9.

4.2. Eigenreactances

The expression analogous to Eq. (6) for eigenreactances X_e is based on specifying the tangential magnetic field component H_x at the input reference plane instead of the electric field component E_y. The component H_z is unconstrained. If the standing wave magnetic field component at the input is $H_x = \cos(\pi x/a)$, then it can be shown that

$$X_e = \frac{8\pi}{\mu a \lambda_g} \int \{\varepsilon E^* \cdot E - \mu H^* \cdot H\}\, dV \tag{9}$$

Once again this expression is quite general. If, in addition, it is assumed that there is no field variation in the y direction as discussed by Schwinger and Saxon (1968),

$$E_y = \frac{\lambda}{2\pi} \sqrt{\frac{\mu}{\varepsilon}} \left\{ \frac{\partial H_z}{\partial x} - \frac{\partial H_x}{\partial z} \right\} \tag{10}$$

In this special case it follows from Eqs. (9) and (10) that

$$X_e = \frac{2\lambda^2}{a\pi\lambda_g} \int \left\{ \left[\frac{\partial H_z}{\partial x} \cdot \frac{\partial H_z}{\partial x} - k^2 H_z^2 \right] + \left[\frac{\partial H_x}{\partial z} \cdot \frac{\partial H_x}{\partial z} - k^2 H_x^2 \right] - \left[\frac{\partial H_Z}{\partial x} \cdot \frac{\partial H_x}{\partial z} - \frac{\partial H_x}{\partial z} \cdot \frac{\partial H_z}{\partial x} \right] \right\} dV \tag{11}$$

This formulation of the eigenreactance in terms of magnetic field components H_x, H_z can also be the basis of a finite element treatment as described by Silvester and Ferrari (1990). However, twice as many potentials are now required for a given triangulation, as was the case for the electric field formulation. The simplest formulation makes use of the differentiation matrices $D^{(i)}$, $i = 1, 2, 3$, defined by Silvester (1978). Since any two matrices $D^{(i)}$ can be derived from the third by using permutation operators, it is actually only necessary to tabulate one of these matrices. The matrix D^1 for the fourth-order case is tabulated in Table 1 for convenience. Most of the entries are zero. These have not been tabulated. Let $h_x(k)$ be the vector of nodal potentials for the magnetic field component H_x for the kth element and $h_z(k)$ be the vector of nodal potentials for the magnetic field component H_z for the kth element. Then an approximate expression for X_e is

$$X_e \cong \frac{2\lambda^2}{a\pi\lambda_g} \left[\sum_k h_x^+(k)[A(k)]h_x(k) + h_x^+(k)[B(k)]h_z(k) + h_z^+(k)[C(k)]h_x(k) + h_z^+(k)[D(k)]h_z(k) \right] \tag{12}$$

TABLE 1. Differentiation Matrix D^1 for the Fourth-Order Case: The Common Divisor Is 3

	1	2	3	4	5	6	7	8	9	10	11	12	13	14	15
1	25	3	3	−1	−1	−1	1	1	1	1	−3	−3	−3	−3	−3
2		22		8	4		−6	−4	−2		16	12	8	4	
3			22		4	8		−2	−4	−6		4	8	12	16
4				18			18	6			−36	−18	−6		
5					18			12	12			−18	−24	−18	
6						18			6	18			−6	−18	−36
7							12				48	12			
8								12				36	24		
9									12				24	36	
10										12				12	48
11															
12															
13															
14															
15															

For the kth element the matrices $A(k)$, $B(k)$, $C(k)$, and $D(k)$ are given by

$$A(k) = \frac{1}{4A} \sum_i \sum_j c_i c_j D^{(i)} \cdot T \cdot D^{(j)'} - \frac{4\pi^2}{\lambda^2} T \tag{13}$$

$$B(k) = \frac{1}{4A} \sum_i \sum_j b_i c_j D^{(i)} \cdot T \cdot D^{(j)'} \tag{14}$$

$$C(k) = \frac{1}{4A} \sum_i \sum_j c_i b_j D^{(i)} \cdot T \cdot D^{(j)'} \tag{15}$$

$$D(k) = \frac{1}{4A} \sum_i \sum_j b_i b_j D^{(i)} \cdot T \cdot D^{(j)'} - \frac{4\pi^2}{\lambda^2} T \tag{16}$$

where A is the area of the kth element, values for the matrix T in the fourth-order case have been tabulated, and the geometric coefficients b_i and c_i ($i = 1, 2, 3$) are those used by Silvester and Ferrari (1990). In the case of the fourth-order elements the dimension of the above matrices is 15×15.

Mathematically, this process bears a strong resemblance to the finite element determination of the characteristic impedance Z_0 or characteristic admittance Y_0 of coaxial line as given by Silvester and Ferrari (1990, pp. 208–213). Instead of solving the two-dimensional Laplace equation, the two-dimensional Helmholtz equation is now being solved. In this reference the results were used to obtain a bound on Z_0. Here the results may be used to obtain a bound on both the even and odd mode eigensusceptances. Note that Eq. (8) for B_e and Eq. (12) for X_e are correct to second order. Furthermore, $B_e = -1/X_e$. Consequently, obtaining a bound on the two-port S matrix parameters becomes possible using the equations given in Section 3 as discussed by Riblet (1995).

5. APPLICATION TO FILTER DESIGN

The first step in filter design usually begins with the generation of a filter prototype based on the electrical specifications. Such a prototype of an N-section waveguide bandpass filter is given in Figure 3. The theory behind such a prototype in the case of waveguide bandpass filters has been discussed by Levy (1967). There are $N + 1$ shunt susceptances B_n separated by N electrical lengths ϕ_n. The next step in the process is to determine mechanical dimensions that correspond to the parameters of the prototype. Since the irises in such filters are separated by roughly a half-wavelength, interaction effects can almost always be neglected. Iris dimensions are found that yield the appropriate susceptance values B_n at the normalized center frequency of the filter. However, as indicated in Figure 1a the electrical center of such an iris does not correspond to the mechanical center. The electrical offset length O_n for each iris must be determined. The parameters ϕ_n, O_n $(n = 1, \ldots, N)$, together allow the cavity lengths to be calculated. In fact, the length ℓ_n of the nth cavity is

$$\ell_n = \frac{\lambda_g}{2\pi}(O_n + \phi_n + O_{n+1}) \tag{17}$$

The parameters B_n, O_n are related to the even and odd mode eigensusceptances B_{en}, B_{on} by simple formulas. For the odd mode circuit the susceptance B_n is shorted out. Consequently, O_n depends only on B_{on}. In fact, O_n is one-half

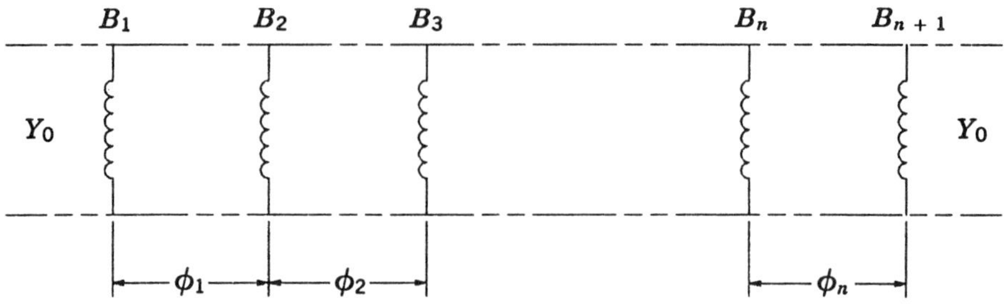

FIGURE 3. Direct coupled filter prototype with shunt inductance coupling.

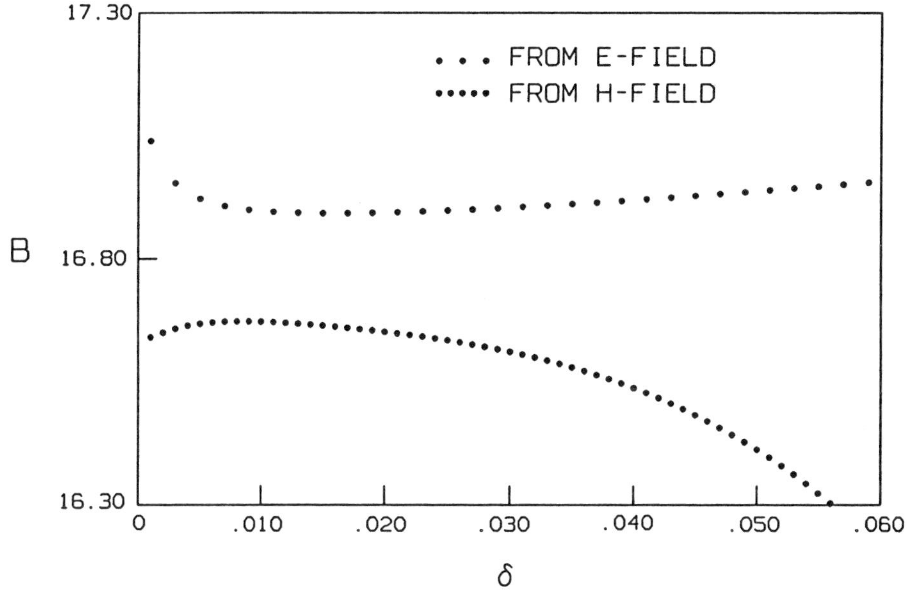

FIGURE 4. Plot of the obstacle susceptance B versus δ for both the eigensusceptance and eigenreactance solutions using the triangulation given in Figure 2.

of the odd mode phase when the reference plane is set to the mechanical center of the iris. The parameter B_n is determined by the requirement that it yield a reflection coefficient magnitude equal to that of the actual iris. In the author's experience the above process yields very accurate designs.

It is desirable for the purpose of filter design to determine the prototype susceptance B_n to an accuracy of several ports per thousand. This can be done with only 40 fourth-order elements in the case of the iris style of Figure 1a. In the case of the iris style of Figure 1b, 64 fourth-order elements are needed. Figure 4 gives a plot of B and Figure 5 a plot of the offset parameter O as a function of δ for the case of an iris of width $W = 0.1638$ and thickness $T = 0.1638$ in a WR75 waveguide. The center frequency is 14.2395 GHz. These values were calculated using double precision arithmetic. Recall that δ is the width of additional fourth-order elements, which have been introduced next to the iris walls where the field components E_y, H_x, and H_z are rapidly varying. It is apparent that both B and O go through a minimum for $\delta \approx 0.016$ inch. This situation is a result of the fact that the expressions (6), (7), and (8) are stationary. The most accurate value for B of 16.892 is obtained at the minimum and the actual value must be slightly less than this minimum value. The consideration of many examples suggests that δ may be taken to be $a/40$, where a is the waveguide width. This example can also be treated using the eigenreactances. In this case a maximum for B of 16.672 is observed in Figure

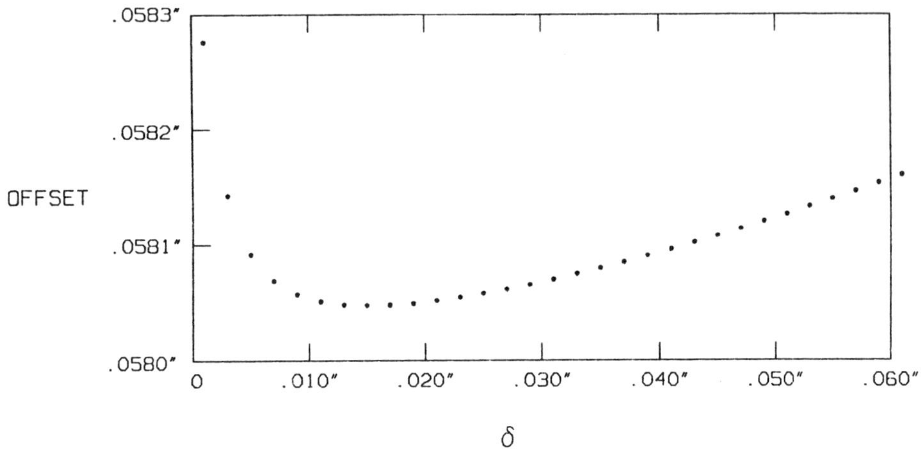

FIGURE 5. Plot of the offset distance O versus δ using the eigensusceptance solutions. Both quantities are in inches.

4 at $\delta \approx 0.009$ in. This maximum value lies below the previous minimum value. The evaluation of many examples suggests that δ may be taken to be $a/100$ in the case of eigenreactances. Consequently, a bound on the actual value is obtained. However, since the field components H_x, H_z are not zero at the iris walls, this method of treating the problem is not as accurate as the method based on the component E_y.

6. APPLICATION TO FILTER ANALYSIS

The ideas developed for triangulating individual irises may be generalized to determine appropriate triangulations of multicavity filters. An accurate analysis can then be carried out with a minimum number of elements. This analysis includes the treatment of all interaction effects and can be used to verify that interaction effects are indeed small. An appropriate triangulation of a seven-cavity filter having the centered irises of Figure 1a is given in Figure 6. Since the electric field component goes through a maximum between any two adjacent irises, 60 fourth-order elements are used between any such irises. A total of 288 fourth-order elements are required to analyze the complete filter.

The analysis of such a filter exceeds the memory capacity of a personal computer unless banded storage is used. The minimum bandwidth is obtained by numbering the nodes starting at the middle of the waveguide input and continuing to the waveguide wall at the input reference plane. The numbering of the nodes then continues until the filter symmetry plane is reached. In the case of the filter of Figure 6, the bandwidth $2M$ has been taken to be 214.

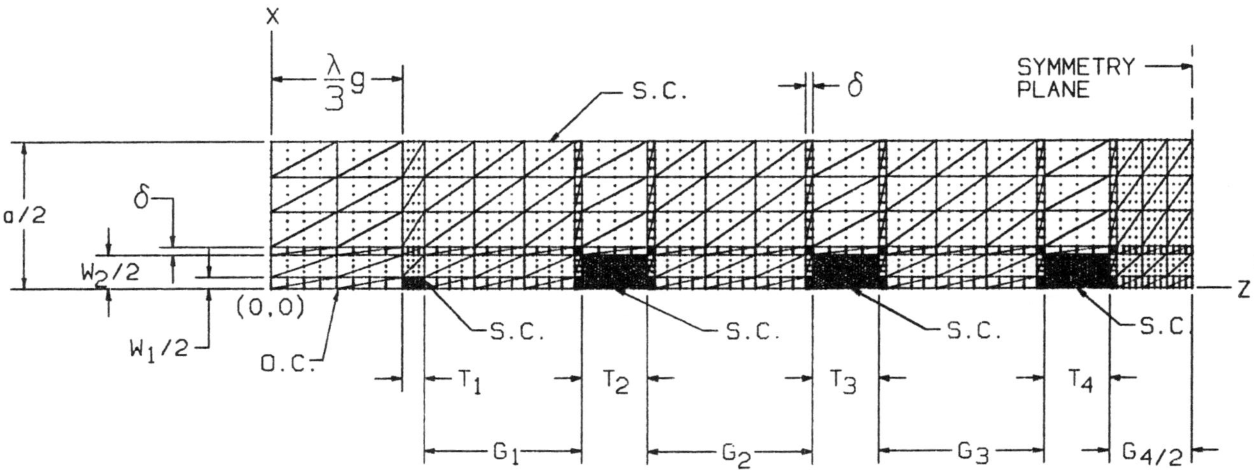

FIGURE 6. Suitable triangulation of a seven-section bandpass filter with fourth-order elements.

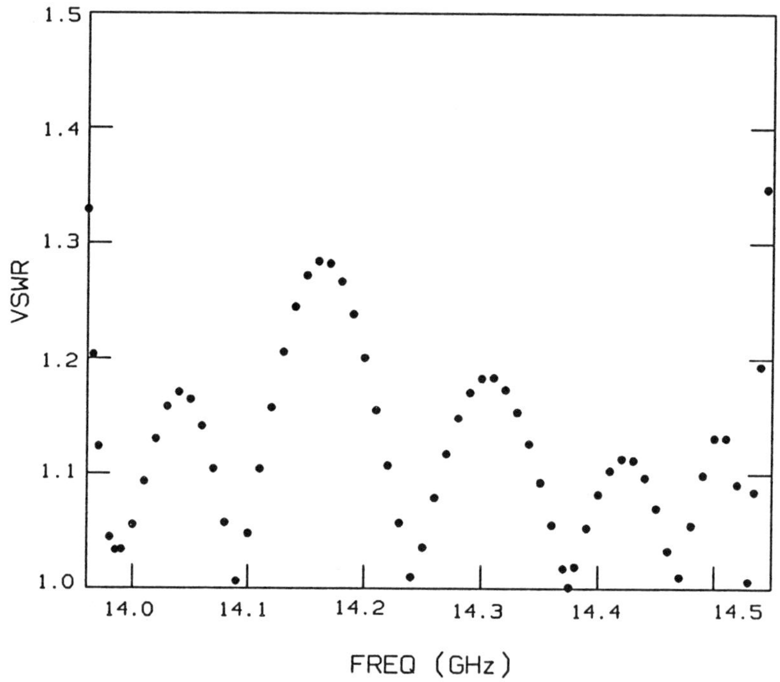

FIGURE 7. Plot of theoretical VSWR versus frequency for a seven-section bandpass filter designed with a passband of 14.0–14.5 GHz.

The analysis of a particular seven-section filter with a design passband of 14.0–14.5 GHz has been carried out. The end irises have a width of $W_1 = 0.0568$ in. while all other irises have a width of $W_2 = 0.1638$ in. The thickness of the irises starting from either end of the filter are $T_1 = 0.0568$ in., $T_2 = 0.1182$ in., $T_3 = 0.1638$ in., and $T_4 = 0.1729$ in., respectively. The gaps between the irises starting from the ends are $G_1 = 0.3944$ in., $G_2 = 0.4307$ in., $G_3 = 0.4308$ in., and $G_4 = 0.4309$ in. Note that the third iris from the end is the same one that was treated as an example in Figures 4 and 5. These dimensions were arrived at using the design process outlined in Section 5. Figure 7 gives a plot of theoretical VSWR versus frequency within the passband. In this example δ was taken to be 0.019 in. The low VSWR given in this figure confirms the accuracy of the design process and the assertion that interaction effects between irises are indeed small.

The program EQSOLV of Silvester and Ferrari (1990) should be modified slightly to avoid complex arithmetic. This is because the FORTRAN command SQRT($S(J, J)$) may under some circumstances yield imaginary numbers in the case of the solution of the two-dimensional Helmholtz equation. Complex numbers can be avoided by defining an additional array SG(J) consisting of the values $+1$ or -1 depending on whether $S(J, J)$ is positive or negative. SG(J) must now multiply the product $S(I, K) * S(J, K)$ where it appears. Furthermore, to minimize memory usage the code for the Cholesky decomposition of S should be written in such a way as to replace S by its upper and lower triangular factors. In the case of banded storage S is no longer a square matrix. The limits on the summations in the Cholesky decomposition no longer depend only on the number of nodes. For certain critical loops these limits should be defined correctly with IF statements prior to entering the loops in order to minimize computation time. If this is done, then the analysis of a seven-section filter, such as that given in Figure 6, will take about 1 minute per frequency point. The FORTRAN code for the Cholesky decomposition of a real banded matrix S is also given in Section 9.

7. TREATMENT OF POWER HANDLING

One of the frequent considerations in waveguide filter design is the question of power handling. This issue may arise with respect to peak power breakdown or multipacting. In either case a knowledge of the traveling wave fields in the filter is needed. The method developed so far has yielded only standing wave solutions for the fields and the filter S matrix parameters via the even and odd mode eigensusceptances. In this section it will be shown that the standing wave solutions together with these eigensusceptances provide sufficient information to determine the traveling wave fields.

Let A be the amplitude of the incident wave and α its phase. Then for a standing wave solution of unit amplitude at the input ($z = 0$)

$$1 = Ae^{\alpha}\{1 \pm |\Gamma|e^{j\theta}\} \tag{18}$$

Here $|\Gamma|$ is the reflection coefficient amplitude and θ its phase. For a reactive termination $|\Gamma| = 1$. The plus sign holds in the case of voltages and the minus sign in the case of currents. This complex equation can be solved for A and α.

$$A^2 = \frac{1}{1 + |\Gamma|^2 \pm 2|\Gamma|\cos\theta} \tag{19}$$

$$\tan\alpha = \frac{\mp|\Gamma|\sin\theta}{1 \pm |\Gamma|\cos\theta} \tag{20}$$

Specializing to the lossless case where $|\Gamma| = 1$,

$$A^2 = \frac{1}{2(1 \pm \cos\theta)} \tag{21}$$

$$\alpha = \left(\frac{-\theta}{2}\right) \tag{22}$$

That is, the amplitude and phase of the incident wave are uniquely determined by the phase θ, which, in turn, is uniquely related to the eigensusceptance B_e by

$$B_e = \tan\left(\frac{-\theta}{2}\right) \tag{23}$$

By renormalizing both the even and odd mode standing wave nodal potentials using the above expressions and then combining them, it is possible to construct the traveling wave solutions. In the case of susceptances where the plus sign in Eq. (21) is appropriate

$$A = \frac{1}{2\cos(\theta/2)} \tag{24}$$

The even and odd mode phases θ_e and θ_o can be obtained directly from the eigensusceptances using Eq. (23). The even and odd mode standing wave potentials must now be multiplied by $\cos(\theta_e/2)$ and $\cos(\theta_o/2)$, respectively, according to Eq. (24). Their phases must be adjusted by $-\theta_e/2$ and $-\theta_o/2$, respectively, according to Eq. (22). The resulting complex normalized potentials are to be combined with a plus sign to obtain the traveling wave solution to the left of the symmetry plane. They are to be combined with a minus sign to obtain the traveling wave solution to the right of the symmetry plane. In the FORTRAN filter analysis program, the array of standing wave nodal potentials has been overwritten with the array of the electric field amplitudes E_y at each node of the filter.

TABLE 2. Maximum Normalized *E* Fields in the Cavities of a Seven-Section Filter

Frequency (GHz)	Cavity #1	Cavity #2	Cavity #3	Cavity #4	Cavity #5	Cavity #6	Cavity #7	Output
13.50	2.2788	0.7491	0.1646	0.0336	0.0069	0.0016	0.0004	0.0001
13.70	3.3310	1.6513	0.5435	0.1670	0.0522	0.0183	0.0068	0.0025
13.80	4.3024	2.8918	1.2761	0.5280	0.2246	0.1065	0.0487	0.0164
13.90	5.7168	6.6784	4.8444	3.2548	2.1776	1.4720	0.8012	0.2727
14.00	2.8631	4.5587	5.0300	5.0544	4.8976	4.5372	2.9407	0.9996
14.12	2.6752	4.0834	3.8663	4.3841	4.0924	3.7703	2.8921	0.9964
14.25	2.8591	3.5792	4.2039	3.8324	4.2110	3.5657	2.8730	0.9998
14.37	2.8432	3.8683	4.0788	4.3495	4.0773	3.8701	2.8424	0.9999
14.50	2.9801	4.4145	4.7910	5.0347	5.0828	4.5042	2.8070	0.9981
14.60	5.3340	6.9956	5.7051	4.2087	3.0255	2.0754	1.1249	0.4066
14.70	4.5232	3.5122	1.8255	0.8880	0.4440	0.2344	0.1112	0.0406
14.80	3.7093	2.2114	0.8946	0.3383	0.1313	0.0558	0.0235	0.0086
15.00	2.7686	1.1876	0.3489	0.0959	0.0268	0.0089	0.0030	0.0010
15.50	1.7851	0.4929	0.0947	0.0170	0.0030	0.0005	0.0001	0.0001

In Table 2 results for the maximum electric field between the irises along the centerline of the filter of Figure 7 are given. These values occur at the center of the cavities where the electric field is a maximum. The cavity numbers have been included in this table. The filter center frequency 14.25 GHz and the filter band edge frequencies 14.0 and 14.5 GHz are included in the table. For these cases the maximum electric field occurs at the center cavities as expected. The amplitude of the electric field at the output is slightly less than 1 as expected for a well-matched filter. Also given are frequencies below the passband and above the passband. For these cases the electric field at the output is small.

In the passband the maximum normalized electric field is close to 5. This means that the peak power handling of this filter must be derated by a factor of 25 from that of standard waveguide. Note that in the table this value of 5 is exceeded at the out-of-band frequencies of 13.90 and 14.60 GHz. In the second cavity the maximum electric field is close to 7 at these two frequencies. At 13.50 and 15.50 GHz the output field amplitudes are 0.000163 and 0.0000678, respectively. This corresponds to a rejection of approximately 80 dB.

8. DISCUSSION OF INSERTION LOSS

Another important issue in waveguide filter design is that of insertion loss. A detailed treatment is beyond the scope of this chapter. However, an approximate treatment is possible once the traveling wave fields in the filter are known. This treatment will be adequate provided that the passband insertion loss is small. The amount of power radiated into the metal is found by integrating the normal component of the Poynting vector $\mathbf{S}_n$ over the metallic surfaces of the filter. This component is

$$\mathbf{S}_n = \tfrac{1}{2} E_t \cdot H_t^* \tag{25}$$

where E_t and H_t are the tangential components of the E and H fields, respectively. Since E_t is very small, it is usual to write it as

$$E_t = Z_m H_t \tag{26}$$

where Z_m is the surface impedance of the metal. An excellent approximation to Z_m is given by

$$Z_m = \sqrt{\frac{\omega\mu}{2\sigma}}\,(1 + j) \tag{27}$$

where σ is the conductivity of the metal.

$$\therefore\ R_e(S_n) = \frac{1}{2}\sqrt{\frac{\omega\mu}{2\sigma}}\,|H_t|^2 \tag{28}$$

Can the eigenmode method, which has been illustrated in this chapter for the case of lossless symmetrical filters, be applied in principle to lossy symmetrical filters as well? The answer to this question is yes. Certainly, the equations given in Section 3 apply to the lossy case. To be sure, the impedance and admittance eigenvalues will now be complex. The advantage of working only with real quantities will have been lost. In fact, the basic quantities of interest may as well be taken to be the even and odd mode reflection coefficients each with a magnitude $|\Gamma_e|$ and $|\Gamma_o|$, respectively, and a phase θ_e and θ_o, respectively. If these are known, then the two-port scattering parameters are determined. Furthermore, it follows from the equations of Section 7 that provided $0 \leqslant |\Gamma_e| < 1$ and $0 \leqslant |\Gamma_o| < 1$ a solution for an incident amplitude A and phase α can always be found, which provides for a unit amplitude standing wave at the input. That is, the same boundary conditions can be applied as in the lossless case. The solutions for the E and H fields will generally be complex, however.

Suppose that solutions for the E and H fields have been obtained. These solutions could be based on the finite element method. Then Montgomery et al. (1987, p. 138) have given equations that relate $|\Gamma|$, θ, the incident power P_i, the average dissipated power P_D, and $W_H - W_E$.

$$P_D = \frac{1}{2} \int \sigma \mathbf{E}^* \cdot \mathbf{E} \, dV \tag{29}$$

$$W_H - W_E = \int \mu \mathbf{H}^* \cdot \mathbf{H} \, dV - \int \varepsilon \mathbf{E}^* \cdot \mathbf{E} \, dV \tag{30}$$

Consequently, P_D and $(W_H - W_E)$ will be assumed to be known in what follows. The equations in question are

$$|\Gamma| \sin \theta = \frac{\omega (W_H - W_E)}{P_i} \tag{31}$$

$$1 - |\Gamma|^2 = \frac{P_D}{P_i} \tag{32}$$

The incident power P_i will be proportional to the square of the amplitude A of the incident signal: that is,

$$P_i = \beta A^2 \tag{33}$$

where the proportionality factor β depends on the waveguide type under consideration. Using Eq. (33) and the expression for A^2 given in Section 7, Eq.

(31) can be rewritten as

$$|\Gamma| \cdot 2\sin\left(\frac{\theta}{2}\right) \cdot \cos\left(\frac{\theta}{2}\right) = \frac{\omega(W_H - W_E)}{\beta}\left[|\Gamma|^2 - 1 + 4|\Gamma|\cos^2\left(\frac{\theta}{2}\right)\right] \tag{34}$$

in the case the plus sign is specified. Consequently, if the E field is specified at the input,

$$\tan\left(\frac{\theta}{2}\right) = \frac{\omega(W_H - W_E)}{2\beta}\left\{\frac{|\Gamma| - 1/|\Gamma|}{\cos^2(\theta/2)} + 4\right\} \tag{35}$$

If the H field is specified at the input,

$$\cot\left(\frac{\theta}{2}\right) = \frac{\omega(W_H - W_E)}{2\beta}\left\{\frac{|\Gamma| - 1/|\Gamma|}{\sin^2(\theta/2)} + 4\right\} \tag{36}$$

A second equation for $|\Gamma|$ is obtained by eliminating P_i from Eqs. (31) and (32):

$$|\Gamma| - \frac{1}{|\Gamma|} = -\frac{2P_D \sin(\theta/2) \cdot \cos(\theta/2)}{\omega(W_H - W_E)} \tag{37}$$

Substituting Eq. (37) into Eq. (35) or (36) results in an equation in $\tan(\theta/2)$ or $\cot(\theta/2)$, respectively. These can be solved for $\tan(\theta/2)$ or $\cot(\theta/2)$ in terms of known quantities. Once θ is known, $|\Gamma|$ follows from Eq. (37).

By considering the lossless case $|\Gamma| = 1$, expressions for the susceptance B_e and reactance X_e can be derived from Eq. (35) and (36), respectively. In particular, in the case of Eq. (35)

$$B_e = -\tan\left(\frac{\theta}{2}\right) = \frac{2\omega(W_E - W_H)}{\beta} \tag{38}$$

In the case of Eq. (36)

$$X_e = -\cot\left(\frac{\theta}{2}\right) = \frac{2\omega(W_E - W_H)}{\beta} \tag{39}$$

By using the proportionality factor β of Eq. (33) for the case of rectangular waveguide, the expressions (6) and (9) of Section 4 can be recovered.

9. FORTRAN SEGMENTS

Several useful FORTRAN segments are listed below. The routine CALCULATE WAVEGUIDE SUSCEPTANCE determines the susceptance from the E-field nodal potentials. The subroutine TRIELM(IE) of Silvester and Ferrari

(1990) is called from this routine, which has been included in the main program sequence. The parameter A is the waveguide width, $C0$ the wavelength, and $C1$ the guide wavelength.

The routine BANDED CHOLESKY DECOMPOSITION is an efficient routine for the Cholesky decomposition of a banded symmetrical real matrix S stored in banded form. The quantity NUMB is the half-bandwidth. The dimension of S is S(MAXNOD,2*NUMB). The array SG with dimension SG(MAXNOD) is needed to avoid complex arithmetic.

```
C******* CALCULATE WAVEGUIDE SUSCEPTANCE *******

      W1=0
      DO 30 K=1,NELMTS
      IE=K
      CALL TRIELM(IE)
      W2=0
      W3=0
      DO 20 I=1,NVE
      W3=0
      DO 10 J=1,NVE
      IROW=NVTX(J,IE)
      W3=W3+(SEL(I,J)-((2*PI/CO)**2)*TEL(I,J))*POTENT(IROW)
   10 CONTINUE
      ICOL=NVTX(I,IE)
      W2=W2+W3*POTENT(ICOL)
   20 CONTINUE
      W1=W1+W2
   30 CONTINUE
      SUSCEP=2*((C1*W1)/(PI*A))


C******* BANDED CHOLESKY DECOMPOSITION: REPLACE S BY ITS
C        UPPER AND LOWER TRIANGULAR FACTORS.
C
        SG(1)=S(1,NUMB)/ABS(S(1,NUMB))
        S(1,NUMB)=SQRT(ABS(S(1,NUMB)))
        S(2,NUMB-1)=S(2,NUMB-1)/(S(1,NUMB)*SG(1))
        S(1,NUMB+1)=S(2,NUMB-1)
        SG(2)=(S(2,NUMB)-S(2,NUMB-1)**2*SG(1))
        SG(2)=SG(2)/ABS(S(2,NUMB)-S(2,NUMB-1)**2*SG(1))
        S(2,NUMB)=SQRT(ABS(S(2,NUMB)-S(2,NUMB-1)**2*SG(1)))
        DO 60 IROW=3,NODES
          IROW1=IROW-1
          ICOL2=NUMB+1-IROW
          IF (ICOL2.LT.1) GOTO 12
```

```
         S(IROW,ICOL2)=S(IROW,ICOL2)/(S(1,NUMB)*SG(1))
         ICOL3=IROW+NUMB-1
         S(1,ICOL3)=S(IROW,ICOL2)
  12     K2=2
         K1=1+NUMB-IROW
         IF (K1.GT.0) GOTO 15
         K2=2+IROW-NUMB
  15   DO 40 ICOL=K2,IROW1
         ICOL1=ICOL-1
         SUM=0.0
             K2=1
             K1=1+NUMB-IROW
             IF (K1.GT.0) GOTO 20
             K2=1+IROW-NUMB
  20     DO 30 K=K2,ICOL1
             K1=K+NUMB
             ICOL2=K1-IROW
             ICOL3=K1-ICOL
           SUM=SUM+S(IROW,ICOL2)*S(ICOL,ICOL3)*SG(K)
  30     CONTINUE
             ICOL2=ICOL+NUMB-IROW
       S(IROW,ICOL2)=(S(IROW,ICOL2)-SUM)/(S(ICOL,NUMB)*SG(ICOL))
             ICOL3 = IROW+NUMB-ICOL
         S(ICOL,ICOL3)=S(IROW,ICOL2)
  40     CONTINUE
           SUM=0.0
               K1=1
               K2=1 + NUMB-IROW
               IF (K2.GT.0) GOTO 45
               K1=1+IROW-NUMB
  45   DO 50 K=K1,IROW1
           ICOL2=K+NUMB-IROW
           SUM=SUM+S(IROW,ICOL2)*S(IROW,ICOL2)*SG(K)
  50   CONTINUE
       SG(IROW=(S(IROW,NUMB)-SUM)/ABS(S(IROW,NUMB)-SUM)
       S(IROW,NUMB)=SQRT(ABS(S(IROW,NUMB)-SUM))
  60   CONTINUE
C      FORWARD ELIMINATE. RTHDSD IS DESTROYED IN THE PROCESS
       RTHDSD(1) = RTHDSD(1)/(S(1,NUMB)*SG(1))
C
       DO 130 IROW=2,NODES
         IROW1=IROW-1
           SUM=0.0
             K1=1
             K2=1+NUMB-IROW
```

```
        IF (K2.GT.0) GOTO 105
        K1=1+IROW-NUMB
  105   DO 110 ICOL=K1,IROW1
        ICOL2=ICOL+NUMB-IROW
          SUM=SUM+S(IROW,ICOL2)*RTHDSD(ICOL)*SG(ICOL)
  110   CONTINUE
  RTHDSD(IROW)=(RTHDSD(IROW)-SUM)/(S(IROW,NUMB)*SG(IROW))
  130 CONTINUE
C       BACK SUBSTITUTE, USING UPPER TRIANGULAR FACTOR
        POTENT(NODES)=RTHDSD(NODES)/S(NODES,NUMB)
        DO 230 K=2,NODES
          IROW=NODES+1-K
          SUM=0.0
          IROW1=IROW+1
          K1=NODES
          K2=NODES+NUMB-IROW
          IF (K2.LT.(2*NUMB+1)) GOTO 215
          K1=NUMB+IROW
  215     DO 220 ICOL=IROW1,K1
            ICOL2=ICOL+NUMB-IROW
              SUM=SUM+S(IROW,ICOL2)*POTENT(ICOL)
  220     CONTINUE
            POTENT(IROW)=(RTHDSD(IROW)-SUM)/S(IROW,NUMB)
  230     CONTINUE
  300     RETURN
```

REFERENCES

Helszajn, J. (1978). *Passive and Active Microwave Circuits.* Wiley, New York.

Levy, R. (1967). Theory of direct-coupled-cavity filters. *IEEE Trans. Microwave Theory Tech.* **MTT-15**, 340–348.

Montgomery, C. G., Dicke, R. H., and Purcell, E. M. (1987). *Principles of Microwave Circuits.* Peter Peregrines, London.

Riblet, G. P. (1995). Eigensusceptances and eigenreactances of symmetrical waveguide inductive obstacles using the finite element method. *IEE Proc. Microwave Antennas Propag.* **142**, 384–388.

Schwinger, J., and Saxon, D. (1968). *Discontinuities in Waveguide.* Gordon & Breach, New York.

Scott, W. R. (1992). Accurate modeling of axisymmetric two-port junctions in coaxial lines using the finite element method. *IEEE Trans. Microwave Theory Tech.* **MTT-40**, 1712–1716.

Silvester, P. P. (1978). Construction of triangular finite element universal matrices. *Int. J. Numer. Methods Eng.* **12**, 237–244.

Silvester, P. P., and Ferrari, R. L. (1990). *Finite Elements for Electrical Engineers.* Cambridge Univ. Press, Cambridge, UK.

CHAPTER SEVEN

The Coupling of Finite Element and Integral Equation Representations for Efficient Three-Dimensional Modeling of Electromagnetic Scattering and Radiation

TOM CWIK, CINZIA ZUFFADA, and VAHRAZ JAMNEJAD

1. INTRODUCTION

Finite element modeling has proved useful for accurately simulating scattered or radiated fields from complex three-dimensional objects whose geometry varies on the scale of a fraction of a wavelength. In order to make this modeling practical for engineering design, it is necessary to integrate the stages of geometry modeling and mesh generation, numerical solution of the fields, and display of field information. The stages of geometry modeling, mesh generation, and field display are commonly completed using commercially available software packages. Algorithms for the numerical solution of the fields need to be written for the specific class of problems considered. Interior problems—simulating fields in waveguides and cavities—have been solved

Finite Element Software for Microwave Engineering, Edited by Tatsuo Itoh, Giuseppe Pelosi, and Peter P. Silvester.
ISBN 0-471-12636-5

successfully using finite element methods because a bounding surface, such as the cavity walls, exactly truncates the problem domain. Exterior problems—simulating fields scattered or radiated from structures—are more difficult to model because of the need to numerically truncate the finite element mesh. To practically compute a solution to exterior problems, the domain must be truncated at some finite surface where the Sommerfeld radiation condition is enforced, either approximately or exactly. Approximate methods attempt to truncate the mesh using only local field information at each grid point, whereas exact methods are global, needing information from the entire mesh boundary. This chapter outlines a method that couples three-dimensional finite element solutions interior to the bounding surface with an efficient integral equation solution that exactly enforces the Sommerfeld radiation condition.

A finite element model of Maxwell's equations furnishes a local solution of fields that is accurate in and around inhomogenous, complex bodies (Silvester and Ferrari, 1990; Jin, 1993). The finite element discretization can be applied to the vector Helmholtz wave equation, derived directly from Maxwell's curl equations, or to a set of equations derived from a scalar–vector potential formulation of the fields. The use of one or the other is directly tied to the choice of three-dimensional vector finite element basis functions. Edge elements (Mur and de Hoop, 1985; Lee and Mittra, 1992) can be naturally used to discretize the Helmholtz equation, providing tangential field continuity across material interfaces, correct discontinuity in normal field components across interfaces, and a natural discretization of tangential field components at geometry surfaces. Nodal elements are used to discretize scalar–vector potential formulations, where auxiliary equations are introduced to enforce boundary conditions at surfaces and to enforce zero divergence of fields in source-free media (Paulsen et al., 1992; Angelini et al., 1993). Formulations based on the use of mixed elements are also possible (Mur, 1993).

Since the finite element formulation of Maxwell's equations applies to an infinite domain, it is necessary to truncate the finite element mesh, enforcing the Sommerfeld radiation condition. Approximate absorbing boundary conditions can be used to provide a computationally efficient enforcement of the radiation condition (Mittra et al., 1989; D'Angelo and Mayergoyz, 1989; Chatterjee et al., 1993). These boundary conditions only require local field information and therefore do not disturb the sparsity of the resulting system of linear equations. Due to the inherent approximations in the enforcement of these boundary conditions, however, problems can arise in their use. Fundamentally, they are most accurate when placed far from the object, requiring additional mesh discretization and resulting computational cost.

An alternative to approximate radiation conditions is the use of an exact expression for the fields on and outside an arbitrary surface that truncates the computational mesh. The problem domain is divided into interior and exterior regions, separated at the mesh boundary. The interior fields are modeled using the finite element representation, while those in the exterior region are modeled using an integral equation formulation. The unknown sources in the integral equation are directly related to the tangential fields on the mesh boundary, and the radiation condition is implicitly enforced exactly through the use of the free-space Green's function. Fields in the two regions are coupled by enforcing boundary conditions on tangential field components at the mesh boundary, thereby producing a unique and exact solution to Maxwell's equations in both regions. One early use of this technique was outlined by Zienkiewicz et al. (1977). To enforce continuity of fields at the mesh boundary, the expansion functions used in the integral equation can be identical or similar to those used at the boundary of the finite element mesh. This approach allows the mesh to conform to the body and results in a near exact enforcement of field continuity between the two regions (Angelini et al., 1993; Jin and Liepa, 1988; Calalo et al., 1989; Gong and Glisson, 1990; Yuan et al., 1990). It also results in the computer storage and solution time being at least proportional to that of a three-dimensional surface integral equation formulation, defeating the computational efficiency of the finite element method and limiting its usefulness.

A second approach decouples the interior finite element mesh from that used for the expansion functions in the integral equation formulation. The advantage of mesh decoupling is that a proper choice of the boundary can lead to an efficient numerical solution. In the unimoment method introduced by Mei (1972), the boundary is the surface of a separable coordinate system (e.g., the surface of a sphere). In the method introduced by Boyse and Seidl (1991) a surface of revolution is proposed that results in eigenfunction series expansion in only one direction (azimuthal). This formulation suggested a node-based tetrahedral expansion in the finite element region, and a Fourier modal azimuthal expansion together with Hermite polynomial functions along the surface of revolution generator for the integral equation expansion. A similar two-dimensional formulation for arbitrary scatterers was given by Cwik (1992). The general three-dimensional formulation in this chapter builds on these previous two works, utilizing vector edge elements to discretize the Helmholtz wave equation, and piecewise triangle functions along the generator to efficiently model the mesh truncation surface. The vector edge elements naturally enforce the boundary condition at perfectly conducting surfaces and do not allow the generation of parasitic fields encountered when applying nodal elements. They also produce fewer nonzero entries in the resultant sparse system of equations relative to nodal elements. This approach is followed, utilizing these specific features:

1. The integral equation mesh is separate from the finite element mesh, and its density can vary according to experienced guidelines of known integral equation solutions, independent of the finite element grid. The integral equation mesh can be as close as possible to the object, limiting the computational expense of both the integral equation and finite element portions of the calculation.
2. The surface that truncates the finite element mesh is a surface of revolution (Boyse and Seidl, 1991). The use of this surface gives an efficient integral equation representation for the exterior fields. Using this

surface decreases the number of unknowns associated with the integral equation portion of the problem. Additionally, the surface needs only enclose the body and can be placed as close as possible to it, thus reducing the size of the problem domain.

3. Vector edge elements are used to discretize the Helmholtz wave equation. As noted above, these functions naturally enforce the proper continuity and discontinuities in vector field components at interfaces. They also naturally enforce the boundary condition at perfectly conducting surfaces and do not allow the generation of parasitic fields encountered when applying nodal elements.

The formulation handles both scattering and radiation problems (e.g., antenna elements or arrays). The radiation problem only differs by the need to model impressed sources by the finite element mesh (Zuffada et al., 1995). In this chapter a detailed description of the scattering formulation will be outlined, followed by an examination of convergence and accuracy, and results for a set of simple scatterers. In Cwik et al. (1996) further results of this approach can be found.

2. THEORETICAL FORMULATION

The scatterer and surrounding space are broken into two regions: an interior part containing the scatterers and free-space region out to a defined surface, and the exterior homogenous part (Figure 1). To efficiently model fields in the exterior region, the surface bounding the interior is prescribed to be a surface of revolution. The following formulation first outlines the interior finite element representation, then the exterior integral equation model, and finally the coupling of fields at the boundary separating the two regions.

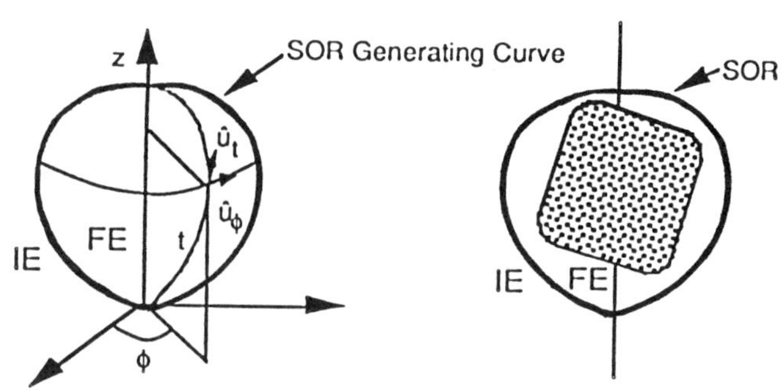

FIGURE 1. Geometry of computational domain showing interior and exterior regions.

2.1. The Finite Element Representation

In the interior region, a finite element discretization of a form of the wave equation is used to model the geometry and fields. Applying a form of Green's theorem—multiplying the wave equation by a testing function **T**, integrating over the volume and using the divergence theorem—a weak form of the wave equation is obtained. This form includes a surface integral that provides for a boundary condition relating the field inside the selected volume to the field on the boundary and thus provides a link to the outside field:

$$\frac{\eta_0}{jk_0}\iiint_V \left[\frac{1}{\varepsilon_r}(\nabla \times \mathbf{H})\cdot(\nabla \times \mathbf{T}^*) - k^2\mu_r \mathbf{H}\cdot\mathbf{T}^*\right]dv - \iint_{\partial V} \mathbf{E}\times\hat{\mathbf{n}}\cdot\mathbf{T}^*\,ds = 0 \tag{1}$$

H is the magnetic field (the **H** equation is used in the development; the formulation of the dual **E** equation is presented in Section 5), **T** is a testing function (the asterisk denotes conjugation), and $\mathbf{E}\times\hat{\mathbf{n}}$ is the tangential component of **E** on the surface S (∂V). In general, ∂V represents all boundaries of the volume, including the surface of revolution and any perfect conductors. The surface integrals over the perfect conductors are identically zero since their integrand includes the tangential electric field $\mathbf{E}\times\hat{\mathbf{n}}$. Equation (1) therefore represents the fields internal to and on the surface S. These fields will be modeled using a set of properly chosen finite element basis functions. In Eq. (1), ε_r and μ_r are the relative permittivity and permeability, respectively, and k_0 and η_0 are free-space wavenumber and impedance, respectively.

2.2. The Combined-Field Integral Equation Representation

In the formulation of the integral equation, fictitious electric ($\mathbf{J} = \hat{\mathbf{n}}\times\mathbf{H}$) and magnetic ($\mathbf{M} = -\hat{\mathbf{n}}\times\mathbf{E}$) surface currents, equivalent to the tangential magnetic and electric fields just on the exterior of the boundary surface, are defined on the boundary. These currents produce fields in the exterior region, which are the scattered fields. The sum of the scattered and the incident fields results in the total field everywhere outside the boundary surface. On the boundary itself, this sum is equal to half the total field. The scattered fields are obtained from the tangential currents via an integral over the boundary using the free-space Green's function kernel. Two equations are obtained for the electric and magnetic fields on the boundary—the electric field integral equation (EFIE) and the magnetic field integral equation (MFIE), respectively. A linear combination of the two with a constant weighting factor α results in the combined field integral equation (CFIE). The general form used in this formulation is

$$Z_M[\mathbf{M}/\eta_0] + Z_J[\mathbf{J}] = V_i \tag{2}$$

where Z_M and Z_J are the integro-differential operators used in defining the CFIE, and V_i represents the incident field.

2.3. Enforcing Boundary Conditions

The previous two sections have outlined field representations for the interior and exterior regions. In the interior region, boundary conditions at any material interface, including perfect conductors, must be enforced by a proper application of the finite element basis functions. At the artificial surface of revolution separating the interior and exterior regions, boundary conditions on the continuity of tangential field components must be enforced.

Initially, four equations are written for the three unknown field quantities of interest. The first unknown is the magnetic field internal to the volume V. The other two are the electric and magnetic surface currents, **J** and **M**, on the boundary.

The four equations are found from the finite element equation (1), the integral equation (CFIE) relating **J** and **M** currents to the incident field (2), and a set of equations enforcing the continuity of **H** and **E** across the boundary. Continuity of the magnetic field across the boundary is enforced in a weak sense

$$\iint_{\partial V} (\hat{\mathbf{n}} \times \mathbf{H} - \mathbf{J}) \cdot (\hat{\mathbf{n}} \times \mathbf{U}^*)\, ds = 0 \tag{3}$$

where **U** is a testing function. This is an essential boundary condition and must be explicitly enforced. Continuity of the electric field across the boundary is made implicit in the finite element equation in the surface integral term $\hat{\mathbf{n}} \times \mathbf{E}$ and is termed a natural boundary condition

$$\iint_{\partial V} (\mathbf{E} \times \hat{\mathbf{n}} - \mathbf{M}) \cdot \mathbf{T}^*\, ds = 0 \tag{4}$$

This equation is combined with Eq. (1) to produce

$$\frac{\eta_0}{jk_0} \int_V \left[\frac{1}{\varepsilon_r} (\nabla \times \mathbf{H}) \cdot (\nabla \times \mathbf{T}^*) - k^2 \mu_r \mathbf{H} \cdot \mathbf{T}^* \right] dv - \int_{\partial V} \mathbf{M} \cdot \mathbf{T}^*\, ds = 0 \tag{5}$$

Equations (5), (3), and (2) constitute the system of equations representing fields in all space in and about the scatterer.

3. NUMERICAL IMPLEMENTATION

The three equations (5), (3), and (2) are discretized using appropriate sets of basis functions. In the interior region, tetrahedral, vector edge elements (Whitney elements) are used. On the boundary surface of revolution, a set of functions with piecewise linear variation along the surface of revolution generator, and with an azimuthal Fourier modal variation, are used.

3.1. The Finite Element Model

An ensemble of elements filling the interior region, excluding any perfect conducting objects, is created using a mesh generator. The elements should accurately represent the magnetic field, the geometry of the scatterer, and the bounding surface of revolution. Since the scatterer is not a body of revolution in general, the finite element mesh will extend out from the scatterer to the surface of revolution. For an accurate model of the fields, tetrahedral, vector edge elements are used to model **H** (Lee and Mittra, 1992)

$$\mathbf{H}(r) = \sum_i h_i \mathbf{W}_i(r) \tag{6a}$$

where

$$\mathbf{W}_{mn}(r) = \lambda_m(r)\nabla\lambda_n(r) - \lambda_n(r)\nabla\lambda_m(r) \tag{6b}$$

and $\lambda(r)$ are the tetrahedral shape functions. Testing functions are also chosen to be the functions $\mathbf{W}(r)$.

These functions are used in the volume integral of Eq. (5). The resultant discretized volume integral is

$$\frac{\eta_0}{jk_0}\int_V \left[\frac{1}{\varepsilon_r}(\nabla \times \mathbf{H})\cdot(\nabla \times \mathbf{W}^*) - k^2\mu_r \mathbf{H}\cdot\mathbf{W}^*\right] dv \Rightarrow \mathsf{KH} \tag{7}$$

where K is the assembled sparse finite element matrix, and H is the vector of complex, finite element basis function amplitudes.

3.2. An Efficient Exterior Integral Equation Model

In the surface of revolution geometry, a cylindrical coordinate system (ρ, ϕ, z) is selected for the exterior region, and orthogonal surface coordinates (t, ϕ) are used on the boundary itself; ϕ is the azimuthal angle variable and t is the contour length variable along the generating curve of the surface of revolution. In the formulation of the integral equation, the equivalent electric and magnetic surface currents (**J** and **M**) are defined just on the outside of the surface through the relations

$$u\mathbf{E} = \mathbf{E}^i - L[\mathbf{J}] + K[\mathbf{M}/\eta_0] \tag{8}$$

$$u\eta_0\mathbf{H} = \eta_0\mathbf{H}^i - K[\mathbf{J}] - L[\mathbf{M}/\eta_0] \tag{9a}$$

in which u is the Heaviside function,

$$u = \begin{cases} 1 & \text{for points outside } \partial V \\ \frac{1}{2} & \text{for points on } \partial V \\ 0 & \text{for points inside } \partial V \end{cases} \tag{9b}$$

and L and K are integro-differential operators given by

$$L[\cdots] = j\eta_0 \iint_{\partial V} (k_0^2[\cdots] + \nabla\nabla'\cdot[\cdots])g(k_0|\mathbf{r} - \mathbf{r}'|)\, ds' \tag{10}$$

$$K[\cdots] = \eta_0 \iint_{\partial V} [\cdots] \times k_0\nabla g(k_0|\mathbf{r} - \mathbf{r}'|)\, ds' \tag{11}$$

In Eqs. (10) and (11) g is the Green's function for unbounded space

$$g(k_0|\mathbf{r} - \mathbf{r}'|) = \frac{e^{-jk_0|\mathbf{r}-\mathbf{r}'|}}{4\pi k_0|\mathbf{r} - \mathbf{r}'|}. \tag{12}$$

From the above equations we can write the following two integral equations on the tangential components of the field, for obtaining the **J** and **M** surface currents. The electric field integral equation, and magnetic field integral equation, respectively, are

$$(\tfrac{1}{2}\eta_0\hat{\mathbf{n}} \times I - K)[\mathbf{M}/\eta_0]|_{\tan} + L[\mathbf{J}]|_{\tan} = \hat{\mathbf{n}} \times \mathbf{M}_i \tag{13a}$$

$$(\hat{\mathbf{n}} \times L)[\mathbf{M}/\eta_0] + (\tfrac{1}{2}\eta_0 I + \hat{\mathbf{n}} \times K)[\mathbf{J}] = \eta_0\mathbf{J}_i \tag{13b}$$

in which, for the sake of symmetry, the source terms (tangential components of the incident field) are given as fictitious surface currents $\mathbf{J}_i$ and $\mathbf{M}_i$. They are presented in a form that is very similar both in terms of dimensions as well as vector orientations. The symbol I represents the unity operator and is introduced for notational consistency.

These two integral equations are linearly combined to obtain the combined-field integral equation

$$\{(1-\alpha)(\tfrac{1}{2}\eta_0\hat{\mathbf{n}}\times I - K) + \alpha(\hat{\mathbf{n}}\times L)\}[\mathbf{M}/\eta_0] + \{(1-\alpha)L + \alpha(\tfrac{1}{2}\eta_0 I + \hat{\mathbf{n}}\times K)\}[\mathbf{J}] = (1-\alpha)\hat{\mathbf{n}}\times\mathbf{M}_i + \alpha\eta_0\mathbf{J}_i \qquad (14)$$

Equation (14) is put in the more compact form given by Eq. (2), where the operators

$$Z_M = \{(1-\alpha)(\tfrac{1}{2}\eta_0\hat{\mathbf{n}}\times I - K) + \alpha(\hat{\mathbf{n}}\times L)\}$$
$$Z_j = \{(1-\alpha)L + \alpha(\tfrac{1}{2}\eta_0 I + \hat{\mathbf{n}}\times K)\} \qquad (15)$$

and source term

$$\mathbf{V}_i = (1-\alpha)\hat{\mathbf{n}}\times\mathbf{M}_i + \alpha\eta_0\mathbf{J}_i \qquad (16)$$

are used. This formulation of the operators follows from the CICERO code development (Medgeysi-Mitschang and Putnam, 1984), although other formulations are also possible.

Using the method of moments, this integral equation is turned into a matrix equation. The unknown currents **M** and **J** are expanded in a finite series of basis functions **U** on the surface of revolution. The testing functions are selected identical to the expansion functions on the surface of revolution. They are written as separable functions of t and ϕ and will have two orthogonal components along the $\hat{\mathbf{t}}$ and $\hat{\boldsymbol{\phi}}$ directions. The azimuthal function is the exponential harmonic $\exp(jn\phi)$ (Fourier harmonics). The variation along the surface of revolution generator is represented by a triangle function $T(t)$ divided by $\rho(t)$, the radial distance from the z-axis. Thus

$$\mathbf{J} = \sum_{n,k} (a^t_{n,k}\mathbf{U}^t_{n,k} - a^\phi_{n,k}\mathbf{U}^\phi_{n,k}) \quad \text{in } \partial V \qquad (17)$$

$$\mathbf{M} = \sum_{n,k} \eta_0(b^t_{n,k}\mathbf{U}^t_{n,k} - b^\phi_{n,k}\mathbf{U}^\phi_{n,k}) \quad \text{in } \partial V \qquad (18)$$

$$\mathbf{V}_i = \sum_{n,k} (c^t_{n,k}\mathbf{U}^t_{n,k} - c^\phi_{n,k}\mathbf{U}^\phi_{n,k}) \quad \text{in } \partial V \qquad (19)$$

and both expansion and testing functions are given as

$$\mathbf{U}^t_{n,k} = \hat{\mathbf{t}}\,\frac{T_k(t)}{\rho(t)}\,e^{jn\phi} \qquad (20)$$

$$\mathbf{U}^\phi_{n,k} = \hat{\boldsymbol{\phi}}\,\frac{T_k(t)}{\rho(t)}\,e^{jn\phi} \qquad (21)$$

$T_k(t)$ is a triangle function spanning the kth annulus on the surface of revolution. Each annulus spans two segments along the generator, each referred to as a strip. Adjacent triangles overlap on one segment. These overlapping triangle functions result in approximations to **J** and **M**, which are piecewise linear in t and a Fourier series in ϕ.

In the moment method, inner products are formed between the testing functions and the integral equations expanded over the basis functions. Here, using the Galerkin method, basis and testing functions are chosen to be identical as indicated above. The complex inner product between any two vectors **U** and **V** is defined as

$$\langle \mathbf{U}, \mathbf{V} \rangle_{\partial V} = \iint_{\partial V} \mathbf{U} \cdot \mathbf{V}^* \, ds \tag{22}$$

The original integral equation is transformed into a set of linear equations for each of the Fourier modes since the Fourier modes are orthogonal and decouple. Thus, in the compact form, it can be written as

$$\sum_m m_m \langle Z_{Mm}[\mathbf{U}_m], \mathbf{U}_n \rangle + \sum_m j_m \langle Z_{Jm}[\mathbf{U}_m], \mathbf{U}_n \rangle = \sum_m v_{im} \langle \mathbf{U}_m, \mathbf{U}_n \rangle \tag{23}$$

where m_m and j_m are the complex unknown amplitudes for each Fourier mode. This is the second equation in the system, representing fields scattered from the object.

3.3. Coupling the Two Representations

The surface integral in Eq. (5) and the first component of the integral in Eq. (3) are termed the coupling integrals, since with a convenient choice of the unknown in the first and of the testing function in the second, they are made to couple interior and exterior field representations. The surface S in these surface integrals is chosen to be that of the surface of revolution. Because the surface of revolution is discretized when using these basis functions, the issue arises of how to represent **W** on S. Indeed, the outer surface of the interior volume is a union of finite element facets. These facets vary, according to the order of finite element representation chosen, from planes to curved surfaces. In general, however, this surface is not identical to the surface of revolution. Similarly, the surface of revolution is obtained by revolving a generating curve around an axis, creating a surface whose cross section is circular. However, for numerical purposes, the generator itself is not necessarily smooth, but is piecewise linear. Thus, practically, only in the limit of fine meshing will the two surfaces coincide with each other.

The finite element function **W** is evaluated approximately on the portion of surface of revolution projected from the triangular facet of the tetrahedron onto a strip. This is accomplished by an orthogonal projection of the tetrahedral facet surface onto the surface of revolution, thus introducing an error that depends on the size of the tetrahedral facet with respect to the curvature of the surface of revolution. The coupling term is given by the integral

$$C = \iint_{\partial V} \mathbf{U} \cdot \mathbf{W}^* \, ds \tag{24}$$

where, for each integral equation basis and finite element testing function, the contributing surface is the union of the projections of a triangular boundary surface onto the proper number of surface of revolution strips (up to two in well-posed cases). Such surfaces are curved triangles, curved quadrilaterals, or curved pentagons. The evaluation of the integrals was done numerically by first inscribing the above irregular surfaces into curved rectangles and then by determining the points inside the region of interest from the knowledge of the simplex coordinates of the original finite element boundary facet and their properties at points inside the facet. These coupling integrals, as well as the discretization of the second surface integral in Eq. (3), complete the discretization of the problem.

4. NUMERICAL SOLUTION OF THE LINEAR SYSTEM

Having introduced the basis and testing functions for the volume as well as the surface unknowns, substitution into the complete set of equations yields

$$\begin{aligned}
&\sum_{p=1}^{P} h_p \langle K_p[\mathbf{W}_p] \cdot \mathbf{W}_q \rangle - \eta_0 \sum_{m=1}^{M} m_m \langle \mathbf{U}_m \cdot \mathbf{W}_q \rangle = 0, \quad q = 1, 2, 3, \ldots, P \\
&-\eta_0 \sum_{p=1}^{P} h_p \langle \mathbf{W}_p \cdot \mathbf{U}_n \rangle + \eta_0 \sum_{m=1}^{M} j_m \langle \mathbf{U}_m \cdot [\mathbf{n} \times \mathbf{U}_n] \rangle = 0, \quad n = 1, 2, 3, \ldots, M \\
&\sum_{m=1}^{M} m_m \langle Z_{Mm}[\mathbf{U}_m] \cdot \mathbf{U}_n \rangle + \sum_{m=1}^{M} j_m \langle Z_{Jm}[\mathbf{U}_m] \cdot \mathbf{U}_n \rangle = \sum_{m=1}^{M} v_{i_m} \langle \mathbf{U}_m \cdot \mathbf{U}_n \rangle, \\
&\qquad n = 1, 2, 3, \ldots, M
\end{aligned} \tag{25}$$

which can be expressed in matrix form as

$$\begin{vmatrix} \mathbf{K} & \mathbf{C} & \mathbf{0} \\ \mathbf{C}^{\dagger} & \mathbf{0} & \mathbf{Z}_0 \\ \mathbf{0} & \mathbf{Z}_M & \mathbf{Z}_J \end{vmatrix} \begin{vmatrix} \mathbf{H} \\ \mathbf{M} \\ \mathbf{J} \end{vmatrix} = \begin{vmatrix} \mathbf{0} \\ \mathbf{0} \\ \mathbf{V}_i \end{vmatrix} \tag{26}$$

where

$$\begin{aligned}
\mathbf{K} &= \langle K_p[\mathbf{W}_p] \cdot \mathbf{W}_q \rangle \\
\mathbf{C} &= -\eta_0 \langle \mathbf{U}_m \cdot \mathbf{W}_q \rangle \\
\mathbf{Z}_0 &= \eta_0 \langle \mathbf{U}_m \cdot [\mathbf{n} \times \mathbf{U}_n] \rangle \\
\mathbf{Z}_M &= \langle Z_{Mm}[\mathbf{U}_m] \cdot \mathbf{U}_n \rangle \\
\mathbf{Z}_J &= \langle Z_{jm}[\mathbf{U}_m] \cdot \mathbf{U}_n \rangle
\end{aligned} \tag{27}$$

K_p is the volume integral operator in Eq. (7); Z_{Mm} and Z_{Jm} are defined in the previous sections. Note that † indicates the adjoint of a matrix, while P and M are the number of finite element and integral equation basis functions, respectively.

It is noted that the overall matrix size is dominated by $\mathbf{K}$ since the surface unknowns are generally a small fraction of those in the volume. Note also that both $\mathbf{K}$ and $\mathbf{C}$ are sparse, $\mathbf{Z}_0$ is tridiagonal, and $\mathbf{Z}_M$ and $\mathbf{Z}_J$ are banded. The overall matrix is very sparse and looks like the scatter plot illustrated in Figure 2 for a small problem generated for graphical demonstration. In particular it is complex, nonsymmetric, and non-Hermitian.

The solution to this matrix equation has been developed using two alternative strategies depending on the application. The two approaches are:

1. Solve the entire system in one step using an iterative algorithm for nonsymmetric systems. In this work the nonsymmetric variant of the

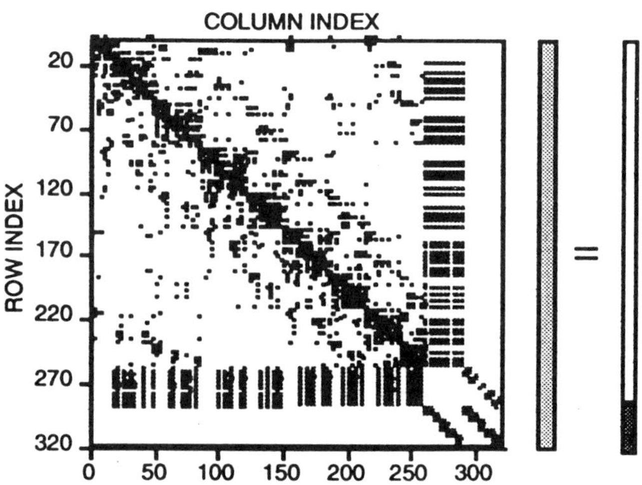

FIGURE 2. Scatter plot figuratively showing structure of system of equations. Darkened spaces indicate nonzero matrix entries.

quasi-minimal residual algorithm (Freund and Nachtigal, 1991; E. Lucas, personal communication, 1993) was applied.

2. Solve the system in two steps by first eliminating $\mathbf{H}$ through the computation of $\mathbf{Z}_K = \mathbf{C}^{\dagger}\mathbf{K}^{-1}\mathbf{C}$ and then by solving the reduced system

$$\begin{aligned} -\mathbf{Z}_K\mathbf{M} + \mathbf{Z}_0\mathbf{J} &= \mathbf{0} \\ \mathbf{Z}_M\mathbf{M} + \mathbf{Z}_J\mathbf{J} &= \mathbf{V}_i \end{aligned} \tag{28}$$

   The first step was accomplished by applying a symmetric variant of the quasi-minimal residual iterative algorithm as well as a direct solver based on an sparse $\mathbf{U}^t\mathbf{D}\mathbf{U}$ factorization. The resulting overall matrix (28) was treated as being dense and the solution of this second system was accomplished via a direct dense LU decomposition, since its size is relatively small.

The choice of the solution method is dependent on a variety of factors, including the number of right-hand side excitation vectors and the efficacy of a preconditioner if iterative algorithms are used to calculate $\mathbf{Z}_K$ in the second method. The computational cost of the first method is mainly due to solving the system iteratively for each right-hand side. The major computational cost in the second method is calculating $\mathbf{Z}_K$; this requires the solution of a system of equations, $\mathbf{K}^{-1}\mathbf{C}$, where $\mathbf{C}$ is a rectangular matrix with a possibly large number of columns for electrically large scatterers. When considering a radiation problem where there are one or a few right-hand sides, or a scattering problem with one or a few excitations, the first method may be preferable. When monostatic radar cross-section calculations are performed and there are upward of thousands of right-hand sides, the second method is more appropriate. This second approach has been implemented on scalable distributed memory computers and is reported in Cwik et al. (1995).

The speed of convergence of iterative methods such as the quasi-minimal residual algorithm depends on the condition number of the matrix. To improve on the condition number of the matrix of Eq. (26) in a one-step solution, or of matrix $\mathbf{K}$ in a two-step solution, standard preconditioning methods have been examined. It is noted that, when preconditioning the original matrix, it is necessary to preserve its valuable properties—in the case of the one-step it is sparsity, whereas for the two-step approach it is sparsity and symmetry. A matrix $\mathbf{M}$ is used as a left preconditioner in the one-step solution

$$\mathbf{M} \cdot \mathbf{A} \cdot \mathrm{x} = \mathbf{M} \cdot \mathbf{b} \tag{29}$$

where $\mathbf{A}$ is the global matrix and $\mathbf{b}$ is the right-hand side. To preserve the sparsity of the matrix, the multiplication on the left-hand side of Eq. (29) is performed every time the matrix is used by the quasi-minimal residual algorithm, so that the algorithm operates on a sparse matrix. It is seen that the solution to the modified problem is the same as that to the original problem.

Alternately, in the two-step method, the matrix on which the quasi-minimal residual algorithm is applied is symmetric. Then the scheme of Eq. (29) is modified into the symmetric scheme

$$\mathbf{M}' \cdot \mathbf{A} \cdot \mathbf{M}' \cdot [(\mathbf{M}')^{-1}\mathbf{x}] = \mathbf{M}' \cdot \mathbf{b} \tag{30}$$

where the matrix $\mathbf{M}'$ is a symmetric preconditioner. Again, to preserve the sparsity of the matrix, the multiplications by $\mathbf{M}'$ are performed every time the matrix is used by the quasi-minimal residual algorithm, so that the algorithm operates on a sparse system.

5. ELECTRIC FIELD FORMULATION

A dual formulation for the electric field within the mesh can be developed by either reformulating the problem beginning with the Helmholtz equation for the electric field in Sections 2 through 4, or by applying duality to the final equations in Section 4. As in the magnetic field formulation, this formulation involves the expansion of the electric field over the mesh as

$$\mathbf{E}(r) = \eta_0 \sum_i e_i \mathbf{W}_i(r) \tag{31}$$

where η_0 is introduced as it was in the expansions in Section 3 to scale the equations properly. The resultant system dual to Eq. (26) then is

$$\begin{vmatrix} \mathbf{K}^E & \mathbf{C} & \mathbf{0} \\ \mathbf{C}^\dagger & \mathbf{0} & -\mathbf{Z}_0 \\ \mathbf{0} & \mathbf{Z}_J & \mathbf{Z}_M \end{vmatrix} \begin{vmatrix} \mathbf{E} \\ \mathbf{J} \\ \mathbf{M} \end{vmatrix} = \begin{vmatrix} \mathbf{0} \\ \mathbf{0} \\ \mathbf{V}_i \end{vmatrix} \tag{32}$$

where $\mathbf{K}^E$ is the dual of Eq. (7)

$$\frac{\eta_0}{jk_0} \int_V \left[\frac{1}{\mu_r} (\nabla \times \mathbf{E}) \cdot (\nabla \times \mathbf{W}^*) - k^2 \varepsilon_r \mathbf{E} \cdot \mathbf{W}^* \right] dv \Rightarrow \mathbf{K}^E \mathbf{E} \tag{33}$$

and other notation remains unchanged.

The electric field formulation has specific advantages when simulating the fields scattered or radiated from particular objects. For example, when considering objects with perfectly conducting flat plates or fins, or objects with electrically thin strips or patches, it is more expedient to apply the perfectly conducting boundary condition using the electric field formulation than the magnetic field formulation. In the electric field formulation, element edges that lie on the perfect conductor are removed from the vector **E**, and the matrix block $\mathbf{K}^E$ is similarly reduced in size. This has the explicit effect of zeroing the coefficients of the electric field lying on the zero thickness perfect conductors, satisfying the boundary condition that the total tangential electric field is zero on the perfect conductor. To accomplish this enforcement of the boundary conditions using the magnetic field formulation, the thin conductor must be modeled as a thin volumetric region. The ensuing mesh generation difficulties of creating the small, often distorted elements make this inconvenient in practice. An alternative strategy is to mesh both faces of the zero thickness conductor with edges that are distinct, though coincident, therefore modeling the tangential magnetic field on both faces of the conductor. This also leads to meshing difficulties in practice.

6. NUMERICAL CONVERGENCE AND RESULTS

Several scatterers have been examined in detail, including dielectric spheres, (coated) metal spheres, finite length (coated) metal cylinders, metal cone-spheres, and metal cubes. Results for the spheres and dielectric cylinders will be presented here; results for the other objects can be found in Cwik et al. (1996). The features of some of the meshes are shown in Table 1. A surrounding shell is used to model a perfectly conducting object coated with a layer of dielectric material. As a special case, by choosing the dielectric coating to be air, the scattering from the perfectly conducting object itself is obtained. Naturally, no elements are required to model fields inside perfect conductors. The choice of metallic objects was based on the existence of results from other codes to be used for comparisons.

When considering the dielectric sphere or cylinder, only the scatterer itself was modeled by the mesh since there is no need to extend the mesh outside the geometry of the scatterer. In the case of the metal sphere, the mesh was a shell conforming to the object with a thickness chosen to fit one tetrahedral element.

TABLE 1. Objects and Their Mesh Densities

Object	Nodes	Elements	Edges
Dielectric sphere 1	865	3,963	5,217
Dielectric sphere 2	2,273	10,240	12,896
Metal sphere (coated)	516	1,716	2,889
Dielectric cylinder	7,076	34,776	43,791

6.1. Numerical Accuracy

The accuracy of this method, compared to analytic solutions or alternative computational methods or measurements, was investigated as a function of several parameters. Some specific parameters are:

- Mesh density
- Number of surface of revolution basis functions along the generator
- Number of surface of revolution azimuthal Fourier modes
- Electric permittivity and magnetic permeability of scatterer

The first three are functions of electrical wavelength. It is noted that these parameters are not independent, but rather are closely related. The canonical object studied for accuracy was the sphere. Both a metal sphere and a dielectric sphere were considered, and in the latter case, meshes of two different densities were investigated. For simplicity the direction of incidence was taken to be the positive z direction with the electric field polarized along x, and the Fourier modes $(-1, 0, +1)$ were considered. Theoretically only the $+1$ and -1 modes are present in the decomposition of this incident plane wave, and no other mode should arise in the scattered field. The quantity of interest was chosen to be the field scattered in the E-plane ($\phi = 0°$). An error measure was constructed by taking the difference between the calculated and exact values, integrated over the range of angles of observation $0° \leqslant \theta \leqslant 180°$. This number was divided by the integral of the analytic solution over the same range to obtain a measure of quality for the solution.

Table 2 illustrates the range of variation for some of the parameters that were considered. The electric permittivity ε_r was fixed at 9 for the study of the dielectric sphere, whereas the metal sphere was treated with a "coating" of free space. The table illustrates that the frequency upper limit was chosen to correspond to the ratio of edges per wavelength close to 10. The choice was made with the intention to investigate the limitations of mesh applicability and to quantify errors in this rather broad range.

TABLE 2. Summary of Edges per Wavelength on Surface of Revolution

Frequency (GHz)	3	5	7	10	15
Metal sphere[a] (16)[b]	50	31	22	15	10
Dielectric sphere 1[c] (16)[b]	17	10	7	—	—
Dielectric sphere 2[c] (24)[b]	25	15	11	—	—

[a]A layer of free space is meshed.
[b]Nominal number of edges along generator.
[c]$\varepsilon_r = 9$ used.

As seen in the following results, high accuracy was obtained for (a) average wavelength to edge ratios between 20 and 40, and (b) wavelength to surface of revolution triangle length ratios between 10 and 20. Lower accuracy can be found for lesser numbers of unknowns per wavelength.

6.2. Radar Cross-Section and Near-Field Results

A set of radar cross-section calculations, showing comparisons with either analytical results or with calculations obtained with the CICERO code, is provided in Figures 3–5. As mentioned previously, results for other objects can be found in Cwik et al. (1996). In the legend our calculation is referred to as PHOEBUS, the name of the software developed from this formulation. Figure 3 illustrates the bistatic radar cross section of a metal sphere (radius = 0.8 cm) coated with a layer of lossy dielectric (thickness = 0.2 cm). Calculations are performed in both the *E*-plane and the *H*-plane and comparisons are made with available analytical results. The frequency of excitation is 15 GHz.

Figure 4 illustrates the monostatic radar cross section of a dielectric cylinder (length = 10.0 cm, radius = 1.0 cm) with $\varepsilon_r = 4$. The comparison is made with

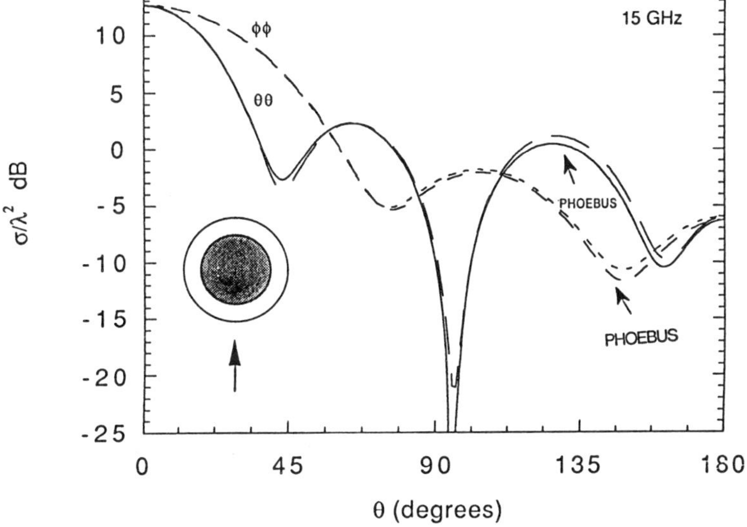

FIGURE 3. Bistatic radar cross section of coated metal sphere (radius = 0.8 cm, coating thickness = 0.2 cm) at 15 GHz.

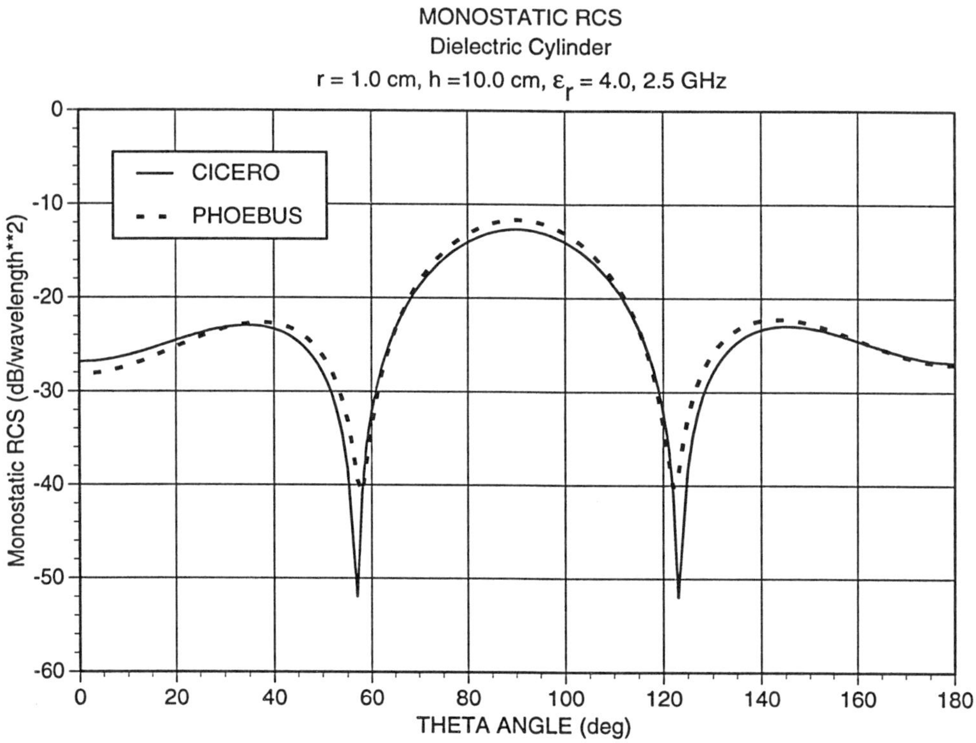

FIGURE 4. Bistatic radar cross section of dielectric cylinder (radius = 1 cm, height = 10 cm) at 2.5 GHz; ϕ–ϕ plane.

the CICERO code for the ϕ–ϕ plane at 2.5 GHz. The discrepancy between the two plots is due to geometry modeling using the piecewise linear finite element and surface of revolution basis functions. This result used finite element tetrahedra with an average edge length of 0.2 cm. By decreasing this edge length the two curves can be made arbitrarily close. The smaller edge length better models the physical geometry but is electrically oversampled (greater than the 20–30 edges per wavelength typically needed). This type of modeling discrepancy is common in the finite element method when using the linear (nonisoparametric) tetrahedron functions.

Figure 5 illustrates the near-field results of the electric and magnetic field on the surface of the dielectric sphere with $\varepsilon_r = 9$. Shown are the $\hat{\mathbf{t}}$ and $\hat{\boldsymbol{\phi}}$ components of both E and H along a line at $\phi = 45°$. The radius of the sphere is 1 cm, and the PHOEBUS results are compared with analytical ones at 5 GHz.

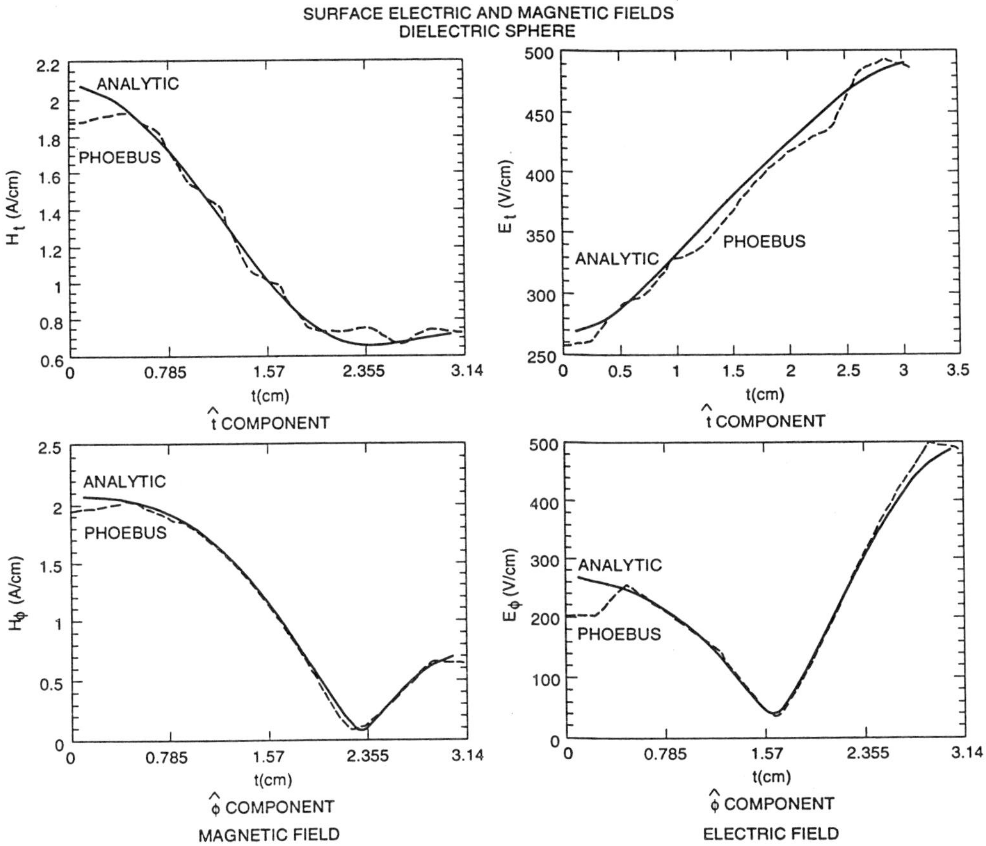

FIGURE 5. Electric and magnetic fields on surface of dielectric sphere (radius = 1 cm) at 5 GHz. Both $\hat{\mathbf{t}}$ and $\hat{\boldsymbol{\phi}}$ components are compared to analytical solutions in a cut along a generator at $\phi = 45°$.

7. CONCLUSIONS

This chapter presents a method for computing the fields of penetrable three-dimensional scatterers of general shape by coupling a finite element solution to an integral equation solution on a surface of revolution. The surface of revolution is chosen to surround the scatterer, resulting in a minimal amount of volume that needs to be discretized. The use of the integral equation provides an exact enforcement of the Sommerfeld radiation condition. Vector edge elements are used to discretize the fields inside the volume, whereas the integral equation is discretized on a decoupled surface mesh, introducing a small set of additional basis functions to the system. Scattered fields for a variety of objects have been investigated, as well as fields inside penetrable scatterers.

ACKNOWLEDGMENTS

The research described in this chapter was carried out at the Jet Propulsion Laboratory, California Institute of Technology, under a contract with the National Aeronautics and Space Administration.

REFERENCES

Angelini, J.-J., Soize, C., and Soudais, P. (1993). Hybrid numerical method for harmonic 3-D Maxwell equations: Scattering by a mixed conducting and inhomogenous anisotropic dielectric medium. *IEEE Trans. Antennas Propag.* **AP-41**(1), 66–76.

Boyse, W. E., and Seidl, A. A. (1991). A hybrid finite element method for near bodies of revolution. *IEEE Trans. Magn.* **MAG-27**, 3833–3836.

Calalo, R., Cwik, T., Ferraro, R., Imbriale, W., Jacobi, N., Liewer, P., Lockhart, T., Lyzenga, G., Mulligan, S., Parker, J., and Patterson, J. (1989). *Hypercube Matrix Computation Task, Research in Parallel Computational Electromagnetics, Report for 1988–1989.* Jet Propulsion Laboratory, California Institute of Technology, Pasadena.

Chatterjee, A., Jin, J.-J., and Volakis, J. (1993). Edge-based finite elements and vector ABC's applied to 3-D scattering. *IEEE Trans. Antennas Propag.* **AP-41**(2), 221–226.

Cwik, T. (1992). Coupling finite element and integral equation solutions using decoupled boundary meshes. *IEEE Trans. Antennas Propag.* **AP-40**(12), 1496–1504.

Cwik, T., Katz, D. Zuffada, C., and Jamnejad, V. (1995). The application of scalable distributed memory computers to the finite element modeling of electromagnetic scattering. *Int. V. Numer. Methods Eng.* (submitted for publication).

Cwik, T., Zuffada, C., and Jamnejad, V. (1996). Efficient coupling of finite element and integral equation representations to model scattering from three-dimensional objects. *IEEE Trans. Antennas Propag.* **AP-40**(A).

D'Angelo, J., and Mayergoyz, I. (1989). On the use of local absorbing boundary conditions for RF scattering problems. *IEEE Trans. Magn.* **MAG-25**, 3040–3043.

Freund, R., and Nachtigal, N. (1991). QMR: A quasi-minimal residual method for non-hermitian linear systems. *Numer. Math.* **60**, 315–339.

Gong, Z., and Glisson, A. (1990). A hybrid equation approach for the solution of electromagnetic scattering problems involving two-dimensional inhomogenous dielectric cylinders. *IEEE Trans. Antennas Propag.* **AP-38**, 60–68.

Jin, J.-J. (1993). *The Finite Element in Electromagnetics.* Wiley, New York.

Jin, J.-J., and Liepa, V. (1988). Application of hybrid finite element method to electromagnetic scattering from coated cylinders. *IEEE Trans. Antennas Propag.* **AP-36**, 50–54.

Lee, J., and Mittra, R. (1992). A note on the application of edge-elements for modeling 3-dimensional inhomogeneously-filled cavities. *IEEE Trans. Microwave Theory Tech.* **MTT-40**(9), 1767–1773.

Lucas, E. (1993), Personal Communications.

Medgeysi-Mitschang, L., and Putnam, J. (1984). Electromagnetic scattering from axially inhomogeneous bodies of revolution. *IEEE Trans. Antennas Propag.* **AP-32**, 797–806.

Mei, K. (1972). Unimoment method of solving antenna and scattering problems. *IEEE Trans. Antennas Propag.* **AP-22**, 760–766.

Mittra, R., Ramahi, O., Khebir, A., Gordon, R., and Kouki, A. (1989). A review of absorbing boundary conditions for two- and three-dimensional electromagnetic scattering problems. *IEEE Trans. Magn.* **MAG-25**(7), 3034–3040.

Mur, G. (1993). The finite-element modeling of three-dimensional electromagnetic fields using edge and nodal elements. *IEEE Trans. Antennas Propag.* **AP-41**(7), 948–953.

Mur, G., and de Hoop, A. (1985). A finite-element method for computing three-dimensional electromagnetic fields in inhomogeneous media. *IEEE Trans. Magn.* **MAG-21**(6), 2188–2191.

Paulsen, D., Boyse, W. E., and Lynch, D. (1992). Continuous potential Maxwell solutions on nodal-based finite elements. *IEEE Trans. Antennas Propag.* **AP-40**(10), 1192–1200.

Silvester, P., and Ferrari, R. (1990). *Finite Elements for Electrical Engineers*, 2nd ed. Cambridge Univ. Press, Cambridge, UK.

Yuan, X., Lynch, D., and Strohbehn, J. (1990). Coupling of finite element and moment methods for electromagnetic scattering from inhomogenous objects. *IEEE Trans. Antennas Propag.* **AP-38**, 386–394.

Zienkiewicz, O., Kelly, D., and Bettess, P. (1977). The coupling of the finite element method and boundary solution procedures. *Int. J. Numer. Methods Eng.* **11**, 355–375.

Zuffada, C., Cwik, T., and Jamnejad, V. (1995). Modeling radiation with an efficient hybrid finite element-integral equation-waveguide mode matching technique. *IEEE Trans. Antennas Propag.* (submitted for publication).

CHAPTER EIGHT

Phased Array Antenna Analysis

JOHN D'ANGELO and ISAAC MAYERGOYZ

1. INTRODUCTION

Numerical analysis of phased array antennas has been the focus of considerable research in recent years. This is due to the practical importance of this problem as well as the formidable challenge it presents for numerical analysts. Indeed, this is a three-dimensional vectorial "wave" problem with a spatially oscillating solution, with complicated interface boundaries, infinite in extent, and geometrically periodic. For this reason, several attempts have recently been made to attack this problem by combining finite elements with other techniques (Jin and Volakis, 1991; Lucas and Fontana, 1995; McGrath and Pyati, 1994; Mias et al., 1994). Our analysis is similar, in some respects, to these publications. However, it also has some salient new features.

This chapter will discuss the application of finite elements to the analysis of phased array antennas. This analysis utilizes "infinite" periodic arrays to approximate large, but finite phased array systems. This "infinite" approximation therefore predicts more accurately the behavior of the antenna elements located at the center of the array than elements located near the edges. This analysis is applicable to various types of antenna elements: waveguide cavities, patch radiators, and printed circuit elements.

Besides the finite element method, this analysis will use three other numerical techniques: the boundary element (integral equation) method, Floquet harmonic expansions, and waveguide mode expansions. These methods are used where they are best applied. The finite element method will be used to

Finite Element Software for Microwave Engineering, Edited by Tatsuo Itoh, Giuseppe Pelosi, and Peter P. Silvester.
ISBN 0-471-12636-5

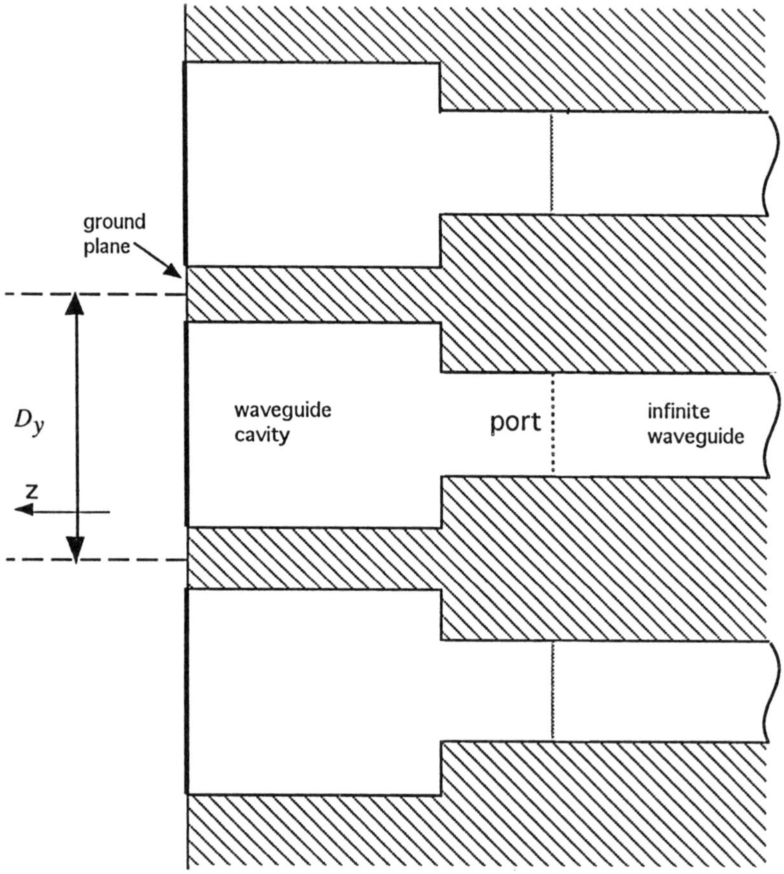

FIGURE 1. *zy* Cross section of a radiating waveguide cavity array in a ground plane.

model the complex shaped, heterogeneous media of the antenna. Both the boundary element and the Floquet harmonic expansion methods will be used to model the exterior infinite domain region. The waveguide modal expansion will be used to both supply the correct excitation of the phased array and calculate the reflection coefficients. As mentioned earlier, our analysis can handle various types of radiators; however, in this chapter we will use a waveguide cavity geometry for descriptive as well as example purposes.

A canonical, side view geometry outline of one waveguide cavity in the phased array is shown in Figure 1. These elements can be placed in either rectangular or skewed periodic arrays as shown in Figure 2. The excitation of the antenna elements will be phased with a constant gradient to produce a Floquet type field in the exterior region. This phase gradient along with the periodic dimension of the array determines the scan angle of the steered main

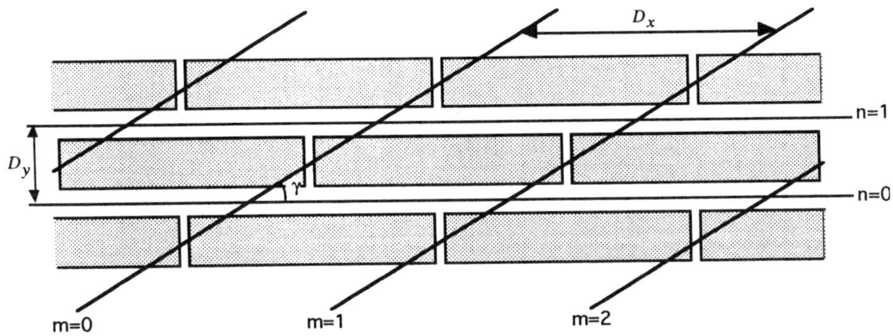

FIGURE 2. *xy* Cross section of radiating waveguide cavity array in a ground plane.

beam. The application of this analysis technique will demonstrate the ability to calculate the reflection coefficient of a cavity antenna as a function of frequency, array dimensions, and array scan angles in both *E* and *H* planes.

Since the antenna elements are electromagnetically coupled to one another, the performance of an antenna element in an array is quite different from its behavior as a single radiating element. Predicting this performance as a function of element spacing, frequency, and scan angle is therefore important to the phased array design.

The following finite element implementation will use vector-based, divergence-free finite element approximation functions, commonly called "edge elements" (Bossavit and Mayergoyz, 1990; D'Angelo and Mayergoyz, 1990). As discussed in Chapter 9, edge elements require higher mesh densities than standard scalar-type finite element approximation functions to achieve comparable accuracy levels. However, because finite element domains here are generally small in physical dimensions, often on the order one cubic wavelength, the additional computational burden of using edge elements is not great. Therefore edge elements, which are straightforward to implement and allow boundary conditions to be readily specified, are used here.

The boundary element integral formulation will allow the modeling of a periodic array of cavities in a ground plane. The periodic Green's function is determined from Floquet theory and is represented by infinite series. Special acceleration techniques will be employed to evaluate these slowly convergent series (Singh et al., 1990).

The alternative use of Floquet harmonic expansions in lieu of boundary elements is also described. Unlike boundary element equations, the Floquet expansion does not result in a fully populated system of equations for the field unknowns on the aperture, thereby lessening the computational requirements. Furthermore, the Floquet expansion provides some flexibility in the choice of the finite element region. In particular, this region can be extended beyond the ground plane in order to include various inhomogeneous design features which may exist there.

Lastly, a modal expansion technique will be used to feed and terminate the antenna. As shown in Figure 1, the waveguide cavity antenna element is terminated by a "port." This antenna is fed by a user selected mode at this port. This mode is determined from a two-dimensional finite element solution of the homogeneous waveguide problem with the port cross section. Homogeneous waveguide modes are also used as basis functions for the unknown reflected field at the port cross section, thereby allowing the direct calculation of the antenna's reflection coefficient. Another advantage of this modal termination is that it eliminates the need for determining or modeling an approximate terminating structure.

2. FINITE ELEMENT FORMULATION

The finite element formulation will be used to solve for the electric field E within the antenna region. Although the method could be reformulated for the magnetic field H as well, this electric field formulation allows boundary conditions on the perfect electric walls to be readily specified.

The basis partial differential equation is the "curl–curl" wave equation obtained directly from Maxwell's equations for source-free regions:

$$\nabla \times \left(-\frac{1}{j\omega\mu} \nabla \times E \right) - j\omega\varepsilon E = 0 \tag{1}$$

The Galerkin-weighted residual method is applied. As a result, Eq. (1) is multiplied (dot product) by an arbitrary vector weighting function, W, and integrated over the solution domain:

$$\int_V W \cdot \left(\nabla \times \left(-\frac{1}{j\omega\mu} \nabla \times E \right) - j\omega\varepsilon E \right) dV = 0 \tag{2}$$

The first term in Eq. (2) is transformed by using the appropriate vector calculus identity:

$$\int_V \left[(\nabla \times W) \cdot \left(-\frac{1}{j\omega\mu} \nabla \times E \right) - \nabla \cdot \left(W \times \left(-\frac{1}{j\omega\mu} \nabla \times E \right) \right) - j\omega\varepsilon W \cdot E \right] dV = 0 \tag{3}$$

The divergence theorem is then used to transform the second term in Eq. (3):

$$\int_V \left[(\nabla \times W) \cdot \left(-\frac{1}{j\omega\mu} \nabla \times E \right) - j\omega\varepsilon W \cdot E \right] dV - \int_S \vec{n} \cdot W \times \left(-\frac{1}{j\omega\mu} \nabla \times E \right) dS = 0 \tag{4}$$

There are three types of boundary surfaces that enclose the finite element domain. The first type, S_{ap}, is the aperture surface. The second type is the perfect conducting surfaces of the waveguide walls. This surface type will be denoted as S_{pec}. The third surface, S_{port}, is the port feed. The surface term in Eq. (4) is then rewritten for the above three boundary surfaces and the Maxwell equation along with a vector identity is used to rearrange terms:

$$\int_V \left[(\nabla \times W) \cdot \left(-\frac{1}{j\omega\mu} \nabla \times E \right) - j\omega\varepsilon W \cdot E \right] dV - \int_{S_{ap}} (\vec{n} \times W) \cdot H \, dS_{ap}$$
$$+ \int_{S_{pec}} (\vec{n} \times W) \cdot H \, dS_{pec} + \int_{S_{port}} W \cdot \vec{n} \times \left(-\frac{1}{j\omega\mu} \nabla \times E \right) dS_{port} = 0 \tag{5}$$

The S_{pec} integral is identically zero because the Direchlet boundary condition for the tangential component of E requires the tangential testing function, $\vec{n} \times W$, to be zero. This term is eliminated from Eq. (5) and the final weak Galerkin form is obtained:

$$\int_V \left[(\nabla \times W) \cdot \left(-\frac{1}{j\omega\mu} \nabla \times E \right) - j\omega\varepsilon W \cdot E \right] dV - \int_{S_{ap}} (\vec{n} \times W) \cdot H \, dS_{ap}$$
$$+ \int_{S_{port}} W \cdot \vec{n} \times \left(-\frac{1}{j\omega\mu} \nabla \times E \right) dS_{port} = 0 \tag{6}$$

Methods to couple the above weak Galerkin form to the boundary element technique and modal expansion technique will be discussed in Sections 4 and 6, respectively.

3. FINITE ELEMENT DISCRETIZATION

As mentioned in the introduction, special vector approximation functions—"edge elements"—are used in this formulation. In the sequel, divergent-free, "zero-order" tetrahedral-shaped edge elements, as shown in Figure 3, are used.

The unknown vector quantity in the element is represented by the following approximation function:

$$E(x, y, z) = \sum_{k=1}^{6} N_k(x, y, z)\, \xi_k \tag{7}$$

In Eq. (7), N_k is the vector approximation function for edge k, and ξ_k is the unknown tangential component of E along this edge. N_k is designed by taking the gradients of the standard, scalar node-based approximation functions. For example, if edge 1 has nodes 1 and 2 as the two end points, then N_1 is

$$N_1(x, y, z) = \lambda_1(x, y, z)\, \nabla\lambda_2(x, y, z) - \lambda_2(x, y, z)\, \nabla\lambda_1(x, y, z) \tag{8}$$

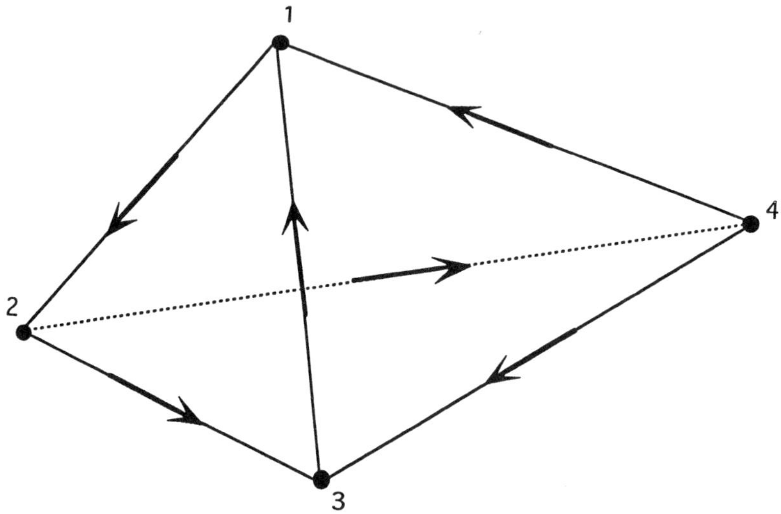

FIGURE 3. Zero-order, tetrahedral-shaped edge element.

where λ_1 and λ_2 are the node-based linear approximation functions. This linearity along with the symmetry of Eq. (8) is the mechanism from which edge elements inherit their divergence-free characteristics.

Using Eq. (8), the curl of E is found as follows:

$$\nabla \times E(x, y, z) = \sum_{k=1}^{6} \nabla \times N_k(x, y, z)\, \xi_k \tag{9}$$

where the curl of the approximation function, N_1, for example, is

$$\nabla \times N_1(x, y, z) = 2\nabla\lambda_1(x, y, z) \times \nabla\lambda_2(x, y, z) \tag{10}$$

The weighting function is identical to the approximation function; that is,

$$W_k(x, y, z) = N_k(x, y, z) \tag{11}$$

The above weighting function and approximation functions are used in the volume integrals of Eq. (6). Similar two-dimensional traces of the above vector functions are determined for surface intergrals S_{ap} and S_{port}.

4. BOUNDARY ELEMENT INTEGRAL EQUATION FORMULATION

The boundary element technique is used to model the open domain exterior to the waveguide cavity and the ground plane. As was done with the finite element formulation, this integral technique is derived directly from Maxwell's equations. The basic equation of analysis, Eq. (12), is the Stratton–Chu formula derived from the source-free curl–curl wave equation for the magnetic field by using the second vector Green's identity.

$$H(r) = \theta \oint_{S'} \left[(\vec{n} \cdot H(r'))\ \nabla' G(r', r) + (\vec{n}' \times H(r')) \times \nabla' G(r', r) \right. \\ \left. + (j\omega\varepsilon_0 \vec{n}' \times E(r'))\ G(r', r) \right] dS' \tag{12}$$

Here, S' is the boundary of the domain, r is an observation point within the domain or on the boundary, and r' is an integration point on the boundary. The coefficient in Eq. (12) is equal to

$$\begin{aligned} \theta &= 1 \quad \text{for } r \text{ outside } S' \\ \theta &= 2 \quad \text{for } r \text{ on } S' \end{aligned} \tag{13}$$

The boundary S' consists of the ground plane and the surface of the aperture. On the ground plane, the first and third terms in the surface integral are zero according to the boundary conditions for perfect electric conductors. The remaining terms are rewritten for the ground plane and aperture surface as follows:

$$H(r) = 2 \oint_{S'_{\text{ap}}} \left[(\vec{n}' \cdot H(r'))\ \nabla' G(r', r) + (\vec{n}' \times H(r')) \times \nabla' G(r', r) \right. \\ \left. + (j\omega\varepsilon_0 \vec{n}' \times E(r'))\ G(r', r) \right] dS'_{\text{ap}} + 2 \oint_{S'_{\text{pec}}} \left[(\vec{n} \times H(r')) \times \nabla' G(r', r) \right] dS'_{\text{pec}} \tag{14}$$

Equation (14) is coupled to the weak Galerkin form, Eq. (6). This coupling is obtained by using the continuity of the tangential components of both the magnetic and electric fields at the aperture. In this way, Eq. (14) can be used to replace the H field in the aperture surface integral of Eq. (6). By substituting Eq. (14) into the aperture surface integral term of Eq. (6), we obtain

$$2 \int_{S_{\text{ap}}} \left[(\vec{n} \times W(r)) \cdot \int_{S'_{\text{ap}}} \left[(\vec{n} \cdot H(r'))\ \nabla' G(r', r) + (\vec{n}' \times H(r')) \times \nabla' G(r', r) \right. \right. \\ \left. + (j\omega\varepsilon_0 \vec{n} \times E(r'))\ G(r', r) \right] dS'_{\text{ap}} + 2 \int_{S'_{\text{pec}}} \left[(\vec{n}' \times H(r')) \right. \\ \left. \left. \times \nabla' G(r', r) \right] dS'_{\text{pec}} \right] \surd\, dS_{\text{ap}} \tag{15}$$

Because r and r' are both in the plane of the aperture, the gradients in Eq. (15) do not have a component in the normal direction to the aperture (only tangential components are present). For this reason, the terms $(\vec{n} \times W(r))$ and $(\vec{n}' \times H(r')) \times \nabla' G(r', r)$ are orthogonal and thus eliminated. Expression (15) can then be rewritten as

$$2\int_{S_{\rm ap}}\int_{S'_{\rm ap}}\left[(\vec{n}\times W(r))\cdot\left(\frac{-1}{j\omega\mu_0}\nabla'\cdot(\vec{n}'\times E(r'))\,\nabla' G(r',r)\right)\right.$$
$$\left.+\,(\vec{n}\times W(r))\cdot(j\omega\varepsilon_0\vec{n}'\times E(r'))\,G(r',r)\right]dS'_{\rm ap}\,dS_{\rm ap} \tag{16}$$

The gradient applied to the Green's function can be switched from the prime to the unprimed coordinate, that is, $\nabla' G = -\nabla G$ and, using the vector identity, $\nabla(\phi\Phi) = \phi\nabla\Phi + \Phi\nabla\phi$, for the first term of Eq. (16), the following is obtained:

$$2\int_{S_{\rm ap}}\int_{S'_{\rm ap}}\left[-(\vec{n}\times W(r))\cdot\nabla\left(\frac{-1}{j\omega\mu_0}\nabla'\cdot(\vec{n}'\times E(r'))\,G(r',r)\right)\right.$$
$$\left.+(\vec{n}\times W(r))\cdot(j\omega\varepsilon_0\vec{n}'E(r'))\,G(r',r)\right]dS'_{\rm ap}\,dS_{\rm ap} \tag{17}$$

By using the vector identity $\nabla\cdot(\phi A) = \nabla\phi\cdot A + \phi\nabla\cdot A$ in the first term of Eq. (17), we find

$$2\int_{S_{\rm ap}}\int_{S'_{\rm ap}}\left[-\nabla\cdot\left(\frac{-1}{j\omega\mu_0}(\vec{n}\times W(r))\,\nabla'\cdot(\vec{n}'\times E(r'))\,G(r',r)\right)\right.$$
$$+\left(\frac{-1}{j\omega\mu_0}\nabla\cdot(\vec{n}\times W(r))\,\nabla'\cdot(\vec{n}'\times E(r'))\,G(r',r)\right)$$
$$\left.+\,(\vec{n}\times W(r))\cdot(j\omega\varepsilon_0\vec{n}'\times E(r'))\,G(r',r)\right]dS'_{\rm ap}\,dS_{\rm ap} \tag{18}$$

A two-dimensional version of the divergence theorem can now be used for the first term in Eq. (18), resulting in a closed line integral around the rim of the aperture; that is,

$$2\int_{S_{\rm ap}}\oint_{\ell_{\rm ap}}-\nabla\cdot\left(\frac{-1}{j\omega\mu_0}(\vec{n}\times W(r))\,\nabla'\cdot(\vec{n}'\times E(r'))\,G(r',r)\right)d\ell_{\rm ap}\,dS_{\rm ap}$$
$$+\,2\int_{S_{\rm ap}}\int_{S'_{\rm ap}}\left[\left(\frac{-1}{j\omega\mu_0}\nabla\cdot(\vec{n}\times W(r))\,\nabla'\cdot(\vec{n}'\times E(r'))\,G(r',r)\right)\right.$$
$$\left.+\,(\vec{n}\times W(r))\cdot(j\omega\varepsilon_0\vec{n}'\times E(r'))\,G(r',r)\right]dS'_{\rm ap}\,dS_{\rm ap} \tag{19}$$

In Eq. (19), the closed line integral is along $\ell_{\rm ap}$, while $\vec{n}_c$ is in the ground plane and normal to the rim as shown in Figure 4.

The dot product in the first term of Eq. (19), $\vec{n}_c\cdot(\vec{n}\times W)$, is zero because the

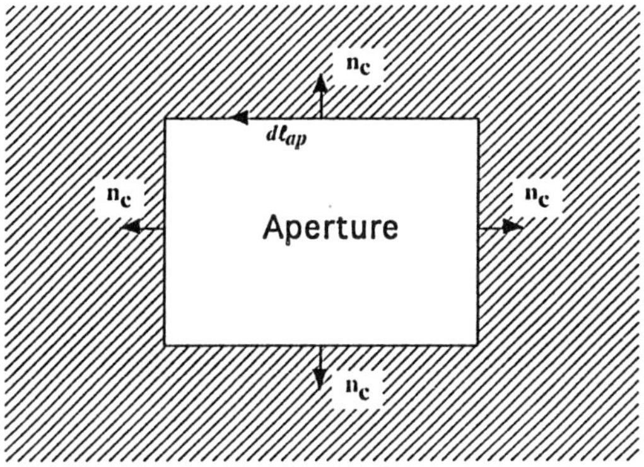

FIGURE 4. Vector directions at aperture rim.

Dirichlet boundary condition for the tangential component of E on the cavity wall implies that tangential components of W are equal to zero. Combining the aperture surface integral with Eq. (6) produces the final finite element/boundary element equation:

$$\int_V \left[(\nabla \times W) \cdot \left(-\frac{1}{j\omega\mu} \nabla \times E \right) - j\omega\varepsilon W \cdot E \right] dV$$

$$+ \int_{S_{\text{port}}} W \cdot \vec{n} \times \left(-\frac{1}{j\omega\mu} \nabla \times E \right) dS_{\text{port}}$$

$$-2 \int_{S_{\text{ap}}} \int_{S'_{\text{ap}}} \left[\frac{-1}{j\omega\mu_0} \nabla \cdot (\vec{n} \times W(r)) \, \nabla' \cdot (\vec{n}' \times E(r')) \, G(r', r) \right.$$

$$\left. + (\vec{n} \times W(r)) \cdot (j\omega\varepsilon_0 \vec{n}' \times E(r')) \, G(r', r) \right] dS'_{\text{ap}} \, dS_{\text{ap}} \tag{20}$$

5. PERIODIC GREEN'S FUNCTION

In this section, we discuss infinite series expressions for periodic Green's functions $G(r', r)$ and various techniques to achieve accelerated convergence for these infinite series.

The spectral form of the free-space Green's function for an infinite periodic structure, as shown in Figure 5, is given by Eq. (21) (Ishimaru, 1991):

$$G(r) = \sum_{m=-\infty}^{+\infty} \sum_{n=-\infty}^{+\infty} \frac{1}{j2D_x D_y k_{z_{mn}}} e^{-jk_{z_{mn}}|z|} e^{-j\vec{k}_{t_{mn}} \cdot \vec{r}} \tag{21}$$

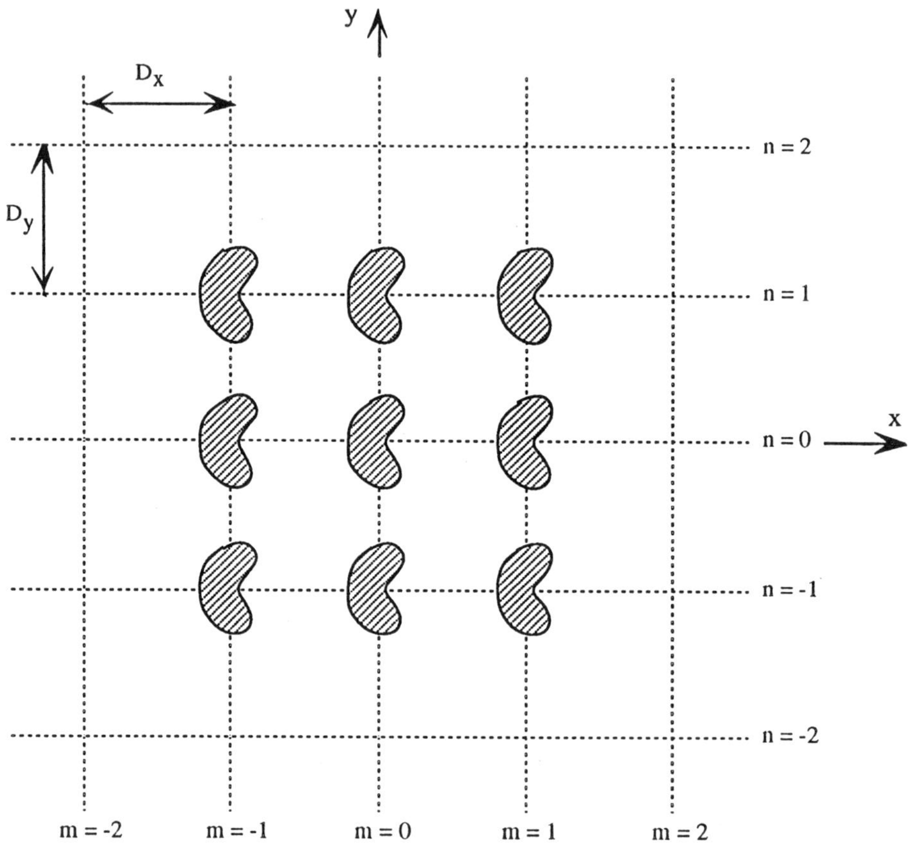

FIGURE 5. Outline of periodic structure.

where, as shown in Figure 5, D_x and D_y are the periodic distances, and

$$
\begin{aligned}
k_{z_{mn}} &= \sqrt{k_0^2 - |\vec{k}_{t_{mn}}|^2} \qquad \text{if } k_0 > |\vec{k}_{t_{mn}}| \\
k_{z_{mn}} &= -j\sqrt{|\vec{k}_{t_{mn}}|^2 - k^2} \quad \text{if } k_0 < |\vec{k}_{t_{mn}}|
\end{aligned} \tag{22}
$$

$$
\vec{k}_{t_{mn}} = \left(\frac{2\pi}{D_x} m + k_0 \sin\theta \cos\phi\right)\vec{x} + \left(\frac{2\pi}{D_y} n + k_0 \sin\theta \sin\phi\right)\vec{y} \tag{23}
$$

In Eq. (23), θ and ϕ are the scan angles of the phased array; that is, the angle directions $\theta = 0$ and $\phi = 0$ would correspond to the direction of propagation of the electromagnetic wave normally from the array; this angle is known as the "boresight" angle.

The direct use of Eq. (21) in the boundary element formulation would be impractical. It would require an inordinate number of terms to converge or numerically may not reach convergence at all. Means of accelerating the convergence by transformations of the above series have been developed (Singh et al., 1990). The two transformations used here are the Kummer's and Poisson transformations.

Kummer's transformation uses series whose terms have the same asymptotic behavior as the original series. Let

$$S = \sum_{n=-\infty}^{n=\infty} f(n) \tag{24}$$

and suppose the $f(n)$ and $f_1(n)$ have the same asymptotics at infinity

$$f(n) \xrightarrow{n \to \infty} f_1(n) \tag{25}$$

then $f_1(n)$ can be added and subtracted from each term in Eq. (24):

$$S = \sum_{n=-\infty}^{n=\infty} [f(n) - f_1(n)] - \sum_{n=-\infty}^{n=\infty} f_1(n) \tag{26}$$

If the sum of the second series is known, then no further transformations are required and Eq. (26) may be used directly. Otherwise, the second series must be converted to a fast convergent series as well. In our case, the second series is converted by a Poisson summation or Fourier transform:

$$S = \sum_{n=-\infty}^{n=\infty} [f(n) - f_1(n)] - \sum_{n=-\infty}^{n=\infty} F_1(2\pi n) \tag{27}$$

By using the above techniques, the following expression for the periodic Green's function can be derived (Singh et al., 1990):

$$\begin{aligned} G(r) = {} & \frac{1}{2A} \sum_{m=-\infty}^{+\infty} \sum_{n=-\infty}^{+\infty} \left[\frac{e^{-jk_{z_{mn}}|z|}}{jk_{z_{mn}}} - \frac{e^{-(\sqrt{|k_{z_{mn}}|^2 + u^2})|z|}}{\sqrt{|k_{z_{mn}}|^2 + u^2}} \right] e^{-j\vec{k}_{t_{mn}} \cdot \vec{r}} \\ & + \frac{1}{4\pi} \sum_{m=-\infty}^{+\infty} \sum_{n=-\infty}^{+\infty} e^{jk_0 D_x \sin\theta \cos\phi + D_y \sin\theta \sin\phi)} \frac{e^{-uR_{mn}}}{R_{mn}} \end{aligned} \tag{28}$$

where u is a smoothing factor equal to $\frac{1}{2}\max(D_x, D_y)$.

Equation (28) is the form of the Green's function that was used in Eq. (20). From our computational experience, the series in Eq. (28) can be approximated by 50 terms (i.e., $m + n$) for most cases.

6. FLOQUET MODAL EXPANSION

This section deals with the modeling of the exterior region to the infinite periodic array by using the Floquet modal expansion. The approach used to couple the Floquet modal expansion with the weak Galerkin form is similar to the method used in Section 4. Namely, the Floquet expansion and the Galerkin form are coupled by making the E and H fields tangentially continuous at the interface between the finite element and the infinite periodic regions.

However, unlike the integral equation method of Section 4, the coupling between the Floquet harmonic expansion and the Galerkin form is accomplished at the interface above the ground plane. This interface is designated as the surface S_{ep} in Figure 6. In addition, side surfaces specified as S_{pb}^{+} and S_{pb}^{-} are also introduced. Equation (5) is modified to include the new surface

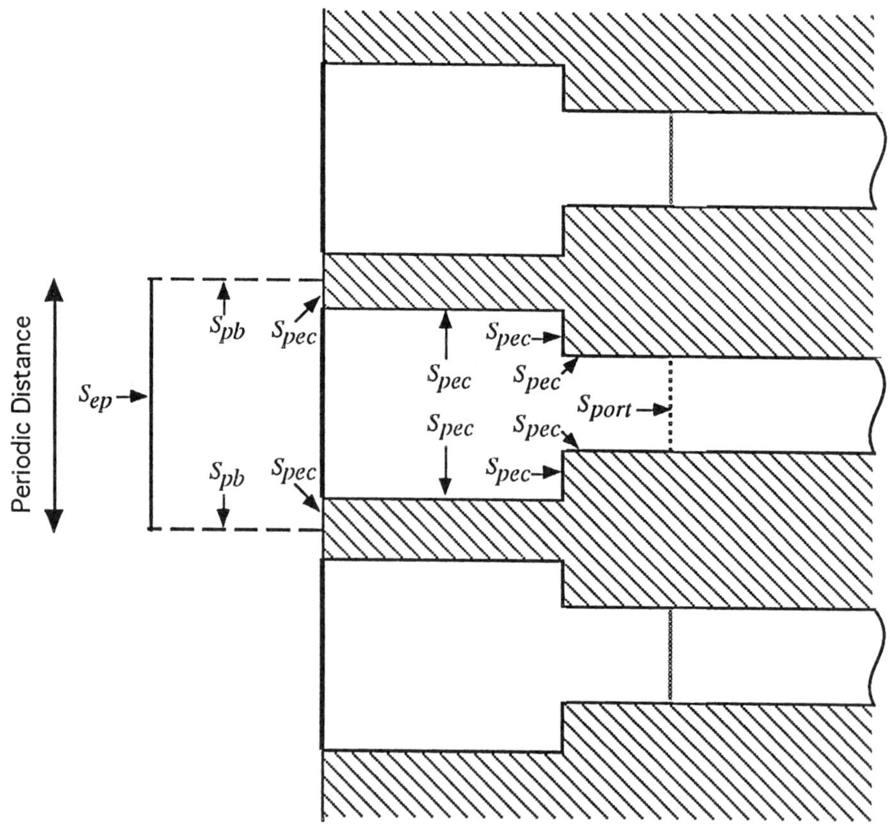

FIGURE 6. Outline of the analysis region and surfaces for the Floquet modal expansion method.

integrals:

$$\int_V \left[(\nabla \times W^*) \cdot \left(-\frac{1}{j\omega\mu} \nabla \times E^{fe} \right) - j\omega\varepsilon W^* \cdot E^{fe} \right] dV$$
$$- \int_{S_{\text{ep}}} (\vec{n} \times W^*) \cdot \left(-\frac{1}{j\omega\mu} \nabla \times E^{fe} \right)_t dS_{\text{ep}}$$
$$- \int_{S_{\text{pec}}} (\vec{n} \times W^*) \cdot \left(-\frac{1}{j\omega\mu} \nabla \times E^{fe} \right)_t dS_{\text{pec}}$$
$$- \int_{S_{port}} (\vec{n} \times W^*) \cdot \left(-\frac{1}{j\omega\mu} \nabla \times E^{fe} \right)_t dS_{port}$$
$$- \int_{S_{\text{pb}}^{+-}} (\vec{n} \times W^*) \cdot \left(-\frac{1}{j\omega\mu} \nabla \times E^{fe} \right)_t dS_{\text{pb}}^{+-} = 0 \tag{29}$$

where underscore t stands for tangetial component. Other differences between Eqs. (5) and (29) include the conjugating of the weights and designation of the fields in the finite element region as E^{fe}.

As previously described, the S_{pec} integral is identically zero because of Dirichlet boundary conditions.

The integrals over the side surfaces, S_{pb}^{+} and S_{pb}^{-}, are also eliminated. These integrals are equal and opposite in value due to the Floquet periodic boundary conditions; that is,

$$E^+ = E^- e^{-j\psi} \tag{30}$$

$$W^+ = W^- e^{-j\psi} \tag{31}$$

where E^+, E^-, W^+ and W^- are the electric fields and the testing functions, respectively, on the S_{pb}^{+} and S_{pb}^{-} surfaces, while the skewed periodic phase term ψ is given by

$$\psi = k_0 \sin\theta \cos\phi (D_x + D_y \cot\gamma) + k_0 \sin\theta \sin\phi D_y \tag{32}$$

Here, k_0 is the wavenumber, θ and ϕ are the scan angles, γ is the skew angle for triangular periodicity, and D_x and D_y are the periodic distances in the x and y directions (γ, D_x, and D_y are shown in Figure 2).

The first step in our method is to represent the electric field in the region above S_{ep} as a Floquet harmonic expansion:

$$E^{pa} = \sum_{m=-\infty}^{\infty} \sum_{n=-\infty}^{\infty} [c_{mn} F_{mn}^{TE} + d_{mn} F_{mn}^{TM}] \tag{33}$$

where E^{pa} is the electric field in the region above S_{ep}, c_m and d_n are the unknown coefficients, and F_{mn}^{TE} and F_{mn}^{TM} are the Floquet harmonics:

$$F_{mn}^{TE} = \frac{\alpha_{mn}\vec{x} - \alpha_m\vec{y}}{\sqrt{A(\alpha_{mn}^2 + \alpha_m^2)}} e^{-j(\alpha_m x + \alpha_{mn} y)} e^{-j\beta_{mn} z} \tag{34}$$

$$F_{mn}^{TM} = \frac{\alpha_m\vec{x} + \alpha_{mn}\vec{y}}{\sqrt{A(\alpha_{mn}^2 + \alpha_m^2)}} e^{-j(\alpha_m x + \alpha_{mn} y)} e^{-j\beta_{mn} z} \tag{35}$$

Here

$$\alpha_m = k_0 \sin\theta \cos\phi + \frac{2\pi m}{D_x} \tag{36}$$

$$\alpha_{mn} = k_0 \sin\theta \sin\phi + \frac{2\pi n}{D_y} - \frac{2\pi m}{D_x \tan\gamma} \tag{37}$$

$$\beta_{mn} = \sqrt{k_0^2 - \alpha_m^2 - \alpha_{mn}^2} \tag{38}$$

$$A = D_x D_y \tag{39}$$

The above Floquet harmonics are orthogonal; that is,

$$\begin{aligned} &\int_{S_{ep}} \vec{n} \times F_{mn}^{TE*} \cdot \vec{n} \times F_{m'n'}^{TE} \, dS_{ep} = 1 \quad \text{if } m = m',\ n = n' \\ &\int_{S_{ep}} \vec{n} \times F_{mn}^{TE*} \cdot \vec{n} \times F_{m'n'}^{TE} \, dS_{ep} = 0 \quad \text{otherwise} \end{aligned} \tag{40}$$

$$\begin{aligned} &\int_{S_{ep}} \vec{n} \times F_{mn}^{TM*} \cdot \vec{n} \times F_{m'n'}^{TM} \, dS_{ep} = 1 \quad \text{if } m = m',\ n = n' \\ &\int_{S_{ep}} \vec{n} \times F_{mn}^{TM*} \cdot \vec{n} \times E_{m'n'}^{TM} \, dS_{ep} = 0 \quad \text{otherwise} \end{aligned} \tag{41}$$

As previously mentioned, the E and H fields are forced to be tangentially continuous at the interface S_{ep}; that is,

$$n \times E^{fe} = n \times E^{pa} \tag{42}$$

$$(\nabla \times E^{fe})_t = (\nabla \times E^{pa})_t \tag{43}$$

By using Eqs. (33), (40), (41), and (42) and representing the E^{fe} in terms of edge elements,

$$E^{fe} = \sum_{i=1}^{\text{edges}} N_i(x, y)\xi_i \tag{44}$$

the following equations are obtained:

$$c_{mn} - \int_{S_{\mathrm{ep}}} \vec{n} \times N_i(x, y)\xi_i \cdot \vec{n} \times F_{mn}^{TE*}\, dS_{\mathrm{ep}} = 0 \tag{45}$$

$$d_{mn} - \int_{S_{\mathrm{ep}}} \vec{n} \times N_i(x, y)\xi_i \cdot \vec{n} \times F_{mn}^{TM*}\, dS_{\mathrm{ep}} = 0 \tag{46}$$

Next, the tangential components of the H field at the interface S_{ep} are made continuous. Using Eq. (33), we find that the transverse component of the curl of the electric field in the region above S_{ep} is equal to

$$(\nabla \times E^{pa})_t = -j \sum_{m=-\infty}^{\infty} \sum_{n=-\infty}^{\infty} \left[c_{mn}\beta_{mn}\vec{n} \times F_{mn}^{TE} + d_{mn}\frac{k_0^2}{\beta_{mn}}\vec{n} \times F_{mn}^{TM} \right] \tag{47}$$

By using Eqs. (43) and (47), we transform Eq. (29) as follows:

$$\begin{aligned}
&\int_V \left[(\nabla \times W^*) \cdot \left(-\frac{1}{j\omega\mu} \nabla \times E^{fe} \right) - j\omega\varepsilon W^* \cdot E^{fe} \right] dV \\
&\quad - \int_{S_{\mathrm{port}}} \frac{1}{\omega\mu} (\vec{n} \times W^*) \cdot \left(-\frac{1}{j\omega\mu} \nabla \times E^{fe} \right) dS_{\mathrm{port}} \\
&\quad - \int_{S_{\mathrm{ep}}} \frac{1}{\omega\mu} (\vec{n} \times W^*) \\
&\quad \cdot \sum_{m=-\infty}^{\infty} \sum_{n=-\infty}^{\infty} \left[c_{mn}\beta_{mn}\vec{n} \times F_{mn}^{TE} + d_{mn}\frac{k_0^2}{\beta_{mn}}\vec{n} \times F_{mn}^{TM} \right] dS_{\mathrm{ep}} = 0
\end{aligned} \tag{48}$$

The set of equations (45), (46), and (48) completely describe the coupling of the Galerkin form and the Floquet modal expansion.

The above formulation has an additional advantage of solving for the unknown expansion coefficients directly. This avoids a harmonic decomposition of the computed fields, which may lead to numerical errors.

7. MODELING OF SEMI-INFINITE WAVEGUIDE REGION AND PORT FEED

This section deals with the modeling of the semi-infinite waveguide section and the port feed.

It is clear that the actual electric field in the semi-infinite waveguide region is the superposition of the forcing waveguide mode and the reflected electric field. The reflected electric field can be represented as the homogeneous waveguide modal expansion at the port cross section. These homogeneous waveguide modes can be calculated by using the well-known techniques for the analysis of infinite homogeneous waveguides.

Thus the electric field in the port cross section can be expressed as

$$E^{wg} = \sum_{m=1}^{M} a_m E_m^{TE} + \sum_{n=1}^{N} b_n E_n^{TM} + E_f^{TE,TM} \tag{49}$$

where a_m and b_n are the unknown expansion coefficients, and E_m^{TE} and E_n^{TM} are modes obtained for the finite element form:

$$E_{m,n}^{TE,TM} = \sum_{j=1}^{\text{edges}} N_j^{xy}(x, y)\xi^{\xi}_{jm,n} e^{-j\beta z} + \sum_{j=1}^{\text{nodes}} n_j(x, y)\xi^{z}_{jm,n} e^{-j\beta z} \vec{z} \tag{50}$$

The phase term $e^{-j\beta z}$ in the last equation will be made equal to 1 by fixing the port at a presumed $z = 0$ position. The last term in Eq. (49), $E_f^{TE,TM}$, is the forcing function term, which is made equal to one of the propagating TE or TM modes.

Similar to the technique used in Section 6, the finite element region and the semi-infinite waveguide region are now coupled by setting E and H fields tangentially continuous at the S_{port} surface; that is,

$$\vec{n} \times E^{fe} = \vec{n} \times E^{wg} \tag{51}$$

$$(\nabla \times E^{fe})_t = (\nabla \times E^{wg})_t \tag{52}$$

The tangential component of E is made continuous by equating an edge-element expansion for E^{fe} and Eq. (49):

$$\begin{aligned} \vec{n} \times E^{fe} &= \sum_{i=1}^{\text{edges}} \vec{n} \times N_i(x, y)\xi_i = \vec{n} \times E^{wg} \\ &= \sum_{m=1}^{M} a_m \vec{n} \times E_m^{TE} + \sum_{n=1}^{N} b_n \vec{n} \times E_n^{TM} + \vec{n} \times E_f^{TE,TM} \end{aligned} \tag{53}$$

Next, we shall use the orthogonality properties of the waveguide modes, that is,

$$\begin{aligned} &\int_{S_{\text{port}}} \vec{n} \times E_m^{TE*} \cdot \vec{n} \times E_{m'}^{TE}\, dS_{\text{port}} = 1 \quad \text{if } m = m' \\ &\int_{S_{\text{port}}} \vec{n} \times E_m^{TE*} \cdot \vec{n} \times E_{m'}^{TE}\, dS_{\text{port}} = 0 \quad \text{if } m \neq m' \end{aligned} \tag{54}$$

$$\begin{aligned} &\int_{S_{\text{port}}} \vec{n} \times E_n^{TM*} \cdot \vec{n} \times E_{n'}^{TM}\, dS_{\text{port}} = 1 \quad \text{if } n = n' \\ &\int_{S_{\text{port}}} \vec{n} \times E_n^{TM*} \cdot \vec{n} \times E_{n'}^{TM}\, dS_{\text{port}} = 0 \quad \text{if } n \neq n' \end{aligned} \tag{55}$$

to obtain the following equations:

$$a_m - \int_{S_{\text{port}}} \vec{n} \times N_i(x, y)\xi_i \cdot \vec{n} \times E_m^{TE*} = \begin{cases} -1 & \text{if } m = f \\ 0 & \text{otherwise} \end{cases} \tag{56}$$

$$b_n - \int_{S_{\text{port}}} \vec{n} \times N_i(x, y)\xi_i \cdot \vec{n} \times E_n^{TM*} = \begin{cases} -1 & \text{if } n = f \\ 0 & \text{otherwise} \end{cases} \tag{57}$$

The transverse component of the curl of the electric field in the infinite waveguide region is equal to

$$(\nabla \times E^{wg})_t = -j \sum_{m=1}^{M} a_m \beta_n \vec{n} \times E_m^{TE} - j \sum_{n=1}^{N} b_n \frac{k_0^2}{\beta_n} \vec{n} \times E_n^{TM} - j\alpha_f \vec{n} \times E_f^{TE,TM} \tag{58}$$

"The forcing function coefficient," α_f, in the last term of Eq. (58) is equal to β_f for TE forcing function modes and equal to k_0^2/β_f for TM modes. By imposing tangential H continuity between the semi-infinite waveguide and finite element region, Eq. (58) can be used to replace the $(\nabla \times E^{fe})_t$ in the S_{port} port surface integral of Eq. (48):

$$\begin{aligned} &\int_V \left[(\nabla \times W^*) \cdot \left(-\frac{1}{j\omega\mu} \nabla \times E^{fe} \right) - j\omega\varepsilon W^* \cdot E^{fe} \right] dV \\ &\quad - \int_{S_{\text{port}}} \frac{1}{\omega\mu} (\vec{n} \times W^*) \cdot \left(\sum_{m=1}^{M} a_m \beta_n \vec{n} \times E_m^{TE} + \sum_{n=1}^{N} b_n \frac{k_0^2}{\beta_n} \vec{n} \times E_n^{TM} \right) dS_{\text{port}} \\ &= \int_{S_{\text{port}}} \frac{1}{\omega\mu} (\vec{n} \times W^*) \cdot \alpha_f \vec{n} \times E_f^{TE,TM} \, dS_{\text{port}} \end{aligned} \tag{59}$$

Equations (56), (57), and (59) form a set of equations that couple the weak Galerkin form with the infinite waveguide modal expansion.

8. MATRIX SOLUTION METHODS

Another numerical challenge is the technique required to solve the resulting matrix equations. The solution of discretized linear equations is made difficult by the fact that the equation matrix is indefinite, unsymmetric, and complex. Also, the use of edge elements results in matrices with condition numbers significantly larger than for similar matrices produced in the case of node-based finite elements. An additional complication is the sparse–dense nature of the equations created by the combination of finite elements and boundary element techniques. The same is true when one deals with the combination of finite elements and modal expansions (although the modal expansion's dense matrix is much smaller in size than the corresponding boundary element matrix).

The most robust and straightforward method to solve the matrix equations is the direct, sparse, Gaussian elimination technique. However, a drawback of direct solution methods is the amount of computer memory required. In two dimensions, direct methods, such as the nested-dissection method (George and Liu, 1981), are efficient because the graph of the sparse matrix connectivity is also two-dimensional and the bandwidth (and therefore the computer memory) of the sparse matrix can be significantly reduced. In three dimensions, for geometries that cannot be effectively partitioned, this is not always the case.

The method employed in our work to solve the matrix equations is a transpose-free quasi-minimal residual iterative technique (Freund, 1991). In conjunction with this iterative technique, we have used a "fill-in" preconditioner (Saad, 1988; Boyse and Seidl, 1996). The often-used preconditioners retain the same nonzero structure as in the original matrix equations. However, for matrices obtained from finite element discretizations of the Helmholtz equation, these preconditoners were found to be ineffective, requiring almost the same amount of computational effort as diagonal preconditioners.

Fill-in preconditioners, however, have been shown to significantly reduce computational time. Fill-in preconditioners are calculated by performing an incomplete elimination process on the matrix equations, but in addition to storing the original nonzero locations in the matrixs, the larger terms produced by the elimination process are also retained. A transpose-free quasi-minimal residual method was used here to eliminate the need for storing or evaluating the transpose of the preconditioner.

On an example problem containing 17,395 complex unknowns, a solver using a biconjugate gradient technique and a diagonal preconditioner required 3415 iterations to reach convergence. (The transpose-free quasi-minimal residual method did not converge with the diagonal preconditioner.) The timing of this problem was 1754 seconds on a Hewlett-Packard 715/75 9000 series computer workstation. Using a fill-in preconditioner, the quasi-minimal residual method required 62 iterations with a timing of 542 seconds to determine the fill-in preconditioner and 360 seconds for the iterative solution.

9. TESTING OF THE PORT MODELING

Before proceeding with the radiation analysis examples, the semi-infinite waveguide/port feed modeling as described in Section 7 will first be tested alone on an example waveguide problem.

Shown in Figure 7 is an isometric view of an abrupt waveguide taper. The finite element/port feed method described above has been compared to a modal solution (Huting and Webb, 1991) for this test problem.

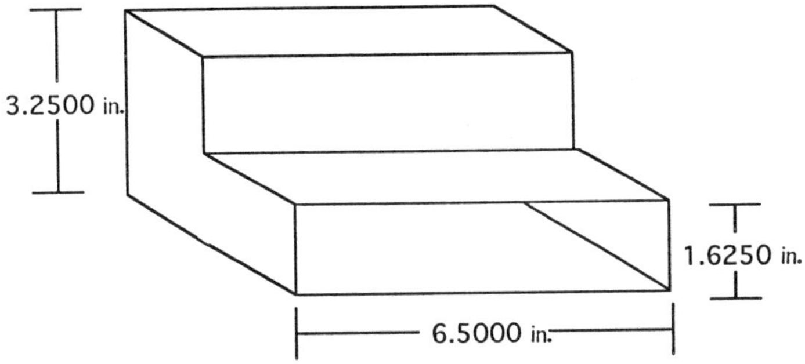

FIGURE 7. Isometic view of an abrupt waveguide taper.

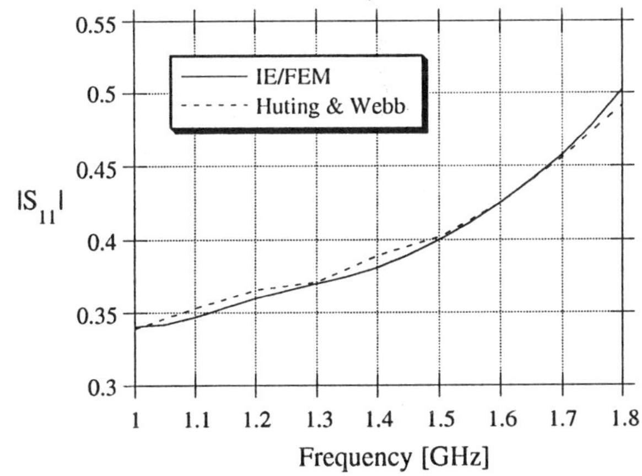

FIGURE 8. Comparison of S_{11} calculation for the abrupt waveguide taper. Integral equation/finite element method versus modal solution.

In this example, port #1 (the larger port) is excited and port #2 remains passive. The frequency is varied from 1.0 to 1.8 GHz and the S_{11} parameter is compared against the modal solution. Figure 8 illustrates the comparison.

10. PERIODIC ARRAY TEST EXAMPLES

The full analysis code has been tested here for an infinite array of periodically placed cavity apertures. The method described above has been compared to a modal solution method by Farrel and Kuhn (1968). Figure 9 shows a subsection of the infinitely periodic square array used in this example. The dimensions a and b describe the size of the aperture, while A and B are the

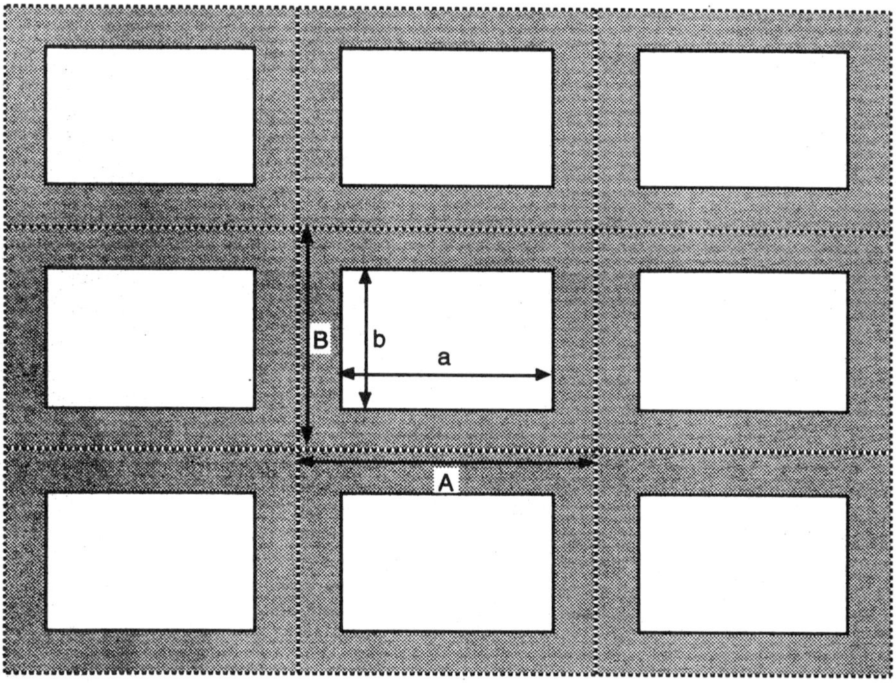

FIGURE 9. Subsection of square infinite periodic array.

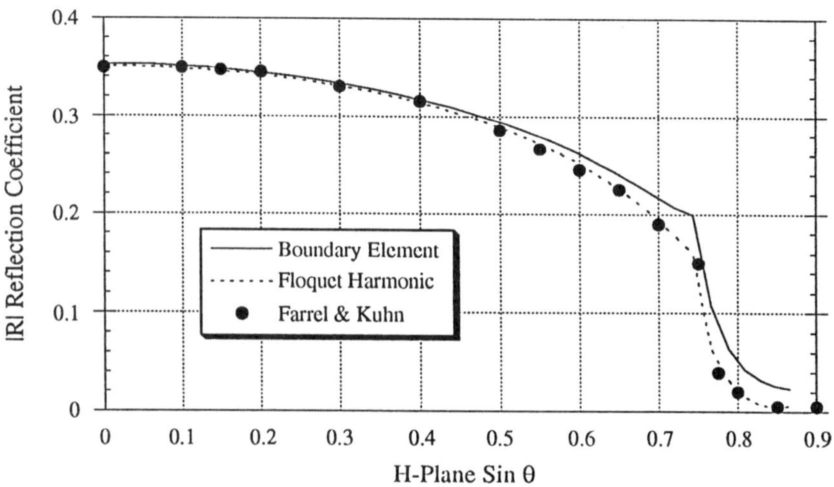

FIGURE 10. Comparison of finite element/boundary element method and finite element/Floquet harmonic method and modal solution for thin wall case.

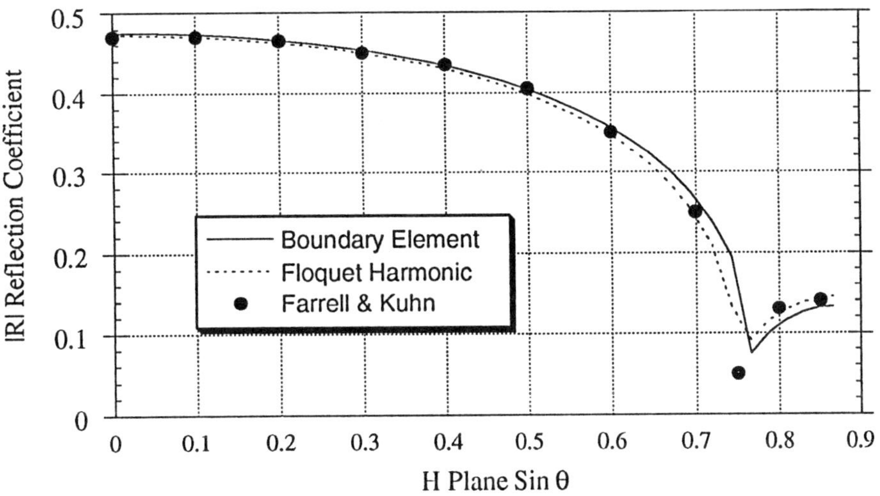

FIGURE 11. Comparison of finite element/boundary element method and finite element/Floquet harmonic method and modal solution for thick wall case.

periodicities of the array. Note that although rectangular apertures are used here, this FE/BE method is capable of modeling arbitrarily shaped apertures.

Two examples from Farrel and Kuhn (1968) will be used for comparison. The first example is called the "thin wall" case, where the periodic distances are equal to the waveguide dimensions, that is, $a = b = A = B = 0.5714$ wavelengths. Figure 10 shows the comparison. The results using the boundary element method and the Floquet harmonic technique produce nearly equal results.

The second is the "thick wall" case where $a = 0.5354$ wavelengths and $b = A = B = 0.5714$ wavelengths. The excitation mode used in this example is the TE mode with the electric field polarized in the b direction. Again, Figure 11 shows nearly equal results between the methods.

For both examples, the emanating field from the array has been scanned from boresight to 60° in the H-field plane (a direction).

11. CONCLUSION

A numerical approach to the modeling of phased array antenna has been presented. This approach combines the finite element technique with the boundary element (integral) technique and Floquet and waveguide modal expansions. The boundary element technique or Floquet modal expansion is used to model the semi-infinite region above the ground plane, while the waveguide modal expansion is employed to model the regular semi-infinite (feeding) waveguide structures. The finite element discretization is used in the part of the antenna region which is highly irregular geometrically and contains design features with complicated media properties. The developed approach has been numerically tested and favorably compared with other techniques for simple geometric configurations.

ACKNOWLEDGMENTS

The authors would like to thank the Lockheed–Martin Corporation for their support of this effort.

REFERENCES

Bossavit, A., and Mayergoyz, I. (1989). Edge elements for scattering problems. *IEEE Trans. Magn.* **MAG-25**, 2816–2821.

Boyse, W. E., and Seidl, A. A. (1996). A block QMR method for computing multiple simultaneous solutions to complex symmetric systems. *SIAM J. Sci. Comput.* **17**, 263–274.

D'Angelo, J., and Mayergoyz, I. D. (1990). Finite element methods for the solution of RF scattering and radiation problems. *Electromagn. J., Spec. Issue*, Apr.

Farrell, G. F., and Kuhn, D. H. (1968). Mutual coupling in infinite planar arrays of rectangular waveguide horns. *IEEE Trans. Antennas Propag*, **AP-16**, 405–414.

Freund, F. W. (1991). *A Transpose-free Quasi-Minimal Residual Algorithm for Non-Hermitian Linear Systems*, Report. The Research Institute for Advanced Computer Science.

George, A., and Liu, J. W. (1981). *Computer Solution of Large Sparse Positive Definite Systems.* Prentice-Hall, Englewood Cliffs, NJ.

Huting, W. A., and Webb, K. J. (1991). Comparison of mode-matching and differential equation techniques in the analysis of waveguide transitions. *IEEE Trans. Microwave Theory Tech.* **MTT-39**, 280–286.

Ishimaru, A. (1991). *Electromagnetic Wave Propagation, Radiation, and Scattering.* Prentice-Hall, Englewood Cliffs, NJ.

Jin, J. M., and Volakis, J. (1991). A hybrid finite element method for scattering and radiation by microstrip patch antennas and arrays residing in a cavity. *IEEE Trans. Antennas Propag.* **AP-39**, 1598–1604.

Lucas, E. M., and Fontana, T. P. (1995). A 3-D hybrid finite element/boundary element method for the unified radiation and scattering analysis of general infinite periodic arrays. *IEEE Trans. Antennas Propag.* **AP-43**, 145–153.

McGrath, D. T., and Pyati, V. P. (1994). Phased array antenna analysis with the hybrid finite element method. *IEEE Trans. Antennas Propag.* **AP-42**, 1625–1630.

Mias, C., Webb, J. P., and Ferrari, R. L. (1994). Finite element analysis of electromagnetic plane wave scattering from an infinite double periodic structure. *COMPEL. Int. J. Comput. Math. Electr. Electron. Eng.* **13**, Suppl. A, 393–398.

Saad, Y. (1988). Preconditioning techniques for nonsymmetric and indefinite linear systems. *J. Comput. Appl. Mathe.* **24**, 89–105.

Singh, S., Richards, W. F., Zinecker, J. R., and Wilton, D. R. (1990). Accelerating the convergence of series representing the free space Green's function. *IEEE Trans. Antennas Propag.* **AP-38**, 1958–1962.

CHAPTER NINE

Large-Scale Finite Element Scattering Analysis on Massively Parallel Computers

JOHN D'ANGELO and ISAAC MAYERGOYZ

1. INTRODUCTION

This chapter describes the implementation of finite element formulations for large radio frequency (RF) scattering problems on massively parallel, distributive memory message passing computers. Included are some results for cases run on the Intel Paragon and the IBM SP2.

The finite element method used here is a frequency domain technique coupled to absorbing boundary conditions for infinite exterior modeling. This formulation is based on decoupled vector Helmholtz techniques with a node-based finite element basis functions (D'Angelo and Mayergoyz, 1991), resulting in a computationally efficient method for the solution of large scattering problems. A large scattering problem is defined here as a scatterer with a surface area on the order of 1000 square wavelengths or greater.

Since nearly all of the computational effort in this method is spent in the sparse matrix solver, this chapter will concentrate on the implementation of an iterative equation solver on distributed memory parallel machines. This chapter will also focus more on practical implementation details than theoretical analysis. Topics discussed include the division of the finite element mesh into subdomains, the communication strategy, preconditioning, and changes from standard single-processor methods.

Finite Element Software for Microwave Engineering, Edited by Tatsuo Itoh, Giuseppe Pelosi, and Peter P. Silvester.
ISBN 0-471-12636-5

As previously mentioned, this method employs node-based finite elements. Unlike the phased array antenna analysis method described in Chapter 8, which utilizes edge-based finite elements, the use of node-based methods, particularly with decoupled vector Helmholtz-type equations, has proved much more efficient for large RF problems. (Edge elements are less efficient because they satisfy the normal continuity of the unknown field in the least-squared sense—higher mesh densities are required to reduce this "weak error.")

However, in this formulation, node-based elements have difficulties of their own. Inaccuracies in the radar cross section result for problems containing perfect electric conductor (PEC) geometries with sharp corners. A means of improving the results for this problem type with combined node and edge elements is also discussed.

The next sections will discuss the finite element formulation, the combined node–edge finite element basis functions, the decomposition of the finite element mesh into subdomains, an outline of the parallel quasi-minimal residual iterative method, the parallel linear algebra functions (matrix multiplies, dot products, and SAXPYs), and results.

2. FINITE ELEMENT FORMULATION

The finite element method described here is used by multiple groups of researchers. The method was first described in detail by Lynch and Paulsen (1990) and has been further refined in recent publications (Boyse et al., 1992). For lack of a better title, the method will be called the "gauge enforcement method."

The gauge enforcement method can be cast in either a four-component formulation containing a vector potential plus a scalar potential unknown to represent the electric or magnetic fields, or the method can be developed for field values directly. The formulation described here will directly solve for the field values, thereby saving an additional unknown at the expense of explicitly enforcing normal discontinuity of the field at material interfaces (i.e., jump in normal E for dielectric interfaces, jump in normal H for magnetic permeability interfaces).

The method can be developed for either the E field or for the H field. For brevity, we will cover only the H field formulation. Equation (1) is the "curl–curl" wave equation derived directly from Maxwell's equations for source-free regions:

$$\nabla \times \frac{1}{\varepsilon_r} \nabla \times \mathbf{H} - k_0^2 \mu_r \mathbf{H} = 0 \tag{1}$$

where ε_r is the relative dielectric constant and μ_r is the relative magnetic permeability of the materials. Two gauge or divergence conditions in addition must be satisfied:

$$\nabla \cdot \mathbf{D} = \rho \tag{2}$$

$$\nabla \cdot \mathbf{B} = 0 \tag{3}$$

Equation (2) is satisfied by the tangential continuity of the unknown magnetic fields. Equation (3) is explicitly enforced by the addition of another term to Eq. (1):

$$\nabla \times \frac{1}{\varepsilon_r} \nabla \times \mathbf{H} - \nabla \left(\frac{1}{\varepsilon_r \mu_r} \nabla \cdot \mu_r \mathbf{H} \right) - k_0^2 \mu_r \mathbf{H} = 0 \tag{4}$$

Equation (4) is the basic equation of the gauge enforcement method. H in Eq. (4) is the total magnetic field consisting of the scattered and incident components; that is, $H = H^s + H^i$. From our experience, better accuracy is obtained when directly solving for the scattered field H^s. Splitting the magnetic field into scattered and incident field components and subtracting the wave equation for the incident magnetic field, that is,

$$\nabla \times \nabla \times H^i - k_0^2 H^i = 0 \tag{5}$$

result in the following equation for the scattered magnetic field:

$$\begin{aligned} &\nabla \times \frac{1}{\varepsilon_r} \nabla \times H^s - \nabla\left(\frac{1}{\varepsilon_r \mu_r} \nabla \cdot \mu_r H^s\right) - k_0^2 \mu_r H^s \\ &\quad = \nabla \times \left(1 - \frac{1}{\varepsilon_r}\right) \nabla \times H^i + \nabla\left(\frac{1}{\varepsilon_r \mu_r} \nabla \cdot \mu_r H^i\right) - k_0^2 (1 - \mu_r) H^i \end{aligned} \tag{6}$$

The terms containing H^i are placed on the right-hand side of the equation, indicating them as forcing functions terms. Equation (6) is cast into finite element equations by applying the Galerkin weighted residual method. Multiplying Eq. (6) by an arbitrary vector weight, W, integrating over the solution region, and integration of the second order terms by parts result in the following:

$$\begin{aligned} &\int_\Omega \left[(\nabla \times W) \cdot \left(\frac{1}{\varepsilon_r} \nabla \times H^s \right) + (\nabla \cdot \mathrm{W}) \left(\frac{1}{\varepsilon_r \mu_r} \nabla \cdot \mu_r H^s \right) - k_0^2 \mu_r H^s \right. \\ &\qquad \left. - \nabla \cdot \left(W \left(\frac{1}{\varepsilon_r \mu_r} \nabla \cdot \mu_r H) \right) - \nabla \cdot \left(W \times \frac{1}{\varepsilon_r} \nabla \times H^s \right) \right] d\Omega \\ &\quad = \int_\Omega \left[(\nabla \times W) \cdot \left(\left(1 - \frac{1}{\varepsilon_r} \right) \nabla \times H^i \right) - (\nabla \cdot W) \left(\frac{1}{\varepsilon_r \mu_r} \nabla \cdot \mu_r H^i \right) - k_0^2 (1 - \mu_r) H^i \right. \\ &\qquad \left. - \nabla \cdot \left(W \times \left(1 - \frac{1}{\varepsilon_r} \right) \nabla \times H^i \right) \right] d\Omega \end{aligned} \tag{7}$$

Equation (7) is further simplified by using the divergence theorem on the last terms of the right- and left-hand sides. In addition, the term

$$\nabla \cdot \left(W \left(\frac{1}{\varepsilon_r \mu_r} \nabla \cdot \mu_r H \right) \right)$$

is made equal to zero on magnetic material interfaces by explicit enforcement of the jump condition, or by weak enforcement by the use of combined edge and node basis functions:

$$\begin{aligned} &\int_\Omega \left[(\nabla \times W) \cdot \left(\frac{1}{\varepsilon_r} \nabla \times H^s \right) + (\nabla \cdot W) \left(\frac{1}{\varepsilon_r \mu_r} \nabla \cdot \mu_r H^s \right) - k_0^2 \mu_r H^s \right] d\Omega \\ &\quad + \int_\Gamma \left[W \cdot \left(n \times \frac{1}{\varepsilon_r} \nabla \times H^s \right) \right] d\Gamma \\ &\quad = \int_\Omega \left[(\nabla \times W) \cdot \left(\left(1 - \frac{1}{\varepsilon_r} \right) \nabla \times H^i \right) - (\nabla \cdot W) \left(\frac{1}{\varepsilon_r \mu_r} \nabla \cdot \mu_r H^i \right) \right. \\ &\quad \left. - k_0^2 (1 - \mu_r) H^i \right] d\Omega + \int_\Gamma \left[W \cdot \left(n \times \left(1 - \frac{1}{\varepsilon_r} \right) \nabla \times H^i \right) \right] d\Gamma \end{aligned} \tag{8}$$

Two types of surface boundaries are encountered. The first is the outer boundary where the finite element equations are truncated and the absorbing boundary conditions are applied. This boundary will be specified as Γ_{out}. Techniques used to apply absorbing boundary conditions are discussed in D'Angelo and Mayergoyz (1991). The second type of boundary is the PEC surface within the solution domain, specified as Γ_{pec}; here the boundary conditions are

$$n \times \frac{1}{\varepsilon_r} \nabla \times \mathbf{H} = 0 \tag{9a}$$

$$n \cdot \mathbf{H} = 0 \tag{9b}$$

Rewritting Eq. (9a) in terms of the scattering and incident field components results in

$$n \times \frac{1}{\varepsilon_r} \nabla \times H^s = -n \times \frac{1}{\varepsilon_r} \nabla \times H^i \tag{10}$$

Using Eq. (10) in Eq. (8) and the fact that only free-space materials exist at the outer boundary (i.e., $\varepsilon_r = \mu_r = 1$), we find

$$\int_\Omega \left[(\nabla \times W) \cdot \left(\frac{1}{\varepsilon_r} \nabla \times H^s \right) + (\nabla \cdot W) \left(\frac{1}{\varepsilon_r \mu_r} \nabla \cdot \mu_r H^s \right) - k_0^2 \mu_r H^s \right] d\Omega$$

$$+ \int_{\Gamma_{\text{out}}} [W \cdot (n \times \nabla \times H^s)] \, d\Gamma_{\text{out}}$$

$$= \int_\Omega \left[(\nabla \times W) \cdot \left(\left(1 - \frac{1}{\varepsilon_r} \right) \nabla \times H^i \right) - (\nabla \cdot W) \left(\frac{1}{\varepsilon_r \mu_r} \nabla \cdot \mu_r H^i \right) \right.$$

$$\left. - k_0^2 (1 - \mu_r) H^i \right] d\Omega + \int_{\Gamma_{\text{pec}}} W \cdot (n \times \nabla \times H^i) \, d\Gamma_{\text{pec}} \tag{11}$$

Equation (11) is the final weak Galerkin form of the equations. Equation (11) is discretized by a standard finite element approximation function in the majority of the solution region and at sharp PEC corners by the special node/edge-based approximation functions. An absorbing boundary condition, based on the Engquist–Majda method, is used and is coupled to Eq. (11) through the surface integral Γ_{out}.

3. NEW FINITE ELEMENT APPROXIMATION FUNCTIONS

As previously mentioned, new types of finite element approximation functions are used in this formulation. These elements are developed to help solve the problem of defining the normal direction at sharp corners.

When using the formulations of Section 2 with standard node-based elements, it was discovered that inaccurate results for a radar cross section (RCS) would consistently be obtained when the obstacle had perfect electric conductors with sharp corners. This error would occur if the RCS were dominated by the sharp corner; that is, valid results would be obtained for normal incidence on a flat plate, but near-edge incidence would cause appreciable error. This error with node-based formulations was also observed by other researchers (Mur, 1993).

The source of error was found to be in the ambiguous direction of the normal component. Using standard node-based elements, as shown in Figure 1, the normal direction is often the average or the weighted average of the normals determined from the elements that have an edge on the corner. The prescribed boundary conditions for the H field and E field ($\vec{n} \cdot H = 0$ and $\vec{n} \times E = 0$), in turn, are also approximated.

A method is developed here to help solve this sharp-corner problem. An approximation function is created, which combines the attributes of both node- and edge-based elements. Edge-based elements depend only on the tangential directions of the field components at an element edge and therefore do not

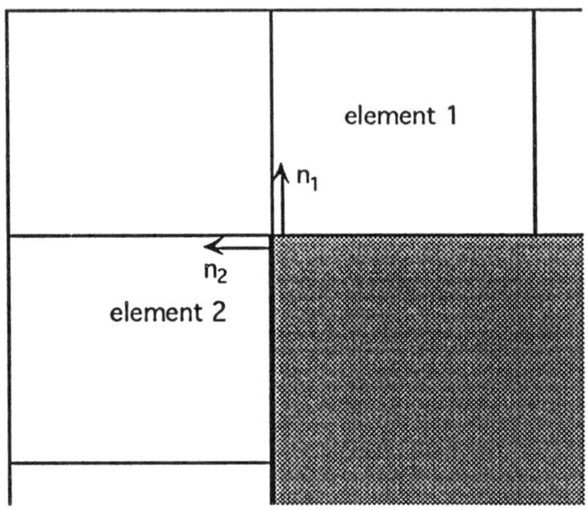

FIGURE 1. Ambiguous normal directions at a corner.

suffer the ambiguity of the normal direction at corners. These newly developed hybrid node/edge elements are used only at the sharp corners. (As mentioned in the previous section, hybrid elements can also be used to provide a jump condition in the normal component of the unknown field quantity at material interfaces.)

As shown in Figure 2, edge unknowns are used at the sharp edges and standard node-based elements are used everywhere else. The mixing of node and edge unknowns occurs in a single element. The placement of the edge unknowns can exist at any node of an element that is touching the element edge.

The magnetic field (or electric field) in an element is defined as follows:

$$H = \sum_{j=1}^{3} \sum_{i=1}^{\text{nodes}} n_i(x, y, z) D_{ij} h_{ij} + \sum_{i=1}^{\text{edges}} N_i(x, y, z) h_i \tag{12}$$

The first term of Eq. (12) is the approximation function for the node-based unknowns; n_i is a standard, scalar finite element shape function, h_{ij} are the three component unknown coefficients, and D_{ij} are the associate vector directions. Note that D_{ij} vector directions need not be aligned to the Cartesian axes but can be rotated and aligned to the flat PEC surfaces to impose the boundary conditions. N_i quantities are the vector edge-element shape functions for the edge unknowns and h_i quantities are the unknown coefficients.

The N_i functions are derived for the three edges at an element vertex. A well-known method of describing the vector field components utilizing covariant and contravariant quantities is employed here (Stratton, 1941; Crowley et al., 1988). A first-order hexahedral element, shown in Figure 2, is used as an example (the method is applicable to any element order or shape).

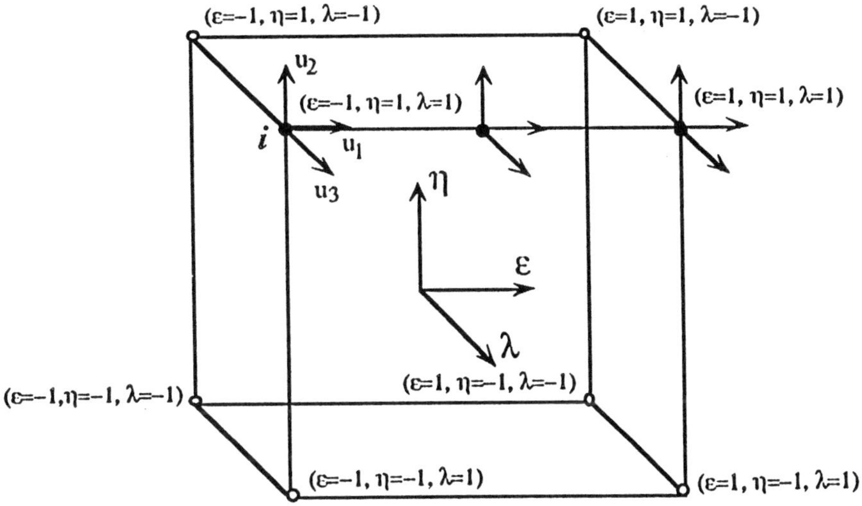

FIGURE 2. Canonical element for hybrid node/edge approximation functions.

A point where an edge unknown is placed in this example is at node i in Figure 2. The finite element here is defined by the local coordinates ($\varepsilon = -1, 1$; $\eta = -1, 1$; $\lambda = -1, 1$). The vector unknowns placed at node i will have tangential continuity with the neighboring elements that have a common face, and the vector directions are determined from a set of unitary and reciprocal unitary vectors.

Using standard finite element techniques, the mapping of the general Cartesian coordinates (x, y, z) to a local coordinate system $(\varepsilon, \eta, \lambda)$ is readily obtained. This mapping is facilitated by the node-based approximation function defined in the local coordinate system:

$$x = \sum n_i(\varepsilon, \eta, \lambda)x_i, \quad y = \sum n_i(\varepsilon, \eta, \lambda)y_i, \quad z = \sum n_i(\varepsilon, \eta, \lambda)z_i \tag{13}$$

A system of unitary vectors (directional vectors for a curvilinear coordinate system) are used here. The unit vectors are U_1, U_2, and U_3 in Figure 2. With respect to the Cartesian coordinate system, these unit vector directions are

$$\begin{aligned} U_1 &= \frac{\partial x}{\partial \eta}\vec{x} + \frac{\partial y}{\partial \eta}\vec{y} + \frac{\partial z}{\partial \eta}\vec{z} \\ U_2 &= \frac{\partial x}{\partial \eta}\vec{x} + \frac{\partial y}{\partial \eta}\vec{y} + \frac{\partial z}{\partial \eta}\vec{z} \\ U_3 &= \frac{\partial x}{\partial \lambda}\vec{x} + \frac{\partial y}{\partial \lambda}\vec{y} + \frac{\partial z}{\partial \lambda}\vec{z} \end{aligned} \tag{14}$$

The derivatives in Eq. (14) are readily obtained from Eq. (13). A second set of reciprocal unitary vectors can be obtained from U_1, U_2, and U_3:

$$U^1 = \frac{U_2 \times U_3}{U_1 \cdot (U_2 \times U_3)}, \quad U^2 = \frac{U_3 \times U_1}{U_1 \cdot (U_2 \times U_3)}, \quad U^3 = \frac{U_1 \times U_2}{U_1 \cdot (U_2 \times U_3)} \tag{15}$$

Using Eq. (15) in (12) the vector edge approximation function at node i is obtained:

$$\begin{aligned} N_{i1}(x, y, z) &= n_i(x, y, z)U^1 \\ N_{i2}(x, y, z) &= n_i(x, y, z)U^2 \\ N_{i3}(x, y, z) &= n_i(x, y, z)U^3 \end{aligned} \tag{16}$$

where n_i is a standard, scalar finite element nodal shape function (written here in terms of the Cartesian coordinates). Constructed in the same manner, the vector approximation functions are made up of one of the two tangential directions on the faces that meet at node i. For example, the scalar shape function n_i is nonzero on the faces where $\varepsilon = -1$, $\eta = 1$, $\lambda = 1$ and the vector direction U^1 is perpendicular to face $\varepsilon = -1$ and has continuous tangential components on faces $\eta = 1$ and $\lambda = 1$.

The approximation function can also be used in the opposite manner: that is, knowing the vector **H** at node i, the coefficients h_i for the edge unknowns can be calculated by the orthogonal condition of the unitary and reciprocal unitary vector:

$$U^i U_j = \delta_{ij} \tag{17}$$

4. DOMAIN DECOMPOSITION

The core procedure of implementing this finite element code on a parallel computer is dividing the mesh into subdomains. This procedure is commonly called domain decomposition. In this implementation, the mesh is divided in an element-wise manner where an element can occur in a single subdomain. Only nodes and their corresponding unknowns are shared between subdomains. The mesh information must be divided so that it can be processed by the individual parallel processors.

This method of subdividing the mesh is called "recursive bisection." Here, the finite element mesh is repeatedly divided into two subdomains with approximately equal numbers of elements. This is illustrated with the two-dimensional mesh of Figure 3. In this example, the first split or bisection is along an approximate vertical line and divides the mesh into two parts. This line is nearly in the middle, allowing the left and right domains to have nearly

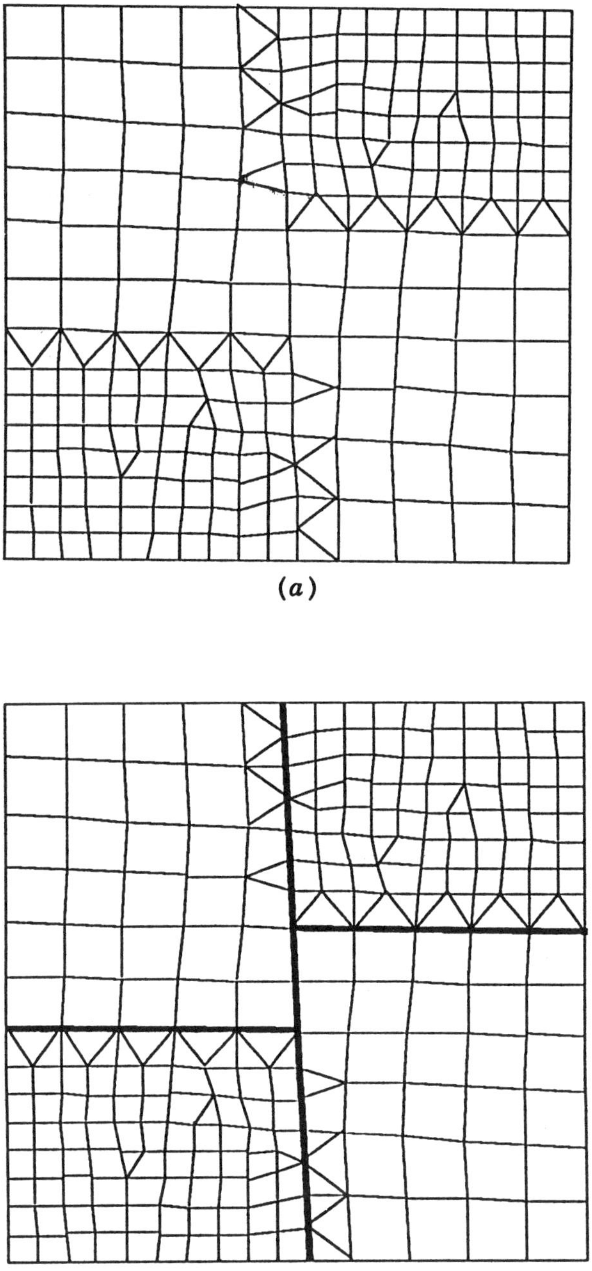

(a)

(b)

FIGURE 3. (a) Sample mesh. (b) Sample mesh with subdomain dividing lines. (*Continues on next page.*)

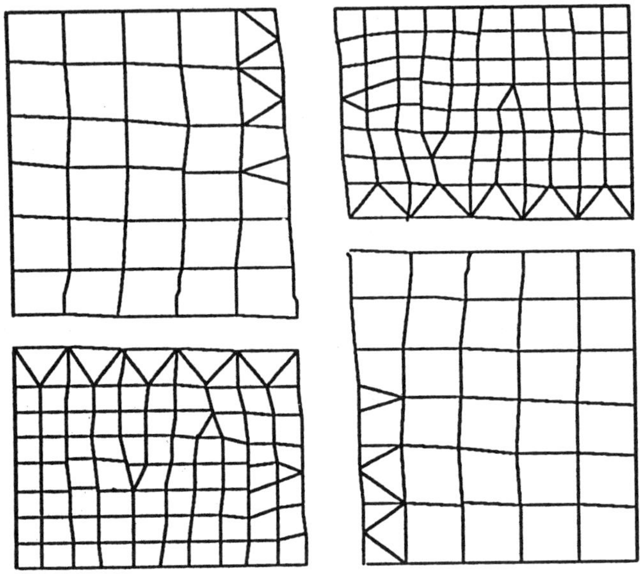

(c)

FIGURE 3. (c) Sample mesh with domains separated.

equal numbers of elements. The second bisection in both domains is along horizontal lines; on the left subdomain the division is nearer the bottom and on the right the division is near the top due to the variation in element density.

In the illustration of Figure 3, two levels of subdivision are used, requiring four processors. If another, third level of subdivision is used, then eight processors are required; with four levels, 16 processors are needed, and so on. The number of processors is therefore 2^l, where l is the number of levels. Note, however, that the incomplete binary tree can be (and often is) used so that the total number of processors need not be an order of 2.

This bisection algorithm is based on the geometric element positions. The mesh is broken approximately along Cartesian axial planes with the algorithm determining which plane to use to provide the best split. The mesh division planes need not be straight lines but can be skewed or jagged-edged to best split the mesh. The recursive subdomaining presented here is one of the many types of subdivision. Another interesting subdomaining method (Simon, 1991) uses an eigenvalue/eigenvector solver on the integer matrix of the mesh connectivity to optimally obtain the division breaks. In this method, the zero contour line of the second mode eigenvector defines the break point. Although eigenvector solvers require a substantial number of operations, by far exceeding solution of the matrix equation themselves, this solver is accelerated by using integer matrices.

5. QMR ALGORITHM

The assembly of the finite element matrices in this approach is not discussed because it is "embarrassingly" parallel (i.e., no communication is required). The communication between processors is located primarily in the matrix solution section of the code. An iterative quasi-minimal residual (QMR) (Freund, 1992) solver is used here. Of all the iterative solvers surveyed—generalized mean residual (GMRES), biconjugate gradient (BICG), biconjugate gradient-stabilized (BICG-STAB), and conjugate gradient squared (CGS)—QMR has proved the most reliable solver for this indefinite complex problem type. Iterative solvers are best suited for large, three-dimensional finite methods (finite elements, finite differences, and finite volumes) because of their reduced computer memory and computational requirements.

This implementation uses a simple diagonal preconditioner. The diagonal preconditioner was chosen because it is easily made parallel. In addition, the diagonal preconditioner allows for matrix splitting between the real and complex sections of the matrix equations, thereby saving an appreciable amount of computer memory. Another preconditioning method that is readily implemented in parallel is "successive over-relaxation" (SOR). However, SOR was found to be ineffective for this problem. Other preconditioners, which are commonly used on single processor computers, such as incomplete factorization types (Saad, 1988), are difficult to implement in a parallel distributed memory environment. More promising parallel proconditioners are of the Schwarz type (Lions, 1988).

Most iterative equation solvers use the same basic linear algebra subroutines (BLAS) such as SAXPY ($a(i) = \Sigma_{j=1}^{n} \alpha b(j) + a(i)$), SDOT, and SCOPY, and all use a "sparse matrix–vector multiplier." Therefore, although the QMR algorithm is illustrated here, the techniques presented are well suited to other types of iterative solvers.

The algorithm for the QMR solver is given below. In this algorithm, Greek letters are scalar quantities, small letters are vectors, and capital letters are matrices. Communication between processors is required in only six places, as designated by the → symbols and only three places within the iteration loop itself.

**** QMR ALGORITHM ****

(1) Start:

$x_0 = 0$

→ Distribute right-hand side, b, to all processors

$M_{ii} = A_{ii}$ (Diagonal Preconditioner)

→ Distribute Preconditioner M_{ii}

$\tilde{v}_1 = M^{-1}b$

$v_0 = p_0 = p_{-1} = 0$

$\rightarrow \beta_1 = (\tilde{v}_1^T \tilde{v}_1)^{1/2}$

$c_0 = c_{-1} = 1$
$s_0 = s_{-1} = 0$

(2) For $k = 1, 2, \ldots$ do:

$\upsilon_k = \tilde{\upsilon}_k / \beta_k$
$z_k = M^{-1} \upsilon_k$
$\rightarrow y_k = A z_k$
$z_k = M^{-1} z_k$
$\rightarrow \alpha_k = \upsilon_k^T z_k$
$\upsilon_{k+1} = z_k - \alpha_k \upsilon_k - \beta_k \upsilon_{\mu-1}$
$\rightarrow \beta_{k+1} = (\tilde{\upsilon}_{k+1}^T \tilde{\upsilon}_{k+1})^{1/2}$
$\theta_k = \overline{s_{k-2}} \omega_{k-1} \beta_k$
$\eta_k = c_{k-1} c_{k-2} \omega_{k-1} \beta_k + \overline{s_{k-1}} \omega_k \alpha_k$
$\tilde{\zeta} = c_{k-1} \omega_k \alpha_k - s_{k-1} c_{k-2} \omega_{k-1} \beta_k$
$|\zeta_k| = (|\tilde{\zeta}_k|^2 + \omega_{k+1}^2 |\beta_{k+1}|^2)^{1/2}$

$$\zeta_k = \begin{cases} |\zeta_k| \tilde{\zeta}_k / |\tilde{\zeta}_k|, & \text{if } \tilde{\zeta}_k \neq 0 \\ |\zeta_k|, & \text{if } \tilde{\zeta}_k \neq 0 \end{cases}$$

$c_k = \tilde{\zeta}_k / \zeta_k$
$s_k = \omega_{k+1} \beta_{k+1} / \zeta_k$
$p_k = (\upsilon_k - \eta_k p_{k-1} - \theta_k p_{k-2}) / \zeta_k$
$\tau_k = c_k \tilde{\tau}_k$
$\tilde{\tau}_{k+1} = -s_k \tilde{\tau}_k$
$x_k = x_{k-1} + \tau_k p_k$;

Communication is required for matrix–vector multiplication $y_k = Az_k$, dot products $\beta_1 = (\tilde{\upsilon}_1^T \tilde{\upsilon}_1)^{1/2}$ and $\beta_{k+1} = (\tilde{\upsilon}_{k+1}^T \tilde{\upsilon}_{k+1})^{1/2}$, inner product $\alpha_k = \upsilon_k^T z_k$, and summation and distribution of the right-hand vector and the preconditioner. The strategy for each of the above communication steps is briefly described here.

The sparse matrix–vector operation for unstructured matrices is where iterative solvers spend the most computational effort, often as much as 80–90% of the computational requirement of the total analysis code, and its implementation is therefore described in detail. The division of matrix A into submatrices is illustrated in Figure 4. In the submatrices, rows of the matrix that correspond to unknowns at the subdomain interfaces are separate. This interface-overlap region is shown as shaded regions. Each submatrix contains its contributions to that row from finite elements in its subdomain. Figure 4 is used for illustration only since the actual matrix division and overlap for the recursive bisection method is more intricate because of multiple connectivity between domains. The vector z_k is also divided according to its respective subdomain boundaries; however, common vector terms at the interfaces are duplicate between processors.

The matrix terms in the overlapped region, however, are not duplicate—each submatrix contains terms corresponding to finite elements solely in its

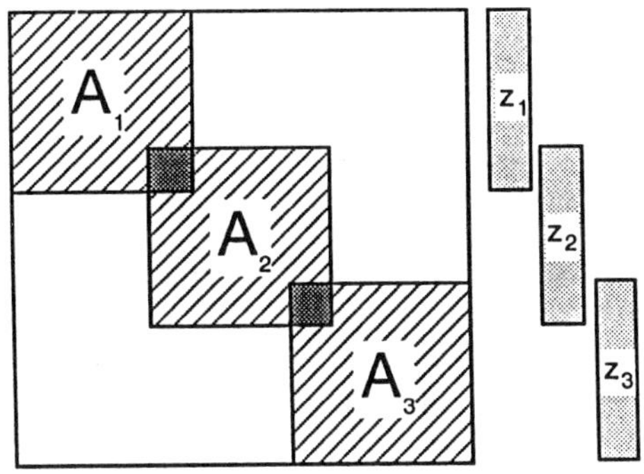

FIGURE 4. Example matrix.

subdomain. Communication between processors is performed when the matrix is used in a matrix–vector product. The distributive law, $(a + b + c + d)x = ax + bx + cx + dx$, is used to combine the resulting vector terms at the boundaries for each row in the overlap section.

This operation, $ax + bx + cx + dx$, requires a proper sequence of multiplication, communication, addition, and redistribution so that each processor obtains the correct values. This sequence is performed as follows. Each node on the interface has with it a binary tree describing its hierarchical division between domains. An example binary tree is shown in Figure 5. This binary tree represents the subdomains to which a particular node is connected. The domains are numbered in binary format; the domains with the larger binary number on the right are called the "masters" of the binary fork and the lower value domains on the left are the "slaves." The matrix–vector multiply sequence starts with each submatrix being multiplied by the vector simultaneously in all processors. Resulting terms on the subdomain interface are then communicated: terms on the slave processors are sent to the masters. This is redundantly performed, level by level, starting from the bottom level of the binary tree to the top. This communication is illustrated in Figure 5. Terms

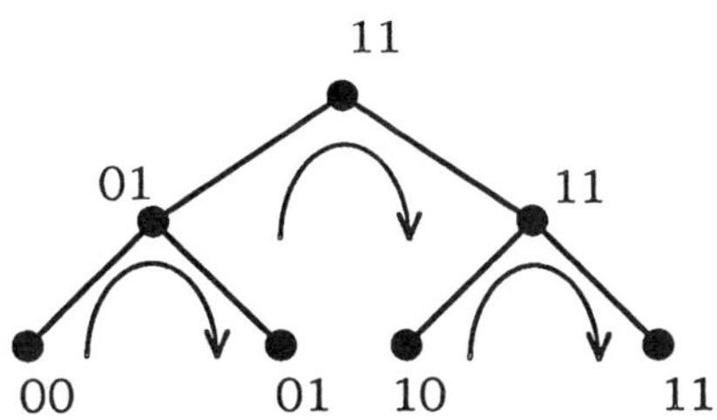

FIGURE 5. Sample binary tree with communication shown from slaves to masters. Master of the binary fork is shown at the fork junction.

from processor 00 are sent to processor 01 and terms from processor 10 are sent to processor 11. This communication is performed simultaneously. Each master processor adds the terms it receives from the slave processor to the corresponding values it has calculated for the interface unknowns. On the next level of the binary tree, communication between processor 01 and processor 11 is performed; again, processor 11 adds the terms from processor 01. When the highest level of the binary tree is reached, redistribution of the terms follows.

For redistribution, as shown in Figure 6, communication is reversed. Communication starts from the highest level of the binary tree to the lowest, and the master processors send information to the slaves. For the first level of communication in this example, processor 11 sends the result, the total sum for the matrix row, to processor 01. On the next level, processor 01 sends the result to processor 00 and processor 11 to processor 10.

Note that each interface node has its own binary tree; the binary trees need not be full, and not all trees have the same number of levels. This binary tree structure is calculated during the recursive bisection procedure and is stored in the database of the subdomain mesh. Also note that the communication is not performed node by node, but in groups of nodes. Multiple interface nodes to be sent from one processor to another are grouped together first and then sent. This grouping avoids an extremely large number of communication calls if a node by node method were used.

The communication for dot and inner products is similar to the strategy of matrix–vector multiplication with an exception: the terms in the arrays corresponding to the interface nodes are duplicate and the strategy is modified to avoid redundant terms. As with the matrix–vector multiplication, the binary tree is used to communicate between domains. The terms in the arrays are duplicate by virtue of prior communication, either by distribution of terms, such as the right-hand side vector, or by the matrix–vector multiplication. Both the dot and inner product operations produce scalar results. This technique used here accumulates this scalar term from the individual processors and, for each communication, subtracts the duplicate term.

As with the matrix–vector multiply sequence, communication starts at the bottom of the binary tree. The resulting scalar term from the dot or inner

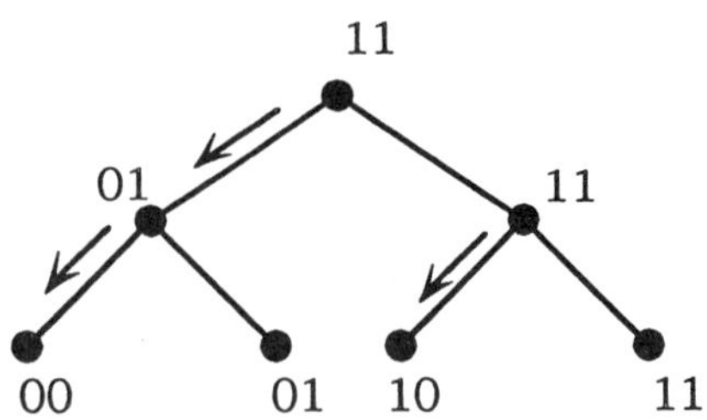

FIGURE 6. Sample binary tree with redistribution communication.

product is performed and the slave sends its result to the master. To avoid redundant terms, however, the slave subtracts the dot or inner product of the interface nodes. This subtraction method is used to avoid an "if" statement in the tight loop of the dot or inner product, thereby increasing the computational rate. The master processor then adds the two resulting scalar terms and the process is repeated. Again, this is performed level by level until the top level of the binary tree is reached. Redistribution then begins and the resulting scalar term of the dot or inner product is sent to all processors.

6. RADAR CROSS-SECTION RESULTS

The first example demonstrating the feasibility of determining the scattering from a large object is a square flat metallic plate, 25 wavelengths per side. This problem consists of a surface area of 1250 square wavelengths (the plate has a finite thickness) and was modeled by 40,302 second-order elements, 201,656 nodes, and 570,316 unknowns. The mesh was terminated one wavelength away by an Engquist–Majda absorbing boundary condition. Computer timings for this problem will be discussed in Section 7. Figure 7 shows the monostatic scattering comparison against a geometric theory of diffraction (GTD) method (Ruck et al., 1970). The results are from an elevation scan from 0° to 60°, VV polarization. The radar cross section was calculated from the finite element solution by the equivalent sources method described in Baucke et al. (1993).

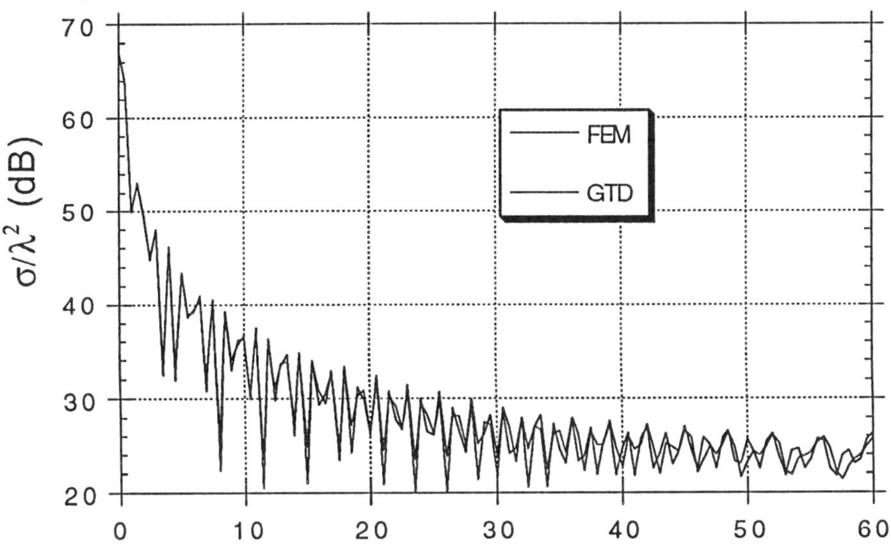

FIGURE 7. FEM and GTD comparison of monostatic scattering from a square metallic plate, 25 wavelengths per side, elevation cut, VV polarization.

7. PARALLEL RESULTS

Results for our parallel QMR matrix solution implementation are shown here. The first set of results is for the square 25 wavelength plate with 570,316 complex unknowns, as discussed in Section 6. The following results are from the Intel Paragon using the NX communication library.

Number of Processors	Computational Time (sec)	Computational Speed (Mflop)
64	431	194
128	288	291

The above timings were obtained with the native communication library. With the "parallel virtual machine" (PVM) (Grant and Skjellum, 1992), a communication package that can be used on parallel machines as well as a cluster of UNIX workstations on a local-area network, the same problem required 965 seconds and obtained a computational rate of 147 Mflops on the Paragon.

The next set of results is for a PEC plate with the dimensions of 3.5 by 2.0 wavelengths. This mesh was overdiscretized with eight second-order elements per wavelength. This resulted in a large mesh of 49,024 elements, 210,466 nodes, and 628,716 complex unknowns. This example was run on the IBM SP2 with the MPI communication library (Gropp et al., 1994).

Number of Processors	Computational Time (sec)	Computational Speed (Mflop)
16	156	475
32	92	800
64	67	1100

The above example had a speed of over 1 gigaflop with 64 processors.

8. CONCLUSIONS

The use of massively parallel computers is the future of computer-aided engineering. Although their use by the computational community is relatively recent and the numerical implementations preliminary, massively parallel, distributed memory computers are already exceeding the computational rates of supercomputers for finite element analysis. As new parallel architectures and methods are developed, larger and more complex problems will be solved. This, in turn, will allow design engineers, the end-users of this technology, to dramatically increase their prediction and optimization capabilities.

The implementation discussed here was the author's first attempt at using parallel computers. Future work will concentrate on the use of combined, or hybrid, iterative and direct solution methods for the solution of very large matrix equations. It is hoped that these hybrid methods will decrease the analysis "through-put" time by requiring less communication between processors and accelerating the convergence rate of the iterative solver.

REFERENCES

Baucke, R. C., D'Angelo, J., and Crabtree, G. D. (1993). Far-field transforms for finite element frequency domain solutions. *IEEE Trans. Antennas Propag.* **AP-41**, 511–514.

Boyse, W. E., Lynch, D. R., Palusen, K. D., and Minerbo, G. N. (1992). Nodal-base finite element modeling of Maxwell's equations. *IEEE Trans. Antennas Propag.* **AP-40**, 642–651.

Crowley, C., Silvester, P. P., and Hurwitz, H. (1988). Covariant projection elements for 3D vector field problems. *IEEE Trans. Magn.* **HAG-24**, 397–400.

D'Angelo, J., and Mayergoyz, I. D. (1991). Three-dimensional RF scattering by the finite element method. *IEEE Trans. Magn.* **MAG-27**, 3827–3832.

Freund, R. W. (1992). Conjugate gradient type methods for linear systems with complex symmetric coefficient matrices. *SIAM, J. Sci. Stat. Comput.* **13**, No. 1.

Grant, J. R., and Skjellum, A. (1992). *The PVM Systems: An In-depth Analysis and Documenting Study: Concise Edition*, Tech. Rep. UCRL-JC-112016. Lawrence Livermore National Laboratory, UCRL, Livermore, CA.

Gropp, W., Lusk, E., and Skjellum, A. (1994). *Using MPI: Portable Parallel Programming with the Message-Passing Interface.* MIT Press, Cambridge, MA.

Lions, P. L. (1988). On the Schwarz alternating method. I. In *Proceedings of the First International Symposium on Domain Demposium Methods for Partial Differential Equations,* (R. Glowinski, G. H. Golub, G. Meurant, and J. Periaux, eds.). SIAM, Philadelphia.

Lynch, D. R., and Paulsen, K. D. (1990). Time domain Integration of Maxwell equations on finite elements. *IEEE Trans. Antennas Propag.* **AP-38**(12), 1933–1942.

Mur, G. (1993). The finite element modeling of three-dimensional electromagnetic fields using edge and nodal elements. *IEEE Trans. Antennas Propag.* **AP-41**, 948–953.

Ruck, G. T., Barrick, D. E., Stuart, W. D., and Krichbaum, C. K. (1970). *Radar Cross Section Handbook.* Plenum, New York.

Saad, Y. (1988). Preconditioning techniques for nonsymmetric and indefinite linear systems. *J. Comput. Appl. Math.* **24**, 89–105.

Simon, H. D. (1991). Partitioning of unstructured problems for parallel processing. *Comput. Syst. Eng.* **2/3**, 135–148.

Stratton, J. A. (1941). *Electromagnetic Theory.* McGraw-Hill, New York.

CHAPTER TEN

Antenna Element and Array Simulation with Commercial Software

ROBERT L. EISENHART

1. INTRODUCTION

While the rest of this book discusses the internal mathematics and manipulation of the finite element modeling algorithms, this chapter looks at what the technique can do from a real life problem-solving point of view, particularly from the aspect of requiring very little knowledge about the internal workings of the software. This is analogous to using a car without understanding what's going on in the engine. The reader is then allowed to concentrate on the problem area of antenna design and simply regard the computer programming as a helpful tool rather than another technical speciality that must be mastered before worthwhile results can be obtained.

The chapter will focus on the use of the commercial finite element software program called High-Frequency Structure Simulator (HFSS), available from Hewlett-Packard. After a brief description of the software, the discussion will address the application to single antenna elements. A technique called waveguide simulation will then be explained for use in modeling large antenna arrays. Some examples will be given for each of the above cases.

Finite Element Software for Microwave Engineering, Edited by Tatsuo Itoh, Giuseppe Pelosi, and Peter P. Silvester.
ISBN 0-471-12636-5

2. HIGH-FREQUENCY STRUCTURE SIMULATOR GENERAL DESCRIPTION

This program can best be described as a computerized automatic network analyzer that characterizes computer modeled hardware. "Structure" in the name refers to the fact that the software representation of the circuit configuration to be analyzed is as identical as possible to the actual three-dimensional, real, physical component. There are no approximations or equivalencies as capacitors or resistors. The simulation refers to the electromagnetic field distribution in that structure under a given excitation and is accomplished by use of the techniques described in the rest of this book. The importance to the user is that the necessary finite element meshing is initially and subsequently adapted automatically until the accuracies prescribed are met. The analysis is first run at a single frequency to establish the mesh but then can be repeated at other frequencies without regenerating the mesh. The primary result of the analysis is the scattering matrix describing the mode voltage relationships at the defined ports. The procedure is similar to working with hardware: it is designed, built (modeled), and tested, resulting in data (scattering parameters) as if it had been measured rather than simulated. The user interface is quickly and easily understood by anyone with microwave circuit design and test experience.

Advantages of this program include arbitrarily shaped structures, unlimited number of ports and modes, and a mixture of dielectrics and conductors and loss characteristics for the materials. Most of these items are used at one time or another in antenna modeling. Using the program is essentially one of drawing the geometry, specifying the material properties and boundary conditions, and then defining the ports. Results of the simulation come in many forms depending on how the information is desired. At the ports the data are presented as S-parameters along with the impedance of the associated transmission lines. In the free-space regions, the data can be seen directly as the electric fields, or interpreted as the far-field radiation patterns. This process is shown in Figure 1 as steps to using the program.

There are many ancillary features to the program, which provide additional information about the component being characterized. These include propagation constants for all modes in the ports, full electric (or magnetic) field description throughout the component, time-varying animated movement of any mode within the component (singly or in combination), and, of particular significance to the array modeling to be discussed later, the ability to separate all the modes in a given port as if they were in independent ports with independent incident and reflected waves, and to provide the resulting S-parameters. Additionally, the program provides fictitious magnetic walls that allow modeling of pieces of circuits, which could never be measured as real hardware. Last but not least is the radiation boundary (or absorbing boundary condition) to simulate the free space around an object. All these features will be used in the following discussions on single elements and array characterization.

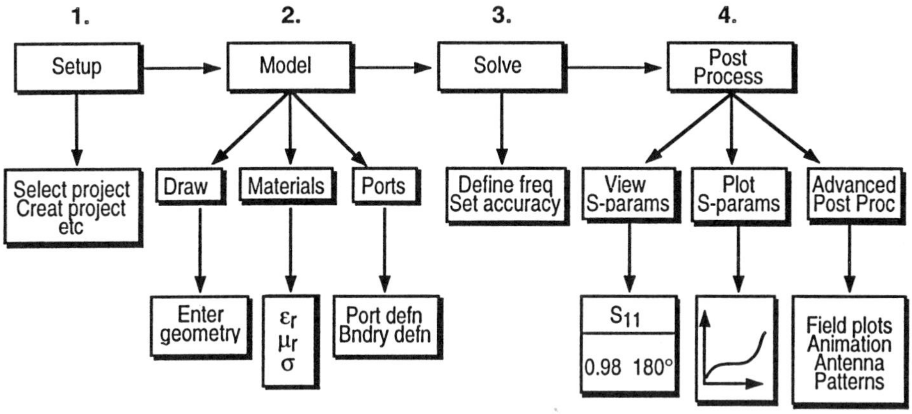

FIGURE 1. Development sequence for setup and use of the High-Frequency Structure Simulator. This procedure follows a logical approach and is easy to use.

3. ANTENNA SINGLE ELEMENT CHARACTERIZATION

The two parts to antenna characterization relate to the two "effective" ports of the element. These are the feed port and the free-space port. The feed port is the least complex and provides the connection to the system. The free-space port represents the radiation environment for the element. Both of these ports will be discussed with the objective of using the High-Frequency Structure Simulator program to develop any desired characteristics.

3.1. Feed Port Description for an Antenna Element

The primary issue at the feed port is the impedance match to the transmission line, which will either deliver or receive the energy from the antenna. There are a lot of choices for the transmission line and it is often determined by mechanical and configurational criteria. Is it low loss, or will it fit? How does the impedance of the line compare generally to the type of antenna element to be used? Many elements will be driven by a coaxial line, which makes the feed very easy to model. Others may use waveguide, microstrip, or stripline. Whatever the approach, the procedure in using the HFSS is similar: build the simulation model as exactly as possible to the real thing rather than approximating anything. The finite element modeling accounts for all of the little variations that would not normally be included in a theoretical analysis. This is why it is possible to get close comparison to measured data quickly. The infinite ground plane or the current distribution no longer need to be assumed. Also, the program accounts for wall thickness of apertures and higher order mode coupling in close quarters in various types of waveguides. In waveguide slots, for example, the internal mutual coupling is accounted for when analyzing the performance.

The impedance of the antenna is the parameter desired at the feed port. As with hardware testing, a useful reference plane for the impedance description will be far enough from any excitation of higher order modes so that they will have no external influence. The characteristic impedance and propagation constant of the feed transmission line will be calculated as part of the simulation. This is particularly helpful when using oddball lines like a ridged waveguide or asymmetrical suspended air-stripline. The reflection coefficient of the feed port is calculated and displayed as S_{11}. This parameter can be interpreted at other positions along the transmission line with the postprocessing features.

3.2. Free-Space Port for an Antenna Element

The free-space side of the antenna is more interesting than the feed port. An absorbing boundary is placed around the antenna to simulate the far-field condition of no reflected energy. The electric field values incident on this surface from the radiating element are subsequently used in the postprocessing to calculate gain and pattern data. This is done by near- to far-field transformations so that the data values as determined on the close-in absorbing boundary can be interpreted as spherical waves at a distance very much greater than the wavelength. The free-space region is usually half or full space depending on whether the antenna is above a ground plane. This boundary condition is part of the modeling for the analysis at the feed port as well as for determining the radiation characteristics.

Therefore a single element is characterized by impedance information on the feed side and field distribution on the radiation side. The following section will explore some of these aspects with a variety of antenna elements.

4. SINGLE ELEMENT EXAMPLES

Three types of elements will be presented to show how the simulation software can be utilized.

4.1. Conical Horn

A very simple low-gain horn is used to demonstrate the ease of designing with this program. A horn was needed for an application, which required a match less than 0.1 reflection but with a gain of at least 10 dB at 975 MHz. It was to be inexpensive and easy to build. The simplest approach was to put a cone directly on the end of a square waveguide and see how bad the match would be. The 20-inch diameter configuration shown in Figure 2 was simulated. There was some concern over the abrupt transition to the 7.7-inch square waveguide but it didn't create any problem. The reflection was less than 0.075 and the patterns were smooth down to -10 dB from the peak on the main lobe. The design task was completed without ever cutting any metal. Figure 2 also shows

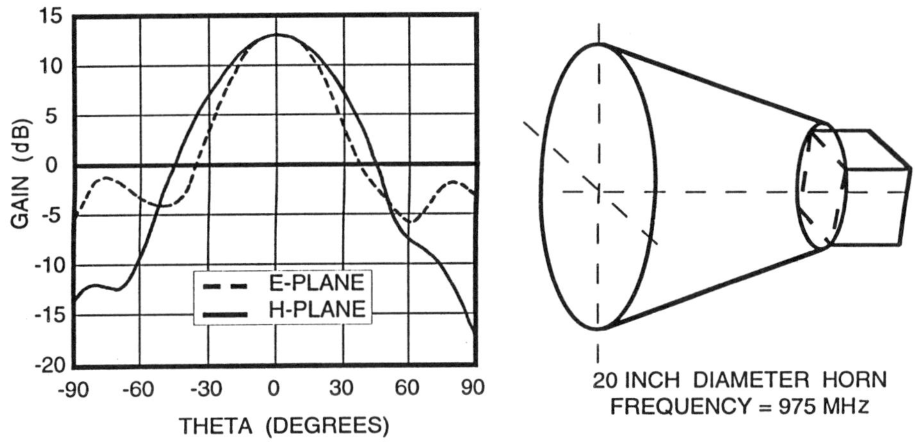

FIGURE 2. Configuration of the horn and the patterns predicted. The patterns matched well to those measured from the hardware.

the predicted *E*- and *H*-plane patterns, which matched closely to those measured later. The plot in the rectangular format of the pattern is available directly from the program.

4.2. Monopole Above a Ground Plane

A whip antenna over a ground plane, driven by a coax line, is used as the second example. The coax feed is taken out through the ground and an absorbing box is placed around the antenna for the simulation. The predicted theta pattern and input match are shown in Figure 3. The line impedance was

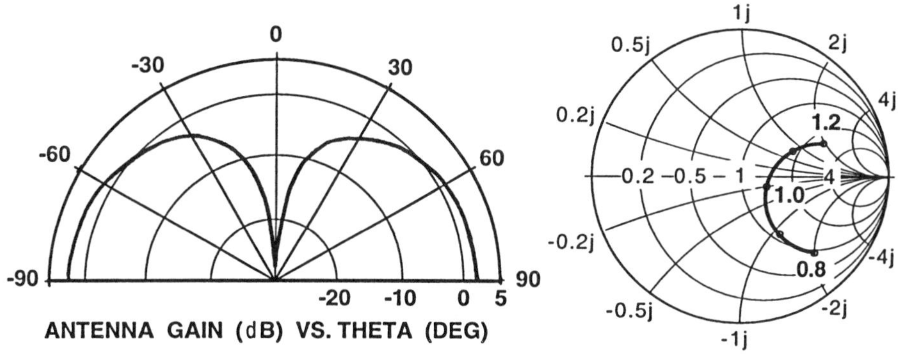

FIGURE 3. *E*-plane pattern and input match for the monopole above a ground plane. The Smith chart is normalized to 50 ohms.

50 ohms. Both the polar plot of the pattern and the Smith chart showing the input match come directly from the postprocessing of the program.

4.3. Waveguide Slot

The example for the radiating slot was taken directly from Elliott (1981) for an offset longitudinal shunt slot. The design goal was a 0.3 normalized slot conductance at a resonant frequency of 9375 MHz. The slot offset from centerline was 0.152 inch with a rounded slot length of 0.625 inch. The slot width was 0.0625 inch cut through a wall of 0.050 inch, driven by standard x-band 0.9 × 0.4 inch waveguide. The model has the slot in shunt with the through port of the waveguide. The configuration is shown in Figure 4. The dashed lines in the figure represent the absorbing boundary walls placed outside the radiating slot. Excellent correlation between the predicted perform-

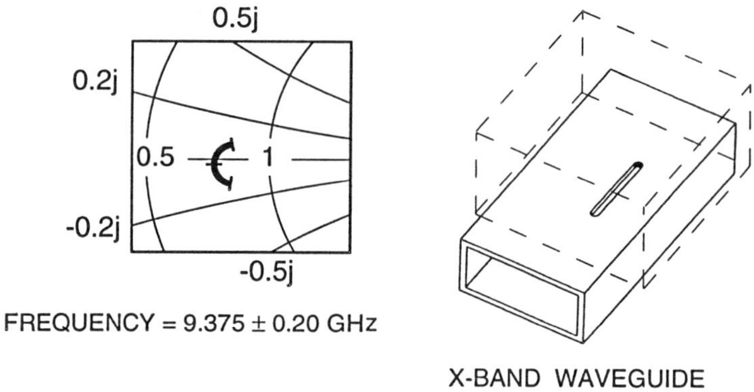

FIGURE 4. Longitudinal slot configuration and the simulated normalized impedance for a shunt slot. The dashed lines indicate the absorbing region placed above the slot to simulate free space.

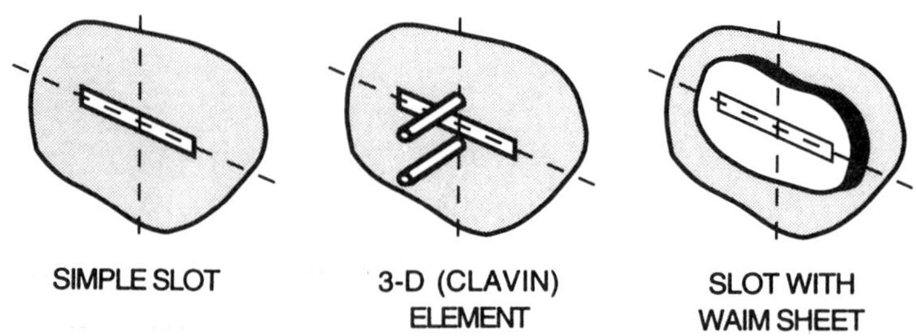

FIGURE 5. Different slot configurations that can be analyzed as single ele-ments or in large arrays. Three-dimensional (3-D) elements pose no problem for the software.

ance and that in Elliott was noted. Figure 4 also shows a Smith chart cutout with the predicted normalized input impedance.

Figure 5 shows two modifications to a radiating slot, which can easily be incorporated into any model using this program. These changes, however, are very difficult to include in more standard theoretical analyses.

5. ARRAY CHARACTERIZATION USING A WAVEGUIDE SIMULATOR

The technique of "waveguide simulation" for planar array characterization was introduced over 30 years ago by Brown and Carberry (1963) and Hannan et al. (1963). It received much attention in the following years as evidenced by Balfour (1965), Hannan and Balfour (1965), and then a little later by Wheeler (1972), Diamond and Knittel (1972), and Gustincic (1972). Along with building small arrays, the simulation approach provides an opportunity to verify the design of a large array at lesser expense than testing a full-up array. However, it requires the use of relatively complex hardware for the generation of each data point (scan angle), followed by careful data interpretation. In addition, broadside and the *E*-plane scan cannot be simulated. For array design, time and money can be used up quickly if any number of design cycles are required.

Nothing much changed through the years for this simulation approach until computers got big enough and fast enough to be applied. Computer simulations are being applied to most problems these days and can work on the waveguide simulated array as well, as discussed by Eisenhart and Park (1995) and Sanford et al. (1995). The computer modeling approach also provides much additional information and insight that are not available through the hardware measurement.

6. HOW THE WAVEGUIDE SIMULATOR WORKS

The following discussion is only to give a conceptual description of how the waveguide simulator works so that the reader can appreciate the application of the computer modeling and the unique features it offers. Thorough general development with all of the details is available in the references.

The original operation of the waveguide simulation technique is based on the imaging of all the elements of an infinite array into a single waveguide "radiation" cell with the metal (electric) walls of a rectangular waveguide. Picture yourself inside a rectangular box with all mirrored walls. You would see a doubly infinite array of images of yourself as you view in all directions, the result of multiple reflections. This simulation works for large arrays and assumes uniform excitation. The imaging accounts for all mutual coupling effects. Only certain scan angle/frequency conditions can be simulated and the maximum scan value is simulated with the fundamental mode of a single cell simulator. Complex higher order modes must be generated for intermediate scan values by using multicell simulators. A general simulator configuration is shown in Figure 6, using open-ended waveguides as the radiating elements. The simulator size is 3×2, which supports many modes and therefore can be used to characterize many scan angles. Hardware testing requires the construction of special higher order mode transitions for each of the scan angles desired, to be tested one mode at a time. Computer simulation, however, allows simultaneous multiple scan values since it can deal independently with the higher modes in a single simulation.

Another advantage of the computer simultation is the ability to use magnetic walls for the radiation cell. Figure 7 shows how many new combinations of scanning modes can be represented by simply changing the boundary conditions of the walls of the unit radiation cell. The lower right-hand corner in the figure shows the all-electric wall configuration, the only one that can be used for hardware testing. The other eight configurations are additionally available for computer simulation, shown as the unit cells in Figure 7. Note particularly that the broadside case ($\theta = 0°$) and the whole E-plane ($\phi = 90°$) can also be simulated. The 4×4 vector sets shown in Figure 7 for each mode configuration show the phasing of the array elements to excite the respective modes. (The radiation cells in this case are unit cells and therefore contain only a single element.)

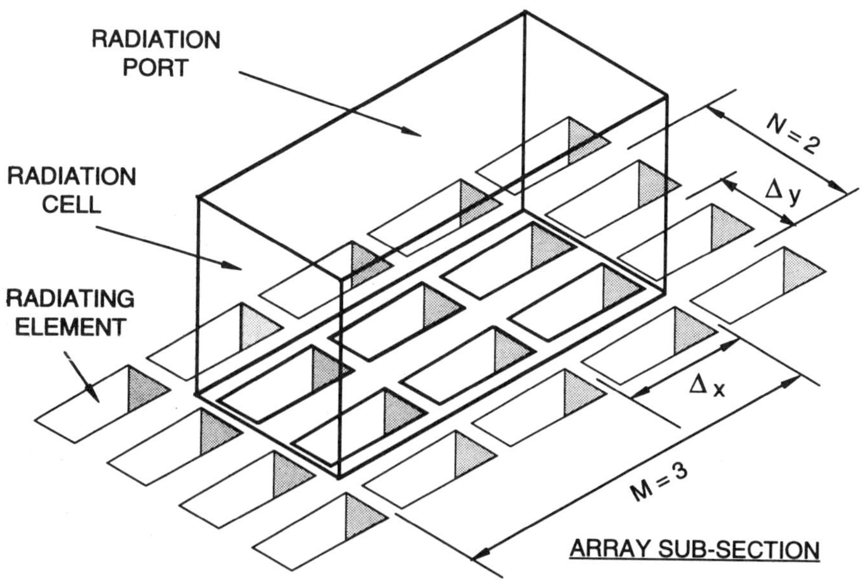

FIGURE 6. Waveguide simulation multicell, which will allow simulation of many different scan angles. Although a rectangular lattice is shown, the procedure can also be adapted to triangular lattices.

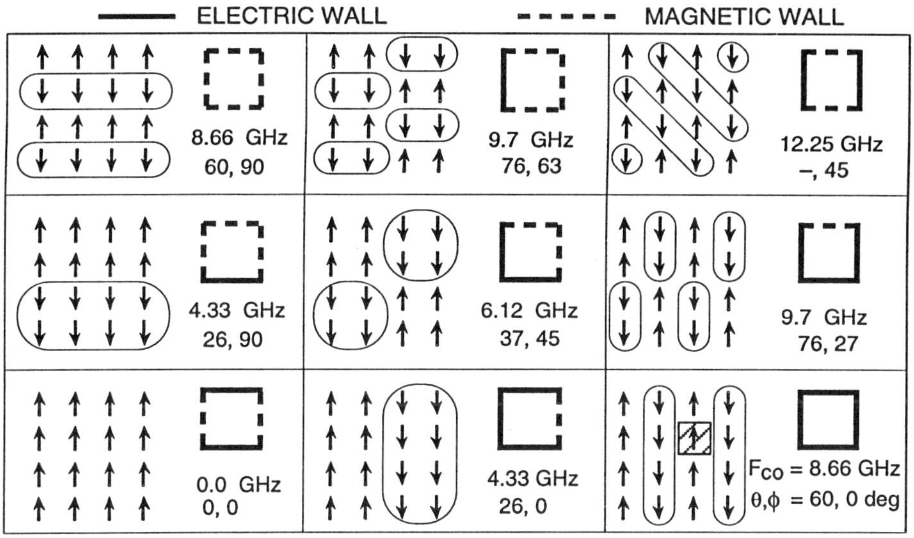

FIGURE 7. Radiation cell configurations for array element imaging. The frequencies indicate the cutoff for each mode and the angles are the scan from broadside and from the *H*-plane. Values are at 10 GHz.

The only difference between the waveguide simulation and the mirrored box is that for the simulation the box is the same order of size as wavelength, and phase/position become important. There is then a relationship between the wavelength (λ_0), box dimensions ($M\Delta_x, N\Delta_y$) radiating element spacing (Δ_x, Δ_y), and the scanning angle (θ_{mn}), which can be represented as

$$\theta_{mn} = \sin^{-1}\left[\left(\frac{m\lambda_0}{2M\Delta_x}\right)^2 + \left(\frac{n\lambda_0}{2N\Delta_y}\right)^2\right]^{1/2} \tag{1}$$

where m and n represent different mode indices. This relationship is for the all-metal case and simply describes the angle of propagation of a pair of waves (or four when off the principal axes) with respect to the waveguide axis in a rectangular waveguide. A real solution cannot exist for $m = n = 0$ in this case with metal walls. As a consequence, there is always some finite scan angle in the *H*-plane since $m \geqslant 1$. Larger radiation cells where $M \gg 1$ can reduce this angle but never eliminate it. Since computer modeling can also utilize magnetic walls as well as electric, cells can be defined in which the TE_{0n} modes propagate, including the TE_{00}, which is another way of writing the TEM mode. This explains the *E*-plane scanning discussed above and produces scanning at additional angles to the electric-wall-only simulations.

7. DESIGN PROCEDURE FOR A TYPICAL ARRAY

Most of these concepts should become clearer by going through an example design. H-plane scanning performance will be used as the goal in the following discussion. The first step is the definition of the array lattice dimensions, from which the radiation cell is established. These dimensions are usually set by system performance requirements relating the frequency and the scanning angles. Choosing a good lattice design would not serve the purpose of this discussion since the objective is to show how the computer simulation can point out problems with a design as well as good performance. Therefore the simplest of configurations will be used as the basis of the following discussion. Consider the rectangular lattice of Figure 6 with $\Delta_x = 1$ inch, $\Delta_y = 0.2$ inch, and frequency at 8 GHz ($\lambda_0 = 1.475$ inch). This would not be a particularly good scanning array lattice because the lattice spacing is large with respect to lambda so that grating lobes come in quickly with scanning. (The y-axis spacing is not relevant since only the H-plane scanning will be considered.) *This is precisely why this lattice has been chosen so that interpretation of these undesirable effects can be demonstrated through the computer simulation.* (An array with a triangular lattice can also be modeled and is shown as an example later in Section 11.)

The second step is to use the unit cell to determine the element match to free-space for the zero-degree scan. Walls are placed at the symmetry planes, which are easily located for a rectangular lattice. The zero-degree scan condition requires the TEM mode in the radiation cell, which is established with a cell wall configuration of top and bottom electric and side walls magnetic. This is best done using the unit cell as shown in Figure 8. The feed

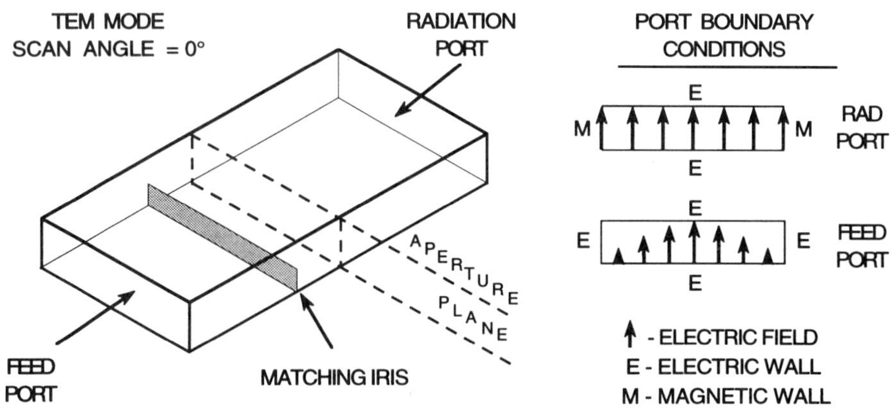

FIGURE 8. Unit cell simulation to match the element at zero-degree scan. A capacitive iris is used for this case.

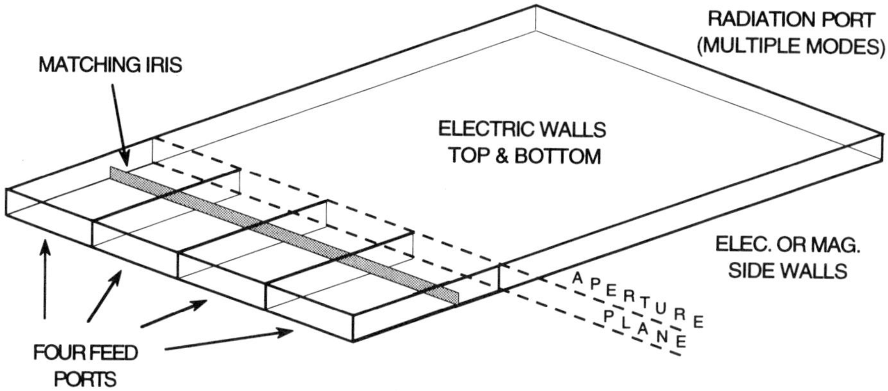

FIGURE 9. The full-size model uses four elements into the radiating cell. Two wall conditions are simulated to create 11 scanning angles. The same matching iris from the unit cell simulation is used to provide a match at broadside.

port waveguide goes directly into the radiation port through the aperture plane. At this plane, however, the sides of the model are changed from the electric walls of the feed guide to magnetic walls of the radiation guide. The field profiles change accordingly as shown and the scan angle $=0°$ for the TEM radiation mode. Initial simulation of the input match at the feed port results in a small mismatch that is slightly inductive at the aperture plane. A capacitive iris is then properly placed to create a good (mag $S_{11} = 0.0058$) match. Now any scanning pattern will be maximum at broadside. Higher order mode coupling between the iris and the aperture is accounted for automatically.

The third step is to model higher order mode cells in order to determine scanning performance off broadside, in this case in the H-plane. Various size cells could be used to simulate different scan angles: however, the computer program will handle multiple higher order modes simultaneously so that a larger multi-element cell can be used to produce many scanning results in one run. A 4×1 radiation cell ($M = 4$) will be used. This size will support 11 propagating modes, six with one boundary condition and five with another, modeled by two simulation runs. This will also demonstrate both single scanning modes and multiple or coupled scanning modes, another way of describing the grating lobe condition. The model is shown in Figure 9. Note that this configuration is just a multiple of the single element model with the expanded radiation cell. The electric or magnetic wall conditions are indicated on the figure. The only change for the second simulation is to make the right-hand side wall electric rather than magnetic. All modes have $n = 0$ [see Eq. (1)], since there is no variation in the y-direction.

Figure 10 shows the electric field profiles of all the propagating modes for this model, using the magnetic walls on the sides of the radiation cell. These

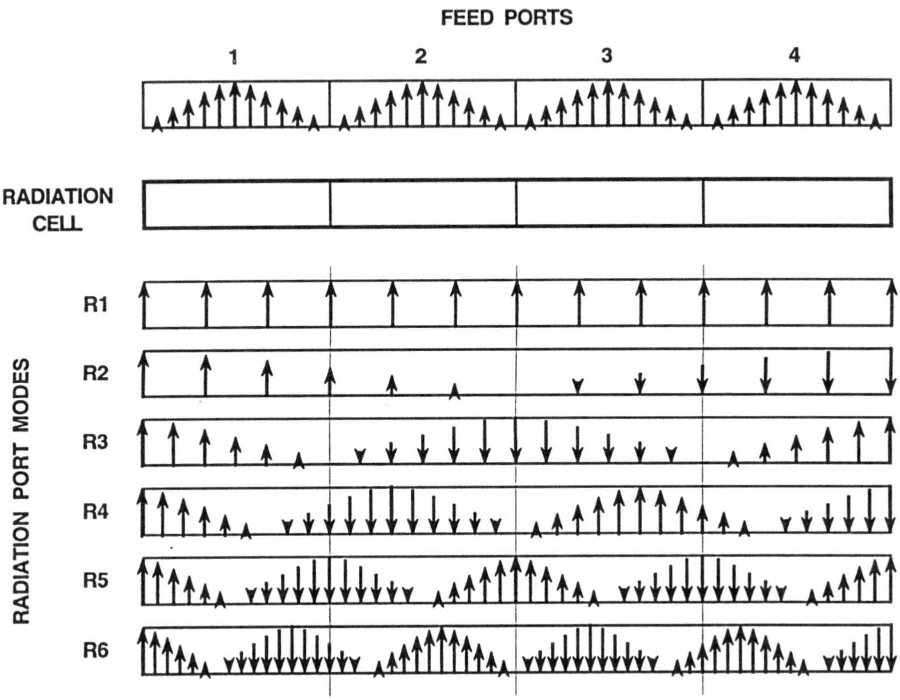

FIGURE 10. Mode electric field profiles for the six modes in the radiation cell. These modes exist for the case with the magnetic side walls.

six modes correspond to $m = 0$ through $m = 5$, integer values. These profiles can provide insight into the coupling between the feed ports and the individual modes in the radiating port.

8. MODE RELATIONSHIPS

Prior to considering the results of the waveguide simulation, it would be instructional to review some of the underlying mathematics. Equation (1) simplifies to

$$\sin \theta_{mn} = \left(\frac{m\lambda_0}{2M\Delta_x} \right) \tag{2}$$

or

$$\sin \theta = 0.1844m \tag{2a}$$

for $n = 0$, $M = 4$, $\Delta_x = 1$ inch, and $\lambda_0 = 1.475$ inch. Since $|\sin\theta| \leqslant 1$, only values of $m \leqslant 11$ are valid and represent propagating modes. The scanning angles that can be simulated with this size cell are easily determined and are as follows:

Mode	m	θ	Mode	m	θ	Mode	m	θ
1	0	0	5	2	21.6	9	4	47.5
2	0.5	5.3	6	2.5	27.5	10	4.5	56.1
3	1	10.6	7	3	33.6	11	5	67.2
4	1.5	16.1	8	3.5	40.2			

Half values for m are also included because they represent the modes with one side wall electric and one magnetic, and $m = 0$ is the TEM mode, where both side walls are magnetic.

The second condition of interest is the onset of multiple or grating lobes (modes in this case) in the radiation cell. This happens when more than one lobe (mode) is created by the same feed guide excitation and can occur when $2\Delta_x > \lambda_0$. Consider the basic scan relationship shown in Figure 11. The effective phase centers of the radiating waveguides are shown in the center of the guides. The distance d is related to the scan angle as

$$\sin\theta = \frac{d}{\Delta_x} \tag{3}$$

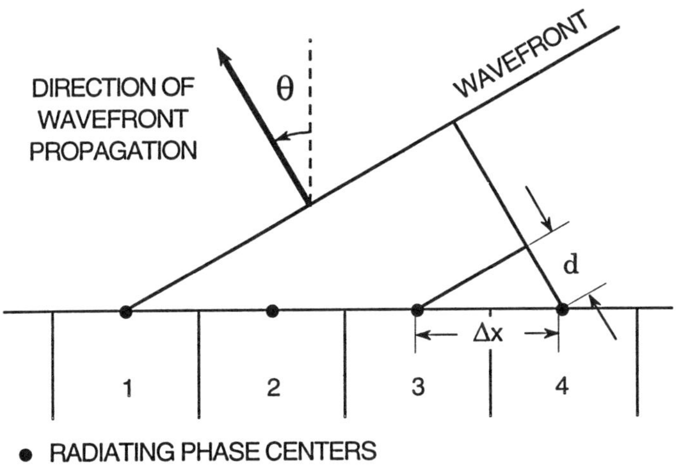

FIGURE 11. This figure shows the relationship between the scan angle and the equivalent phase length at the element phase centers, which is necessary to create the scan.

For a wavefront to propagate in the direction shown, the distance d also represents the phase delay at guide 3 relative to 4, which in radians is

$$\varphi = \frac{2\pi d}{\lambda_0} \pm 2\pi i \tag{4}$$

which can be satisfied by more than one d. For example, if

$$\varphi = \frac{2\pi d_1}{\lambda_0} + 2\pi = \frac{2\pi d_2}{\lambda_0}, \quad \text{then } \lambda_0 = d_1 - d_2 \tag{5}$$

Substituting for both d's and understanding that θ_2 is on the opposite side of the axis as θ_1, then

$$\frac{\lambda_0}{\Delta_x} = \sin\theta_1 + \sin\theta_2 \tag{6}$$

The threshold for the second mode to occur is when $\theta_2 = 90°$. Substituting in Eq. (6) and soving for m, we find

$$m \leqslant 2M(1 - \Delta_x/\lambda_0) = 2.576 \tag{7}$$

Therefore the first six modes, or all modes with $m \leqslant 2.5$, are single mode cases. For $m \geqslant 3$, there will be multiple modes in the radiation cell and they will couple to each other. Consider further this coupling relationship. Substituting for both angles in Eq. (6) with Eq. (2) and simplifying gives

$$m_1 + m_2 = 2M \tag{8}$$

This shows which modes will coexist: that is, for m's greater than 2.5 and less than 5.5 there will be a corresponding value for a second mode. This will be pointed out in the simulation data.

9. DATA INTERPRETATION

The *Active Element Pattern* represents the element pattern, as modified by the mutual coupling effects, to be superimposed on the array pattern to determine the total effective radiating pattern over the scan volume. It can be interpreted as the effective area of the element and is calculated as

$$A_e \propto [1 - \Gamma(\theta)^2] \cos \theta \quad (9)$$

where $\cos \theta$ is the projected area term and $\Gamma(\theta)$ is the array reflection coefficient as determined from the simulation. As previously mentioned, this function can only be determined for discrete values of θ from a given simulation. However, many configurations can be analyzed, which will then provide sufficient points to understand the behavior with scan. The Active Element Pattern is typically determined experimentally, particularly for complex three-dimensional elements.

Determination of the Active Element Pattern is the main objective of this analysis. From Eq. (9) it is seen that $\Gamma(\theta)$ is the unknown, and this is precisely what is produced by the waveguide simulator. How is it specifically generated? Since there are six modes on the radiation port, this port behaves like six independent ports. Adding the four feed ports then makes a model with effectively ten ports. The simulation S-parameter matrix will therefore be a 10×10, showing the interrelationship between all these modes. This scattering matrix in Figure 12 is essentially shown as it is provided by the simulation program. This data set is for the same wall configuration that is represented by the field profiles in Figure 10. An additional data line showing the propagation constant (radians/meter) is included at the bottom of the chart for reference and is also provided by the program. By reciprocity, the match of the array for all these modes when used as a transmitter is the same as when receiving. Therefore, if the S_{ii} for an incident mode in the radiation cell is known, this will be the match in the feed ports for the phasing to generate that same mode at the associated scan angle. All that is needed then is the reflection coefficients for these higher order modes in the radiation port.

There is much information to be derived from Figure 12, most of which would not be available from waveguide simulation with hardware. First, consider the values desired for the Active Element Pattern. R1, R2, and R3 are the three propagating modes for this configuration. Note that S_{R1R1} is the same as when a single cell was used for matching for $\theta = 0°$. The second and third scan angles also show excellent match. Note particularly that there is no coupling between these modes and other modes in the radiation port. (The small numbers are all numerical noise, below 40 B.) Next consider column R4. There are two significant values in the radiation port modes. The first, S_{R4R4}, represents the reflection of the array to the incoming wave. The second value, S_{R6R4}, represents the scattering into the companion mode as related by Eq. (8). Note then that S_{R4R6} is also the same cross-coupling value as expected.

The value for S_{R5R5} is a special case since the two modes of the coupled mode pair occur at the same angle. Therefore it is reasonable to expect something irregular, as this is sort of a singularity. The total reflection indicated is due to the alignment of the peak of the field profile (as seen in Figure 10, mode R5) with the metal walls of the feed guides at the aperture

	RADIATION PORT MODES									
	R1	R2	R3	R4	R5	R6	F1	F2	F3	F4
R1	0.0058	0.0013	0.0006	0.0017	0.0010	0.0001	-0.4999	-0.5001	-0.5000	-0.4999
R2	0.0013	0.0281	0.0002	0.0004	0.0014	0.0003	0.6530	0.2706	-0.2705	-0.6530
R3	0.0006	0.0002	0.1085	0.0012	0.0004	0.0023	0.4973	-0.4975	-0.4965	0.4969
R4	0.0017	0.0004	0.0012	0.3413	0.0008	0.4166	0.2277	-0.5507	0.5503	-0.2282
R5	0.0010	0.0014	0.0004	0.0008	1.0000	0.0005	-0.0009	-0.0016	0.0005	0.0008
R6	0.0001	0.0003	0.0023	0.4166	0.0005	0.7600	-0.1352	0.3255	-0.3261	0.1352
F1	-0.4999	0.6530	0.4973	0.2277	-0.0009	-0.1352	0.0517	0.0530	-0.0257	0.0103
F2	-0.5001	0.2706	-0.4973	-0.5507	-0.0016	0.3255	0.0530	-0.1080	0.0693	-0.0257
F3	-0.5000	-0.2706	-0.4965	0.5503	0.0005	-0.3261	-0.0257	0.0693	0.1133	0.0531
F4	-0.4999	-0.6530	0.4969	-0.2282	0.0008	0.1352	0.0103	-0.0257	0.0531	0.0533
Γ	167.67	164.79	155.85	139.67	113.20	64.88	113.20	113.20	113.20	113.20

FIGURE 12. Scattering matrix for the six-mode/four-feed simulation using the magnetic side walls. The highlighted values are the reflection coefficients for the scanning modes in the radiating cell. The data shown for the feed guide are not directly needed to get the Active Element Pattern but it is interesting to see the coupling values and relate them to the profiles of Figure 10.

plane. If electric walls had been used in the simulation of the radiation cell, then the mode profile would be shifted sideways by 90° and would have lined up perfectly with the feed guides, resulting in a very low reflection. This unique case will be discussed more in Section 10.

Consider next the coupling between the radiation port modes and the feed port modes. The magnitude of the coupling coefficient is proportional to the integral of the product of the mode electric field profiles at the aperture plane. In simpler terms, the profiles that match up well will have a higher coupling value than the ones that do not, and the sign is determined by the up or down nature for that particular port. Considering the column for R1, the values for S_{F1R1} through S_{F4R1} are identical. This is also to be expected since the profile shown in Figure 10, mode R1, is uniform cross the aperture plane. Looking at any of the other combinations results in the same correlation between the field profiles and the coupling factors. One last example would be for R4, where the appearance of the field profile is roughly +low, −high, +high, and −low in a relative sense. Comparing these observations to the values in Figure 12, column R4 shows good correlation as well.

The importance of this visual correlation goes way beyond the confirmation of the mode-matching mathematics. All these data are provided by the simulation program and are extremely useful in developing an understanding of what is actually happening within whatever component is being analyzed. In particular, when there is a problem that needs to be figured out, using the field patterns as a diagnostic tool is invaluable.

It might be expected by some readers that the scanning match could be determined by consideration of only the S-parameters at the feed ports. This is indeed true; however, it is much more complicated and requires knowledge of the phase portion of the S-parameters as well (which I have left out because of space limitations). The method takes into account the mutual coupling terms between the feed ports and the necessary excitation at these ports to create each particular mode. This would be viewed as a "looking out" versus "looking in" approach.

If the active element scanning performance is all that is desired, it can be determined from the radiation port modes alone (strictly looking in). For this case, the model can be simplified by removing the feed port boundary conditions and substituting waveguide terminations at the ports. The S-matrix will now only be 6×6, requiring less calculation time and memory for the computer. This technique is used in the triangular array example discussed in Section 11.1.

CAUTION: It is absolutely necessary to include all propagating modes in the port analyses or the answers will be in error. However, if you also include unnecessary evanescent modes it will slow down the computations and add nothing to the data. The easiest way to ensure the right number is to run a "ports only" analysis with a large number of modes. This is a quick computation that gives the propagation constants for all the included modes. A simple review of the answers tells which are propagating and which can be discarded.

10. SIMULATION RESULTS

The goal was to determine the Active Element Pattern from the simulation. Figure 13 consolidates the relevant data from three simulations. It is presented to show how the data dovetail together to present a more complete picture and also to show some redundancy in the data. The configuration that supports the most modes has the two magnetic walls for the radiation cell because it includes the TEM mode. The complementary configuration is the magnetic/electric wall case, which supports the half-integer m values. The electric/electric wall case is totally redundant to the first case and simply corroborates that data. The electric/electric case does show one interesting data point at $m = 4$. This was the total reflection case for the magnetic/magnetic where the peaks of the scanning mode lined up with the walls between the feeds. Now the peaks line up perfectly with the center of the feed waveguides and the coupling is very strong. In fact, since the mode field profiles match so well, the slight reflection is due to the matching capacitive iris, which is not needed for this special case.

	m	Γ	θ	mm	me	ee	RhoMag	AEP (dB)
SINGLE MODES	0	167.67	0	0.0058			0.0058	0
	0.5	166.95	5.3		0.0098		0.0098	0.019
	1	164.79	10.6	0.0281		0.0277	0.0281	0.075
	1.5	161.12	16.1		0.0592		0.0592	0.189
	2	155.85	21.6	0.1085		0.1083	0.1085	0.368
	2.5	148.78	27.5		0.2021		0.2021	0.702
MULTIPLE MODES	3	139.67	33.6	0.3413		0.3409		
	*			0.4166		0.4162		
	3.5	128.06	40.2		0.4321			
	*				0.4748			
	4	113.2	47.5	1		0.1985		
	4.5	93.55	56.1		0.634			
	*				0.4748			
	5	64.88	67.2	0.76		0.7606		
	*			0.4166		0.4162		

mm – MAGNETIC / MAGNETIC WALLS
me – MAGNETIC / ELECTRIC WALLS
ee – ELECTRIC / ELECTRIC WALLS

* – COUPLED MODES
AEP – ACTIVE ELEMENT PATTERN

FIGURE 13. Results consolidation for three simulations showing the H-plane Active Element Pattern data out to 27.5° scan. Coupled mode data representing the grating lobe occurrence are also shown out to 67.2°.

The scanning angle of the mode can be determined by Eq. (2a) as previously shown. It is easily confirmed from the Γ value provided by the program as

$$\theta_i = \cos^{-1}\left(\frac{\Gamma_i}{\Gamma_0}\right) \tag{10}$$

where Γ_0 is for the 0° scan case.

The resulting scan data for the Active Element Pattern for this case is not very interesting in itself because it is well behaved over the relative small scan range applicable. However, many interesting aspects of the problem were discussed during the process definition. A more practical array with realistic problems focuses more on the performance in the example in Section 11.

Once this waveguide simulation technique is fully understood, it allows the flexibility to investigate a wide variety of array elements and lattice configurations without expensive hardware fabrication. Wide angle impedance matching (WAIM) techniques can also be considered simply by altering the model accordingly. Particularly significant is the ability to model complex three-dimensional radiating elements, which are not tractable with other theoretical approaches. This technique is basically independent of the type of feed line, radiating element, and other tuning in the immediate vicinity of the array elements.

11. TRIANGULAR ARRAY EXAMPLE

The objective of this example is to demonstrate the power of the computer approach, by predicting blindzones in simulation data as compared to some approximate theoretical results from Stark (1974). The performance of the array configuration considered is extremely sensitive to the waveguide width of the radiating elements, providing an excellent vehicle for the demonstration.

11.1. Configuration Definition

This discussion focuses on the triangular array shown in Figure 14. The imaging is a little more complex than for a rectangular array, but still possible, as discussed in Balfour (1965). The dimensions of the array were borrowed from Stark so that performance comparison could be made directly to his predictions. The array lattice spacing is pretty typical for this type of array. Note the dashed lines in front of the array face. These outline the radiation cell or simulating waveguide that contains the effect of all the array element images. The largest cell used is shown and it provided eight scan points in the H-plane. Smaller radiating cells (in width) were also used to provide additional scan angles. Three arrays were simulated, with the waveguide widths $W = 0.858$ in.,

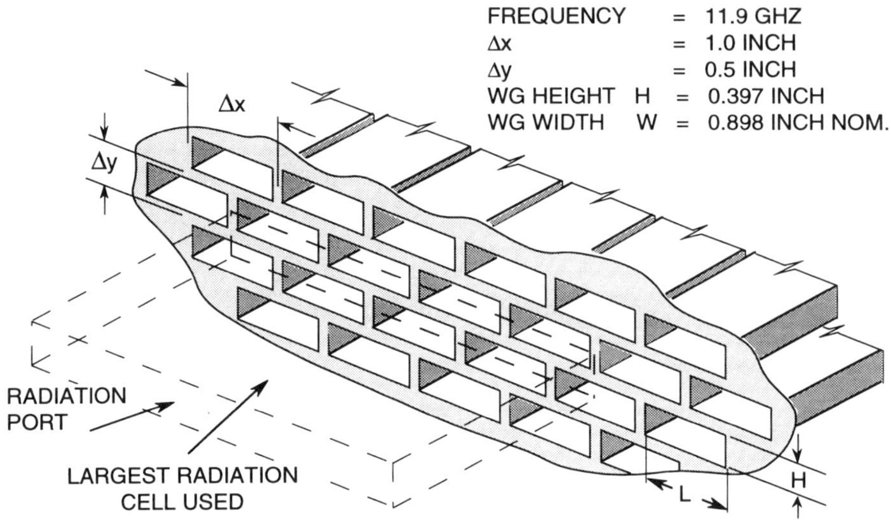

FIGURE 14. Triangular array under consideration. The dimensions were borrowed from the literature so that direct comparisons to results could be made.

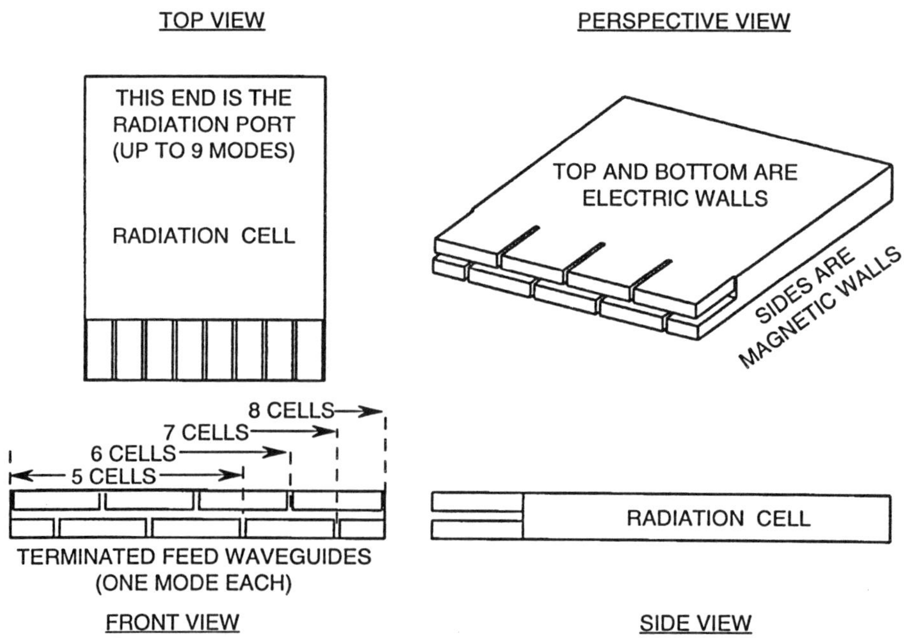

FIGURE 15. A 5-to-8 cell configuration for H-plane scanning. Changing the radiation cell size produces a wide variety of modes with their associated scan angles.

0.898 in., and 0.939 in. and the other dimensions fixed. What we looked for was the "blindzone" or high reflection in a scanning region of the H-plane scan and how it shifted with change in W. This "blindzone" shows up in the Active Element Pattern when scan phasing is such that the mutual coupling constructively provides a high reflection excitation at the element being analyzed. As is necessary for this approach to work, all mutual coupling effects are accounted for through imaging.

In this example a different approach is used in order to simulate a large number (22) of closely spaced scan points. Four different multi-element simulator cells, all with the magnetic/magnetic side wall boundaries, are used. The model is shown in Figure 15. Note that since the solution matrices would be quite large, the feed ports have been terminated with effective loads as suggested in Section 8.

11.2. Triangular Array Data

Simulation data for the seven-cell case with $W_2 = 0.898$ inch is shown in Figure 16. This S-matrix is representative data for the four simulations and different widths and shows only the modes of the radiation cell because of the terminated feed ports. The mode reflection coefficients are highlighted along the diagonal. This example was not matched at 0° scan and also shows a wide range of reflection values for these modes.

The relevant data were collected from each of the four simulations and consolidated into Figure 17. It is easily seen how the results from the different cases complement each other to create a very compete data set, showing scan performance out to 60°. The Active Element Pattern is then calculated and shown in the right-hand column.

Next, the data for each of the different widths is plotted in Figure 18. This shows the accumulated results of multiple simulations (four cells per curve) for

(8 iterations)	Port 1 (Radiation Port)						
	Mode 1	Mode 2	Mode 3	Mode 4	Mode 5	Mode 6	Mode 7
Port 1 Mode 1	0.2200	0.0152	0.0031	0.0053	0.0218	0.0052	0.0086
Port 1 Mode 2	0.0152	0.3116	0.0187	0.0058	0.0034	0.0062	0.0081
Port 1 Mode 3	0.0031	0.0187	0.6551	0.0187	0.0149	0.0095	0.0303
Port 1 Mode 4	0.0053	0.0058	0.0187	0.9739	0.0293	0.0340	0.0159
Port 1 Mode 5	0.0218	0.0034	0.0149	0.0293	0.9764	0.0136	0.0386
Port 1 Mode 6	0.0052	0.0062	0.0095	0.0340	0.0136	0.9518	0.0233
Port 1 Mode 7	0.0086	0.0081	0.0303	0.0159	0.0386	0.0233	0.9386
Gamma (Re)	0.00	0.00	0.00	0.00	0.00	0.00	0.00
Gamma (Im)	249.4	246.9	239.2	225.7	205.5	176.0	131.1

FIGURE 16. S-parameter simulation results for the seven-cell configuration.

8 cell	7 cell	6 cell	5 cell	Scan Angle		Act Ele Pat (dB)
Scan Angle	Scan Angle	Scan Angle	Scan Angle	S-ang	Rho Mag	(1-Rho^2)*cos(S-ang)
0	0	0	0	0	0.22	-0.215
7.1				7.1	0.289	-0.412
	8.1			8.1	0.312	-0.488
		9.5		9.5	0.35	-0.628
			11.4	11.4	0.302	-0.502
14.4				14.4	0.545	-1.669
	16.4			16.4	0.655	-2.614
		19.3		19.3	0.801	-4.708
21.8				21.8	0.902	-7.618
			23.4	23.4	0.914	-8.208
	25.2			25.2	0.974	-13.331
29.7		29.7		29.7	0.994	-19.833
	34.5			34.5	0.976	-14.080
			36.5	36.5	0.978	-14.562
38.3				38.3	0.964	-12.558
		41.4		41.4	0.957	-11.998
	45.1			45.1	0.952	-11.796
48.1				48.1	0.952	-12.036
			52.5	52.5	0.959	-13.107
		55.7		55.7	0.953	-12.863
	58.2			58.2	0.939	-12.053
60				60	0.834	

FIGURE 17. Data summary for eight-iteration simulation with five, six, seven, and eight multicell configurations. Note how the data from different configurations complement each other to form a near continuous scan prediction.

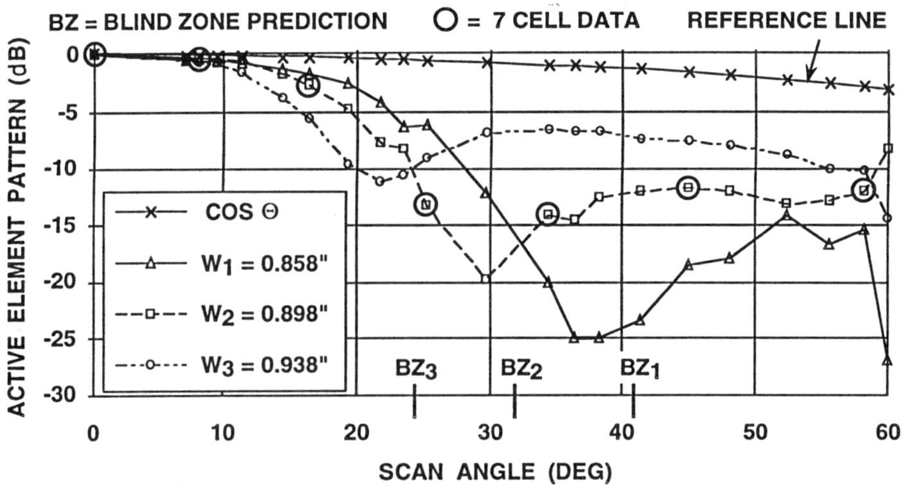

FIGURE 18. Simulated scanning performance for the waveguide array. Performance is shown to be very sensitive to element dimensions and correlates well with theory based on a two-mode approximation.

the three waveguide widths shown. Note that less than a $\pm 5\%$ change in width results in significant performance changes in the Active Element Pattern. The three marks at the bottom of the graph are the predicted scan angles for high reflection or blindzone (BZ) centers for the three configurations as determined by the approximate theory of Stark (1974). The simulation results closely correlate with Stark yet are likely to be more accurate because *all* higher order waveguide modes are being included in the determination of $\Gamma(\theta)$, rather than just the two-mode approximation of Stark. The reference line is also shown to indicate the roll-off of the pattern due to array projected area reduction with scanning.

The power of this technique has been demonstrated to show how significant performance characteristics of scanned arrays can easily be determined. The question naturally arises as to what are the computer requirements and program running times. Figure 19 shows how quickly the computer program converged to useful data. Six iterations show the data trends quite well for the W_2 configuration. Figure 18 used eight iterations for all three widths to get smoother data. One "width" curve at eight iterations typically took 15 hours as the total for all four-cell simulations, but only 5 hours at six iterations.

An idea of the running time and RAM requirements are shown in Figure 20. The software was run on a Hewlett-Packard model 750 computer with 256 MB of RAM. The run time and memory requirements are not unreasonable considering the power of the analysis tool, resulting in savings in cost and design time.

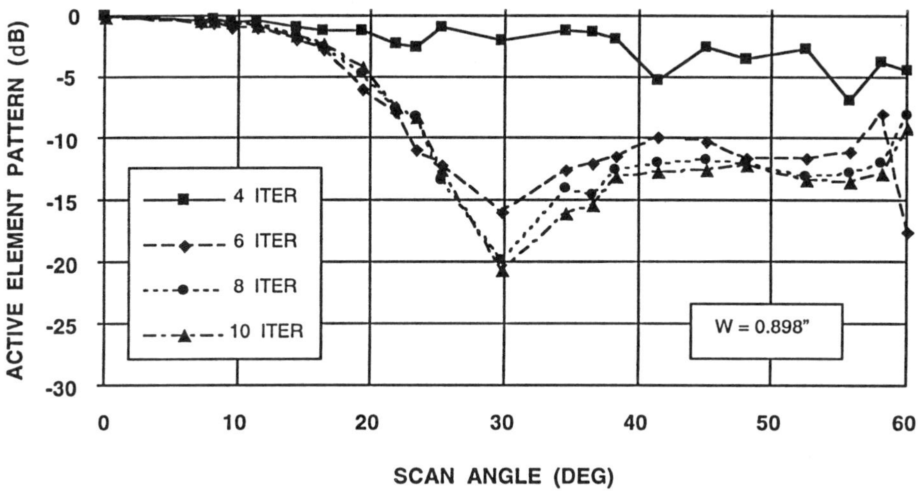

FIGURE 19. Simulation convergence with number of iterations. Useful trends are apparent after only six iterations of mesh adapting.

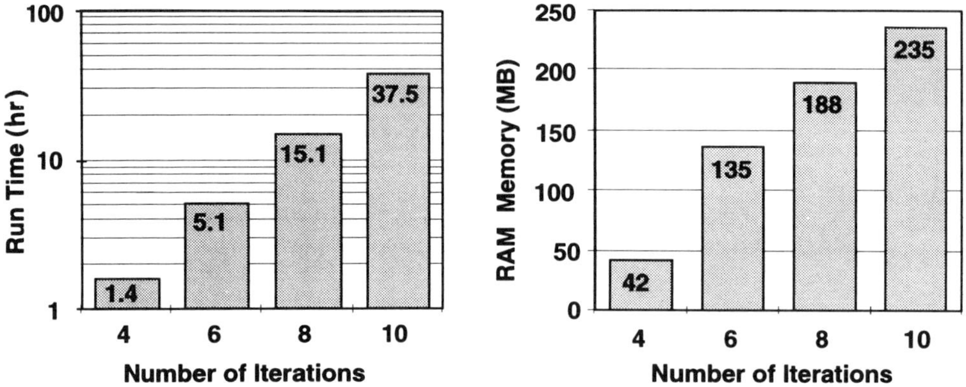

FIGURE 20. The simulation time and memory requirements are shown as a function of the number of adaptive iterations for the finite element meshing for the example triangular array.

12. CONCLUSIONS

Computer simulation capability exists that can accurately predict single antenna element performance and large array scanning performance for fast and inexpensive design. The computer modeling is directly applicable to complex three-dimensional radiating elements, which are not amenable to other theoretical analysis.

REFERENCES

Balfour, M. A. (1965). Phased array simulators in waveguide for triangular arrangement of elements. *IEEE Trans. Antennas Propag.* **AP-13**, 475–476.

Brown, B. R., and Carberry, T. F. (1963). A technique to simulate the self and mutual impedance of an array. *IEEE Trans. Antennas Propag.* **AP-11**, 377–378.

Diamond, B. L., and Knittel, G. H. (1972). A new procedure for the design of a waveguide element for a phased-array antenna. In *Phased Array Antennas*, (A. A. Oliner and G. H. Knittel, eds.), pp. 149–156. Artech House, Dedham, MA.

Eisenhart, R. L., and Park, P. K. (1995). Phased array scanning performance simulation. *IEEE AP-S Int. Symp. Dig.*, pp. 2002–2005.

Elliott, R. S. (1981). *Antenna Theory and Design*, pp. 407–414. Prentice-Hall, Englewood Cliffs, NJ.

Gustincic, J. J. (1972). The determination of active array impedance with multielement waveguide simulators. *IEEE Trans. Antennas Propag.* **AP-20**, 589–595.

Hannan, P. W., and Balfour, M. A. (1965). Simulation of a phased-array antenna in waveguide. *IEEE Trans. Antennas Propag.* **AP-13**, 342–353.

Hannan, P. W., Meier, P. J., and Balfour, M. A. (1963). Simulation of phased array antenna impedance in waveguide. *IEEE Trans. Antennas Propag. (Commun.)* **AP-11**, 715–716.

Sanford, J. R., Johansson, M., and Mukhamedov, L. (1995). Method performs planar-array calculations. *Microwaves & RF*, January, pp. 106–116.

Stark, L. (1974). Microwave theory of phased array antennas—A review. *Proc. IEEE* **62**(12), 1661–1701.

Wheeler, H. A. (1972). A survey of the simulator technique for designing a radiating element in a phased-array antenna. In *Phased Array Antennas* (A. A. Oliner and G. H. Knittel, eds.), pp. 132–148. Artech House, Dedham, MA.

CHAPTER ELEVEN

A Finite Element Method for Electromagnetic Subsurface Tomography

MARC SAILLARD, PATRICK VINCENT, and DANIEL MAYSTRE

1. INTRODUCTION

The aim of the study described in this chapter is to provide a precise numerical tool for calculating the near field scattered by a set of parallel homogeneous cylindrical bodies, with arbitrary cross sections, illuminated by a time-harmonic source of finite extension. The solution of this direct diffraction problem is a first step toward solving an inverse problem of vital importance: the electromagnetic cross-borehole subsurface tomography.

Most of the previous attempts at solving this problem have been using approximate methods (Chew, 1990, and references herein; Torres-Verdin and Habashy, 1994, 1995). On the other hand, a rigorous finite-difference formalism has been developed by Druskin and Knizhnerman (1988). From the experimental point of view, a very recent paper (Peden and Brew, 1995) has shown results from a laboratory scale model. Working in situ, Sato et al. (1995) have focused on the interest of measuring the cross-polarized component of the electromagnetic field. In acoustics, closely related problems have been investigated (see, e.g., Gelius and Stamnes, 1990).

After a presentation of the practical problem, a detailed description of the rigorous theory is given. The first step of this theory is to show that the

Finite Element Software for Microwave Engineering, Edited by Tatsuo Itoh, Giuseppe Pelosi, and Peter P. Silvester.
ISBN 0-471-12636-5

scattered field can be expressed as a sum of space frequency components, thanks to a Fourier transform with respect to the direction of the cylinder's axis. Each component obeys a system of two coupled boundary integral equations, which are solved with the help of a boundary finite element method. Particular attention has been paid to the integration of the singular part of the kernels of the integral equations. The accuracy of the results has been checked against classical criteria and comparisons with numerical data provided by other authors have been achieved. Lastly, two numerical examples show the generality of the method, including the case of high contrast of conductivities.

2. THE PRACTICAL PROBLEM OF TOMOGRAPHY

The physical problem is outlined in Figure 1. Two electromagnetic antennas, a source S and a receiver R, are placed in two parallel boreholes B_S and B_R, respectively. These antennas consist of monochromatic electric or magnetic dipoles, protected from the drilling mud by a plastic cover. Note that some of the currently used antennas are line sources. This kind of antenna can be represented by a set of electric dipoles. On the other hand, current loop antennas behave like a single magnetic dipole.

The frequency can vary from a few hertz to 1 MHz, depending on the nature of the subsoil. Indeed, a compromise must be found between the penetration of the electromagnetic wave and the resolution. The two boreholes are about a few tens of meters in distance.

The numerical solution of this three-dimensional general problem is very difficult to envisage today. Therefore simplifications are needed. A preliminary

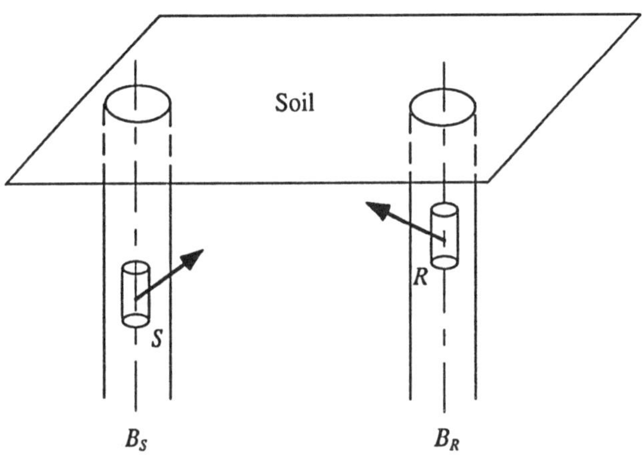

FIGURE 1. The physical problem.

study (Ngakosso, 1994) has shown that, in this frequency range, the influence of both boreholes, the drilling mud, and the plastic cover of the antennas can be neglected. Moreover, the strong attenuation of the electromagnetic field in most subsoils leads to a small contribution from the reflection on the earth–air interface.

It is also assumed that the subsoil consists of homogeneous inclusions buried in a homogeneous medium. Finally, each inclusion is represented by a cylindrical body perpendicular to the vertical plane Π containing the axes of the boreholes. The cross-section is the intersection of the three dimensional object with Π. Indeed, it can be conjectured that the main contribution to the diffracted field at R comes from the part of the scatterer close to Π, at least if the size of the buried object is large enough (typically a few meters).

3. FINITE ELEMENT METHOD: THEORY

3.1. The Simplified Physical Problem

The simplified physical problem is schematized in Figure 2. A magnetic or electric dipole source of wavelength $\lambda_0 = 2\pi/k_0$ in vacuum, located in the xz plane, illuminates a set of N isotropic homogeneous cylinders parallel to the y-axis, of arbitrary bounded cross sections Ω_i of boundaries C_i $(1 \leqslant i \leqslant N)$, with infinite length in the y-direction. Each cylinder will be characterized by its complex permittivity ε_i and its complex magnetic permeability μ_i, the homogeneous isotropic embedding medium of cross section Ω having the

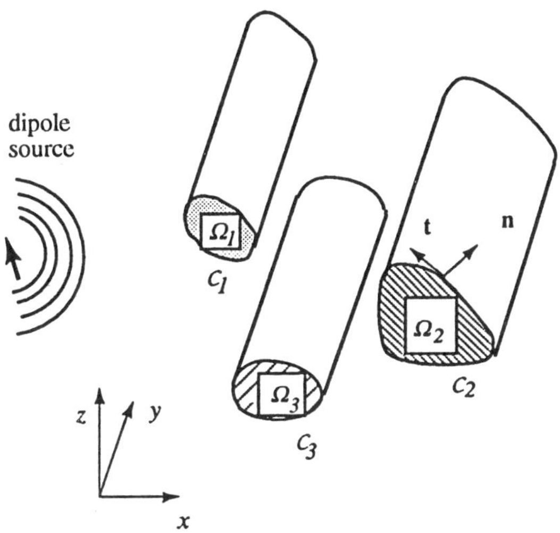

FIGURE 2. The simplified physical problem.

electric constants ε and μ (ε_0 and μ_0 denote the corresponding constants in vacuum). C denotes the entire set of C_i.

Using a time dependence in $\exp(j\omega t)$, the complex amplitudes of the total electric and magnetic fields will be denoted by $\mathbf{E}$ and $\mathbf{H}$. The incident field $(\mathbf{E}^i, \mathbf{H}^i)$ being defined as the field generated by the dipole in an infinite homogeneous medium of electric constants ε and μ, the scattered field $(\mathbf{E}^d, \mathbf{H}^d)$ is defined outside the cylinders as the difference between the total and incident fields. At any point of C_i the unit normal $\mathbf{n}$ is directed toward the exterior of Ω_i and the tangential vector $\mathbf{t}$ is given by

$$\mathbf{t} = \mathbf{n} \times \mathbf{e}_y \tag{1}$$

with $\mathbf{e}_x$, $\mathbf{e}_y$, $\mathbf{e}_z$ representing unit vectors of the coordinate axes.

3.2. Decomposition of the Field in Uncoupled Space-Frequency Components

When the source is an electric dipole of electric moment $\mathbf{P}$, the incident field is given by

$$\mathbf{E}^i = \frac{\mathbf{P}}{j\omega\varepsilon}(\nabla^2 + k^2)\frac{e^{-jkR}}{4\pi R} \tag{2}$$

where R is the distance between the observation point M ($\mathbf{OM} = \mathbf{r} + y\mathbf{e}_y$) and the dipole located at point S of the xz plane ($\mathbf{OS} = \mathbf{r}_s$)

$$R = \sqrt{|\mathbf{r} - \mathbf{r}_s|^2 + y^2}$$

$k = k_0 v$ is the wavenumber in the embedding medium ($v = \sqrt{\varepsilon\mu}/\sqrt{\varepsilon_0\mu_0}$ complex index). Using the decomposition of an outgoing spherical wave in outgoing conical waves (Gradshteyn and Ryzyk, 1983)

$$\frac{e^{-jkR}}{R} = -\frac{j}{2}\int_{-\infty}^{+\infty} H_0^{(2)}(\tilde{k}|\mathbf{r} - \mathbf{r}_s|)e^{j\beta y}\, d\beta \tag{3}$$

with $\tilde{k} = \sqrt{k^2 - \beta^2}$, $\mathrm{Re}(\tilde{k}) + \mathrm{Im}(\tilde{k}) \geqslant 0$, the longitudinal components of the fields can be expanded in Fourier integrals of y. Denoting by $\hat{f}(x, \beta, z)$ the Fourier transform of $f(x, y, z)$

$$f(x, y, z) = \int_{-\infty}^{+\infty} \hat{f}(x, \beta, z)e^{j\beta y}\, d\beta \tag{4}$$

we get

$$\hat{E}_y^i(x, \beta, z) = \frac{-1}{8\pi\varepsilon\omega}\left\{P_y\tilde{k}^2 H_0^{(2)}(\tilde{k}|\mathbf{r}-\mathbf{r}_s|) - j\frac{\tilde{k}\beta}{|\mathbf{r}-\mathbf{r}_s|}\mathbf{P}\cdot(\mathbf{r}-\mathbf{r}_s)H_1^{(2)}(\tilde{k}|\mathbf{r}-\mathbf{r}_s|)\right\} \quad (5)$$

$$\hat{H}_y^i(x, \beta, z) = \frac{-j}{8\pi}\left\{\frac{\tilde{k}}{|\mathbf{r}-\mathbf{r}_s|}\mathbf{e}_y\cdot(\mathbf{P}\times(\mathbf{r}-\mathbf{r}_s))H_1^{(2)}(\tilde{k}|\mathbf{r}-\mathbf{r}_s|)\right\} \quad (6)$$

In the same way, the longitudinal components of the incident field generated by a magnetic dipole of moment **M** are given by

$$\hat{E}_y^i(x, \beta, z) = \frac{-\omega\mu}{8\pi}\left\{\frac{\tilde{k}}{|\mathbf{r}-\mathbf{r}_s|}\mathbf{e}_y\cdot(\mathbf{M}\times(\mathbf{r}-\mathbf{r}_s))H_1^{(2)}(\tilde{k}|\mathbf{r}-\mathbf{r}_s|)\right\} \quad (7)$$

$$\hat{H}_y^i(x, \beta, z) = \frac{-j}{8\pi}\left\{M_y\tilde{k}^2 H_0^{(2)}(\tilde{k}|\mathbf{r}-\mathbf{r}_s|) - j\frac{\tilde{k}\beta}{|\mathbf{r}-\mathbf{r}_s|}\mathbf{M}\cdot(\mathbf{r}-\mathbf{r}_s)H_1^{(2)}(\tilde{k}|\mathbf{r}-\mathbf{r}_s|)\right\} \quad (8)$$

Thus the incident field can be considered as a sum of terms having a space frequency β in y. For each of these terms, the Fourier transforms $\hat{\mathbf{E}}_T(x, \beta, z)$ and $\hat{\mathbf{H}}_T(x, \beta, z)$ of the transverse components (in the xz plane) can be derived from the longitudinal components. In Ω the following formulas apply:

$$\tilde{k}^2\hat{\mathbf{E}}_T = j\beta\nabla\hat{E}_y + j\omega\mu\mathbf{e}_y\times\nabla\hat{H}_y \quad (9)$$

$$\tilde{k}^2\hat{\mathbf{H}}_T = j\beta\nabla\hat{H}_y - j\omega\varepsilon\mathbf{e}_y\times\nabla\hat{E}_y \quad (10)$$

Similar equations give the transverse components of the field inside Ω_i, $\tilde{k}$ being replaced by $\tilde{k}_i$ such that $\tilde{k}_i = \sqrt{k^2 v_i^2 - \beta^2}$, $\mathrm{Re}(\tilde{k}_i) + \mathrm{Im}(\tilde{k}_i) \geqslant 0$, $v_i = \sqrt{\varepsilon_i\mu_i}/\sqrt{\varepsilon_0\mu_0}$. Finally, for each space frequency β, the problem is reduced to the determination of the Fourier transforms $\hat{E}_y(x, \beta, z)$ and $\hat{H}_y(x, \beta, z)$ of the fields. Since $\hat{E}_y(x, \beta, z)\exp(j\beta y)$ and $\hat{H}_y(x, \beta, z)\exp(j\beta y)$ satisfy, in the three-dimensional space, Helmholtz equations with wavenumbers k and k_i, $\hat{E}_y$ and $\hat{H}_y$ satisfy, in the xz space, Helmholtz equations with wavenumbers $\tilde{k}$ and $\tilde{k}_i$:

$$\nabla^2\hat{E}_y + \kappa^2\hat{E}_y = 0 \quad (11)$$

$$\nabla^2\hat{H}_y + \kappa^2\hat{H}_y = 0 \quad (12)$$

with $\kappa = \tilde{k}$ in Ω and $\kappa = \tilde{k}_i$ in Ω_i.

Moreover, since the y-component of the field is tangential to the surface of the cylinders, $\hat{E}_y$ and $\hat{H}_y$ must be continuous across C_i:

$$\hat{E}_y^+ = \hat{E}_y^- \quad (13)$$

$$\hat{H}_y^+ = \hat{H}_y^- \quad (14)$$

the superscripts + and − denoting the limits of the fields on both sides of C_i in Ω and Ω_i, respectively. A second boundary condition can be deduced from the continuity of the components on **t** of the electric and magnetic fields. Using Eqs. (9) and (10) and the similar equations obtained in each cylinder Ω_i, these boundary conditions take the following form:

$$\frac{1}{\tilde{k}^2}\left\{j\beta\frac{d\hat{E}_y^+}{dt} - j\omega\mu\frac{d\hat{H}_y^+}{dn}\right\} = \frac{1}{\tilde{k}_i^2}\left\{j\beta\frac{d\hat{E}_y^-}{dt} - j\omega\mu_i\frac{d\hat{H}_y^-}{dn}\right\} \tag{15}$$

$$\frac{1}{\tilde{k}^2}\left\{j\beta\frac{d\hat{H}_y^+}{dt} + j\omega\varepsilon\frac{d\hat{E}_y^+}{dn}\right\} = \frac{1}{\tilde{k}_i^2}\left\{j\beta\frac{d\hat{H}_y^-}{dt} + j\omega\varepsilon_i\frac{d\hat{E}_y^-}{dn}\right\} \tag{16}$$

In conclusion, we are led, for each space-frequency component of the fields to a boundary-value problem in a two-dimensional space defined by Eqs. (11)–(16) and the radiation conditions on $\hat{E}_y^d$ and $\hat{H}_y^d$ at infinity, the incident field being given by Eqs. (5) and (6) for an electric dipole and Eqs. (7) and (8) for a magnetic dipole. It is interesting to note that this problem is very close to the problem of scattering by the same structure illuminated by a plane wave in conical diffraction. The solution of the physical problem will be deduced from the elementary solution of the space-frequency components of the field by the following Fourier integral:

$$E_y^d(x, y, z) = \int_{-\infty}^{+\infty} \hat{E}_y^d(x, \beta, z)e^{j\beta y}\, d\beta \tag{17}$$

$$H_y^d(x, y, z) = \int_{-\infty}^{+\infty} \hat{H}_y^d(x, \beta, z)e^{j\beta y}\, d\beta \tag{18}$$

3.3. The Integral Representation of the Field, Integral Equations

In order to solve the space-frequency elementary problem, a rigorous integral formalism based on Green's function will be used. The classical integral theory (Vincent et al., 1995) leads to the solution of four coupled boundary integral equations with unknowns $\hat{E}_y$, $\hat{H}_y$ and their normal derivatives on C. It has been established by Maystre and Vincent (1972) that, in the problem of scattering by cylinders in the classical (nonconical) mounting, the size of the set of integral equations can be reduced by a factor of 2 using adequate unknown functions. Of course, for computational considerations, this reduction in size is very interesting. It will be shown that the same reduction holds in the conical mounting. Since 1972, many papers have proposed other integral theories reducing by a factor of 2 the number of integral equations in various kinds of scattering problems (e.g., by Maystre, 1978; Marx, 1982; Kleinman and Martin, 1988).

The main step of this integral theory lies in the choice of the unknown functions on C. It is well known that the choice of the longitudinal components of the electric and magnetic fields is not sufficient to represent the fields at any point in the form of integrals since the Green's theorem needs also the normal derivatives of these functions to represent the field. If one wants to find a couple of functions defined on C and satisfying this requirement, a natural way is to adopt two potentials, one for the longitudinal component of the electric field and one for the longitudinal component of the magnetic field, generating in Ω a field equal to the scattered field in the space-frequency problem. Indeed, if these two potentials exist, it is possible to express the scattered fields in Ω in an integral form. Going to the boundary of Ω will allow one to express in an integral form the longitudinal components of the fields, their normal derivatives and their tangential derivatives on C. The use of boundary conditions (13)–(16) will enable one to express the values of the same limits inside Ω_i, and thus the longitudinal components of the fields inside Ω_i in the form of integrals, using Green's theorem.

Let us define these potentials in a rigorous manner. With this aim, we define, for a given value β of the space frequency, a function $F(x, z)$ which satisfies a radiation condition at infinity and

$$\text{in } \Omega, \quad F(x, z) = \hat{E}_y^d(x, \beta, z) \tag{19}$$

$$\forall i, \text{ in } \Omega_i, \quad \nabla^2 F + \tilde{k}^2 F = 0 \tag{20}$$

$$\forall i, \quad F \text{ is continuous across } C_i \tag{21}$$

It is worth noting that definitions (19) and (21) permit one to deduce the internal limit of F from the value of $\hat{E}_y^d$ on C_i. Hence F satisfies a boundary-value problem inside Ω_i with Helmholtz equation (20). The question of the uniqueness of the solution of this problem will not be considered here, and thus it will be assumed that the optical index v is not purely real.

We denote by G the two-dimensional Green's function of homogeneous space for Eq. (20)

$$G(M, M') = \frac{j}{4} H_0^{(2)}(\tilde{k} MM') \tag{22}$$

Since F satisfies the same Helmholtz equation in Ω and Ω_i and is continuous across C_i, it can be expressed in an integral form as a function of the jump ϕ of the normal derivative of F across C_i. The value of F at any point Q of space is given by

$$F(Q) = \int_C G(Q, M')\phi(M')\, ds' \tag{23}$$

with

$$\phi = \frac{dF^+}{dn} - \frac{dF^-}{dn} \tag{24}$$

Since F is continuous, the limit of $F(Q)$ when Q tends to a point M of C can be written

$$F^{+}(M) = \int_C G(M, M')\phi(M')\, ds' \tag{25}$$

In order to simplify the notations, Eq. (25) will be written in an operator form:

$$F^{+} = G\phi \tag{26}$$

Due to the discontinuity of the normal derivative of F on C, the value of the normal derivative of F in Ω will be determined by writing (Maystre and Vincent, 1972)

$$\frac{1}{2}\left(\frac{dF^{+}(M)}{dn} + \frac{dF^{-}(M)}{dn}\right) = \int_C \frac{dG(M, M')}{dn}\phi(M')\, ds' \tag{27}$$

the normal derivative of G being taken with respect to M.

From Eqs. (24) and (27), it comes out that

$$\frac{dF^{+}}{dn} = \frac{\phi}{2} + \frac{dG}{dn}\phi \tag{28}$$

On the other hand, since F is continuous across C, the tangential derivative of F is continuous across C as well and is given by

$$\frac{dF^{+}}{dt} = \frac{dG}{dt}\phi \tag{29}$$

Bearing in mind that F identifies with $\hat{E}_y^d$ in Ω, the limits on C of the space-frequency components of the total longitudinal electric field and its derivatives are given by

$$\hat{E}_y^{+} = e^i + G\phi \tag{30}$$

$$\frac{d\hat{E}_y^{+}}{dn} = \frac{\phi}{2} + \frac{de^i}{dn} + \frac{dG}{dn}\phi \tag{31}$$

$$\frac{d\hat{E}_y^{+}}{dt} = \frac{de^i}{dt} + \frac{dG}{dt}\phi \tag{32}$$

with

$$e^i(M) = \hat{E}_y^i(x, \beta, z) \tag{33}$$

In order to find the expressions of the space-frequency components of the longitudinal magnetic field and its derivatives, it suffices to define a function F' satisfying Eqs. (19)–(21), but with $\hat{H}_y^d$ instead of $\hat{E}_y^d$. That way, it can be found that

$$\hat{H}_y^+ = h^i + G\psi \tag{34}$$

$$\frac{d\hat{H}_y^+}{dn} = \frac{\psi}{2} + \frac{dh^i}{dn} + \frac{dG}{dn}\psi \tag{35}$$

$$\frac{d\hat{H}_y^+}{dt} = \frac{dh^i}{dt} + \frac{dG}{dt}\psi \tag{36}$$

with

$$h^i(M) = \hat{H}_y^i(x, \beta, z) \tag{37}$$

and

$$\psi = \frac{dF'^+}{dn} - \frac{dF'^-}{dn} \tag{38}$$

In order to find the integral expressions of the space-frequency elements of the longitudinal components of the fields inside a cylinder C_i, it is necessary to express the internal limits of these fields inside Ω_i. This goal can be reached from Eqs. (15) and (16) by remarking that the tangential derivatives of $\hat{E}_y$ and $\hat{H}_y$ are continuous across C. This property entails that

$$\frac{d\hat{E}_y^-}{dn} = \frac{\tilde{k}_i^2\varepsilon}{\tilde{k}^2\varepsilon_i}\frac{d\hat{E}_y^+}{dn} - \frac{\beta}{\omega\varepsilon_i}\left(1 - \frac{\tilde{k}_i^2}{\tilde{k}^2}\right)\frac{d\hat{H}_y^+}{dt} \tag{39}$$

$$\frac{d\hat{H}_y^-}{dn} = \frac{\tilde{k}_i^2\mu}{\tilde{k}^2\mu_i}\frac{d\hat{H}_y^+}{dn} + \frac{\beta}{\omega\mu_i}\left(1 - \frac{\tilde{k}_i^2}{\tilde{k}^2}\right)\frac{d\hat{E}_y^+}{dt} \tag{40}$$

and thus from Eqs. (13), (14), (30)–(32), and (34)–(36),

$$\hat{E}_y^- = e^i + G\phi \tag{41}$$

$$\hat{H}_y^- = h^i + G\psi \tag{42}$$

$$\frac{d\hat{E}_y^-}{dn} = \alpha_1\left(\frac{\phi}{2} + \frac{de^i}{dn} + \frac{dG}{dn}\phi\right) + \alpha_2\left(\frac{dh^i}{dt} + \frac{dG}{dt}\psi\right) \tag{43}$$

$$\frac{d\hat{H}_y^-}{dn} = \alpha_3\left(\frac{\psi}{2} + \frac{dh^i}{dn} + \frac{dG}{dn}\psi\right) + \alpha_4\left(\frac{de^i}{dt} + \frac{dG}{dt}\phi\right) \tag{44}$$

with

$$\alpha_1 = \frac{\tilde{k}_i^2 \varepsilon}{\tilde{k}^2 \varepsilon_i}, \quad \alpha_2 = -\frac{\beta}{\omega \varepsilon_i}\left(1 - \frac{\tilde{k}_i^2}{\tilde{k}^2}\right), \quad \alpha_3 = \frac{\tilde{k}_i^2 \mu}{\tilde{k}^2 \mu_i}, \quad \alpha_4 = \frac{\beta}{\omega \mu_i}\left(1 - \frac{\tilde{k}_i^2}{\tilde{k}^2}\right) \tag{45}$$

the α_p depending on the cylinder C_i where the point is located.

Now the field inside C_i can be written, using Green's theorem:

$$\hat{E}_y(Q) = \int_{C_i} \left\{ \frac{dG_i(Q, M')}{dn'} \hat{E}_y^-(M') - G_i(Q, M') \frac{d\hat{E}_y^-}{dn}(M') \right\} ds' \tag{46}$$

$$\hat{H}_y(Q) = \int_{C_i} \left\{ \frac{dG_i(Q, M')}{dn'} \hat{H}_y^-(M') - G_i(Q, M') \frac{d\hat{H}_y^-}{dn}(M') \right\} ds' \tag{47}$$

G_i being the Green's function of the homogeneous space having the electrical constants of the cylinder C_i.

Now the Fourier transform of the fields have been expressed at any point of space in terms of the two functions ϕ and ψ. It is worth noting that the expressions of the field in Ω [Eq. (23) and the same for the magnetic field] and Ω_i [Eqs. (46) and (47)] do not ensure the boundary conditions (13)–(16) to be fulfilled. Indeed, we have used the Green's theorem inside Ω_i using the limits of the fields and their normal derivatives on C_i provided by Eqs. (41)–(44). However, it is well known that in a boundary-value problem, the field and its normal derivative cannot be imposed together, except if they satisfy a condition of compatibility. The condition of compatibility on the electric field is quite simple: it suffices to write that the limit of $\hat{E}_y(Q)$ when Q tends to a point M of C_i is actually equal to $\hat{E}_y^-(M)$. Calculating this limit (Dautray and Lions, 1985), the compatibility leads to the following equation:

$$\frac{1}{2}\hat{E}_y(M) = \int_{C_i} \left\{ \frac{dG_i(M, M')}{dn'} \hat{E}_y^-(M') - G_i(M, M') \frac{d\hat{E}_y^-}{dn}(M') \right\} ds' \tag{48}$$

The same equation for the magnetic field gives a second equation:

$$\frac{1}{2}\hat{H}_y(M) = \int_{C_i} \left\{ \frac{dG_i(M, M')}{dn'} \hat{H}_y^-(M') - G_i(M, M') \frac{d\hat{H}_y^-}{dn}(M') \right\} ds' \tag{49}$$

Inserting in Eqs. (48) and (49) the expressions of the limits given by Eqs. (41)–(45) leads to a system of two coupled integral equations with two unknown functions ϕ and ψ. Denoting by $G'(M, M')$ the function defined by

$$G'(M, M') = \begin{cases} G_i(M, M') & \text{if } M \in C_i \text{ and } M' \in C_i, i \in (1, N) \\ 0 & \text{otherwise} \end{cases} \tag{50}$$

the two integral equations take, in the operator form,

$$\left\{\left(\frac{1}{2}-\frac{dG'}{dn'}\right)G+\alpha_1 G'\left(\frac{1}{2}+\frac{dG}{dn}\right)\right\}\phi+\alpha_2 G'\frac{dG}{dt}\psi$$

$$=-\frac{e^i}{2}+\frac{dG'}{dn'}e^i-G'\left(\alpha_1\frac{de^i}{dn}+\alpha_2\frac{dh^i}{dt}\right) \tag{51}$$

$$\left\{\left(\frac{1}{2}-\frac{dG'}{dn'}\right)G+\alpha_3 G'\left(\frac{1}{2}+\frac{dG}{dn}\right)\right\}\psi+\alpha_4 G'\frac{dG}{dt}\phi$$

$$=-\frac{h^i}{2}+\frac{dG'}{dn'}h^i-G'\left(\alpha_3\frac{dh^i}{dn}+\alpha_4\frac{de^i}{dt}\right) \tag{52}$$

3.4. Uniqueness of the Solution: Irregular Frequencies

A Dirichlet problem can be associated with each elementary diffraction problem, characterized by the space frequency β of the incident conical wave. This new problem, schematized in Figure 3 with one scatterer C_1 only, consists in finding a function F_D satisfying a Helmholtz equation with wavenumber $\tilde{k}$ in Ω_1 and a Dirichlet boundary condition:

$$\nabla^2 F_D+\tilde{k}^2 F_D=0 \quad \text{with } F_D=0 \text{ on } C_1 \tag{53}$$

Physically, F_D can be interpreted as the axial component of the electric field of a TM mode of a cylindrical cavity with perfectly conducting walls. This cavity has the same shape as the actual scatterer and is filled with the embedding medium ε, μ.

The Green's theorem applies in Ω_1 and yields

$$\forall M\in C_1,\quad F_D(M)=0=\int_{C_1} G(M,M')\phi_D(M')\,ds' \tag{54}$$

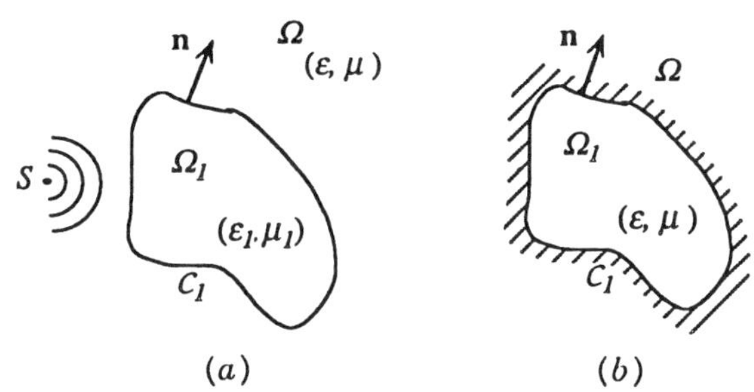

FIGURE 3. (a) Diffraction problem with conical incident wave. (b) Associated interior problem, without incident field.

where ϕ_D denotes the opposite of the normal derivative of F_D inside C_1. Taking the limits of both tangential and normal derivatives of F_D inside C_1, one gets

$$0 = \int_{C_1} \frac{dG}{dt}(M, M')\phi_D(M')\, ds' \tag{55}$$

$$-\phi_D(M) = 2\int_{C_1} \frac{dG}{dn}(M, M')\phi_D(M')\, ds' \tag{56}$$

As a consequence of Eqs. (54)–(56), if a nonzero solution of the associated interior problem exists, the solution of the actual diffraction problem is not unique. Indeed, if $\{\phi, \psi\}$ is a solution of the pair of the two coupled integral equations [Eqs. (51) and (52)], $\{\phi + a\phi_D, \psi + b\phi_D\}$, where a and b are two arbitrary complex numbers, is also a solution. This property has been demonstrated previously for the acoustic problem by Kleinman and Martin (1988) and for the three-dimensional electromagnetic case by Martin and Ola (1993). The eigenvalues of the associated Dirichlet interior problem form a discrete set of values of $\tilde{k}^2$ (thus of β) called irregular frequencies. These values are located in the interval characterized by $|\mathrm{Re}(\beta)|$ smaller than $\mathrm{Re}(k)$.

In the practical problem dealt with here, since the subsoil is generally very lossy, the imaginary part of the irregular frequencies is large. This property not only ensures the uniqueness of the solution for real values of β, but it also permits one to perform an accurate integration along the β real axis, as in Eqs. (17) and (18).

3.5. Boundary Elements Method

In this section, the solution of the integral equations will be reduced to that of a linear system of equations, using a boundary elements method. First, the shape of the cylinders will be represented by polygonal ones. Figure 4 shows an example of this representation for one of the cylinders. Thus the N cylinders are represented by M linear segments $\Gamma_j (1 \leqslant j \leqslant M)$ with middle M_j and length d_j.

As usual in the boundary elements method, the functions ϕ and ψ are represented by the following sums:

$$\phi = \sum_{j=1}^{M} \phi_j f_j \tag{57}$$

$$\psi = \sum_{j=1}^{M} \psi_j f_j \tag{58}$$

and here the functions f_j are defined by

$$f_j(M) = \begin{cases} 1 & \text{if } M \in \Gamma_j \\ 0 & \text{otherwise} \end{cases} \tag{59}$$

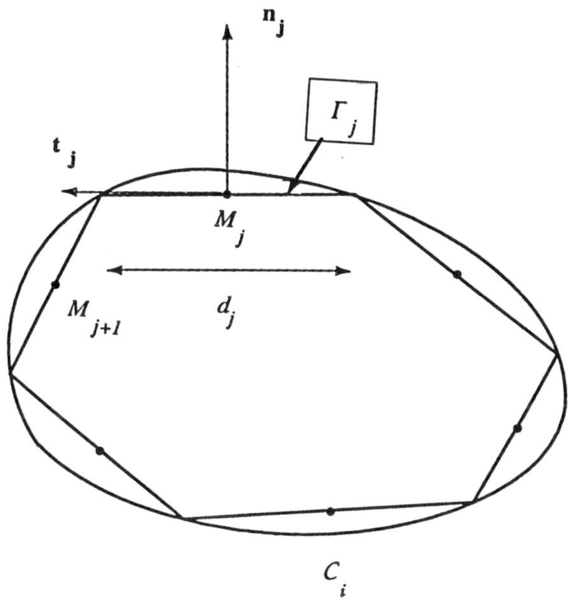

FIGURE 4. Representation of a cylinder.

Then all the terms of Eqs. (51) and (52) will be projected on a set of test functions identical to the set f_j. This means that not only the final results, obtained after a complete integration of the terms, but also the intermediate results of integration [e.g., $G\phi$ in the second term $dG'/dn'\, G\phi$ of Eq. (51)] will be projected on the set of test functions f_j. That way, each integration will associate with a sum of functions f_j another sum of functions f_j and each operator O contained in the integral equation will reduce to a complex matrix of size M. Indeed, if O applies to the function $\tau(M')$ to give $\rho(M)$,

$$\rho(M) = \int_C O(M, M')\tau(M')\, ds' \tag{60}$$

the representation of τ and ρ will lead to the following equation

$$\sum_{j=1}^{M} \rho_j f_j = \sum_{j'=1}^{M} \tau_{j'} \int_C O(M, M') f_{j'}(M')\, ds' \tag{61}$$

and after projecting the above equation on the test functions

$$\rho_j = \sum_{j'=1}^{M} \tau_{j'} \iint_{C,C} O(M, M') f_j(M) f_{j'}(M')\, ds\, ds' \tag{62}$$

and finally

$$\rho_j = \sum_{j'=1}^{M} O_{j,j'}\tau_{j'} \tag{63}$$

with

$$O_{j,j'} = \iint_{M\in\Gamma_j, M'\in\Gamma_{j'}} O(M, M')\, ds\, ds' \tag{64}$$

The two integral equations contain three kinds of integral operators. The first kind (G, G') is directly deduced from the Green's functions. They present a logarithmic singularity when M and M' tend to each other. The second kind of operator (dG/dn, dG'/dn') is deduced from the Green's functions by taking the normal derivative. In contrast with what could be expected at a first glance, they do not contain singularities when M and M' tend to each other. Finally, the third kind of operator (dG/dt) is deduced from the Green's function by taking the tangential derivative. It behaves like a principal value when M' tends to M. Let us consider the first kind of operator. We have to perform integrations of the following form:

$$G_{i,j} = \iint_{\Gamma_i,\Gamma_j} G(M, M')\, ds\, ds' \tag{65}$$

Using the expression of G given by Eq. (22), and bearing in mind (Abramowitz and Stegun, 1965) that

$$\text{if } z \to 0, \quad \frac{j}{4} H_0^{(2)}(z) \approx \frac{1}{2\pi}(\ln(z/2) + \gamma) + \frac{j}{4} \tag{66}$$

γ being the Euler constant, Eq. (65) is written in the form

$$G_{i,j} = \iint_{\Gamma_i,\Gamma_j} \left(G(M, M') - \frac{1}{2\pi}\ln(MM') \right) ds\, ds' + \frac{1}{2\pi} \iint_{\Gamma_i,\Gamma_j} \ln(MM')\, ds\, ds' \tag{67}$$

The first term on the right-hand side of Eq. (67) is not singular and can be calculated by taking the value of the integrand when M and M' identify with M_i and M_j:

$$\iint_{\Gamma_i,\Gamma_j} \left(G(M, M') - \frac{1}{2\pi}\ln(MM') \right) ds\, ds' = \left[G(M_i, M_j) - \frac{1}{2\pi}\ln(M_i M_j) \right] d_i d_j \tag{68}$$

Of course, when i is equal to j, the right-hand side of Eq. (68) must be considered as the limit of the expression between brackets when M_i and M_j tend to each other. In order to calculate the second term in closed form, it is interesting to write MM'^2 in the form

$$MM'^2 = (a + bw_i + cw_j)(\bar{a} + \bar{b}w_i + \bar{c}w_j) \tag{69}$$

with

$$w_i = \mathbf{t}_i \cdot \mathbf{M}_i\mathbf{M} \tag{70}$$

and

$$a = \begin{cases} \dfrac{\mathbf{M}_i\mathbf{M}_j \cdot [c\mathbf{t}_j - b\mathbf{t}_i + (b\mathbf{t}_j - c\mathbf{t}_i)(\mathbf{t}_i \cdot \mathbf{t}_j)]}{1 - (\mathbf{t}_i \cdot \mathbf{t}_j)^2} & \text{if } |\mathbf{t}_i \cdot \mathbf{t}_j| \neq 1 \\ -\mathbf{n}_i \cdot \mathbf{M}_i\mathbf{M}_j + j\mathbf{t}_i \cdot \mathbf{M}_i\mathbf{M}_j \quad \text{if } \mathbf{t}_i = \mathbf{t}_j \\ -(\mathbf{t}_i \cdot \mathbf{M}_i\mathbf{M}_j + j\mathbf{n}_i \cdot \mathbf{M}_i\mathbf{M}_j) \quad \text{if } \mathbf{t}_i = -\mathbf{t}_j \end{cases} \tag{71}$$

$$\bar{c} = b = \sqrt{-\mathbf{t}_i \cdot (\mathbf{t}_j - j\mathbf{n}_j)} \tag{72}$$

The second term on the right-hand side of Eq. (67) can be calculated in closed form, remarking that

$$\iint_{\Gamma_i,\Gamma_j} \ln(MM')\, ds\, ds' = \mathrm{Re}\left\{\frac{1}{4bc} \cdot [\mathscr{L}(a+b'+c') + \mathscr{L}(a-b'-c') - \mathscr{L}(a-b'+c') - \mathscr{L}(a+b'-c')]\right\} \tag{73}$$

with

$$b' = b\frac{d_i}{2}, \quad c' = c\frac{d_j}{2}, \quad \mathscr{L}(x) = x^2[2\ln(x) - 3]$$

The second kind of function in the kernel, for instance dG/dn, given by

$$\frac{dG}{dn} = \frac{j\tilde{k}}{4}\mathbf{n}(M) \cdot \frac{\mathbf{MM'}}{MM'} H_1^{(2)}(\tilde{k}MM') \tag{74}$$

is not singular since when M tends to M', the scalar product $\mathbf{n}(M) \cdot \mathbf{MM'}/MM'$ tends to zero, compensating the behavior in $1/MM'$ of the Hankel function. However, at edges, or even when two cylinders are very close to each other, this kernel may have very strong variations and thus it is much more reliable to operate as for the first kind of operator and to write

$$\left(\frac{dG}{dn}\right)_{i,j} = \iint_{\Gamma_i,\Gamma_j} \frac{j\tilde{k}}{4}\mathbf{n}(M)\cdot\frac{\mathbf{MM'}}{MM'}\left(H_1^{(2)}(\tilde{k}MM') - \frac{2j}{\pi\tilde{k}MM'}\right)ds\,ds'$$
$$-\frac{1}{2\pi}\iint_{\Gamma_i,\Gamma_j}\frac{\mathbf{n}(M)\cdot\mathbf{MM'}}{MM'^2}ds\,ds' \qquad (75)$$

Remarking that

$$\mathbf{n}(M)\cdot\mathbf{MM'} = \mathbf{n}_i\cdot\mathbf{M}_i\mathbf{M}_j + w_j\mathbf{n}_i\cdot\mathbf{t}_j \qquad (76)$$

and bearing in mind Eq. (69), the last term of Eq. (75) can be calculated in closed form:

$$\iint_{\Gamma_i,\Gamma_j}\frac{\mathbf{n}(M)\cdot\mathbf{MM'}}{MM'^2}ds\,ds'$$
$$= -\mathrm{Im}\{c[\mathscr{K}(a+b'+c')+\mathscr{K}(a-b'-c')-\mathscr{K}(a-b'+c')-\mathscr{K}(a+b'-c')]\} \qquad (77)$$

with

$$\mathscr{K}(x) = x\ln(x)$$

The first term on the right-hand side of Eq. (75) can be replaced by its value (or its limit) when M and M' identify with by M_i and M_j.

The integral containing the tangential derivative of the Green's function can be achieved in the same way, remarking that, this time, the integral containing the singular part can be calculated using the following equation:

$$\iint_{\Gamma_i,\Gamma_j}\frac{\mathbf{t}(M)\cdot\mathbf{MM'}}{MM'^2}ds\,ds'$$
$$= \mathrm{Re}\{c[\mathscr{K}(a+b'+c')+\mathscr{K}(a-b'-c')-\mathscr{K}(a-b'+c')-\mathscr{K}(a+b'-c')]\} \qquad (78)$$

3.6. Basic Properties of the Field

The results obtained by the numerical computation can be checked using analytical relations deduced from the symmetry properties of the Maxwell equations. Two theorems are used: energy balance and reciprocity. The first one is easily implemented for incident plane waves and lossless systems, allowing fast control of the solution of the linear systems. The second can be used for dipole sources; it takes into account the computation of the inverse Fourier transform.

To express the energy balance, let us define a fictitious circular surface with radius going to infinity and let us divide the total field into incoming and outgoing waves. The flux of the Poynting vector of outgoing waves across the boundary of this surface must be equal to that of the incoming ones. Using cylindrical coordinates (r, y, θ) for points located on this boundary, the asymptotic expression of the field deduced from the behavior of Hankel functions yields:

$$\hat{\mathbf{E}}_y^d(r, \beta, \theta) \approx g_e(\theta) \frac{e^{-j\tilde{k}r}}{\sqrt{r}} \tag{79}$$

where $g_e(\theta)$ can be obtained through the Kirchhoff–Helmholtz integral (23) as a function of one of the solutions ϕ of the integral equations (51) and (52):

$$g_e(\theta) = \frac{e^{j\pi/4}}{\sqrt{8\pi\tilde{k}}} \int_{C_i} \phi(x', z') e^{j\tilde{k}(x' \cos\theta + z' \sin\theta)} \, ds' \tag{80}$$

A similar expression holds for the angular behavior $g_h(\theta)$ of the magnetic field as a function of ψ.

Assuming an incident plane wave with wavevector $\mathbf{k}^i = \tilde{k}\cos\theta_0 \mathbf{e}_x - \beta\mathbf{e}_y + \tilde{k}\sin\theta_0\mathbf{e}_z$ and polarization angle δ_0 between the electric field and the normal to the $(\mathbf{k}^i, \mathbf{e}_y)$ plane, the energy balance can be written as the classical optical theorem (Petit and Zolla, 1994):

$$\Phi_1 + \Phi_2 = 0 \tag{81}$$

with

$$\Phi_1 = \frac{k}{2\tilde{k}} \int_0^{2\pi} \left(\sqrt{\frac{\varepsilon}{\mu}} |g_e(\theta)|^2 + \sqrt{\frac{\mu}{\varepsilon}} |g_h(\theta)|^2 \right) d\theta \tag{82}$$

$$\Phi_2 = \sqrt{\frac{2\pi}{\tilde{k}}} \operatorname{Re} \left\{ \left(\sqrt{\frac{\varepsilon}{\mu}} g_e(\theta_0) \sin\delta_0 - \sqrt{\frac{\mu}{\varepsilon}} g_h(\theta_0) \cos\delta_0 \right) e^{-j\pi/4} \right\} \tag{83}$$

This criterion is usually satisfied within a 1% accuracy.

Consider now two diffraction problems differing only through their electric and magnetic current source distributions: $(\mathbf{J}_1^e, \mathbf{J}_1^h)$ for the first one and $(\mathbf{J}_2^e, \mathbf{J}_2^h)$ for the second one. The corresponding electromagnetic fields are denoted by $(\mathbf{E}_1, \mathbf{H}_1)$ and $(\mathbf{E}_2, \mathbf{H}_2)$, respectively. A classical calculation (Van Bladel, 1964), based on a Wronskian integration taking into account the Helmholtz equation, yields

$$\iiint_{\Omega} (\mathbf{E}_1 \cdot \mathbf{J}_2^e + \mathbf{H}_1 \cdot \mathbf{J}_2^h)\, dU = \iiint_{\Omega} (\mathbf{E}_2 \cdot \mathbf{J}_1^e + \mathbf{H}_2 \cdot \mathbf{J}_1^h)\, dU \tag{84}$$

For dipole sources located at $\mathbf{r}_1$ and $\mathbf{r}_2$, the currents are represented by Dirac distributions:

$$\mathbf{J}_1^e = \mathbf{P}_1 \delta(\mathbf{r} - \mathbf{r}_1), \quad \mathbf{J}_1^h = \mathbf{M}_1 \delta(\mathbf{r} - \mathbf{r}_1)$$
$$\mathbf{J}_2^e = \mathbf{P}_2 \delta(\mathbf{r} - \mathbf{r}_2), \quad \mathbf{J}_2^h = \mathbf{M}_2 \delta(\mathbf{r} - \mathbf{r}_2)$$

where $\mathbf{P}_{1,2}$ and $\mathbf{M}_{1,2}$ describe the electric and magnetic moments, respectively. Then, Eq. (84) takes the simple form

$$\mathbf{E}_1(\mathbf{r}_2) \cdot \mathbf{P}_2 + \mathbf{H}_1(\mathbf{r}_2) \cdot \mathbf{M}_2 = \mathbf{E}_2(\mathbf{r}_1) \cdot \mathbf{P}_1 + \mathbf{H}_2(\mathbf{r}_1) \cdot \mathbf{M}_1 \tag{85}$$

4. NUMERICAL IMPLEMENTATION

4.1. Reconstruction of the Solution

The first part of this section is devoted to the reconstruction of the solution of the original problem from the superposition of elementary two-dimensional problems. This reconstruction is achieved through the inverse Fourier transform of the transfer function, defined by Eqs. (17) and (18), using an adequate contour of integration in the complex plane.

Bearing in mind that incident waves with $|\text{Re}(\beta)| > \text{Re}(k)$ represent evanescent waves, and that their damping increases with $|\text{Re}(\beta)|$, it appears that truncating the path of integration with a negligible loss of accuracy is possible. The appropriate truncation mainly depends on the frequency, or more exactly, on the distance from the source to the boundary of the nearest scatterer in terms of wavelengths. The smaller this latter is, the longer the path. For instance, at a frequency $f = 1$ Hz, a dipole source located 25 m from the cylinder requires a very long path for $\text{Re}(\beta)$: $[-200\,\text{Re}(k), 200\,\text{Re}(k)]$. The same problem at $f = 1000$ Hz is accurately solved with a ten times smaller interval. The ratio of the computation times is also close to 10. Even though one gets accurate results, this rigorous tool is not well suited for low frequencies. In the 1–100 Hz range, its main interest is to permit one to check the validity of approximate theories.

The existence of irregular frequencies when $-\text{Re}(k) < \text{Re}(\beta) < \text{Re}(k)$ has already been discussed from a theoretical point of view in Section 3.4. Numerically, as long as the conductivity σ of the embedding medium is higher than 10^{-5} S/m, the numerical integration remains accurate in the frequency range used in this problem (1 Hz to 1 MHz). Nevertheless, in a very unusual situation, if the ratio σ/f expressed in S/m·Hz is smaller than about 10^{-11}, one must move the corresponding part of the contour of integration away from the real axis of the β complex plane. In a recent paper by Ngakosso et al. (1995), this way to overcome the problem has been justified theoretically and its numerical efficiency has been verified.

The conductivities σ_i of the buried bodies are not linked to the irregular frequencies but the propagation constants of the modes propagating along the y-axis direction inside these cylindrical bodies are functions of these parameters. In most cases, the losses are large enough to ensure high imaginary parts of the propagation constants and thus accurate numerical integration. But, once again, if the ratio σ_i/f is smaller than about 10^{-11}, the influence of the modes has to be taken into account in the inverse Fourier transform. For instance, a tunnel can probably support guided waves with low leakage. Since, in that case, the associated poles in the β complex plane are close to the real axis, one has to integrate a function with very sharp peaks, except if the contour is moved away from these poles. Fortunately, a simple argument tells us that the poles stay in the two quarters of the β complex plane defined by

$$\mathrm{Re}(\beta)\,\mathrm{Im}(\beta) < 0$$

Indeed, a mode cannot propagate without any damping in a lossy medium. Bearing in mind that the time dependence is in $\exp(j\omega t)$, a y dependence in $\exp(j\beta y)$ describes a wave propagating toward $y = -\infty$ if $\mathrm{Re}(\beta) > 0$. Such a wave is damped provided $\mathrm{Im}(\beta) < 0$.

As a conclusion, if one of the media is almost lossless, the path of integration is transformed into a step function, an imaginary part with the same sign as the real part being added to the real values of β. Numerically, it has been noticed that $|\mathrm{Im}(\beta)|$ greater than $\mathrm{Re}(k)/20$ permits one to smooth the transfer function.

4.2. Tests of Validity

Two kinds of tests are made. On one hand, tests based on the basic properties of the electromagnetic field described in Section 3.6 are used; on the other hand, purely numerical tests to make comparisons with numerical results from other theories have been achieved.

To check that the Lorentz reciprocity relation is numerically satisfied, let us consider the two following diffraction problems, depicted in Figure 5:

1. Calculate at S_2 the magnetic field $\mathbf{H}_1$ scattered by a given obstacle when a magnetic dipole with moment $\mathbf{M}_1 = \mathbf{e}_z$ is located at S_1.
2. Calculate at S_1 the magnetic field $\mathbf{H}_2$ scattered by the same obstacle when a magnetic dipole with moment $\mathbf{M}_2 = \mathbf{e}_x$ is located at S_2.

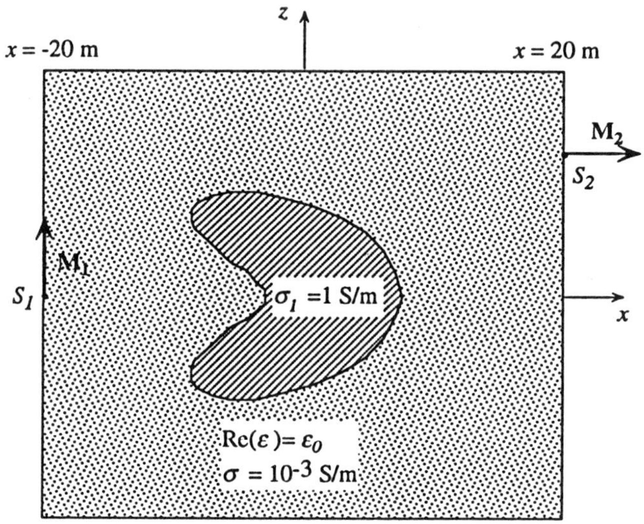

FIGURE 5. Test based on reciprocity: description of the two diffraction problems.

According to the reciprocity relation given by Eq. (85), the magnetic fields $\mathbf{H}_1$ and $\mathbf{H}_2$ are linked by the relation

$$\mathbf{e}_x \cdot \mathbf{H}_1 = \mathbf{e}_z \cdot \mathbf{H}_2$$

The computation has been made for a unique cylinder with the following parameters: the frequency is $f = 1$ MHz, the points S_1 and S_2 are located at $(-20\text{ m}, 0, 0)$ and $(20\text{ m}, 0, z_2)$, respectively; the subsoil is characterized by the real part of its permittivity $\text{Re}(\varepsilon) = \varepsilon_0$, its conductivity $\sigma = 10^{-3}$ S/m and its permeability $\mu = \mu_0$, while the corresponding parameters of the scatterer are $\text{Re}(\varepsilon_1) = 3\varepsilon_0$, $\sigma_1 = 10^{-2}$ S/m and $\mu_1 = \mu_0$. The shape of the boundary is given by the parametric equations:

$$x(t) = 2(-1 + \cos t + \cos 2t)$$
$$z(t) = 2(1.5 \sin t + 0.25 \sin 2t)$$

In Figure 6, the real part and the imaginary part of $\mathbf{e}_x \cdot \mathbf{H}_1$ and of $\mathbf{e}_z \cdot \mathbf{H}_2$ (A/m) have been plotted when z_2 is varied from -20 m to 20 m. The reciprocity relation is satisfied within less than 1%.

The results of the finite element method have also been compared with those published by Torres-Verdin and Habashy (1994). To show the good agreement, the diffraction problem described in Figure 10 (p. 1065) of this reference has been solved here. It is represented in Figure 7: two cylinders with square cross

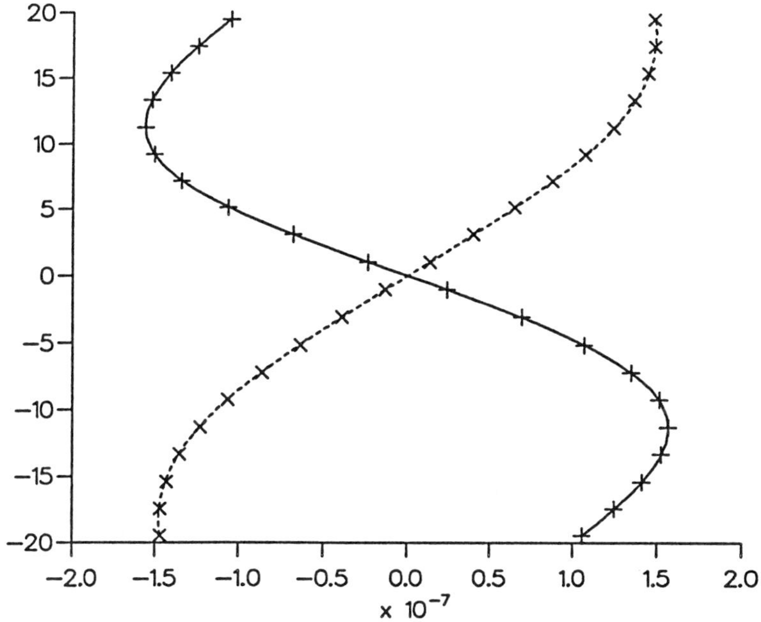

FIGURE 6. Test based on reciprocity: $\mathrm{Re}(\mathbf{e}_x \cdot \mathbf{H}_1)$ (solid line), $\mathrm{Im}(\mathbf{e}_x \cdot \mathbf{H}_1)$ (dashed line), $\mathrm{Re}(\mathbf{e}_z \cdot \mathbf{H}_2)$ (+) and $\mathrm{Im}(\mathbf{e}_z \cdot \mathbf{H}_2)$ (×) versus the depth z_2 of S_2.

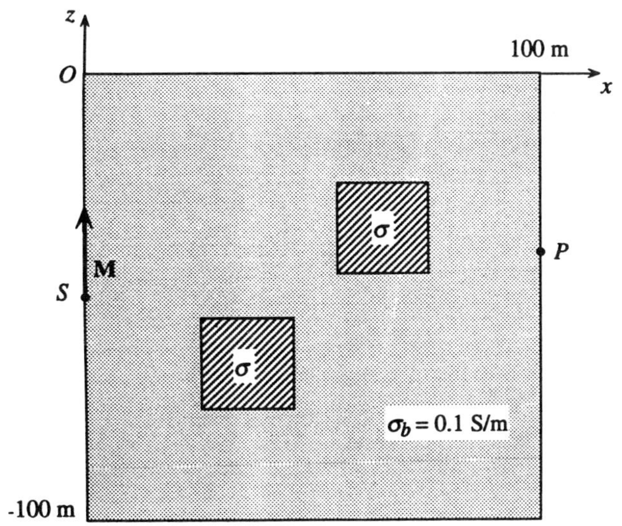

FIGURE 7. Comparison with other theories: description of the diffraction problem.

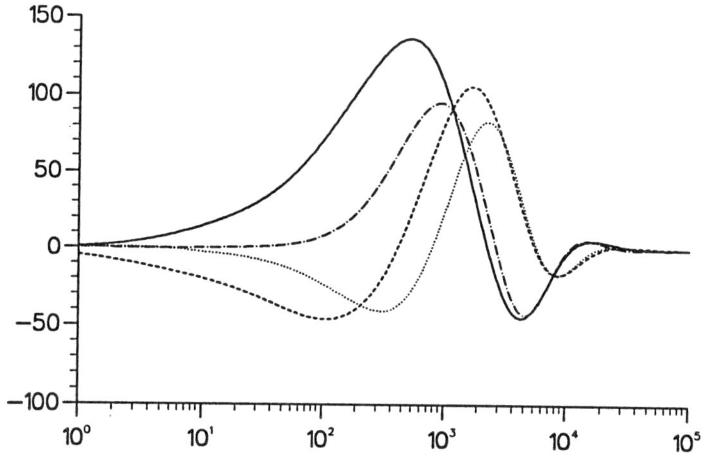

FIGURE 8. Comparison with other theories: vertical component of the diffracted magnetic induction field **B** at P versus frequency (in femtotesla). To be compared with Figure 10 (p. 1065) of the paper by Torres-Verdin and Habashy (1994). Solid line: $\sigma = 100$ S/m, real part. Dashed line: $\sigma = 100$ S/m, imaginary part. Dot–dash line: $\sigma = 1$ S/m, real part. Dotted line: $\sigma = 1$ S/m, imaginary part.

sections 20 m × 20 m and conductivity σ are embedded in a homogeneous background with conductivity $\sigma_b = 0.1$ S/m. A vertical magnetic dipole with moment $\mathbf{M} = \mathbf{e}_z$ is located at $S(0, 0, -50\text{ m})$.

Figure 8 gives the z component of the magnetic field at $P\,(100\text{ m}, 0, -40\text{ m})$ when the frequency varies from 1 Hz to 100 kHz, for both $\sigma = 1$ S/m and $\sigma = 100$ S/m. For the lower ratio σ/σ_b, the agreement with the rigorous finite-difference method by Druskin and Knizhnerman (1988) is excellent, as well as with the nonlinear approximation by Torres-Verdin and Habashy (1994). When the contrast of conductivities becomes high, the approximation fails but the two rigorous methods still give the same result.

The computations have been performed on an IBM Risc6000 workstation (15 Mflops), and the computation time varies from 1 min 15 s when the frequency f is greater than 1 kHz to 10 min at $f = 1$ Hz.

5. NUMERICAL RESULTS

In this section, two examples of the capabilities of the method are given. The first one deals with a vertical magnetic dipole, while the second one is concerned with a vertical electric dipole. In this latter case, the incident field is scattered by two bodies with different conductivities.

First, let us consider a bone-shaped body with conductivity $\sigma = 1$ S/m, embedded in a homogeneous background with conductivity $\sigma_b = 0.1$ S/m, in which a vertical magnetic dipole oscillating with frequency $f = 2$ kHz has been placed at S (-60 m, 10 m). The profile of the scatterer is described by the parametric equations:

$$x = 5(\sin\theta + 0.5\sin 3\theta), \quad z = 2.5\cos\theta, \quad 0 \leqslant \theta \leqslant 2\pi$$

Figure 9 gives the maps of the modulus of the two transverse components of the total magnetic field B_x and B_z around and inside the buried body.

The source is located 10 m from the left side of the maps. Because of its conductivity, the field is damped inside the object. Concerning B_z, the best transmission occurs through the thin middle part and generates a small secondary lobe in the central part of the shadow area. The two other secondary lobes of this area result from the diffraction by the ends. The representation of B_x is more complicated. Here, the incident field is minimum in the thin central part, leading to the splitting of the former central secondary lobe into two smaller ones around $z = +10$ m and $z = -10$ m.

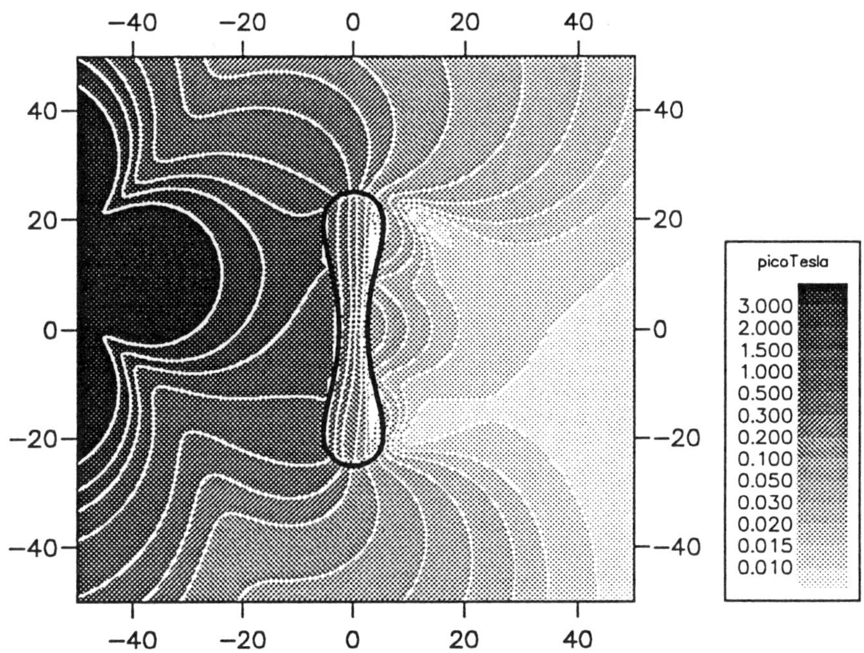

(*a*)

FIGURE 9. Map of the modulus of the transverse components of the magnetic field. The dark solid line represents the boundary of the scatterer. (a) Vertical component. (*Continues on next page.*)

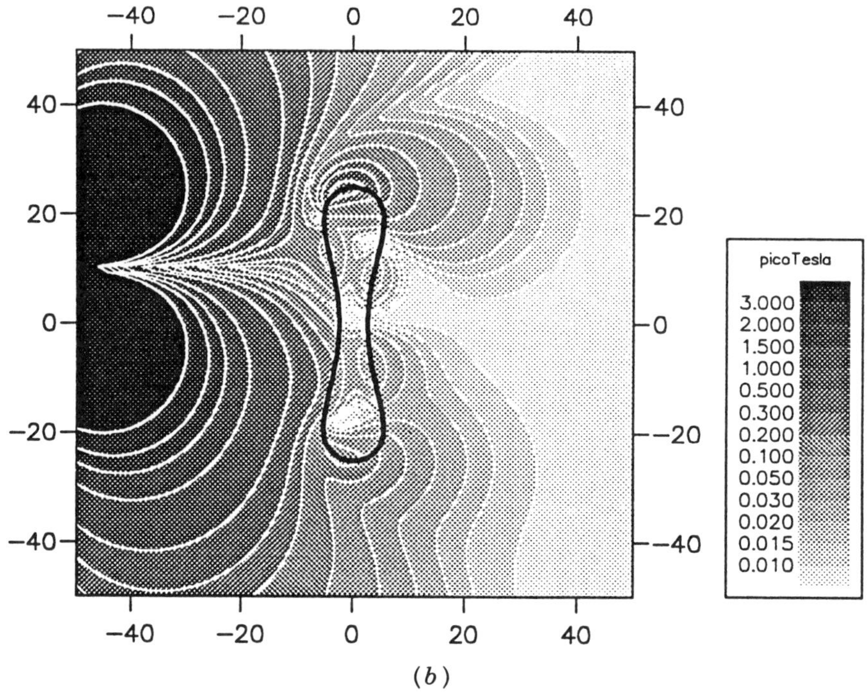

FIGURE 9. (b) Horizontal component.

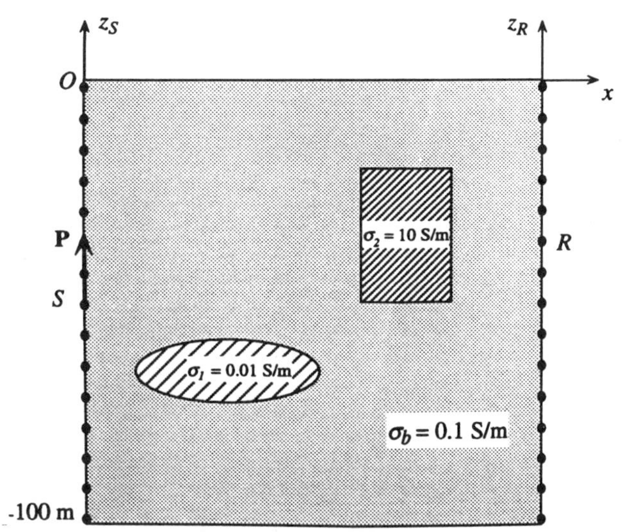

FIGURE 10. Description of the problem.

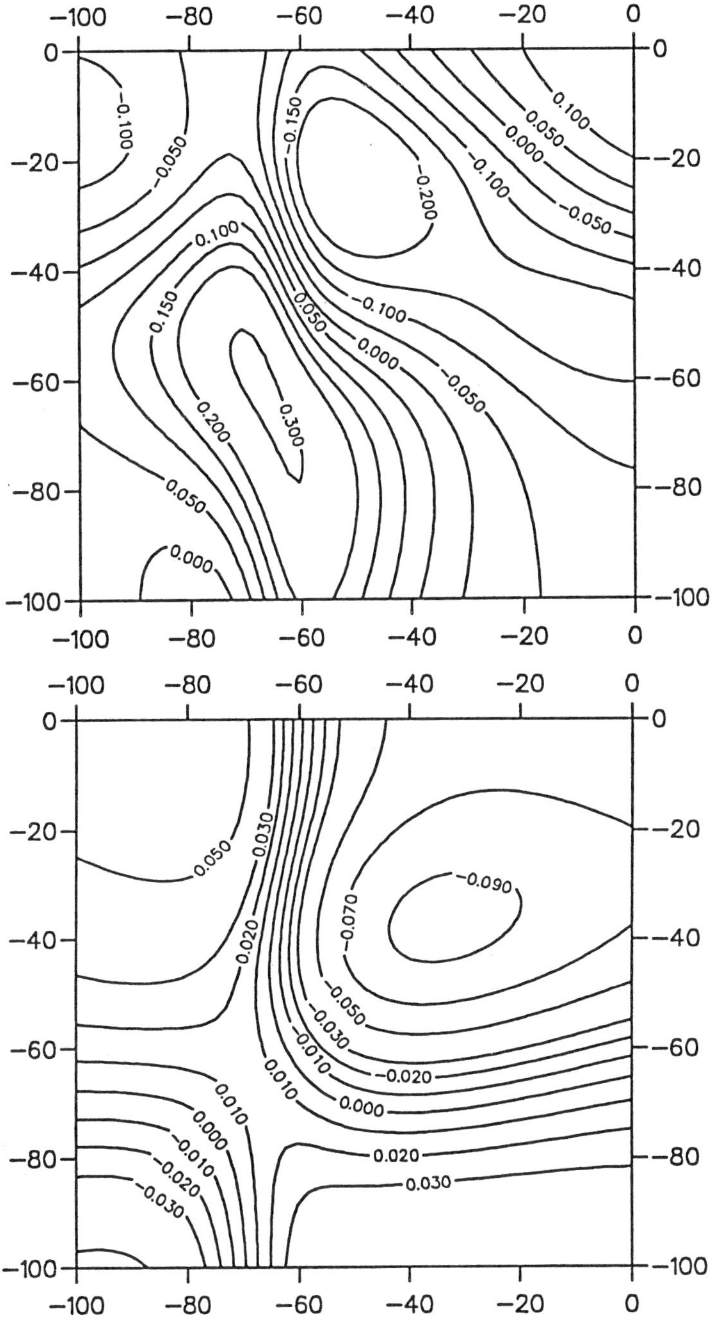

FIGURE 11. Real part (top) and imaginary part (bottom) of E_z when the depths of the source z_S and of the receiver z_R are varied from -100 m to 0. The frequency f of the vertical electric dipole is 0.1 kHz.

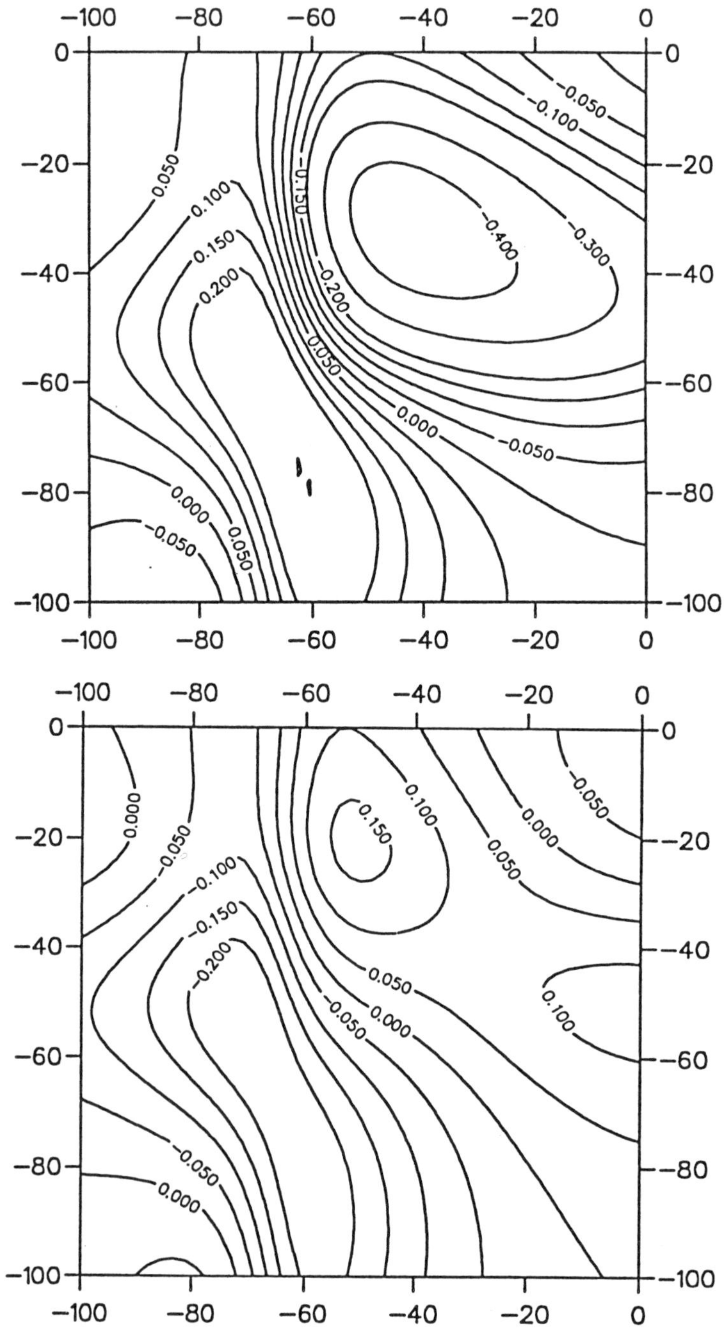

FIGURE 12. Same as Figure 11 with $f = 1\,\text{kHz}$.

The second example gives a typical result of what can really be measured in situ. Calculations are made at various depths of the receiver R for different positions of the source S along its borehole ($x = -100$). Plotting a component of the diffracted field as a function of the two depths z_S and z_R of the source and of the receiver gives a diagram called a tomogram. It contains the information available for the solving of the inverse problem.

Examples are given for the problem depicted in Figure 10. The two parallel boreholes have been drilled in a medium with conductivity $\sigma_b = 0.1$ S/m, 100 m from each other. Two cylindrical bodies, one with elliptic cross section (1/2 x-axis: 20 m, $\frac{1}{2}$ z-axis: 7 m) centered at O_1 (-70 m, -65 m) and conductivity $\sigma_1 = 0.01$ S/m, one with rectangular cross section (20 m wide, 30 m high) centered at O_2 (-30 m, -35 m) and conductivity $\sigma_2 = 10$ S/m, take place between the boreholes. Fifty positions of the source and 100 positions of the receiver have been investigated, from $z_S = z_R = 0$ to -100 m, with a 2-m step for z_S and a 1-m step for z_R. Under these conditions, using 100 finite elements to describe the whole boundary, about 1 hour of computation time was needed for each frequency on an IBM Risc6000 workstation. Figures 11 and 12 show both the real part and the imaginary part of the vertical component of the electric field E_z for two usual frequencies, 0.1 and 1 kHz.

6. CONCLUSION

The combination of the integral formalism with the use of boundary finite elements permits one to elaborate powerful and versatile numerical tools. As applied to subsurface tomography, this rigorous method gives reliable results in a large frequency range, even for high contrasts of conductivities. The code can handle several cylinders with arbitrary cross sections. It can also deal with any kind of source, provided the Fourier transform of the incident field is known.

The computations have been performed on a workstation with reasonable computation times and memory requirements. This makes such a tool valuable for testing the domain of validity of approximate theories. The present model could be extended in order to take into account the air–ground interface, assuming the surface roughness is cylindrical as well.

The solution of the corresponding inverse problem will require the knowledge of a functional derivative of the diffracted field with respect to the shapes and the conductivities of the cylinders (Roger, 1982; Chew and Liu, 1993), combined with a Newton–Kantorovitch process and regularization (Roger, 1981). It is known, from the experience of the laboratory in this field (Roger and Maystre, 1980), that another requirement to deal with the inverse problem is to have at hand an accurate code for solving the direct problem. Obviously, the boundary finite element method fulfills this prerequisite.

ACKNOWLEDGMENTS

The authors thank the Bureau des Recherches Géologiques et Minières (Orléans, France) for financial support.

REFERENCES

Abramowitz, M., and Stegun, I. A. (1965). *Handbook of Mathematical Functions*, p. 360. Dover, New York.

Chew, W. C. (1990). *Waves and Fields in Inhomogeneous Media*, Chapter 9. Van Nostrand-Reinhold, New York.

Chew, W. C., and Liu, Q.-H. (1993). Using CG-FFHT method to solve borehole inverse problem at induction frequencies. *Proc. PIERS*, Pasadena, 1993, p. 284.

Dautray, R., and Lions, J. L. (1985). *Analyse mathématique et calcul numérique pour les sciences et techniques*, Vol. 2, p. 682. Masson, Paris.

Druskin, V., and Knizhnerman, L. (1988). A spectral semi-discrete method for the numerical solution of 3-D non-stationary problems in electrical prospecting. *Izv. Acad. Sci. USSR, Phys. Solid Earth* (*Engl. Transl.*) **8**, 63–74.

Gelius, L.-J., and Stamnes, J. J. (1990). Diffraction tomography: Potentials and problems. In *Scattering in Volumes and Surfaces* (M. Nieto-Vesperinas and J. C. Dainty, eds., pp. 91–109. Elsevier, Amsterdam.

Gradshtein, I. S., and Ryzyk, I. M. (1983). *Table of Integral, Series and Products*, p. 710. Academic Press, New York.

Kleinman, R. E., and Martin, P. A. (1988). On single integral equations for the transmission problem of acoustics. *SIAM J. Appl. Math.* **48**, 307–325.

Martin, P. A., and Ola, P. (1993). Boundary integral equations for the scattering of electromagnetic waves by a homogeneous dielectric obstacle. *Proc. R. Soc. Edinburgh, Sect. A: Math.* **123**, 185–208.

Marx, E. (1982). Single integral equation for wave scattering. *J. Math. Phys.* **23**, 1057–1065.

Maystre, D. (1978). A new general integral theory for dielectric coated gratings. *J. Opt. Soc. Am.* **68**, 490–495.

Maystre, D., and Vincent, P. (1972). Diffraction d'une onde électromagnétique plane par un objet cylindrique non infiniment conducteur de section arbitraire. *Opt. Commun.* **5**, 327–330.

Ngakosso, E. (1994). Modélisation de la diffraction d'un rayonnement dipolaire en Electromagnétisme: Application à la tomographie. Thèse de Doctorat, Université Aix-Marseille III.

Ngakosso, E., Saillard, M., and Vincent, P. (1995). Theoretical and numerical study of conical diffraction by cylindrical objects. *JEWA* **9**, 479–502.

Peden, I. C., and Brew, J. (1995). A laboratory scale model for the study of subsurface scattering in low-loss media with applications to ground penetrating radar. *J. Appl. Geophys.* **33**, 109–118.

Petit, R., and Zolla, F. (1994). The method of fictitious sources as applied to the conical diffraction by a homogeneous rod. *JEWA* **8**, 1–18.

Roger, A. (1981). Newton-Kantorovitch algorithm applied to an electromagnetic inverse problem. *IEEE Trans. Antennas Propg.* **AP-29**, 232–238.

Roger A. (1982). Reciprocity theorem applied to the computation of functional derivatives of the scattering matrix. *Electromagnetics* **2**, 69–83.

Roger, A., and Maystre, D. (1980). Inverse scattering method in electromagnetic optics: application to diffraction gratings. *J. Opt. Soc. Am.* **70**, 1483–1495.

Sato, M., Ohkubo, T., and Niitsuma, H. (1995). Cross-polarization borehole radar measurements with a slot antenna. *J. Appl. Geophys.* **33**, 53–61.

Torres-Verdin, C., and Habashy, T. M. (1994). Rapid 2.5-dimensional forward modeling and inversion via a new nonlinear scattering approximation. *Radio Sci.* **29**, 1051–1079.

Torres-Verdin, C., and Habashy, T. M. (1995). A two-step linear inversion of two-dimensional electrical conductivity. *IEEE Trans. Antennas Propag.* **AP-43**, 405–415.

Van Bladel, J. (1964). *Electromagnetic Fields*, p. 234. McGraw-Hill, New York.

Vincent, P., Saillard, M., and Ngakosso, E. (1995). Rigorous computation of the near field diffracted by a cylinder from dipole sources. *IEEE Trans. Magn.* **MAG-31**, 1614–1617.

CHAPTER TWELVE

Absorbing Boundary Conditions for Finite Element Mesh Truncation

Ü. PEKEL and R. MITTRA

1. INTRODUCTION

The finite element method (FEM) in electromagnetics was originally employed in the analysis of closed-region problems, where one can formulate the boundary value problem using a Dirichlet, Neumann, or mixed type of boundary condition. The numerical advantages of FEM over integral equation techniques, such as the sparsity and bandedness of the pertaining system matrices, and the ability to handle medium inhomogeneities and/or anisotropies in the analysis without additional computational cost, have paved the way for FEM to become a versatile tool for solving electromagnetic modeling problems with complex geometries.

However, when analyzing electromagnetic radiation or scattering problems in unbounded regions with FEM, the infinite domain must be truncated to enable one to perform the analysis of the open-region problem with an inherently finite amount of computational resources. The truncation of the FEM mesh that covers the problem domain is, in general, achieved either by explicitly coupling the exterior domain solution to the interior domain FEM solution with the help of a surface integral equation that plays the role of a boundary condition, or through the imposition of some form of (approximate)

Finite Element Software for Microwave Engineering, Edited by Tatsuo Itoh, Giuseppe Pelosi, and Peter P. Silvester.
ISBN 0-471-12636-5

absorbing boundary condition (ABC) at the truncation boundaries. Clearly, the major challenge in the construction of all types of ABCs has been to make them as nearly transparent as possible to the outgoing waves, and to bring them in as close as possible to the radiating or scattering object without unduly sacrificing the numerical accuracy.

The procedure of direct coupling of FEM for the interior domain and the method of moments (MoM) for the exterior domain, or the use of the mode matching concept, lead to the boundary element method (BEM), as outlined by Paulsen et al. (1988) and Jin (1993), in which nonlocal, but exact boundary conditions are employed to truncate the FEM mesh at the expense of sacrificing the sparsity of the system matrices. A related *coupling procedure*, in which a surface integral equation for the fields in the exterior domain is directly imposed as a boundary condition for the interior domain fields, serves as basis for the hybrid finite element method (HFEM), as explained by Lynch et al. (1986), Jin and Liepa (1988), Yuan (1990), and Jin et al. (1991), which is essentially a combination of FEM and BEM. This approach, however, does not mitigate the problem generally encountered with the use of nonlocal boundary conditions, since the coupling among all the unknowns on the boundary surface inevitably gives rise to a dense system matrix.

The disadvantage of working with dense system matrices has provided considerable incentive to workers in this area to formulate local, but approximate, absorbing boundary conditions (ABCs) that enable one to truncate the mesh without spoiling the sparsity of the system matrices, as demonstrated by Engquist and Majda (1977), Bayliss et al. (1982), Liao et al. (1984), and Higdon (1986). In these approaches, a certain approximate relationship is enforced between the unknowns and their normal derivatives at a boundary surface located at a given distance away from the object. As the boundary surface is taken to *infinity*, the approximate relationship becomes an exact one. To make up for the approximate nature of the ABCs without destroying the bandedness property of the FEM matrix, an approach that uses an exact relationship between the unknowns and their normal derivatives has been developed by Keller and Givoli (1989). The approximate approaches have been extended in the relatively recent past to the fully three-dimensional vector case by Peterson (1988), Webb and Kanellopoulos (1989), and Chatterjee et al. (1993). However, it has been shown by Mittra and Ramahi (1989) that all types of approximate ABCs are susceptible, to a varying extent, to numerical shortcomings in the case of evanescent wave (or cutoff mode) incidence, or, in the case of propagating waves, incidence angles that are close to the grazing incidence limit, because of the inherently approximate nature of the ABCs at a finite distance from the object.

A third category of mesh truncation techniques, in which the interior and exterior domain solutions are coupled on either a single separable surface or two geometrically conforming surfaces, are referred to as the unimoment method, introduced by Mei (1974), and the bymoment method, introduced for two-dimensional applications by Cangellaris and Lee (1990) and extended to three-dimensional problems by Pekel and Lee (1995), respectively. In the unimoment method, eigenfunction expansions in the appropriate coordinate system are employed to represent the fields in the exterior region; the interior region field solutions are determined from the FEM solution of a number of Dirichlet-type boundary value problems. The two sets of solutions are coupled on the separable surface by enforcing the continuity condition of the tangential field components across that surface. In the bymoment method, which has been introduced to overcome the difficulties associated with the unimoment method, the exterior fields are expressed in terms of an appropriate series of expansion functions on the first boundary surface, which geometrically conforms to that of the scattering object. The interior region solutions are found from the FEM solution of a number of Neumann-type boundary value problems, with the same set of expansion functions being utilized in the formulation of the boundary conditions. The two solutions are finally coupled on the second boundary surface by an application of Green's theorem over the problem domain.

A number of numerical techniques have recently been introduced for the purpose of FEM mesh truncation. One of these techniques, referred to as the measured equation of invariance (MEI) and developed by Mei et al. (1994), incorporates a numerical boundary condition, which is designed to pass, with little or no reflection, the fields that are generated by a certain set of test currents, referred to as *metrons* (measures), located on the surface of the scatterer. The (linear) measured equations, through which the boundary conditions are enforced in this approach, are specified to be local, excitation-independent, and applied at distances very close to the scattering object. Various numerical investigations have been performed recently on the validity of this technique, as outlined in Gordon et al. (1993) and Jevtic and Lee (1994, 1995). Recent studies have shown that MEI is really one form of numerical absorbing boundary condition (NABC), and a rigorous theory for the NABCs has been developed by Stupfel and Mittra (1995). Thc NABCs have been applied by Boag et al. to open structures (1994a) and to two-dimensional problems (1994b), and by Boag and Mittra to general three-dimensional problems (1995). The NABC is essentially a linear numerical relationship that relates the value of the field at an edge (node) located at the boundary surface to those at a selected number of neighboring edges (nodes). The relationship in question is derived by imposing the constraint that the linear equations be satisfied, to within a certain tolerance, by all of the outgoing waves originating from the interior domain. The equations that are associated with this relationship are used *in lieu* of the contributions originating from the surface integral term of the FEM formulation.

As an additional recent development in this area, the perfectly matched layer (PML) concept has been introduced by Berenger (1994) with the objective of developing an absorbing boundary condition for the finite difference time domain (FDTD) method. In its original formulation, the PML approach was based on the splitting of the field components into two subcomponents, which are weighted with different constitutive parameters. The original formulation has been compared and linked by Mittra and Pekel (1995) to alternate coordinate-scaling or lossy mapping approaches of space, developed by Chew and Weedon (1994) and Rappaport (1995), respectively, and has in the meantime been modified and extended by Pekel and Mittra (1995a–d) to the FEM frequency domain (FEFD) analysis of radiation and scattering problems. In these applications, the PML-type medium has been employed in conjunction with analytical and numerical ABCs serving as its termination to obtain an improved boundary condition for the FEFD method at a limited increase in computational cost. The governing equations that characterize this particular approach neither require the splitting of the field components of interest nor involve negative conductivity parameters. Alternately, a related approach, in which a uniaxial anisotropic medium is used to synthesize a perfectly matched absorber (PMA), has been introduced by Z. S. Sacks, D. M. Kingsland, R. Lee, and J. F. Lee (private communication). The underlying equations of this approach can be derived from the original PML equations through component-scaling, which is found to lead to the presence of negative conductivity values in the diagonal material property tensors in the direction of propagation.

The objective of this chapter is to discuss the application of the three-dimensional NABCs alluded to above and to investigate, in a detailed manner, the performance of the PML concept in the context of mesh truncation in FEFD applications.

2. A THREE-DIMENSIONAL APPLICATION OF THE NUMERICAL ABSORBING BOUNDARY CONDITION

In this section, the numerical absorbing boundary condition (NABC) approach is applied in conjunction with a three-dimensional edge-based finite element frequency domain (FEFD) formulation to the analysis of the problem very low frequency (VLF) electromagnetic radiation and scattering. In this approach, the singular value decomposition (SVD) scheme is employed to derive linear relationships that link the field values at the edges located on the boundary surface to the field values at neighboring edges. These linear relationships are satisfied, within a specified tolerance, by all of the outgoing waves originating from the interior domain and are used *in lieu* of the contributions of the surface integral term in the FEFD formulation. The NABC scheme appears to be one of the few viable mesh truncation techniques at VLF, since numerical experiments have shown that conventional ABC terminations at the boundaries lead to results that resemble those obtained when using perfect magnetic conductors (PMCs) as termination, as demonstrated by Ko et al. (1994), in which case one has to resort to impractically large domain sizes to ensure sufficient numerical accuracy.

A thorough treatment of the underlying theory of the NABC concept can be found in a paper by Stupfel and Mittra (1995). In this section, we directly follow the method presented by Boag et al. (1994a) for open structures, and its extension to two-dimensional structures, as described by Boag et al. (1994b), to arrive at an NABC scheme developed by Boag and Mittra (1995), which is suitable for three-dimensional FEFD formulations. In contrast to the two-dimensional application, however, the number of edges linked at the boundaries through the NABC scheme is kept constant, while the coefficients that characterize the scheme are determined through an SVD-based algorithm due to Golub and Van Loan (1989).

2.1. Formulation of the Three-Dimensional NABC Scheme

We begin with the formulation of the NABC scheme by considering a general three-dimensional problem, the two-dimensional frontal view of whose geometry is depicted in Figure 1. We implicitly assume that the time-harmonic variation is of the form $e^{+j\omega t}$. A three-dimensional FEFD scattered field formulation employing hexahedral edge elements developed by Mahadevan, Irvine, Pekel, and Mittra serves as the basis in the analysis. It is assumed that all the inhomogeneities in the problem are confined to the region enclosed by the surface S, and that the boundary surface C, which encloses the surface S, is convex.

To be able to derive a linear relationship between the field values at the boundary edges and the neighboring ones, we consider a set of L edges with center points $\mathbf{r}_l$ and tangential unit vectors $\mathbf{t}_l$, where $l = 0, 1, \ldots, L-1$, and the subscript 0 denotes the edge for which the linear relationship is to be obtained. It should be noted that a *point-matching* interpretation is used in the determi-

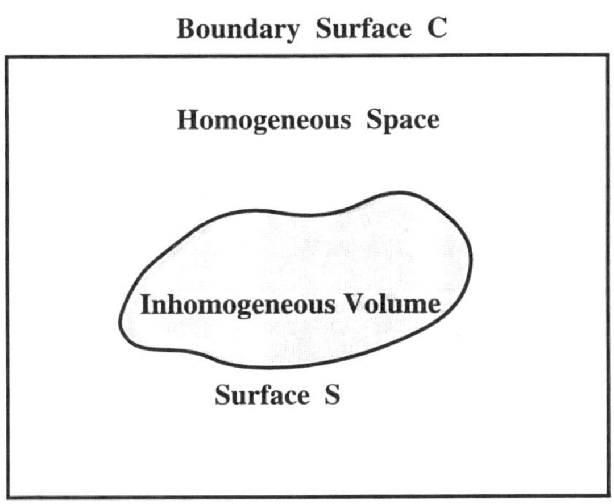

FIGURE 1. Geometry of a scatterer or source surrounded by a homogeneous domain.

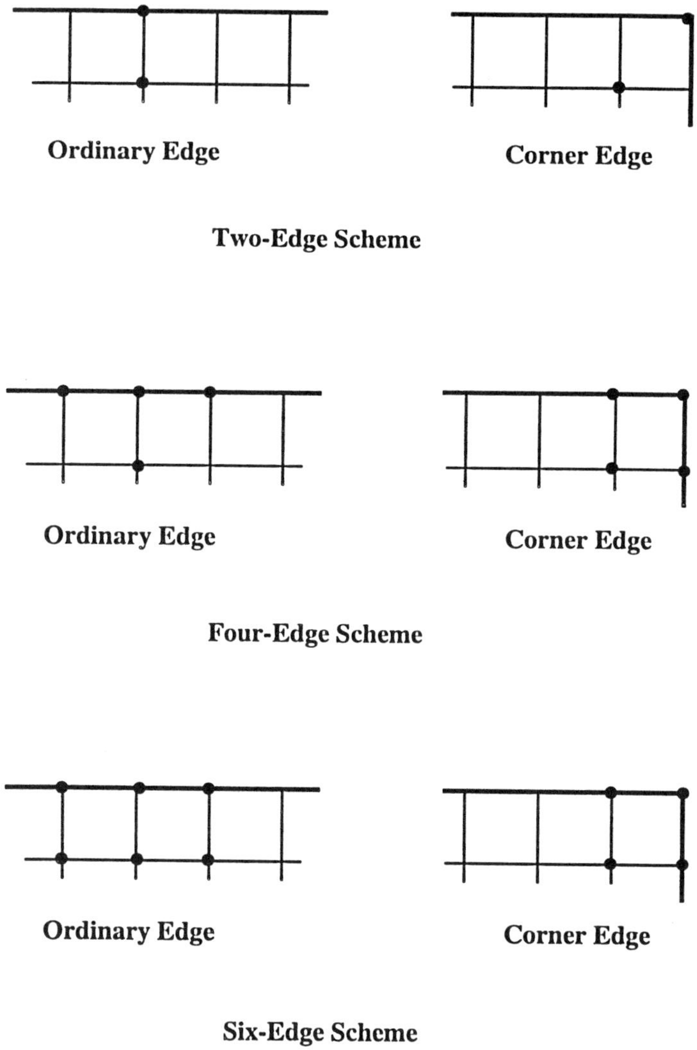

FIGURE 2. The NABC schemes with $L = 2$, $L = 4$, and $L = 6$.

nation of the coefficients that characterize the NABC scheme when edge elements are used; that is, the tangential field components are sampled at the centers of the edges. The specific edge schemes that are used in this section, namely, the $L = 2$, $L = 4$, and $L = 6$ cases, are explained, again using a two-dimensional frontal view, as in Figure 2. We note that one must distinguish between *ordinary* edges and *corner* edges in the respective arrangements. Since orthogonal hexahedral edge elements are employed in this study, the NABCs only link parallel edges, and not orthogonal ones, thereby *simulating* absorbing boundary conditions derived from the scalar wave equations. Other options may be considered for general edge elements.

With the above constraints at hand, we can set up the NABC relationship in the following manner:

$$\mathbf{t}_0 \cdot \mathbf{E}(\mathbf{r}_0) = \sum_{l=1}^{L-1} c_l \mathbf{t}_l \cdot \mathbf{E}(\mathbf{r}_l) \tag{1}$$

where the unknown coefficients c_l that characterize the NABC are to be determined. The selection of an *optimal* value for the number L of edges to be linked among each other through the NABC equations depends on the required numerical accuracy and the stability of the procedure of evaluation of the coefficients. The field in the homogeneous region external to S (including the boundary surfaces) can be calculated from the identity

$$\mathbf{E}(\mathbf{r}) = \int_S \bar{\bar{\mathbf{G}}}(\mathbf{r}, \mathbf{r}_s) \cdot \mathbf{J}_{\mathrm{eq}}(\mathbf{r}_s)\, ds \tag{2}$$

where $\bar{\bar{\mathbf{G}}}$ denotes the pertinent dyadic Green's function, and $\mathbf{J}_{\mathrm{eq}}$ is the equivalent electric surface current flowing on the surface S. Although the current $\mathbf{J}_{\mathrm{eq}}$ is not known *a priori*, it is clear that if Eq. (1) were satisfied by the vectorial quantity $\bar{\bar{\mathbf{G}}}(\mathbf{r}, \mathbf{r}_s) \cdot \mathbf{t}_i$, with $i = 1, 2$ denoting the two directions tangential to the surface S at the point $\mathbf{r}_s$, then this equation would also be satisfied by any field that obeys Eq. (2). Consequently, we can set up a matrix equation of the form

$$\mathbf{Ac} \cong \mathbf{b} \tag{3}$$

by enforcing the condition that Eq. (1) be approximately satisfied by the quantity $\bar{\bar{\mathbf{G}}}(\mathbf{r}, \mathbf{r}_s) \cdot \mathbf{t}_n$ for $n = 1, 2, \ldots, N$. In Eq. (3), $\mathbf{c}$ is a column vector with the $L - 1$ unknown coefficients, and $\mathbf{A}$ is an N by $L - 1$ matrix whose nlth entry is given by $A_{nl} = \mathbf{t}_l \cdot \bar{\bar{\mathbf{G}}}(\mathbf{r}, \mathbf{r}_s) \cdot \mathbf{t}_n$. Also, the right-hand side vector $\mathbf{b}$ has N entries, which are defined by $\mathbf{b}_n = \mathbf{t}_0 \cdot \bar{\bar{\mathbf{G}}}(\mathbf{r}, \mathbf{r}_s) \cdot \mathbf{t}_n$, where we assume that such an approximate equation is formulated for possibly all edges on the boundary surface C, with L edges linked through the NABC at the point $\mathbf{r}_s$ and excited by N *test sources* located on the surface S.

Equation (3) represents an overdetermined system of equations (i.e., the number N of test sources is larger than the number L of edges to be related among each other) and can formally be solved for $\mathbf{c}$ in the least-square sense by the use of a QR-based decomposition of $\mathbf{A}$. However, this procedure may result in a severe increase in the condition number of the system matrix when the above equations replace the corresponding FEM equations, leading to slower convergence of the CG-based solution scheme and lower numerical accuracy. To prevent the possibility of such numerical instabilities, we return to the physical interpretation of the NABC equations. We realize that if the set of *test sources* is sufficiently dense over the surface S, the number of degrees of freedom present in the field in the vicinity of $\mathbf{r}_0$ (i.e., the location of the edge of interest), denoted by N_{dof}, can be measured by evaluating the number of singular values of the matrix $\mathbf{A}$ that are larger than a relative threshold value, as shown in Golub and Van Loan (1989). It is evident that there is no particular need to construct NABC equations that are more accurate than the conventional FEM equations in the interior domain. A measure for the accuracy in the interior domain can be found from Δ_{FEM}, namely, the rms error that occurs in the conventional FEM equations by the same set of *test fields* $\bar{\bar{\mathbf{G}}}(\mathbf{r}, \mathbf{r}_s) \cdot \mathbf{t}_n$ for $n = 1, 2, \ldots, N$. A representative value of Δ_{FEM} can be calculated by considering only those edges located one layer away from the boundary surface C in all three directions, instead of calculating the error in the entire problem domain, thereby lowering the additional computational cost. Equation (3) can thus be *solved* by applying an SVD-based algorithm on $\mathbf{A}$ as

$$\mathbf{c} = \mathbf{U}^H \boldsymbol{\Sigma}^+ \mathbf{V}\mathbf{u} \tag{4}$$

where $\mathbf{A} = \mathbf{U}\boldsymbol{\Sigma}\mathbf{V}^H$, and the matrices $\mathbf{U}$ and $\mathbf{V}$ are orthogonal, and the matrix $\boldsymbol{\Sigma}$ is diagonal. It should be noted that the superscript H in Eq. (4) denotes the Hermitian conjugate of the original matrix, and $\boldsymbol{\Sigma}^+$ represents the rank-deficient pseudo-inverse of the matrix $\boldsymbol{\Sigma}$, which is obtained by discarding those singular values σ_l for which the inequality $\sigma_l/\sigma_{\max} < \Delta_{\mathrm{FEM}}$ is valid. The number of the retained singular values corresponds to the effective rank of $\mathbf{A}$, and also represents the value of N_{dof} present in the field at the location $\mathbf{r}_0$. Clearly, by deliberately discarding smaller singular values, we sacrifice, to some extent, the numerical accuracy of the solution of Eq. (3); however, the trade-off is that we are able to ensure its stability.

At this point, we can return to the unsettled issue of selection of an *optimal* value for the parameter L, namely, the number of edges to be incorporated in the NABC scheme. Since the parameter N_{dof} varies over the boundary surface C, the selection of the *optimal* value of the parameter L is directly determined by the largest value of N_{dof} evaluated at a number of representative edges. In this study, all the boundary edges have been considered in the determination of L, although it is known that in practice the largest values of N_{dof} materialize in those regions of the mesh with the largest radii of curvature.

The NABC expression given in Eq. (1) is used to replace the conventional FEM equations incorporating the contributions from the surface integral in the FEM formulation at the boundary edges. With this new FEM system matrix at hand, we can now solve for the field distribution in the entire problem domain.

2.2. Numerical Results

In this section, we begin by considering the problem of radiation from a linear antenna located in a homogeneous, lossy dielectric medium, namely, seawater, and investigate this problem by using a three-dimensional finite element frequency domain (FEFD) approach in conjunction with the above-mentioned numerical absorbing boundary conditions (NABCs). The analysis of the linear antenna is then extended to the case in which a perfectly conducting inhomogeneity in the form of a prism is present in the problem of domain of interest. The frequency of operation is chosen to be 1.0 kHz, and the linear antenna is assumed to have a uniform current distribution along its entire length. The FEFD approach in question is based on a scattered electric field formulation and employs a nonuniform, but orthogonal, mesh structure that comprises hexahedral, first-order edge elements with 12 edges (or unknowns) per element, as developed by Mahadevan et al. (1996).

Linear Antenna Radiating in a Homogeneous Surrounding Medium of Seawater. The geometry that is considered in the first case is shown in Figure 3, where the linear antenna is radiating in the homogeneous surrounding seawater medium. It is noted that the linear antenna, which is assumed to have a length of 10.0 m, is modeled in the FEFD approach by a number of consecutive edge elements whose constitutive parameters are identical to those of the surrounding seawater medium, namely, $\varepsilon_r = 72.0$, $\sigma = 4.0$, and $\mu_r = 1.0$. The *source elements*, and thus the antenna model, have a width and depth of 1.0 m. The four x-directed edges of each of these *source elements*, in contrast to the four y-directed and four z-directed edges, are assigned a known, uniform electric field strength in order to obtain a uniform current distribution along the length of the linear antenna.

During the process of evaluation of the complex coefficients that are associated with the NABCs for the case of the homogeneous problem domain,

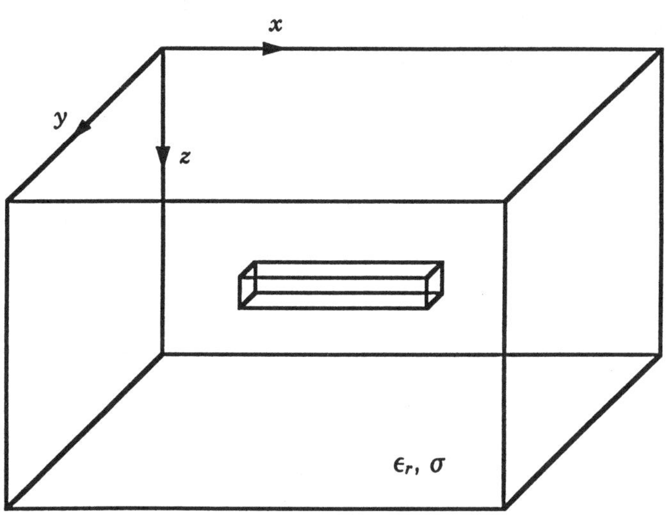

FIGURE 3. Linear antenna radiating in a surrounding homogeneous medium of seawater.

a sufficient number of Hertzian dipole sources are placed as test sources along the central axis of the linear antenna model. While the sufficiency criterion is based partly on the geometry and overall size of the primary source, it is primarily dependent on the number of edges that are incorporated in the particular NABC scheme under consideration. Clearly, the number of edges used in the formulation of the NABCs, and hence the number of complex coefficients that are associated with them, directly govern the performance of the NABCs for a specific size of the problem domain, and the resulting accuracy of the calculated fields. In this section, two-, four-, and six-edge schemes are considered in the NABC formulations, as was described earlier in greater detail.

In the first numerical application, the size of the homogeneous problem domain is selected as $82\,\text{m} \times 41\,\text{m} \times 41\,\text{m}$ in the x-, y-, and z-directions, respectively, with the 10-m linear antenna residing in the center of the domain. The orthogonal, but nonuniform mesh that is applied to analyze this problem leads to 58,632 unknown electric field values. The relatively large size of the problem domain implies that the two-edge scheme, instead of the four- or six-edge scheme, is sufficient to formulate the NABCs along the six boundary surfaces of the domain, since the field components along the various edges on these surfaces become less and less coupled as the boundary surfaces are taken farther away from the primary source. As expected, among all the schemes of this study, the two-edge scheme is the computationally least expensive one, since it does not require the explicit solution of an overdetermined system of linear equations, which must be set up in the four- and six-edge schemes by specifying a number of test sources (i.e., Hertzian dipoles) that are larger than the number of edges to be incorporated in the scheme under consideration. In the two-edge scheme, a simple *optimization rule* based on mean-square-error minimization can be applied to determine the second complex coefficient in accordance with the freely selected first one, although by again employing a comparatively larger number of test sources directed along the length of the linear antenna. With the first coefficient c_0 selected as unity, we can determine the second one, denoted by c_1, as

$$c_1 = \frac{\mathbf{A}^H \mathbf{b}}{\mathbf{A}^H \mathbf{A}} \tag{5}$$

where the superscript H again denotes the Hermitian conjugate of the original matrix, and $\mathbf{b}$ is the right-hand side vector in Eq. (3) associated with the NABC coefficients.

The variations of the dominant field components E_x, E_z, and H_y radiated by the linear antenna in the homogeneous seawater medium as calculated by the FEFD approach in conjunction with the two-edge NABC scheme are plotted in Figure 4. The calculated field values are compared to the respective exact ones, where all curves are obtained by sampling the field values along a line at a distance of 18.0 m underneath the antenna, or 2.5 m (i.e., 1.25 layers)

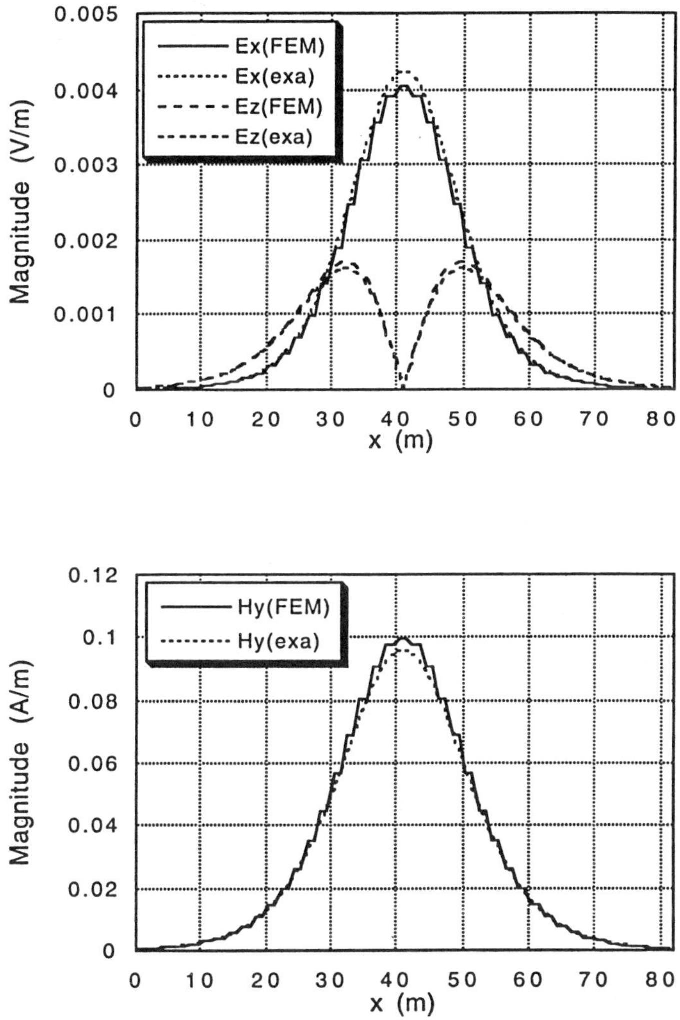

FIGURE 4. Field amplitudes produced by linear antenna radiating in homogeneous medium of seawater, calculated at a distance of 18.0 m underneath antenna.

above the bottom boundary surface. It is clearly seen that the agreement between the calculated and exact field variations is good, which implies that the two-edge NABC scheme is sufficient in terms of interrelating the radiated field components along the element edges on the boundary surfaces for this particular size of the problem domain.

Next, we lower the size of the homogeneous problem domain to 58 m × 29 m × 29 m, with the orthogonal, nonuniform mesh, which requires 21,804 unknown electric field values along the element edges. It is known that as the size of the domain is gradually reduced, it becomes necessary to incorporate a growing number of edges in the formulation of the NABCs. However, in the application of the four-edge, six-edge, and also higher-order NABC schemes, the solution of the overdetermined system of linear equations will, in contrast to the two-edge scheme with its simple *optimization rule*, require the use of a numerically stable algorithm that is geared toward solving such systems. Moreover, we also need a clear and well-defined criterion for *optimally* choosing the particular number of edges to be incorporated in the NABC formulation, such that the condition-number of the resulting FEM matrix is kept at an acceptable level, and the accuracy of the resulting field solutions is sufficiently high. In this section, a singular value decomposition (SVD) algorithm is employed to achieve both goals while solving the overdetermined system of equations.

The implementation of the SVD-based criterion that is used to select the number of edges to be included in the NABC formulation and to determine the tolerance threshold to be applied in the subsequent evaluation of the NABC coefficients can be described in the following manner. First, the fields radiated by the test sources are evaluated along the element edges, which lie on the six surfaces that are located exactly one layer of elements underneath the respective six main boundary surfaces. The calculated test fields for each edge on the six surfaces in question are then multiplied with the corresponding FEM matrix entries situated in the rows that are associated with the particular edge at which the test fields were calculated. The partial results of these multiplications are then added in order to obtain a measure of the error for each edge on those six surfaces, denoted by the parameter Δ_{FEM} in the preceding subsection. This error measure is directly linked to the finite accuracy of the FEM matrix; obviously, if the entries in the FEM matrix were completely accurate, the resulting sum for each edge would be equal to zero. The rms average of these nonzero error measures for the edges located on the six surfaces in question is divided by the rms average of the calculated test fields at those edges to obtain a normalized, global error measure. This measure serves as the SVD tolerance threshold and is used together with the highest order NABC scheme under consideration to perform an initial SVD analysis of all the overdetermined systems associated the edges on the boundary surfaces. The optimal number of edges (i.e., the parameter L) included in the main NABC analysis procedure is determined by taking the highest rank number (or, equivalently, N_{dof}) of the overdetermined systems encountered in this initial SVD analysis phase and incrementing it by one. While the reduced number of edges eventually included in the NABC scheme may lead to a slight decrease in the numerical accuracy of the calculated fields, since some of the smaller singular values are deliberately discarded in this process, it results, as mentioned above, in a numerically stable process of determination of the NABC coefficients.

The E_x, E_z, and H_y field components, as obtained by using FEM in conjunction with an *optimal* four-edge NABC scheme for the domain size

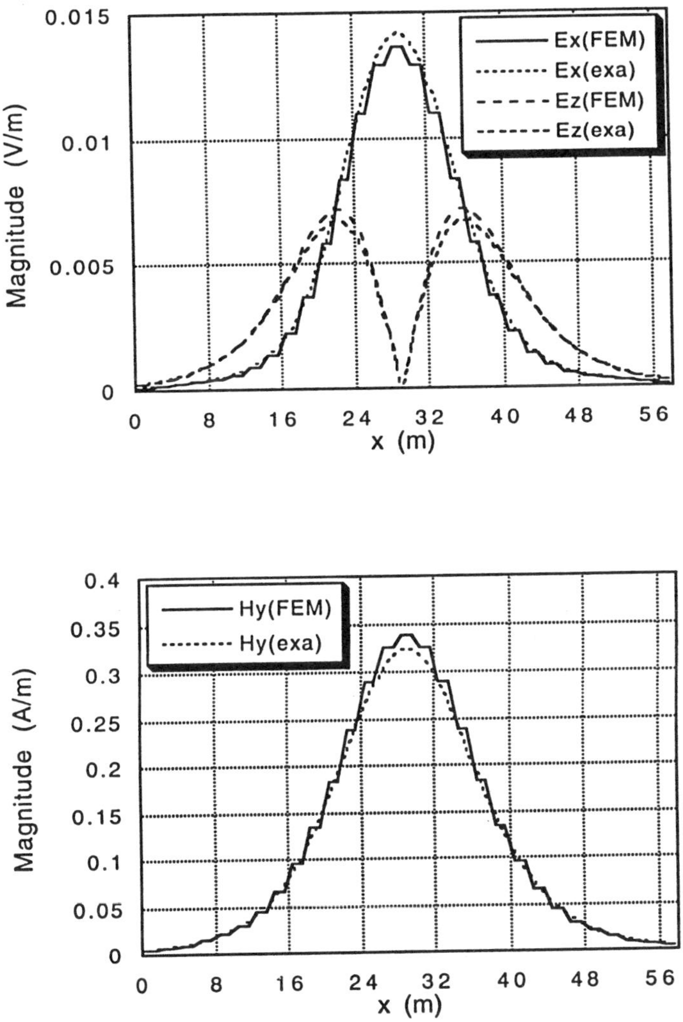

FIGURE 5. Field amplitudes produced by linear antenna radiating in homogeneous medium of seawater, calculated at a distance of 12.0 m underneath antenna.

under consideration, are plotted and compared to their analytical counterparts in Figure 5, where the field points are located at a distance of 12.0 m underneath the linear antenna, or 2.5 m (i.e., 1.25 layers) above the bottom boundary surface. Again, the agreement between the calculated and exact field variations appears to be good, which implies that the *optimal* four-edge NABC scheme is sufficient for this particular domain size.

For a homogeneous domain of a smaller size of 42 m × 21 m × 21 m, the FEFD approach can be applied in conjunction with an *optimal* six-edge NABC

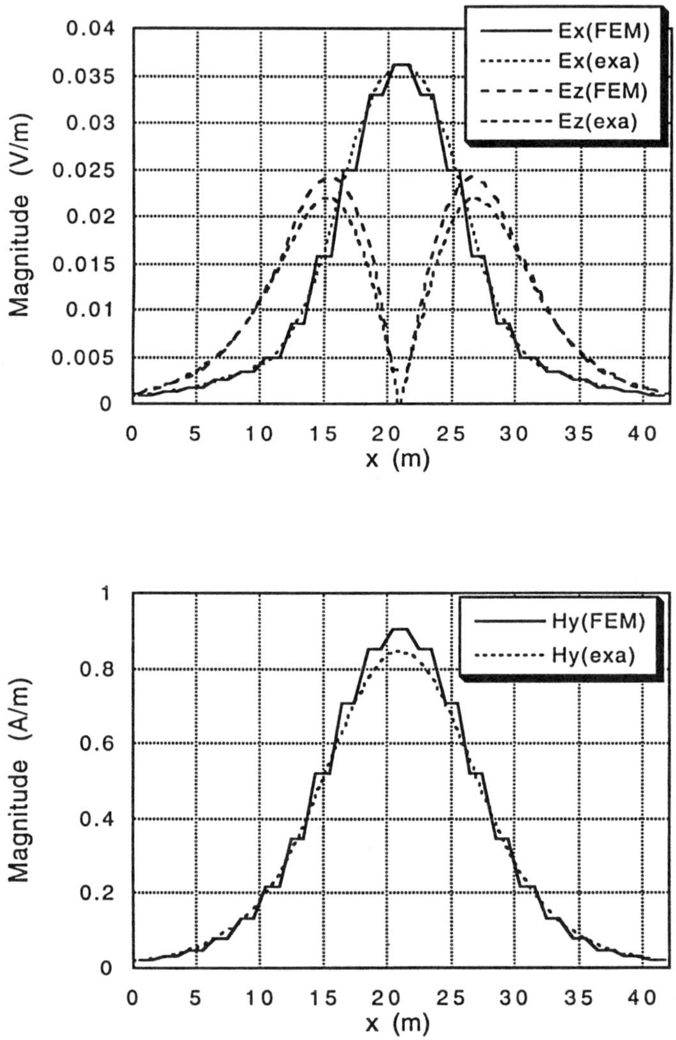

FIGURE 6. Field amplitudes produced by linear antenna radiating in homogeneous medium of seawater, calculated at a distance of 8.0 m underneath antenna.

scheme, as predicted by the above outlined *optimization procedure.* The calculated field components and the exact ones, which are calculated along a line of field points that lie 8.0 m underneath the antenna (and 2.5 m, i.e., 1.25 layers above the bottom boundary surface) are plotted in Figure 6. The agreement between the two sets of results is seen to be good, which indicates that the six-edge scheme for the NABCs is effective for this particular problem domain. We note that as the size of the problem domain is steadily reduced, a larger number of edges must be incorporated in the formulation of the NABCs to ensure a proper absorption of the radiated fields at the six boundary surfaces. However, a substantially reduced domain size clearly implies a considerably lower CPU time requirement in the computation of the field solutions.

Linear Antenna in Seawater Radiating in the Presence of a PEC Inhomogeneity. In this numerical application of the NABCs, a small, perfectly conducting prism is added to the originally homogeneous surrounding medium of seawater, while retaining the same linear antenna model with the same dimensions. The size of the problem domain is chosen to be 50 m × 21 m × 39 m in the x-, y- and z-directions, respectively. The linear antenna again resides at the center of the domain, as shown in Figure 7. The dimensions of the conducting prism are given by 2 m × 1 m × 2 m, and it is located on the same y-plane as the linear antenna. The center of the prism is 16 m off the center of the linear antenna in the x-direction, and 9.5 m away from it in the z-direction.

During the application of the NABCs in this particular problem geometry, the test sources are again placed along the length of the linear antenna; however, one test source is placed at the center of the conducting inhomogeneity (i.e., prism) as well, in order to incorporate the effect of its presence in the domain in the formulation of the NABCs. The total field components of interest, which have been calculated with the help of the *optimal* four-edge NABC scheme, are plotted in Figure 8 along a line that lies at a distance of 8.0 m directly underneath the linear antenna (or, equivalently, 0.5 m above the prism). We observe that the overall magnitudes of the field components of interest in this numerical application agree with the ones of the comparatively

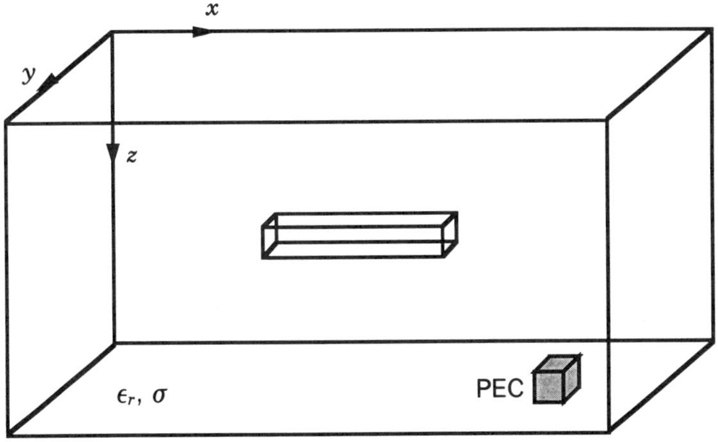

FIGURE 7. Linear antenna radiating in seawater in the presence of a PEC prism.

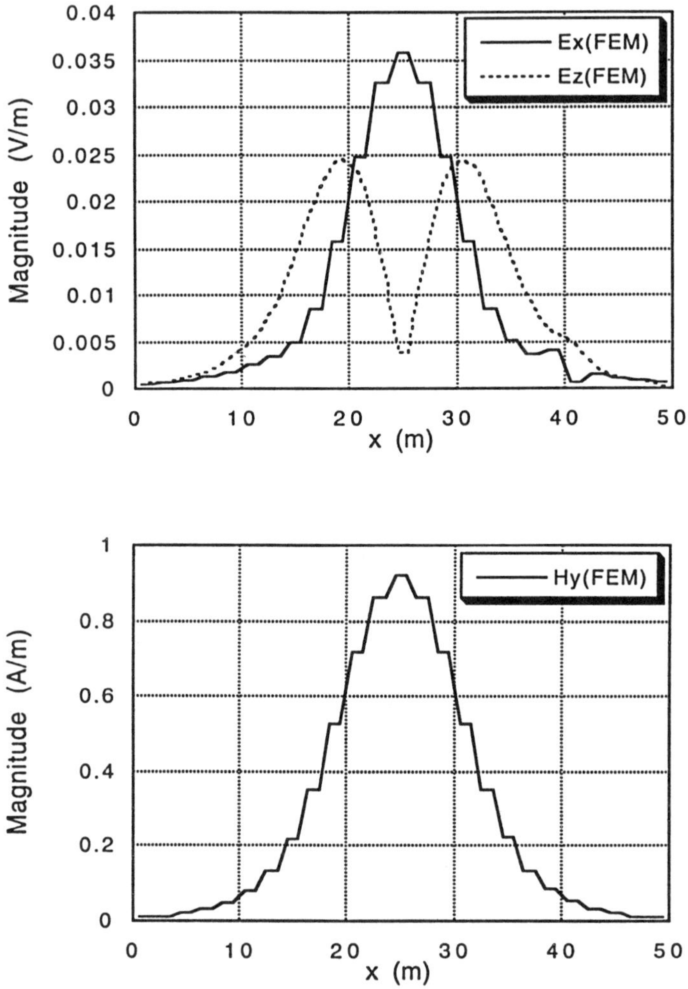

FIGURE 8. Field amplitudes produced by linear antenna radiating in seawater in the presence of the PEC prism, calculated at a distance of 8.0 m underneath antenna.

smaller homogeneous domain of the preceding application, since the field components are calculated at the same distance from the linear antenna. However, it is clearly seen that in the presence of the conducting prism, centered at $x = 41.0$ m, the E_x component exhibits a local minimum and the E_z component a local maximum immediately above the prism, as one would expect from the nature of the implied boundary conditions for the tangential and normal electric fields at the surface of the conducting prism. Although the effect of the inhomogeneity on the H_y component is not as evident, we can, nevertheless, note a slight buildup in this component in the neighborhood of the prism, due to the fact that a mainly x-directed current is induced in this object when it is illuminated by an x-directed linear antenna. The *perturbations* that are observed in the field components of interest due to the presence of the conducting object obviously imply that the object (i.e., the prism) is electrically detectable, as found by Ko et al. (1994). This numerical example also demonstrates the effectiveness of the NABCs for cases in which perfectly conducting inhomogeneities are present in the problem domain.

2.3. Concluding Remarks

In this section, we have demonstrated the validity of the numerical absorbing boundary condition (NABC) as a tool for mesh truncation in a three-dimensional finite element method frequency domain (FEFD) analysis scheme, which employs hexahedral edge elements. The NABC is based on a number of linear relationships, which are formulated on the boundary of the mesh in order to enforce an approximate absorption condition that is to be satisfied by all outgoing waves originating in the interior domain. The calculation of the complex coefficients that characterize the NABC is accomplished through the use of an SVD scheme, which utilizes a global error measure associated with the FEM solution in the interior domain as a criterion of accuracy. The NABC is found to be particularly useful in the very low frequency (VLF) range of operation, where conventional ABCs do not work properly and lead to inaccurate results. The NABCs also offer a major computational advantage over the application of approximation mesh truncation techniques at VLF, such as PEC and/or PMC surface terminations, since the latter techniques generally require the use of problem domain of impractically large sizes.

3. THE PERFECTLY MATCHED LAYER AS A MESH TRUNCATION TECHNIQUE FOR THE FINITE ELEMENT METHOD

Recently, a novel concept referred to as *perfectly matched layer* (PML) has been introduced by Berenger (1994) for the purpose of effectively establishing absorbing boundary conditions for the finite difference time domain method (FDTD). The PML concept is based on the phenomenon of reflectionless absorption of electromagnetic waves by a non-Maxwellian medium, and it has been validated and extended to three-dimensional Cartesian FDTD meshes by Katz et al. (1994). However, in its orginal formulation, the PML approach requires the decomposition (i.e., splitting) of the components of the electric and magnetic fields into two constituent parts, thereby resulting overall in 12 field components in which two different constitutive parameters are assigned to the two subcomponents of each of the original six field components. In this chapter, a frequency domain interpretation of the PML concept is presented in which the somewhat artificial decomposition of the original, physical field components is avoided, but in which distributed dependent sources are utilized in the equations that characterize the PML medium. The constitutive parameters of the medium will be seen to retain negative imaginary parts, or, equivalently, the electric and magnetic conductivities of the PML medium will turn out to be non-negative real numbers. The resulting equations for the PML medium will be shown to be identical to the ones that were presented in a paper based by Chew and Weedon (1994) on the concept of longitudinal scaling of Maxwell's equations, or, equivalently, the lossy anisotropic mapping of space, as described by Rappaport (1995). Finally, another equivalent version developed by Pekel and Mittra (1995a–d), which utilizes *transverse scaling*, will be employed to obtain a suitable weak formulation for the finite element method (FEM) analysis of radiation and scattering problems.

3.1. The PML Equations

The aim in formulating the governing equations of the PML medium is to ensure that a plane wave that is incident from free space to the semi-infinite PML region at an arbitrary angle is completely absorbed there, without giving rise to a reflected wave in the free-space region. Equivalently, it might be said that the aim is to ensure the complete transmission of the incident plane wave from the free-space region into the PML region, as indicated in Figure 9.

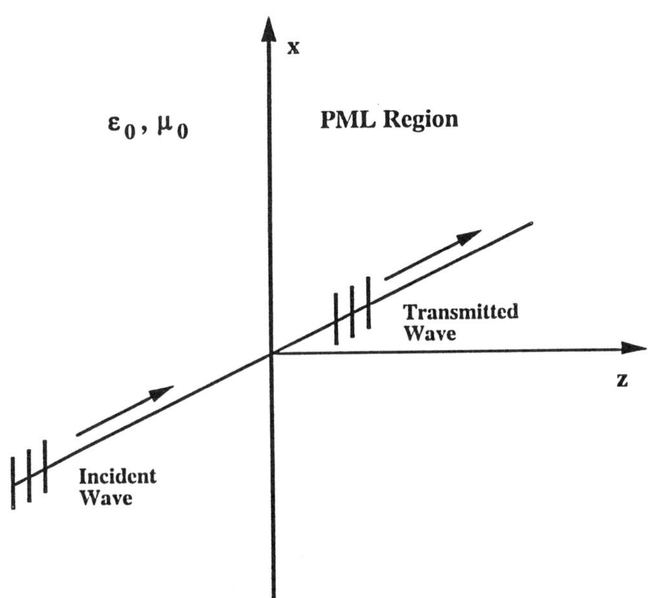

FIGURE 9. Reflectionless absorption of an incoming plane wave at a free-space to PML medium interface.

The original, split-form version of the PML equations in the time domain are given by Berenger (1994) and Katz et al. (1994) in the following form:

$$\mu_0 \frac{\partial H_{xy}}{\partial t} + \sigma_y^* H_{xy} = -\frac{\partial(E_{zx} + E_{zy})}{\partial y}, \quad \mu_0 \frac{\partial H_{xz}}{\partial t} + \sigma_z^* H_{xz} = \frac{\partial(E_{yx} + E_{yz})}{\partial z} \tag{6a}$$

$$\mu_0 \frac{\partial H_{yz}}{\partial t} + \sigma_z^* H_{yz} = -\frac{\partial(E_{xy} + E_{xz})}{\partial z}, \quad \mu_0 \frac{\partial H_{yx}}{\partial t} + \sigma_x^* H_{yx} = \frac{\partial(E_{zx} + E_{zy})}{\partial x} \tag{6b}$$

$$\mu_0 \frac{\partial H_{zx}}{\partial t} + \sigma_x^* H_{zx} = -\frac{\partial(E_{yx} + E_{yz})}{\partial z}, \quad \mu_0 \frac{\partial H_{zy}}{\partial t} + \sigma_y^* H_{zy} = \frac{\partial(E_{xy} + E_{xz})}{\partial y} \tag{6c}$$

$$\varepsilon_0 \frac{\partial E_{xy}}{\partial t} + \sigma_y E_{xy} = \frac{\partial(H_{zx} + H_{zy})}{\partial y}, \quad \varepsilon_0 \frac{\partial E_{xz}}{\partial t} + \sigma_z E_{xz} = -\frac{\partial(H_{yx} + H_{yz})}{\partial z} \tag{7a}$$

$$\varepsilon_0 \frac{\partial E_{yz}}{\partial t} + \sigma_z E_{yz} = \frac{\partial(H_{xy} + H_{xz})}{\partial z}, \quad \varepsilon_0 \frac{\partial E_{yx}}{\partial t} + \sigma_x E_{yx} = -\frac{\partial(H_{zx} + H_{zy})}{\partial x} \tag{7b}$$

$$\varepsilon_0 \frac{\partial E_{zx}}{\partial t} + \sigma_x E_{zx} = \frac{\partial(H_{yx} + H_{yz})}{\partial x}, \quad \varepsilon_0 \frac{\partial E_{zy}}{\partial t} + \sigma_y E_{zy} = -\frac{\partial(H_{xy} + H_{xz})}{\partial y} \tag{7c}$$

We note in the above equations that after the field components are split, their respective subcomponents are *weighted* by different constitutive parameters, raising the question whether the PML medium can be described by conventional 3×3 constitutive parameter tensors. To answer this question we eliminate all the split fields ($H_{xy}, E_{xy}, \ldots$) from Eqs. (6) and (7) and replace them by the conventional ones ($H_x, E_x, \ldots$) instead, by using simple addition and substitution in the above six pairs of equations [e.g., we add and solve Eq. (6a) to get $H_{xz} = -jk_z E_y/(\sigma_2^* + j\omega\mu_0)$]. The unsplit equations read:

$$(j\omega\mu_0 + \sigma_1^*)H_x = -jk_z E_y + jk_y E_z + \frac{\sigma_2^* - \sigma_1^*}{j\omega\mu_0 + \sigma_2^*} jk_z E_y \tag{8a}$$

$$(j\omega\mu_0 + \sigma_2^*)H_y = -jk_x E_z + jk_z E_x + \frac{\sigma_3^* - \sigma_2^*}{j\omega\mu_0 + \sigma_3^*} jk_x E_z \tag{8b}$$

$$(j\omega\mu_0 + \sigma_3^*)H_z = -jk_y E_x + jk_x E_y + \frac{\sigma_1^* - \sigma_3^*}{j\omega\mu_0 + \sigma_1^*} jk_y E_x \tag{8c}$$

$$(j\omega\varepsilon_0 + \sigma_1)E_x = -jk_y H_z + jk_z H_y - \frac{\sigma_2 - \sigma_1}{j\omega\varepsilon_0 + \sigma_2} jk_z H_y \tag{9a}$$

$$(j\omega\varepsilon_0 + \sigma_2)E_y = -jk_z H_x + jk_x H_z - \frac{\sigma_3 - \sigma_2}{j\omega\varepsilon_0 + \sigma_3} jk_x H_z \tag{9b}$$

$$(j\omega\varepsilon_0 + \sigma_3)E_z = -jk_x H_y + jk_y H_x - \frac{\sigma_1 - \sigma_3}{j\omega\varepsilon_0 + \sigma_1} jk_y H_x \tag{9c}$$

where, for convenience, we have used a frequency domain representation, and the subscripts 1, 2, 3 of the electric and magnetic conductivities (σ) can be associated with the x-, y-, and z-directions, respectively. We note that in addition to the regular terms arising from Maxwell's curl equations in an anisotropic medium with a uniaxial, 3×3 conductivity tensor, there are terms involving the differences of the conductivities—for example $(\sigma_1 - \sigma_2)$—appearing on the right-hand sides of the above equations, which must be interpreted as source terms, albeit dependent, that are distributed in the entire semi-infinite PML region. We must therefore conclude that due to the presence of the distributed dependent sources on the right-hand sides of Eqs. (8a) to (9c), the PML medium ceases to be passive, even if the σ values are non-negative.

Next, we observe that if we impose the *PML impedance matching condition*

$$\frac{\sigma_1}{\varepsilon_0} = \frac{\sigma_1^*}{\mu_0}, \quad \frac{\sigma_2}{\varepsilon_0} = \frac{\sigma_2^*}{\mu_0}, \quad \frac{\sigma_3}{\varepsilon_0} = \frac{\sigma_3^*}{\mu_0} \tag{10}$$

then Eqs. (8) and (9) become *dual* equations. Without loss of generality, we can represent the wave propagating in the PML medium as a linear combination of TE-to-z and TM-to-z modes, as defined by Born and Wolf (1964) as

$$\begin{aligned} \mathbf{E}^{\mathrm{TE}} &= (\hat{\mathbf{x}} a_x^e \sin\phi - \hat{\mathbf{y}} a_y^e \cos\phi) e^{-j(k_x x + k_y y + k_z z)} \\ \mathbf{H}^{\mathrm{TE}} &= (-\hat{\mathbf{x}} b_x^e \cos\theta\cos\phi - \hat{\mathbf{y}} b_y^e \cos\theta\sin\phi + \hat{\mathbf{z}} b_z^e \sin\theta) e^{-j(k_x x + k_y y + k_z z)} \end{aligned} \tag{11a}$$

$$\begin{aligned} \mathbf{E}^{\mathrm{TM}} &= (-\hat{\mathbf{x}} a_x^m \cos\theta\cos\phi - \hat{\mathbf{y}} a_y^m \cos\theta\sin\phi + \hat{\mathbf{z}} a_z^m \sin\theta) e^{-j(k_x x + k_y y + k_z z)} \\ \mathbf{H}^{\mathrm{TM}} &= (-\hat{\mathbf{x}} b_x^m \sin\phi + \hat{\mathbf{y}} b_y^m \cos\phi) e^{-j(k_x x + k_y y + k_z z)} \end{aligned} \tag{11b}$$

where the superscripts e and m denote the TE and TM modes, respectively. Obviously, in order to be able to match the fields at the interface of the free-space and the PML region, we must impose the following constraints on the complex coefficients $a_x^{e,m}$, $a_y^{e,m}$, $b_x^{e,m}$, and $b_y^{e,m}$ that are associated with the tangential x- and y-directions in Eqs. (11a) and (11b):

$$a_x^{e,m} = a_y^{e,m}, \quad b_x^{e,m} = b_y^{e,m}, \quad a_x^{e,m} = -\eta_0 b_y^{e,m} \tag{12}$$

Furthermore, as a consequence of the phase-matching condition that has to be satisfied at the interface between the two regions, the tangential wavenumbers k_x and k_y in the PML region must be equal to their counterparts in the free-space region, which implies that

$$k_x = k_0 \sin\theta\cos\phi = k_{x0}, \quad k_y = k_0 \sin\theta\sin\phi = k_{y0} \tag{13}$$

In addition, to ensure a proper decay of the wave in the PML region along the preferred z-direction, and thereby produce the desired absorption effect, the wavenumber k_z in the PML region must have the form

$$k_z = k_0(1 - jk_{ri}) \cos\theta \tag{14}$$

The positive-real number k_{ri} can be chosen in such a way as to achieve a prespecified value of attenuation for a given value of the transmission (= incidence) angle θ.

We now substitute the field representation in the PML region, as postulated in Eqs. (11a) and (11b) in the form of a linear combination of TE-to-z and TM-to-z fields, into Eqs. (8a) to (9c). After some algebraic manipulations, we arrive at the following conclusions:

$$1 - jk_{ri} = 1 - j\frac{\sigma_2}{\omega\varepsilon_0} = 1 - j\frac{\sigma_2^*}{\omega\mu_0} \tag{15}$$

$$\sigma_1 = \sigma_3 = 0, \quad \sigma_1^* = \sigma_3^* = 0 \tag{16}$$

The above equalities are required in order to avoid contradictory conditions arising during the solution of the system of equations. Of course, the σ values must be permuted as the direction of the normal is changed from z to x to y. Also, at the edges of the problem domain, where two faces come together, one may simply *overlap* the two constitutive parameters pertaining to the two preferred directions of propagation, while maintaining a zero conductivity in the third dimension. Likewise, at the corner regions of the problem domain, one may include all three conductivities in the governing equations since one is effectively faced with three directions of propagation. We should point out that suitable combinations of these conductivities must be used in the edge and corner regions in order to effect a smooth transition of the field behavior as one sweeps through the corner region in the process of going from one face to another. A more detailed description of the edge and corner regions in the PML medium will be given in the next section.

With the above constraints at hand, the frequency-domain equations (8a) to (9c) for the PML medium may be rewritten in an *unsplit* form as:

$$j\omega\mu_0 H_x = jk_y E_z - jk_z E_y \frac{1}{1 - j\dfrac{\sigma_2^*}{\omega\mu_0}} \tag{17a}$$

$$j\omega\mu_0 H_y = jk_z E_x \frac{1}{1 - j\dfrac{\sigma_2^*}{\omega\mu_0}} - jk_x E_z \tag{17b}$$

$$j\omega\mu_0 H_z = jk_x E_y - jk_y E_x \tag{17c}$$

$$j\omega\varepsilon_0 E_x = jk_z E_y \frac{1}{1 - j\dfrac{\sigma_2}{\omega\varepsilon_0}} - jk_y E_z \tag{18a}$$

$$j\omega\varepsilon_0 E_y = jk_x H_z - jk_z H_x \frac{1}{1 - j\dfrac{\sigma_2}{\omega\varepsilon_0}} \tag{18b}$$

$$j\omega\varepsilon_0 E_z = jk_y H_x - jk_x H_y \tag{18c}$$

where we have also set $\sigma_1 = \sigma_3 = \sigma_1^* = \sigma_3^* = 0$. The above six equations are effectively identical to the equations that were presented by Chew and Weedon (1994) in a recent investigation of perfectly matched absorbing boundary conditions based on the longitudinal scaling of Maxwell's equations. It is evident that Eqs. (17) and (18) can be obtained from conventional Maxwell's curl equations by introducing the complex spatial variable z', which is related to the spatial variable z by

$$z' = z\left(1 - j\frac{\sigma_2}{\omega\varepsilon_0}\right) = z\left(1 - j\frac{\sigma_2^*}{\omega\mu_0}\right) \tag{19}$$

assuming that the z-direction is the preferred direction of propagation, and the x- and y-directions are the tangential directions in the PML medium. This leads us to conclude that the space-mapping procedure leads to non-Maxwellian fields as well.

Alternately, it is also possible to start the entire procedure outlined above by deleting the dependent source terms in Eqs. (8) and (9) at the outset, such that the above equations are entirely Maxwellian. If we require the field solution in the PML region to still have the same form as before, we obtain a different uniaxial medium, whose constitutive parameters are 3×3 (diagonal) matrices with *negative* σ_3 and σ_3^*, with $\sigma_1 = \sigma_2 > 0$, $\sigma_1^* = \sigma_2^* > 0$, and whose ε_3 and μ_3 are also different from ε_0 and μ_0. This result is consistent with that reported by Z. S. Sacks, D. M. Kingsland, R. Lee, and J. F. Lee (private communication) and leads to the conclusion that a Maxwellian PML medium must be active with negative σ values and scaling of field components in the normal direction. A more detailed treatment of this approach is given in Section 3.6.

A weak form, which is suitable for implementation in the context of finite element frequency domain analysis and is based on the *transversely scaled* version of Eqs. (17) and (18), is presented in the next section.

3.2. Weak Form for FEM

In this section, the *unsplit* equations given in Eqs. (17) and (18) are further modified to a *transversely scaled* version with a view to obtaining a system of equations that can be expressed in terms of a modified *curl-like* operator, as proposed by Pekel and Mittra (1995a–d). This modified curl operator can be implemented directly in finite element method frequency domain (FEFD) applications without the burden of a surface integral term in the weak form.

Rather than trying to implement Eqs. (17) and (18) directly in an FEFD formulation, one can manipulate these equations through a straightforward multiplication to arrive at

$$(j\omega\mu_0 + \sigma_2^*)H_x = jk_{y'}E_z - jk_zE_y \tag{20a}$$

$$(j\omega\mu_0 + \sigma_2^*)H_y = jk_zE_x - jk_{x'}E_z \tag{20b}$$

$$(j\omega\mu_0 + \sigma_2^*)H_z = jk_{x'}E_y - jk_{y'}E_x \tag{20c}$$

$$(j\omega\varepsilon_0 + \sigma_2)E_x = jk_zH_y - jk_{y'}H_z \tag{21a}$$

$$(j\omega\varepsilon_0 + \sigma_2)E_y = jk_{x'}H_z - jk_zH_x \tag{21b}$$

$$(j\omega\varepsilon_0 + \sigma_2)E_z = jk_{y'}H_x - jk_{x'}H_y \tag{21c}$$

where the scaled wavenumbers $k_{x'}$ and $k_{y'}$ are related to the ordinary ones, namely, k_x and k_y, through the equations

$$k_{x'} = k_x\left(1 - j\frac{\sigma_2}{\omega\varepsilon_0}\right) = k_x\left(1 - j\frac{\sigma_2^*}{\omega\mu_0}\right) \tag{22a}$$

$$k_{y'} = k_y\left(1 - j\frac{\sigma_2}{\omega\varepsilon_0}\right) = k_y\left(1 - j\frac{\sigma_2^*}{\omega\mu_0}\right) \tag{22b}$$

It can be shown that Eqs. (8a) to (9c) represent a version of the original PML equations in which the transverse coordinates x and y, rather than the normal coordinate z, have been scaled to arrive at the new transversc coordinates x' and y', defined in the following manner:

$$x' = \frac{x}{\left(1 - j\dfrac{\sigma_2}{\omega\varepsilon_0}\right)} = \frac{x}{\left(1 - j\dfrac{\sigma_2^*}{\omega\mu_0}\right)}, \quad y' = \frac{y}{\left(1 - j\dfrac{\sigma_2}{\omega\varepsilon_0}\right)} = \frac{y}{\left(1 - j\dfrac{\sigma_2^*}{\omega\mu_0}\right)} \tag{23}$$

By switching back from the spectral domain to the spatial domain, one can rewrite Eqs. (8a) to (9c) in the following more compact form, which is also required to eventually obtain a weak form suitable for the FEFD formulation in ordinary PML elements in which one can identify a single direction of propagation:

$$\mathbf{\nabla}' \times \mathbf{H} = j\omega\varepsilon_0\left(1 - j\frac{\sigma_2}{\omega\varepsilon_0}\right)\mathbf{E}, \quad \mathbf{\nabla}' \times \mathbf{E} = -j\omega\mu_0\left(1 - j\frac{\sigma_2^*}{\omega\mu_0}\right)\mathbf{H} \tag{24}$$

where the modified curl operator is defined through the modified operator ∇' as

$$\nabla' = \hat{\mathbf{x}} \frac{\partial}{\partial x'} + \hat{\mathbf{y}} \frac{\partial}{\partial y'} + \hat{\mathbf{z}} \frac{\partial}{\partial z} \tag{25}$$

reflecting the fact that the transverse coordinates x and y, and not the normal coordinate z, have been scaled in Eqs. (8a) to (9c). It is clear from Eqs. (11a) and (11b) that the new partial differential operators $\partial/\partial x'$ and $\partial/\partial y'$ are related to their conventional Cartesian counterparts through the equations

$$\frac{\partial}{\partial x'} = \left(1 - j\frac{\sigma_2}{\omega\varepsilon_0}\right)\frac{\partial}{\partial x} = \left(1 - j\frac{\sigma_2^*}{\omega\mu_0}\right)\frac{\partial}{\partial x} \tag{26a}$$

$$\frac{\partial}{\partial y'} = \left(1 - j\frac{\sigma_2}{\omega\varepsilon_0}\right)\frac{\partial}{\partial y} = \left(1 - j\frac{\sigma_2^*}{\omega\mu_0}\right)\frac{\partial}{\partial y} \tag{26b}$$

For elements that lie at the edges (and not at the corners) of a given three-dimensional problem domain, where one is effectively faced with two directions of propagation, say, x and z, it is necessary to modify the *curl-like* operator to a different form based on the modified differential operator given below:

$$\nabla' = \hat{\mathbf{x}} \frac{\partial}{\partial x} + \hat{\mathbf{y}} \frac{\partial}{\partial y'} + \hat{\mathbf{z}} \frac{\partial}{\partial z} \tag{27}$$

The two conductivity parameters σ_2 and σ_3, which are associated with the z- and x-directions, respectively, must be chosen equal to each other, that is, $\sigma_2 = \sigma_3$ (and, as a consequence of the *PML impedance matching condition*, $\sigma_2^* = \sigma_3^*$). This constraint can be enforced without a loss of generality and leads to a considerable simplification in the governing equations. For elements that lie at the corners of a three-dimensional problem domain, and are therefore associated with three directions of propagation, the *curl-like* operator in fact reduces to the conventional curl operator. One can then write

$$\nabla' = \nabla = \hat{\mathbf{x}} \frac{\partial}{\partial x} + \hat{\mathbf{y}} \frac{\partial}{\partial y} + \hat{\mathbf{z}} \frac{\partial}{\partial z} \tag{28}$$

It should also be noted that in this case the three conductivity parameters σ_2, σ_3, and σ_1, which are associated with the z-, x-, and y-directions, respectively, as shown by Mittra and Pekel (1995), have been chosen to be equal; that is, we have let $\sigma_2 = \sigma_3 = \sigma_1$ (and $\sigma_2^* = \sigma_3^* = \sigma_1^*$) without loss of generality.

Having derived appropriate systems of partial differential equations of an essentially non-Maxwellian nature involving modified curl operators for the

PML region, we can proceed next to obtain a weak form of these equations suitable for implementation in the FEFD approach. Using Eq. (24) with the isotropic right-hand sides as a starting point, we can obtain a modified curl of curl of **E** equation, which reads

$$\nabla' \times \left[\frac{1}{\left(1 - j\dfrac{\sigma_2^*}{\omega\mu_0}\right)} \nabla' \times \mathbf{E} \right] - k_0^2 \left(1 - j\frac{\sigma_2}{\omega\varepsilon_0}\right) \mathbf{E} = 0 \tag{29}$$

The above equation can be weighted with appropriate vectorial weighting functions $\boldsymbol{\phi}_i$, which can ordinarily be chosen from the set of expansion functions in accordance with the Galerkin approach, and then integrated over the volume of the PML region where this equation is to be enforced. It can readily be shown that the conventional vector differential identities (as, for instance, the divergence theorem) remain valid when the conventional curl operator is replaced by the modified one. With the help of these identities, one arrives at the following weak form for the PML region:

$$\begin{aligned} &\int_{V(\mathrm{PML})} \left[\left(1 - j\frac{\sigma_2^*}{\omega\mu_0}\right)^{-1} \nabla' \times \mathbf{E} \cdot \nabla' \times \boldsymbol{\phi}_i - k_0^2 \left(1 - j\frac{\sigma_2}{\omega\varepsilon_0}\right) \mathbf{E} \cdot \boldsymbol{\phi}_i \right] dv \\ &= \oint_{S(\mathrm{PML})} (j\omega\mu_0 \hat{\mathbf{n}}' \times \mathbf{H}) \cdot \boldsymbol{\phi}_i \, ds \end{aligned} \tag{30}$$

It should be noted that the modified outward normal unit vector $\hat{\mathbf{n}}'$ in Eq. (30) is defined in the same manner as the modified differential operator ∇'. Therefore the contribution of the surface integral on the right-hand side of this equation will be zero, since the *scaling* was performed on the *transverse* components, but not on the normal ones. In contrast, in the FEFD implementation of the *longitudinal scaling procedure*, which utilizes scaling in the normal direction, one encounters discontinuities in the normal unit vector across the interface between free space and the PML region, which unavoidably lead to nonzero contributions from the surface integral in question. However, it should also be pointed out that, in the *transverse scaling* approach, such discontinuities do appear in the normal unit vector along internal interfaces inside the PML region between the ordinary PML elements and the corner elements. The corresponding surface integral contributions originating from such internal interfaces can be neglected since they are secondary in nature.

Numerical Results for PEC Surface Termination. To provide numerical evidence for the effectiveness of the above outlined PML equations acting as an absorbing boundary condition (ABC), the radiation from a thin, linear dipole antenna located in free space is investigated by using a three-dimensional FEFD approach in conjunction with these PML equations. The results

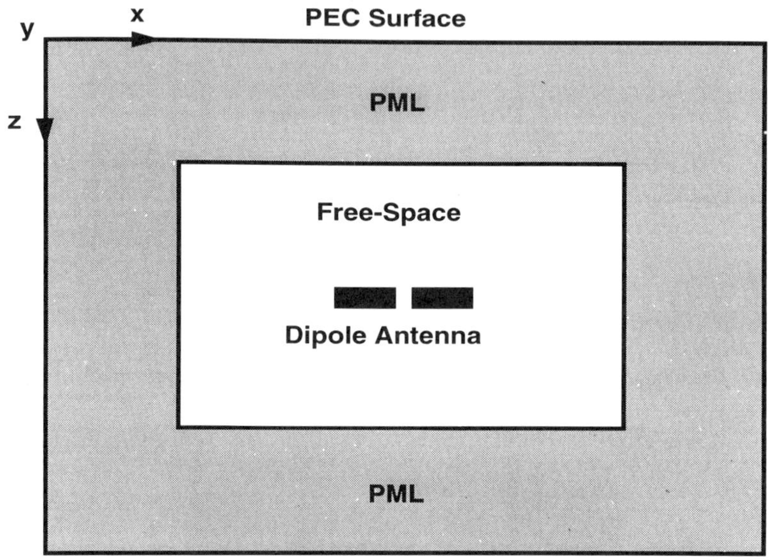

FIGURE 10. Linear dipole antenna radiating in free space surrounded by PML medium with a PEC surface termination.

generated with the PML-based ABC are compared to the exact, analytical results, and also to those obtained by using numerical absorbing boundary conditions (NABCs) applied directly for mesh truncation in the absence of any PML-type material. The frequency of operation is chosen to be 300 MHz, and the dipole antenna is assumed to have a triangular current distribution along its total length of $0.125\lambda_0$ (where the free-space wavelength λ_0 is equal to 1.0 m). The FEFD approach under consideration is based on a scattered electric field formulation and utilizes a nonuniform, but orthogonal, mesh structure that comprises hexahedral, first-order edge elements with 12 edges (or unknowns) per element, as described by Mahadevan et al. (1996).

A two-dimensional frontal view of the specific problem geometry is given in Figure 10, with the linear dipole antenna radiating in free space. The antenna is located at the center of the problem domain, and the total thickness of the PML-type medium surrounding the domain of interest is taken to be identical in all three directions, which leads to overall three orthogonal planes of symmetry. In this subsection, the surrounding layers of the PML-type medium are themselves terminated by PEC surfaces, and no ABCs, either analytical or numerical, are directly employed to reduce the reflections originating from the termination of the outermost layer of the PML, as explained by Pekel and Mittra (1995a).

The size of the problem domain of interest is selected as $0.545\lambda_0 \times 0.445\lambda_0 \times 0.445\lambda_0$, not including the eight layers of the surrounding PML-type

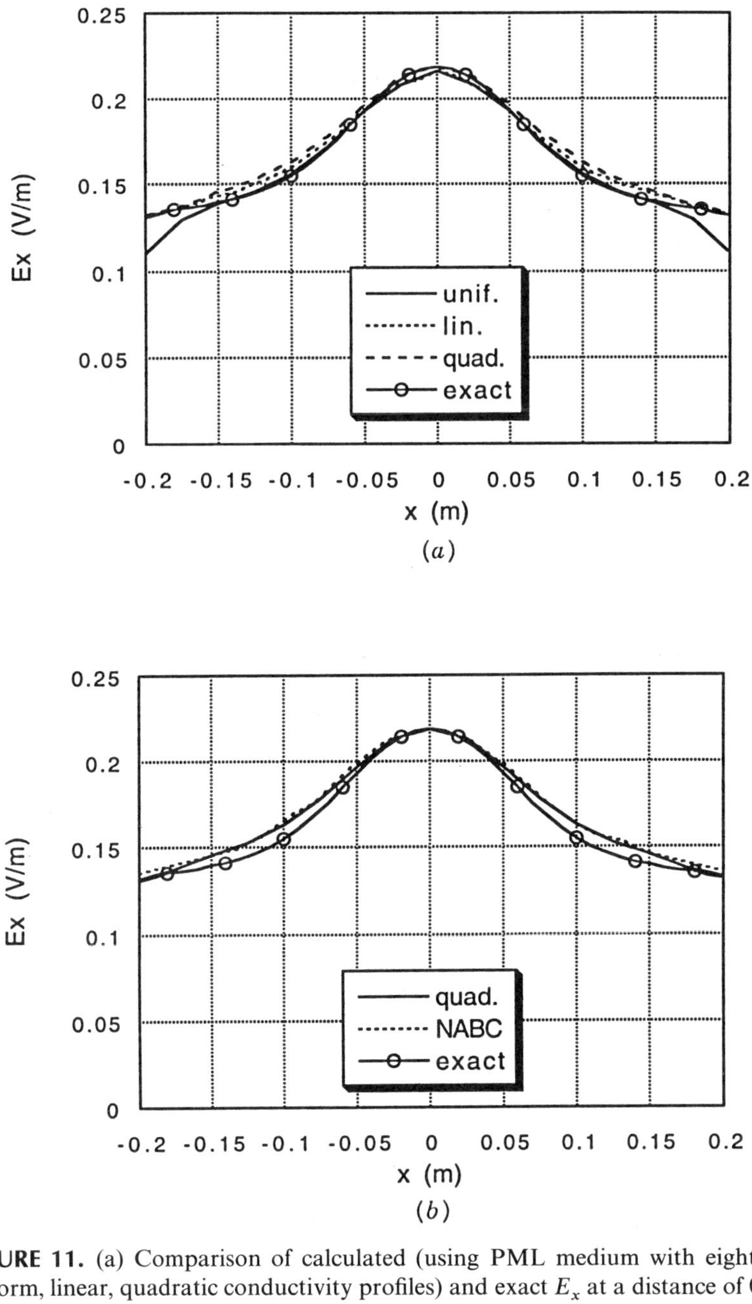

FIGURE 11. (a) Comparison of calculated (using PML medium with eight layers of uniform, linear, quadratic conductivity profiles) and exact E_x at a distance of $0.2\lambda_0$ from antenna. (b) Comparison of calculated (using PML medium with eight layers of quadratic conductivity profile and NABC separately) and exact E_x at a distance of $0.2\lambda_0$ from antenna.

medium with a total thickness of $0.3\lambda_0$ in all three directions. The orthogonal FEM mesh that is applied on this particular problem domain generates a total number of 77,796 unknown electric field values along the element edges. Figure 11a shows the variations of the field component E_x, as computed by using the FEFD approach with uniform (solid curve), linear (dotted curve), and quadratic (dashed curve) conductivity profiles along the eight layers of the PML medium in all three directions. The computed field results are compared to the exact one represented by a solid line with circular markers, all of the curves being obtained along field points that lie on a line parallel to the antenna at a distance of $0.2\lambda_0$ underneath it. We observe that the accuracy of the field values obtained with the quadratic profile is higher than that of the uniform and linear profiles. The performance of the PML-type medium with eight layers and PEC termination appears to be quite satisfactory in terms of its ability to simulate a reasonably reflectionless and absorptive medium and thus provides us with a guideline for synthesizing an effective ABC for FEFD applications.

The corresponding E_x variation computed by using a *two-edge NABC scheme* in the absence of the PML medium is displayed in Figure 11b, together with the PML result obtained for a quadratic conductivity profile and the exact result. It is seen that the numerical accuracies of both techniques are essentially comparable, with the PML result being slightly more accurate. However, because of the nonsymmetric nature of the FEM global matrix associated with the application of the NABCs, one is faced, in general, with larger CPU time requirements in conjunction with this class of boundary conditions.

Next, we consider a case in which the problem domain has the same physical dimensions as the one in the preceding case, but the number of the surrounding PML layers is lowered to four, with a total thickness of $0.2\lambda_0$ in all three directions. This case is investigated with a view to reducing the number of layers in order to ease the intensive memory requirements associated with a large number of layers of PML-type medium. The total number of unknowns is in this case given by 30,116. The computed variations of E_x for uniform, linear, and quadratic conductivity profiles in the PML region are shown and compared to the exact field behavior in Figure 12.

We note that the accuracy of the results worsens, particularly for the quadratic conductivity profile, and this is attributable to the fact that the amount of attenuation provided by the PML medium becomes insufficient as the number of layers is reduced. Consequently, one is faced with the necessity of loosening the discretization in order to compensate for that effect, which, in turn, imposes a severe burden on the numerical accuracy of the results, since the extremely rapid decay of the fields in the PML region cannot be accurately captured by first-order elements in a coarsely discretized mesh. Clearly, this observation suggests that for the PML medium with a PEC termination, a very low number of layers of the PML medium will, in general, not result in a reflectionless, absorptive medium and thus will not provide a satisfactory performance as an ABC in FEFD applications.

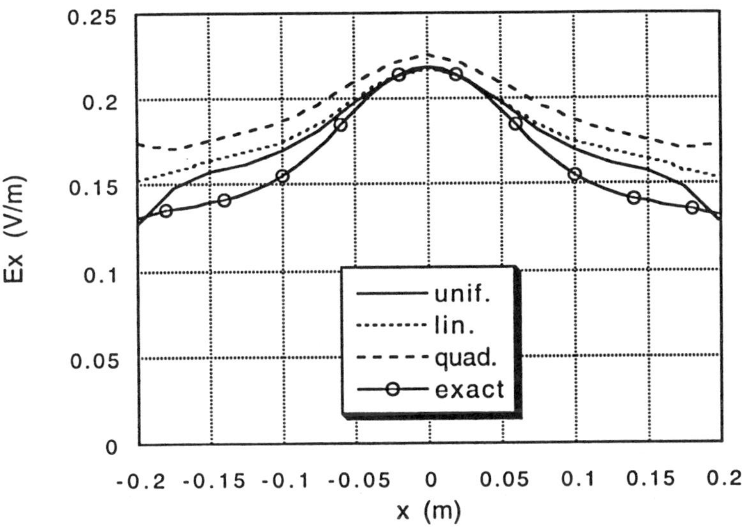

FIGURE 12. Comparison of calculated (using PML medium with four layers and uniform, linear, quadratic conductivity profiles) and exact E_x at a distance of $0.2\lambda_0$ from antenna.

It is conjectured that the accuracy of the PML approach could be increased substantially by doubling the number of layers from, say, 8 to 16, albeit at a substantial increase in computational cost in terms of CPU time and memory requirements that would render it less efficient than the NABCs, despite the symmetry property of the FEM system matrices associated with the PML-based ABC. This naturally leads us to examine the possibility of incorporating an ABC (as opposed to a PEC) termination to the PML medium, as demonstrated by Pekel and Mittra (1995b), becausc this modification has the potential of reducing the required total thickness of the PML medium, and hence increasing the computational efficiency without sacrificing the numerical accuracy.

Numerical Results for ABC/NABC Termination. In this subsection we employ the modified PML equations for the case in which the PML-type medium is terminated with an analytical ABC (of the Engquist–Majda form) or a numerical ABC (NABC) rather than PEC surfaces, and we show some pertinent numerical results in order to validate the application. The combination of the PML medium and the analytical or numerical ABCs as the termination leads to an improved boundary condition for FEFD applications.

The *transversely scaled* PML equations are employed in conjunction with analytical and numerical ABCs serving as termination, as outlined by Pekel and Mittra (1995b), to investigate the test problem of radiation from a thin, linear dipole antenna located in free space, and the FEFD results are com-

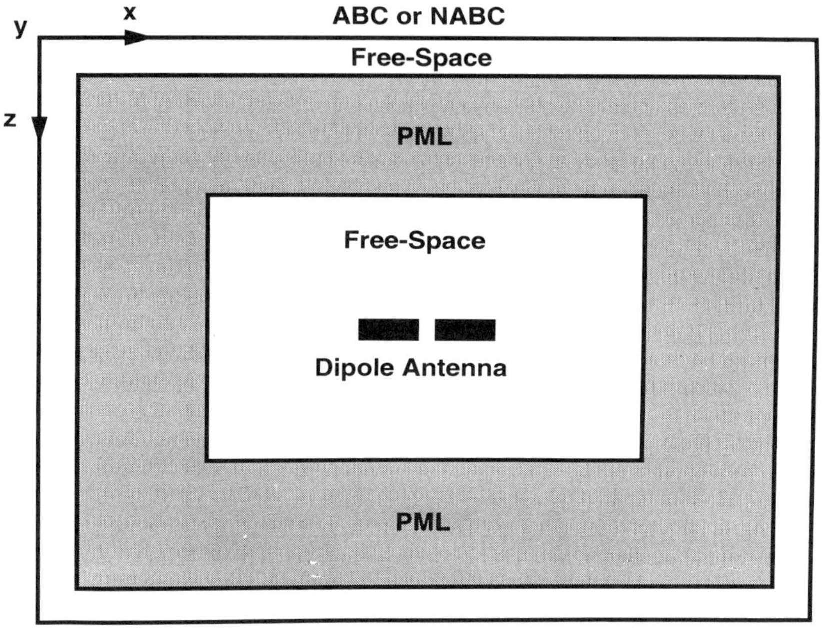

FIGURE 13. Linear dipole antenna radiating in free-space region surrounded by PML medium, and a single layer of free space terminated by analytical or numerical ABC.

pared with those obtained from exact formulas. The frequency of operation is chosen to be 300 MHz, and the dipole antenna is assumed to have a triangular current distribution along it total length of $0.125\lambda_0$ (where the free-space wavelength λ_0 is equal to 1.0 m). The FEFD approach under consideration is based on a scattered electric field formulation and again utilizes a nonuniform, but orthogonal, mesh structure that comprises hexahedral, first-order edge elements with 12 edges (or unknowns) per element.

A two-dimensional frontal view of the dipole antenna problem geometry is given in Figure 13. The antenna is located at the center of the computational domain, and the total thickness of the PML-type medium surrounding the domain of interest is taken to be identical in all three directions. In this subsection, the surrounding layers of the PML-type medium are themselves enclosed by a single layer of free-space when an analytical ABC of the Engquist–Majda form or a numerical ABC (NABC) is employed as the termination. The size of the problem domain of interest is selected to be $0.425\lambda_0 \times 0.225\lambda_0 \times 0.225\lambda_0$ in the *x*-, *y*-, and *z*-directions, respectively, excluding the surrounding PML-type medium.

With the size of the problem domain being specified above, only four layers of the PML medium with a total thickness of $0.14\lambda_0$ in all three directions and an analytical ABC termination are employed to synthesize an ABC. The

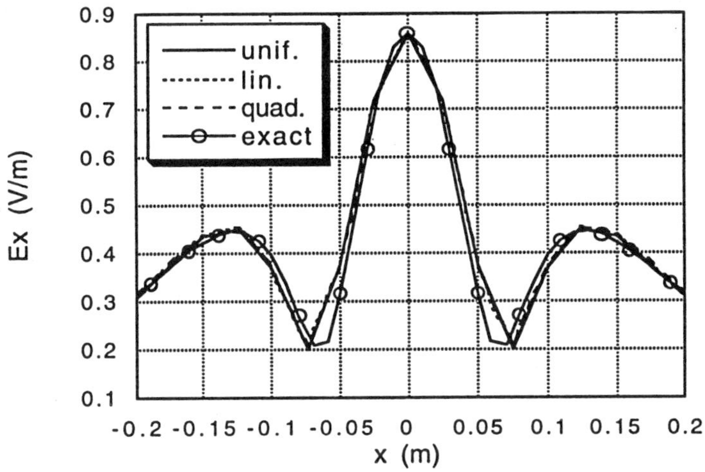

FIGURE 14. Comparison of calculated (PML medium with four layers of uniform, linear, quadratic conductivity profiles with ABC termination) and exact E_x at a distance of $0.1\lambda_0$ from antenna.

orthogonal FEM mesh that is applied on this particular problem domain generates a total number of 32,036 unknown electric field values along the element edges. Figure 14 shows the variations of the field component E_x, as computed by using the FEFD approach with uniform (solid curve), linear (dotted curve), and quadratic (dashed curve) conductivity profiles along the four layers of the PML medium in all three directions. The computed field results are compared to the exact one, represented by a solid line with circular markers, where all of the curves are obtained at a distance of $0.1\lambda_0$ underneath the antenna.

We observe that while the agreement between the calculated and the exact results is good, the accuracy of the field values obtained with the uniform profile is slightly higher than that of the linear and quadratic profiles. In the previous subsection, it was found that for a PEC surface termination, the quadratic profile gave rise to the highest numerical accuracy, since it was a more suitable choice for *matching* the zero conductivity value of the free-space region at one end of the PML medium and the infinite conductivity value of the PEC surface at its other end. In this case, the uniform conductivity profile *matches* the zero conductivity value of the free-space region at either end of the PML medium. We should also point out that the conductivity values used in the case of the ABC termination have been consistently smaller than the ones used with the PEC termination, since in the latter case a considerably larger amount of attenuation has to be provided by the PML medium to effectively absorb the full reflection from the PEC surface. The performance of the PML-type medium with four layers and analytical ABC termination appears to be satisfactory in terms of its ability to simulate a reasonably reflectionless and absorptive medium, which is in contrast to the findings of the previous subsection, where four layers of the PML medium with a PEC termination were found to generally generate inaccurate results. The consequence of this observation is that it is possible to synthesize an improved boundary condition for FEFD applications via the use of the PML medium with an analytical ABC termination, while incurring only a limited increase in computational cost.

Next, we retain the physical size of the problem domain but replace the analytical ABC termination of the preceding numerical example by an NABC termination, with a view to evaluating the overall effectiveness of the PML/NABC combination as a boundary condition for the FEFD approach. We should note that the complex coefficients associated with the NABC have been determined in free space, that is, in the absence of the PML medium, since there is no convenient way to determine the field radiated by the test sources through the box-shaped, layered PML medium. The total number of unknowns is in this case again given by 32,036. The computed variations of E_x for uniform, linear, and quadratic conductivity profiles in the PML region are shown and compared to the exact field behavior in Figure 15, at the same location of the field points as in the preceding numerical example.

We note that the results are, in general, sufficiently accurate, with the numerical accuracy for the uniform profile being once again higher than that of the linear and quadratic profiles. Thus it is possible to synthesize an improved boundary condition for FEFD applications, comprising a relatively low number of layers of the PML medium and a numerical ABC termination,

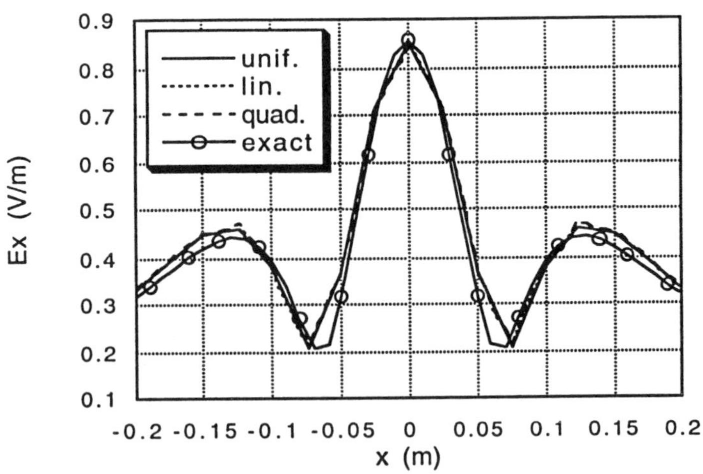

FIGURE 15. Comparison of calculated (PML medium with four layers and uniform, linear, quadratic conductivity profiles with NABC termination) and exact E_x at a distance of $0.1\lambda_0$ from antenna.

with only a small increase in the computational cost. However, it should be stressed that the application of the numerical ABCs requires CPU times that are, in general, larger than those needed when the analytical ABCs are used, since the NABCs inherently lead to nonsymmetric FEM system matrices, whereas the analytical ones preserve the matrix symmetry.

3.3. The PML Concept in Conjunction with Geometries with Discontinuities in Material Properties

In this section, we consider the application of the *transversely scaled* version of the PML equations to radiation problems involving geometries with discontinuities in material properties, which calls for a modification of the underlying equations in order to incorporate the relevant material parameters. We show that a *material-dependent, composite* PML medium can be employed for mesh truncation without introducing numerical instabilities, which have been reported by Mittra et al. (1995) in the PML/time domain analysis of similar configurations. We further show that the number of layers in the PML medium can be kept at a moderate level, say, four, by terminating them with an analytical absorbing boundary condition (ABC) or, at the expense of a lower numerical accuracy, with PEC surfaces. The PML-based boundary condition is demonstrated to yield results that are more accurate than those obtained with a conventional ABC termination.

We modify the *transversely scaled* version of the PML equations to account for the presence of the relative permittivity and permeability parameters ε_r and μ_r. The equations read:

$$\nabla' \times \mathbf{H} = j\omega\varepsilon_0\left(\varepsilon_r - j\,\frac{\sigma_2}{\omega\varepsilon_0}\right)\mathbf{E}, \quad \nabla' \times \mathbf{E} = -j\omega\mu_0\left(\mu_r - j\,\frac{\sigma_2^*}{\omega\mu_0}\right)\mathbf{H} \tag{31}$$

The above *curl-like* operator is based on the *material-dependent* modified differential operator ∇' defined by

$$\nabla' = \hat{\mathbf{x}}\,\frac{\partial}{\partial x'} + \hat{\mathbf{y}}\,\frac{\partial}{\partial y'} + \hat{\mathbf{z}}\,\frac{\partial}{\partial z} \tag{32}$$

where $\hat{\mathbf{z}}$ is the assumed normal direction of propagation, and

$$\frac{\partial}{\partial x'} = \left(1 - j\,\frac{\sigma_2}{\omega\varepsilon_r\varepsilon_0}\right)\frac{\partial}{\partial x} = \left(1 - j\,\frac{\sigma_2^*}{\omega\mu_r\mu_0}\right)\frac{\partial}{\partial x} \tag{33a}$$

$$\frac{\partial}{\partial y'} = \left(1 - j\,\frac{\sigma_2}{\omega\varepsilon_r\varepsilon_0}\right)\frac{\partial}{\partial y} = \left(1 - j\,\frac{\sigma_2^*}{\omega\mu_r\mu_0}\right)\frac{\partial}{\partial y} \tag{33b}$$

implying that the *scaling* is performed in the *transverse* directions. The parameters σ_2 and σ_2^* represent electric and magnetic conductivities, respectively, associated with the direction of propagation in the anisotropic version of the original PML equations introduced by Berenger (1994), and satisfy the *material-dependent PML impedance matching condition*, namely, $\sigma_2/(\varepsilon_r\varepsilon_0) = \sigma_2^*/(\mu_r\mu_0)$.

The above equations are valid for ordinary elements that lie along the material-to-PML region interface. For elements lying at the edges, and not at the corners, of a three-dimensional problem domain and allowing propagation in two directions, say, x and z, the *curl-like* operator must be based on the following *material-dependent* differential operator in which only the y-coordinate is scaled:

$$\nabla' = \hat{\mathbf{x}}\frac{\partial}{\partial x} + \hat{\mathbf{y}}\frac{\partial}{\partial y'} + \hat{\mathbf{z}}\frac{\partial}{\partial z} \tag{34}$$

For elements that lie at the corners of the problem domain and have three possible directions of propagation, the *curl-like* operator reverts to the conventional Cartesian curl operator as

$$\nabla' = \nabla = \hat{\mathbf{x}}\frac{\partial}{\partial x} + \hat{\mathbf{y}}\frac{\partial}{\partial y} + \hat{\mathbf{z}}\frac{\partial}{\partial z} \tag{35}$$

Using Eq. (31) as a starting point, we can obtain a *material-dependent* modified curl–curl of $\mathbf{E}$ equation directly leading to the following *material-dependent* weak form for the PML region:

$$\int_{V(\text{PML})}\left[\left(\mu_r - j\frac{\sigma_2^*}{\omega\mu_0}\right)^{-1}\nabla'\times\mathbf{E}\cdot\nabla'\times\boldsymbol{\phi}_i - k_0^2\left(\varepsilon_r - j\frac{\sigma_2}{\omega\varepsilon_0}\right)\mathbf{E}\cdot\boldsymbol{\phi}_i\right]dv = \oint_{S(\text{PML})}(j\omega\mu_0\hat{\mathbf{n}}'\times\mathbf{H})\cdot\boldsymbol{\phi}_i\,ds \tag{36}$$

where $\boldsymbol{\phi}_i$ denotes the vectorial weighting functions, and the material-dependent, modified outward normal unit vector $\hat{\mathbf{n}}'$ is defined in the same manner as the modified differential operator ∇'. In contrast to the longitudinal (z) scaling case, the contribution of the surface integral on the right-hand side of Eq. (36) is again zero as previously outlined in the material-independent case, due to the fact that the *scaling* was performed on the *transverse* components. The corresponding contributions of the surface integrals at the interfaces between the ordinary PML elements and the corner elements inside the PML region are again secondary in nature and can thus be neglected.

Numerical Results for the Material-Dependent Case. The *material-dependent, transversely scaled* PML equations are employed, in conjunction with analytical ABCs or PEC surfaces serving as the termination, to investigate the

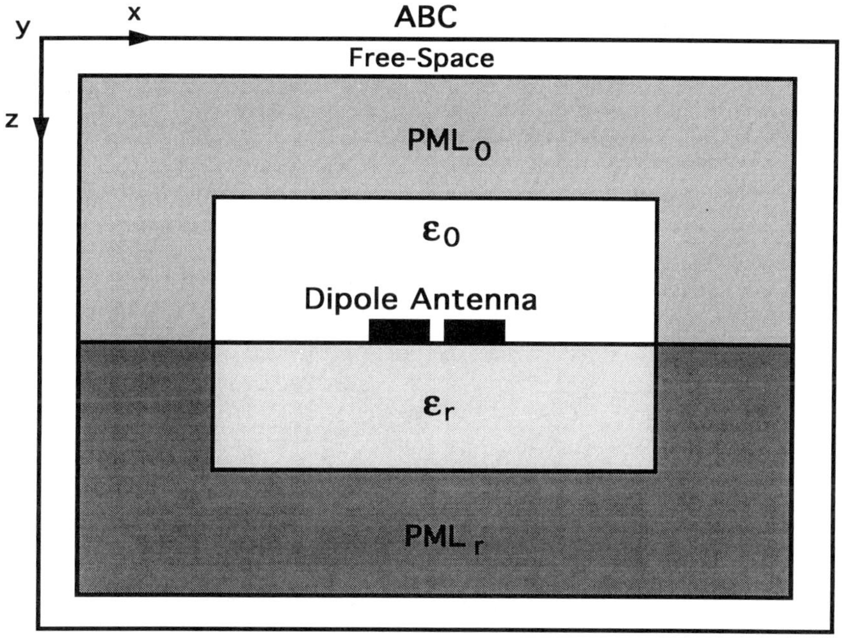

FIGURE 16. Electric dipole antenna residing on the interface between dielectric half-space and free space, surrounded by a composite PML medium and a single layer of free space terminated by an analytical ABC.

problem of radiation from a $0.125\lambda_0$ electric dipole antenna residing on the interface between a semi-infinite dielectric half-space with a relative dielectric permittivity of 2.25 and free space as shown in Figure 16, where $\lambda_0 = 1.0$ m. The FEFD approach utilizes a nonuniform, but orthogonal, mesh structure comprised of hexahedral, first-order edge elements. The dipole antenna is located at the center of the computational domain, where the total thickness of the surrounding *composite* PML medium is again identical in all three directions, as implemented by Pekel and Mittra (1995d). The PML medium is composed of two main sections whose parameters are *matched* to those of the free-space medium above and the dielectric medium underneath the antenna, respectively. The layers of the *composite* PML medium are enclosed by a single layer of free space terminated by either an analytical Engquist–Majda ABC or directly by PEC surfaces without an additional free-space layer. The size of the interior region is given by $0.425\lambda_0 \times 0.325\lambda_0 \times 0.325\lambda_0$ in the x-, y-, and z-directions, respectively.

The PML region has a total thickness (away from the interior region) of $0.14\lambda_0$ and comprises four layers. The total number of unknowns is 32,036 and 19,412 for the ABC and PEC termination cases, respectively. Figure 17a shows the main field component of interest, namely, E_x, calculated along a line parallel to and $0.1\lambda_0$ above the antenna for a *composite* PML medium with a uniform conductivity profile in depth and the two types of termination under consideration. The calculated field variations are compared to the exact one obtained from the pertinent Sommerfeld integral. It is seen that for a moderate number of layers in the PML medium, the ABC termination yields more accurate results than the PEC termination, as stated previously. A comparison

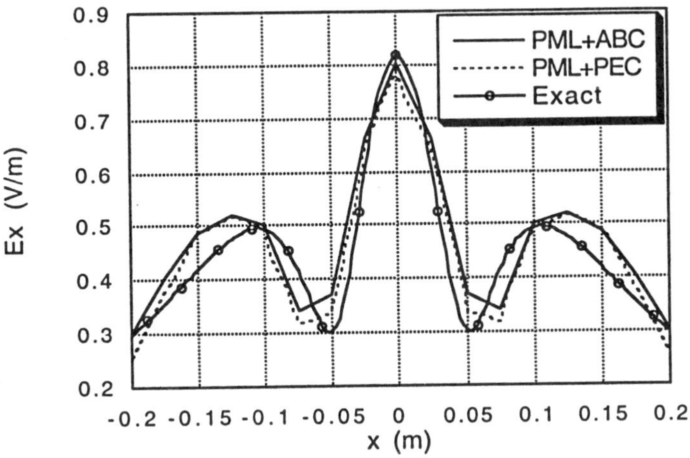

(a)

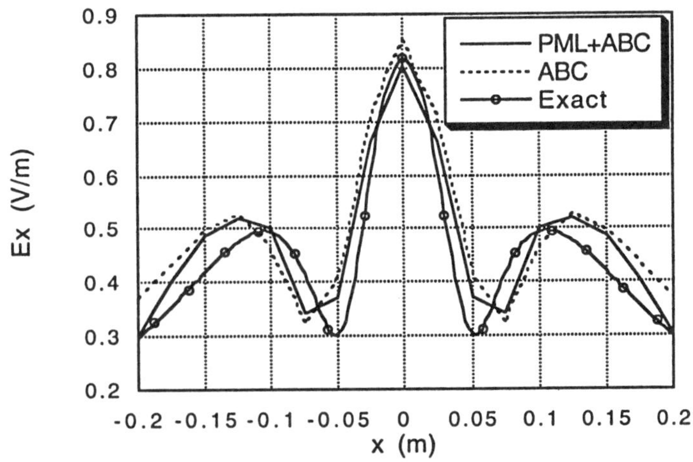

(b)

FIGURE 17. (a) Comparison of calculated (four-layer PML medium with uniform conductivity profile and ABC or PEC termination) and exact E_x variations at a distance of $0.1\lambda_0$ above the antenna. (b) Comparison of calculated (four-layer PML medium with uniform conductivity profile and ABC termination, and ABC termination alone) and exact E_x variations at a distance of $0.1\lambda_0$ above the antenna.

of the PML + ABC and the analytical ABC (in the absence of the PML medium) results computed at the same location of field points is given in Figure 17b.

The presence of the surrounding PML medium clearly improves the accuracy of the calculated fields, albeit at a slightly lower level than comparable applications in free space, which do not involve geometries where very large regions with material property discontinuities have to be truncated to a domain of limited size. A four-layer PML medium with a uniform conductivity profile and an analytical ABC termination appears to be capable of simulating a reasonably reflectionless and absorptive medium without any numerical instabilities despite the composite nature of the PML medium and the additional internal reflections caused by it, and thus constitutes a foundation for an improved boundary condition for the FEFD analysis of problems with discontinuities in material properties.

The improved boundary condition requires an overall smaller problem domain than does the conventional ABC termination, which must be applied considerably farther away from the antenna to attain the same numerical accuracy. The accuracy of the results generated with this boundary condition can steadily be improved by increasing the number of layers in the PML medium from, say, four, albeit again at the expense of a rapid increase in the number of unknowns.

3.4. The PML Concept in the Synthesis of Effective ABCs for the FEFD Analysis of Scattering Problems

In this subsection, we consider the extension of the *transversely scaled* version of the PML equations to scattering problems, which requires a modification of the procedure outlined in the previous sections for radiation problems to be able to incorporate a term associated with the incident field. We show that the number of layers in the PML medium can be held at a moderate level, say, four, by terminating them with an analytical absorbing boundary condition (ABC), while maintaining a satisfactory level of numerical accuracy.

By using the *modified curl* equations and the weak formulation derived in the preceding sections, the FEFD formulation for the scattered electric field inside the PML region can be written for a completely known incident field as proposed by Pekel and Mittra (1995c) in the following manner:

$$\int_{V(\mathrm{PML})} \left[\left(1 - j\frac{\sigma_2^*}{\omega\mu_0}\right)^{-1} \nabla' \times \mathbf{E}^{\mathrm{scat}} \cdot \nabla' \times \boldsymbol{\phi}_i - k_0^2 \left(1 - j\frac{\sigma_2}{\omega\varepsilon_0}\right) \mathbf{E}^{\mathrm{scat}} \cdot \boldsymbol{\phi}_i \right] dv$$
$$= \int_{V(\mathrm{PML})} \left[k_0^2 \left(1 - j\frac{\sigma_2}{\omega\varepsilon_0}\right) \mathbf{E}^{\mathrm{inc}} \cdot \boldsymbol{\phi}_i - \nabla' \times \left(\left(1 - j\frac{\sigma_2^*}{\omega\mu_0}\right)^{-1} \nabla' \times \mathbf{E}^{\mathrm{inc}} \right) \cdot \boldsymbol{\phi}_i \right] dv \tag{37}$$

where the *modified curl* operator has been defined above, and the superscripts scat and inc identify the scattered and incident parts of the electric field, respectively. We should note that since the *scaling* is again performed on the transverse components, the contribution of the surface integral on the right-hand side that would be encountered in the *longitudinal scaling* case is equal to zero. In the non-PML regions, one can use a conventional FEFD scattered field formulation given by Mahadevan et al. (1996):

$$\int_V \left[\frac{1}{\mu_r} \nabla \times \mathbf{E}^{\text{scat}} \cdot \nabla \times \boldsymbol{\phi}_i - k_0^2 \varepsilon_r \mathbf{E}^{\text{scat}} \cdot \boldsymbol{\phi}_i \right] dv + \oint_S \left(\hat{\mathbf{n}} \times \frac{1}{\mu_r} \nabla \times \mathbf{E}^{\text{scat}} \right) \cdot \boldsymbol{\phi}_i \, ds$$

$$= \int_V \left[k_0^2 (\varepsilon_r - 1) \mathbf{E}^{\text{inc}} \cdot \boldsymbol{\phi}_i - j\omega\mu_0 \left(1 - \frac{1}{\mu_r} \right) \mathbf{H}^{\text{inc}} \cdot \boldsymbol{\phi}_i \right] dv \tag{38}$$

where the surface integral on the left-hand side will have to be evaluated only on the outermost surface S enclosing the entire problem domain when a conventional ABC termination is to be implemented.

Numerical Results for Electromagnetic Scattering from a PEC Plate. The *transversely scaled* PML equations have been employed, in conjunction with analytical ABCs serving as termination, to investigate the problem of scattering from a $1\lambda_0 \times 1\lambda_0$ conducting square plate located in free space as indicated in Figure 18, with $\lambda_0 = 1.0\,\text{m}$. The FEFD approach utilizes a nonuniform, but orthogonal, mesh structure comprised of hexahedral, first-order edge elements. The plate is located at the center of the computational domain, and the total thickness of the surrounding PML medium is identical in all three directions. The layers of the PML medium are enclosed by a single layer of free space terminated by an analytical Engquist–Majda ABC. The size of the interior region is given by $1.3\lambda_0 \times 1.3\lambda_0 \times 0.3\lambda_0$ in the x-, y-, and z-directions, respectively.

In this numerical application, the PML region has a total thickness (away from the interior region) of $0.14\lambda_0$, comprising four layers, and the total number of unknowns is 69,984. A plane wave is normally incident on the plate, with a y-polarized incident electric field (pointing into the page). Figure 19a shows the bistatic RCS pattern in the principal plane ($\phi = \pi/2$) for a PML medium with a uniform conductivity profile and an analytical ABC termination. As stated earlier, the uniform profile generally yields more accurate results than the linear or quadratic ones when ABC terminations are used. The calculated RCS pattern is compared to the one obtained with the same ABC termination applied in the absence of the PML medium, and to a method of moments (MoM) reference pattern. A comparison of the bistatic RCS patterns on the orthogonal plane ($\phi = 0$) is given in Figure 19b. The presence of the

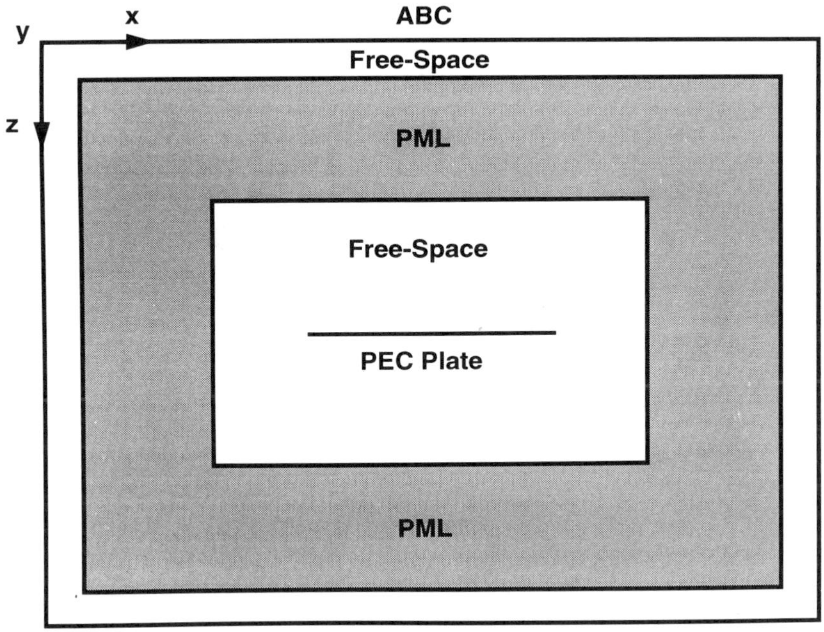

FIGURE 18. PEC plate in free-space region surrounded by PML medium and a single layer of free space terminated by analytical ABC.

surrounding PML medium gives rise to a clear improvement in the accuracy of the RCS patterns in both planes. Thus a four-layer PML medium with a uniform conductivity profile and an analytical ABC termination is capable of simulating a reasonably reflectionless and absorptive medium and leads to an improved boundary condition for the FEFD analysis of scattering problems.

Next, we use the four layers of the PML-type medium alone, without the ABC termination on the single surrounding layer of free space, while retaining the size of the problem domain. An approximate boundary condition that assumes a complete absorption the scattered field in the PML region is indirectly enforced as termination by neglecting the surface integral term in Eq. (38). Figures 20a and 20b show the bistatic RCS patterns in the principal and orthogonal planes, respectively, computed with the FEFD approach utilizing this particular boundary condition (BC), and the deterioration of the patterns when compared to the PML + ABC results is apparent. This leads us to recommend that the PML medium not be terminated by an approximate BC, or, for that matter, a PEC surface, but rather with an ABC termination, especially for the practical case where only a moderate number of layers is employed in the PML medium.

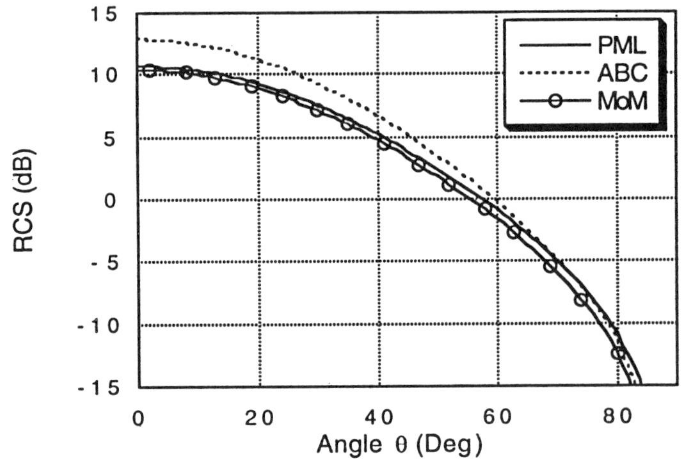

(a)

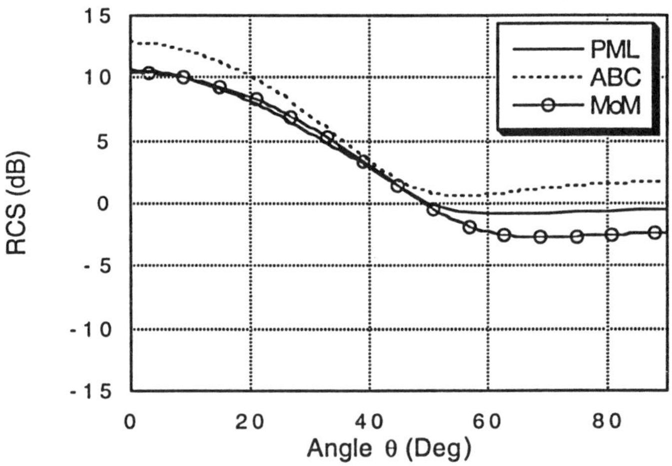

(b)

FIGURE 19. (a) Comparison of calculated RCS patterns (PML medium with four layers of uniform conductivity profile with ABC termination, ABC alone, and MoM) on the principal plane. (b) Comparison of calculated RCS patterns (PML medium with four layers of uniform conductivity profile with ABC termination, ABC alone, and MoM) on the orthogonal plane.

3.5. The Perfectly Matched Absorber (PMA) Approach

As an alternative to the above outlined approach, it is also possible to synthesize a theoretically reflectionless medium by suitably defining the two material properties (viz., permittivity and permeability) as 3×3 complex

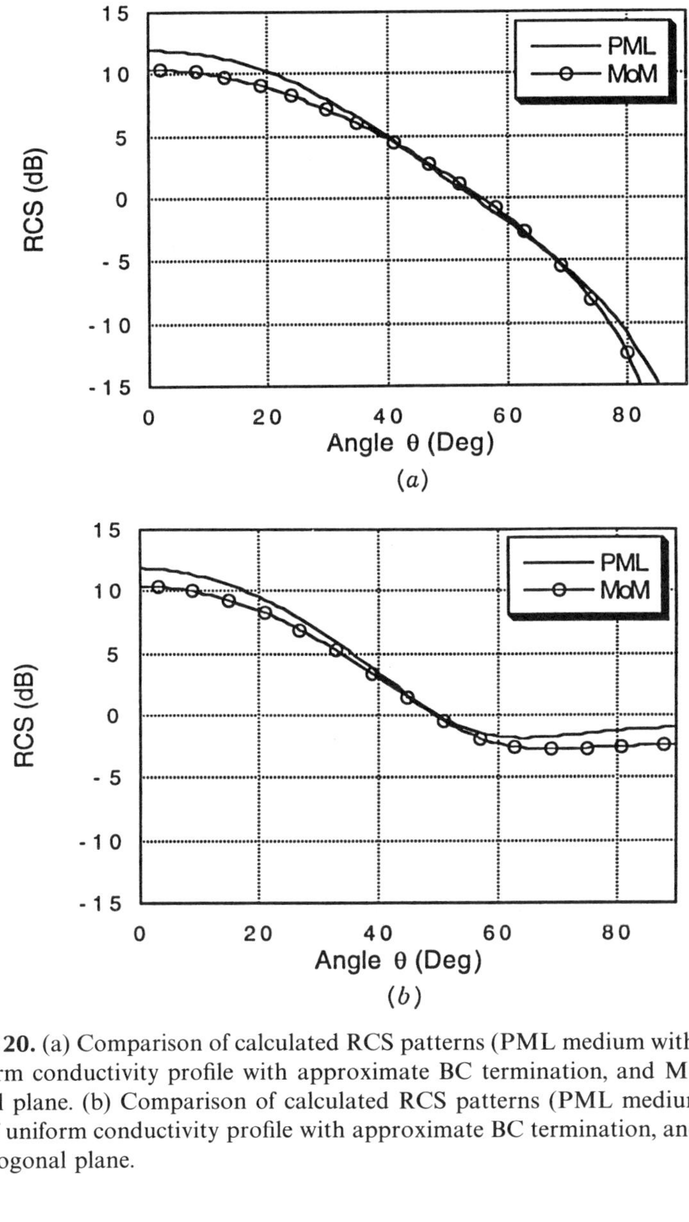

FIGURE 20. (a) Comparison of calculated RCS patterns (PML medium with four layers of uniform conductivity profile with approximate BC termination, and MoM) on the principal plane. (b) Comparison of calculated RCS patterns (PML medium with four layers of uniform conductivity profile with approximate BC termination, and MoM) on the orthogonal plane.

diagonal tensors in such a way as to ensure a reflectionless interface and a complete absorption of the incident wave inside the medium. Then a number of layers of this perfectly matched absorber (PMA) type medium can be used as an absorbing boundary condition (ABC) in FEM analysis, as proposed by Z. S. Sacks, D. M. Kingsland, R. Lee, and J. F. Lee (private communication).

The equations that characterize the PMA-type medium can be derived by considering the time-harmonic form of Maxwell's equations as

$$\nabla \cdot (\bar{\bar{\varepsilon}}\mathbf{E}) = 0, \quad \nabla \cdot (\bar{\bar{\mu}}\mathbf{H}) = 0 \tag{39a}$$

$$\nabla \times \mathbf{E} = -j\omega\bar{\bar{\mu}}\mathbf{H}, \quad \nabla \times \mathbf{H} = j\omega\bar{\bar{\varepsilon}}\mathbf{E} \tag{39b}$$

where the complex diagonal permittivity and permeability tensors $\bar{\bar{\varepsilon}}$ and $\bar{\bar{\mu}}$ are constrained to satisfy the *free-space impedance matching condition* $\bar{\bar{\mu}}\bar{\bar{\varepsilon}}^{-1} = \mu_0/\varepsilon_0$, with ε_0 and μ_0 denoting the free-space permittivity and permeability parameters, respectively. The two tensors for the anisotropic PMA medium can thus be written in the following manner:

$$\bar{\bar{\mu}} = \mu_0 \mathbf{\Lambda} = \mu_0 \begin{bmatrix} a & 0 & 0 \\ 0 & b & 0 \\ 0 & 0 & c \end{bmatrix}, \quad \bar{\bar{\varepsilon}} = \varepsilon_0 \mathbf{\Lambda} = \varepsilon_0 \begin{bmatrix} a & 0 & 0 \\ 0 & b & 0 \\ 0 & 0 & c \end{bmatrix} \tag{40}$$

The dispersion relation for this particular medium can be found through a conventional variable transformation technique, and its solution describes an ellipsoid in k-space, resulting in the following separation constants:

$$k_x = k_0 \sin\theta \cos\phi, \quad k_y = k_0 \sin\theta \sin\phi, \quad k_z = k_0 \cos\theta \tag{41}$$

where it is assumed that the x- and y-directions are transverse to the straight interface, and z is the direction of propagation normal to it, as indicated in Figure 9 albeit for a free-space to PML-medium transition. By performing a reflection–transmission analysis based on a TE-to-z and TM-to-z mode decomposition at the free-space to PMA medium interface, the respective reflection coefficients R^{TE} and R^{TM} have been found by Z. S. Sacks et al. as

$$R^{\mathrm{TE}} = \frac{\cos\theta_i - \sqrt{\dfrac{b}{a}}\cos\theta_t}{\cos\theta_i + \sqrt{\dfrac{b}{a}}\cos\theta_t}, \quad R^{\mathrm{TM}} = \frac{\sqrt{\dfrac{b}{a}}\cos\theta_t - \cos\theta_i}{\cos\theta_i + \sqrt{\dfrac{b}{a}}\cos\theta_t} \tag{42}$$

subject to the phase matching condition given by

$$\sqrt{bc}\,\sin\theta_t = \sin\theta_i \tag{43}$$

where the parameters a, b, and c appear in the definition of the matrix $\mathbf{\Lambda}$ in Eq. (40), and the subscripts i and t denote incidence- and transmission-related parameters, respectively. It is evident that one must choose $\sqrt{bc} = 1$ (i.e., $c = 1/b$) in the phase matching condition so that the reflection coefficient will be independent of the incidence angle. Moreover, one must select the parameters a and b to be equal (i.e., $a = b$), to ensure zero reflection coefficients for both the TE and TM modes at the straight interface. It follows that the parameter matrix $\mathbf{\Lambda}$ used in Eq. (40) is explicitly given by

$$\mathbf{\Lambda} = \begin{bmatrix} \alpha - j\beta & 0 & 0 \\ 0 & \alpha - j\beta & 0 \\ 0 & 0 & \dfrac{\alpha + j\beta}{\alpha^2 + \beta^2} \end{bmatrix} \tag{44}$$

where the parameter a is equal to $\alpha - j\beta$ (i.e., $a = \alpha - j\beta$), and $b = a$, $c = 1/b = (\alpha - j\beta)^{-1}$. It should be noted that for a free-space to PMA medium interface, the parameter α will be selected as unity, namely, equal to the value of the relative permittivity ε_r of free space. Therefore one is faced with the presence of *negative* conductivity values in the direction of propagation in the material parameter tensors that characterize the uniaxial PMA medium. These *negative* conductivities render the medium *active* and ensure a reflectionless transmission at the interface. It should also be mentioned that one can *overlap* the constitutive parameters at the edges and corners of the domain through a simple multiplication of the respective two or three material property tensors.

Finally, we should point out that the PMA equations presented above can be derived directly from the frequency domain version of Berenger's PML equations, albeit through the use of component scaling in addition to coordinate scaling.

3.6. Concluding Remarks

In this section, we have shown that the presence of a surrounding PML medium as part of the absorbing boundary condition leads to a clear improvement in the near-field variations and RCS (far-field) patterns in various problems. However, at present, we should point out that numerical experimentation is required in order to determine *optimal* conductivity values and discretization. A moderate number of layers, say, four, of the PML medium with uniform conductivity profile and analytical ABC termination have been found to yield a reasonably reflectionless and absorptive medium, and thus lead to an improved boundary condition for FEFD analysis. However, we have also demonstrated that four layers of PML medium with a PEC termination, or some form of approximate boundary condition termination, may give rise to a lower numerical accuracy. It is therefore preferable not to terminate the PML medium in an approximate BC or a PEC surface if a relatively low number of layers is employed in the PML medium in order to keep the computational cost at a minimum level.

REFERENCES

Bayliss, A., Gunzburger, M., and Turkel, T. (1982). Boundary conditions for the numerical solution of elliptic equations in exterior regions. *SIAM J. Appl. Math.* **42**, 430–451.

Berenger, J. P. (1994). A perfectly matched layer for the absorption of electromagnetic waves. *J. Comput. Phys.* **114**, 185–200.

Boag, A., and Mittra, R. (1995). A numerical absorbing boundary condition for 3D edge-based finite element analysis of very low frequency fields. *Microwave Opt. Technol. Lett.* **9**, 22–27.

Boag, Am., Boag, Al., Mittra, R., and Leviatan, Y. (1994a). A numerical absorbing boundary condition for finite difference and finite element analysis of open structures. *Microwave Opt. Technol. Lett.* **7**, 395–398.

Boag, Am., Boag, Al., and Mittra, R. (1994b), A numerical absorbing boundary condition for edge-based finite element analysis. *Microwave Opt. Technol. Lett.* **7**, 733–737.

Born, M., and Wolf, E. (1964). *Principles of Optics.* Macmillan, New York.

Cangellaris, A. C., and Lee, R. (1990). The bymoment method for two-dimensional electromagnetic scattering. *IEEE Trans. Antennas Propag.* **AP-38**, 1429–1437.

Chatterjee, A., Jin, J.-M., and Volakis, J. L. (1993). Edge-based finite elements and vector ABCs applied to 3D scattering. *IEEE Trans. Antennas Propag.* **AP-41**, 221–226.

Chew, W. C., and Weedon, W. H. (1994). A 3D perfectly matched medium from modified Maxwell's equations with stretched coordinates. *Microwave Opt. Technol. Lett.* **7**, 509–604.

Engquist, B., and Majda, A. (1977). Absorbing boundary conditions for the numerical simulation of waves. *Math. Comput.* **31**, 629–651.

Golub, G. H., and Van Loan, C. F. (1989). *Matrix Computations*, pp. 592–599. Johns Hopkins Univ. Press, Baltimore, MD.

Gordon, R., Mittra, R., Glisson, A., and Michielssen, E. (1993). Finite element analysis of electromagnetic scattering by complex bodies using an efficient numerical boundary condition for mesh truncation. *Electron. Lett.* **29**, 1102–1103.

Higdon, R. L. (1986). Absorbing boundary conditions for difference approximations to the multidimensional wave equation. *Math. Comput.* **47**, 437–459.

Jevtic, J. O., and Lee, R. (1994). A theoretical and numerical analysis of the measured equation of invariance. *IEEE Trans. Antennas Propag.* **AP-42**, 1097–1105.

Jevtic, J. O., and Lee, R. (1995). How invariant is the measured equation of invariance? *IEEE Microwave Guided Wave Lett.* **5**, 45–47.

Jin, J.-M. (1993). *The Finite Element Method in Electromagnetics.* Wiley, New York.

Jin, J.-M., and Liepa, V. V. (1988). Application of the hybrid finite element method to electromagnetic scattering from coated cylinders. *IEEE Trans. Antennas Propag.* **AP-36**, 55–70.

Jin, J.-M., Volakis, J. L., and Collins, J. D. (1991). A finite element-boundary integral method for scattering and radiation by two- and three-dimensional structures. *IEEE Antennas Propag. Mag.* **33**, 22–32.

Katz, D. S., Thiele, E. T., and Taflove, A. (1994). Validation and extension to three dimensions of the Berenger PML absorbing boundary condition for FDTD meshes. *IEEE Microwave Guided Wave Lett.* **4**, 268–270.

Keller, J., and Givoli, D. (1989). Exact non-reflecting boundary conditions. *J. Comput. Phys.* **82**, 172–192.

Ko, W. L., Harms, P., Pekel, Ü., and Mittra, R. (1994). *Detection of Objects under Seawater*, Techn. Rep. Electromagnetic Communication Laboratory, University of Illinois, Urbana. Submitted to the Naval Research Laboratory, Mapping, Charting, and Geodesy Branch, Stennis Space Center, MS.

Liao, Z. P., Wong, H. L., Yang, B. P., and Yuan, Y. F. (1984). A transmitting boundary for transient wave analysis. *Sci. Sini. Ser. A* (Engl. Ed.) **27**, 1063–1076.

Lynch, D. R., Paulsen, K. D., and Strohbehn, J. W. (1986). Hybrid finite element method for unbounded electromagnetic problems in hyperthermia. *Int. J. Numer. Methods Eng.* **23**, 1915–1937.

Mahadevan, K., Irvine, B., Pekel, Ü., and Mittra, R. (1996). Edge-based finite element analysis of singly- and doubly-periodic scatterers using absorbing and periodic boundary conditions. *Electromagnetics* **16**, 1–16.

Mei, K. K. (1974). Unimoment method for solving antenna and scattering problems. *IEEE Trans. Antennas Propag.* **AP-22**, 760–766.

Mei, K. K., Pous, R., Chen, Z., Liu, Y. W., and Prouty, M. D. (1994). Measured equation of invariance: A new concept in field computations. *IEEE Trans. Antennas Propag.* **AP-42**, 320–328.

Mittra, R., and Pekel, Ü. (1995). A new look at the perfectly matched layer (PML) concept for the reflectionless absorption of electromagnetic waves. *IEEE Microwave Guided Wave Lett.* **5**, 84–86.

Mittra, R., and Ramahi, O. (1989). Absorbing boundary conditions for the direct solution of partial differential equations arising in electromagnetic scattering problems. In *Finite Element Finite Difference Methods in Electromagnetic Scattering* (M. A. Morgan, ed.), Vol. II, pp. 133–173. Elsevier, New York.

Mittra, R., Pekel, Ü., and Veihl, J. (1995). A theoretical and numerical study of Berenger's perfectly matched layer (PML) concept for mesh truncation in time and frequency domains. *Int. Conf. Workshop Approx. Numer. Methods Solution Maxwell Equations, 3rd*, Oxford.

Paulsen, K. D., Lynch, D. R., and Strohbehn, J. W. (1988). Three-dimensional finite, boundary, and hybrid element solutions of the Maxwell's equations for lossy dielectric media. *IEEE Trans. Microwave Theory Tech.* **MTT-36**, 682–693.

Pekel, Ü., and Lee, R. (1995). A 3-D application of the bymoment method for electromagnetic scattering. *J. Electromag. Waves Appl.* **9**, 973–992.

Pekel, Ü., and Mittra, R. (1995a). A finite element method frequency domain application of the perfectly matched layer (PML) concept. *Microwave Opt. Technol. Lett.* **9**, 117–122.

Pekel, Ü., and Mittra, R. (1995b). Mesh truncation in the finite element frequency domain method with a perfectly matched layer (PML) applied in conjunction with analytic and numerical absorbing boundary conditions. *Microwave Opt. Technol. Lett.* **9**, 176–180.

Pekel, Ü., and Mittra, R. (1995c). An application of the perfectly matched layer (PML) concept to the finite element method frequency domain analysis of scattering problems. *IEEE Microwave Guided Wave Lett.* **5**, 258–260.

Pekel, Ü., and Mittra, R. (1995d). A finite element frequency domain (FEFD) formulation with perfectly matched layer (PML) for mesh truncation. *Jt. IEEE AP-S Symp. URSI Radio Sci. Meet.*, Newport Beach, California.

Peterson, A. F. (1988). Absorbing boundary conditions for the vector wave equation. *Microwave Opt. Technol. Lett.* **1**, 62–64.

Rappaport, C. M. (1995). Perfectly matched absorbing boundary conditions based on anisotropic lossy mapping of space. *IEEE Microwave Guided Wave Lett.* **5**, 90–92.

Stupfel, B., and Mittra, R. (1995). A theoretical study of numerical absorbing boundary conditions. *IEEE Trans. Antennas Propag.* **AP-43**, 478–487.

Webb, J. P., and Kanellopoulos, V. N. (1989). Absorbing boundary conditions for the finite element solution of the vector wave equation. *Microwave Opt. Technol. Lett.* **2**, 370–372.

Yuan, X. (1990). Three-dimensional electromagnetic scattering from inhomogeneous objects by the hybrid moment and finite element method. *IEEE Trans. Microwave Theory Tech.* **MTT-38**, 1053–1058.

CHAPTER THIRTEEN

FEM Applications to Conformal Antennas

JOHN L. VOLAKIS, JIAN GONG, and TAYFUN OZDEMIR

1. INTRODUCTION

Antennas are the most well known and widely used electromagnetic devices, and new applications in mobile communications and radar guidance for automated highways present us with continuing demands for new designs. Printed and integrated antenna (Pozar, 1992; James and Hall, 1989) of all sorts are very popular because of their conformality, ease of fabrication, light weight, and low cost. By and large, the moment method has been the simulation method of choice for printed antenna analysis. However, the large computational requirements of this full matrix technique coupled with the need to employ complicated Green's functions make it less attractive. Printed antennas are often coupled to complex feeding structures and are typically placed in finite cavities for heat dissipation, to suppress surface wave power loss, and for reduced coupling to adjacent radiating elements. These environments are indeed quite stressing for traditional integral formulations, and the need to model finite and inhomogeneous or anisotropic substrates/superstrates, lumped and distributed loads, as well as multicompartment cavities make the finite element method (FEM) very attractive for modeling printed antennas such as those shown in Figure 1. Basically, the FEM is best suited for modeling the inhomogeneous cavity without any compromise in the geometrical characteristics of the resonant device, its feed structure, or the material loading. Thick

Finite Element Software for Microwave Engineering, Edited by Tatsuo Itoh, Giuseppe Pelosi, and Peter P. Silvester.
ISBN 0-471-12636-5

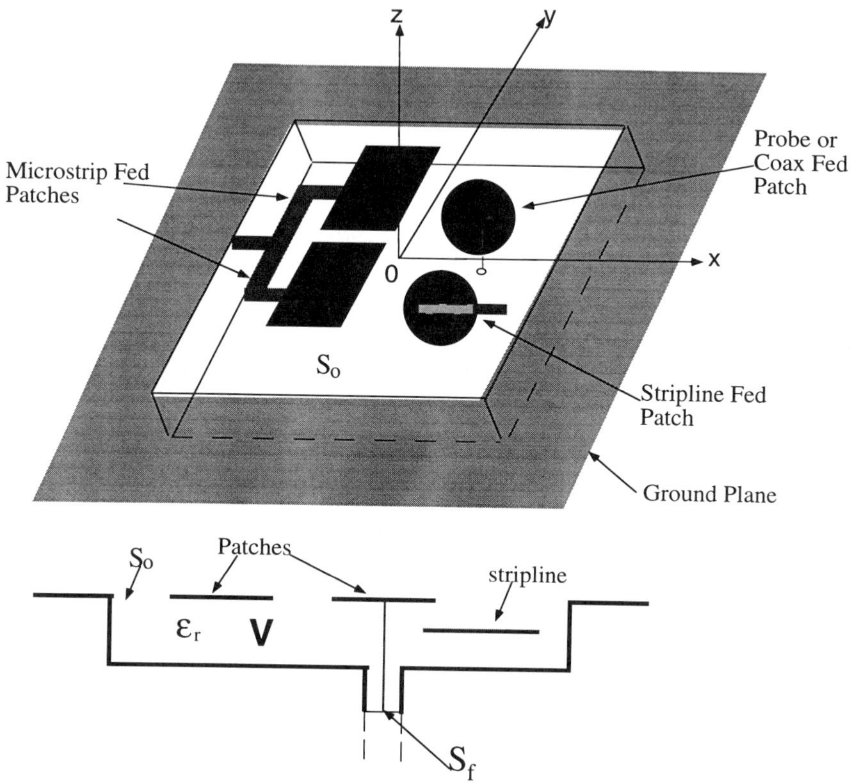

FIGURE 1. Geometry of cavity-backed microstrip antennas.

substrates or superstrates (isotropic or not) are readily included in the formulations, and, as will be seen, a variety of feeds can be implemented with great ease. Of course, the truncation of the FEM mesh is a challenging task and must be performed accurately for input impedance calculations. Use of the boundary integral (Jin and Volakis, 1991b; Volakis et al., 1994a; Cwik, 1992; Yuan et al., 1990; Antilla and Alexopoulos, 1992) for mesh truncation yields the most robust formulation at the expense of higher computational requirements. However, more recently, absorbing boundary conditions (Chatterjee et al., 1993) and use of artificial absorbers (Ozdemir and Volakis, 1994; Gong and Volakis, 1995a) have yielded very promising results. The latter, of course, are associated with fully sparse matrices and are also applicable to antennas on curved and treated platforms.

Other chapters have dealt with the details of the FEM formulation and its implementation for scattering calculations. Thus this chapter begins with only a brief account of the FEM formulation giving particular emphasis on the treatment of anisotropic materials and feed modeling, including current probe, voltage gap, coax cable, and aperture coupled feeds. Also, details relating to the implementation of lumped and distributed loads are included. The boundary integral, absorbing boundary condition, and artificial absorber termination schemes are discussed next, along with the resulting system of equations. A large portion of the chapter, however, is devoted to the presentation of various applications of the finite element formulation for antenna analysis and design. Pattern, resonance frequency, and input impedance calculations are presented for various patch antennas on planar and nonplanar surfaces using different mesh elements, feed models, and mesh truncation schemes. Indeed, these example applications serve to demonstrate the versatility and robustness of the finite element method. Some of the results have already appeared in the literature, but several of them, such as those dealing with high contrast superstrates, distributed loading, finite substrate, and finite curvature effects, are new.

2. FORMULATION

Consider the antenna configuration shown in Figure 1, where the cavity is recessed in a ground plane and is intended to house the radiating elements and their feeding structure such as coaxial cables, striplines, microstrip lines, or aperture-coupled microstrip lines. Any inhomogeneous material filling can be handled by the FEM formulation, including resistive cards and lumped or distributed loads, often used for frequency and bandwidth control. Thin anisotropic layers have been found very useful for truncating the finite element meshes and this aspect of their use will be addressed later.

The goal is to seek the solution of the vector wave equation

$$\nabla \times \{\bar{\bar{\mu}}_r^{-1} \cdot (\nabla \times \mathbf{E})\} - k_0^2 \bar{\bar{\varepsilon}}_r \cdot \mathbf{E} = -jk_0 Z_0 \mathbf{J}^i - \nabla \times (\bar{\bar{\mu}}_r^{-1} \cdot \mathbf{M}^i) \qquad (1)$$

in which $\mathbf{E}$ represents the total electric field, $(\bar{\bar{\varepsilon}}_r, \bar{\bar{\mu}}_r)$ denote the relative permittivity and permeability tensors of the computational domain, k_0 is the free-space wavenumber, and Z_0 is the free-space intrinsic impedance. In addition, $\mathbf{J}^i$ and $\mathbf{M}^i$ denote the impressed electric and magnetic sources, respectively. As is well known, the standard finite element (FE) solution scheme is to consider the weak form of Eq. (1), instead of solving it directly. This can be achieved by extremizing the pertinent functional (Volakis et al., 1994a, b)

$$F(\mathbf{E}) = \frac{1}{2} \iiint_V \{(\nabla \times \mathbf{E}) \cdot \bar{\bar{\mu}}_r^{-1} \cdot (\nabla \times \mathbf{E}) - k_0^2 \bar{\bar{\varepsilon}}_r \cdot \mathbf{E} \cdot \mathbf{E}\}\, dv$$
$$+ \iiint_V \mathbf{E} \cdot (jk_0 Z_0 \mathbf{J}_i + \nabla \times \bar{\bar{\mu}}_r^{-1} \cdot \mathbf{M}_i)\, dv$$

$$+ jk_0 Z_0 \iint_{S_0 + S_f} \mathbf{E} \cdot (\mathbf{H} \times \hat{n})\, ds$$

$$+ \frac{1}{2} jk_0 Z_0 \left\{ \iint_{S_r} \frac{1}{R} (\hat{n} \times \mathbf{E}) \cdot (\hat{n} \times \mathbf{E})\, ds + \frac{d}{Z_L s} \iiint_{V_l} \mathbf{E} \cdot \mathbf{E}\, dv \right\} \tag{2}$$

where R denotes the resistivity or impedance of the surface S_r, S_0 represents the aperture excluding the metallic portions, S_f stands for the junction opening to possible guided feeding structures, and $\mathbf{H}$ is the corresponding magnetic field on S_0 and S_f whose outer normal is given by $\hat{n}$. Also, V_s is the source volume and V_l is the volume occupied by the load Z_L, whose length and cross section are givem by d and s, respectively. It is noted that S_r must be closed when it satisfies an opaque impedance boundary condition but can be open (i.e., a finite plate) if it satisfies the penetrable resistive sheet condition

$$\hat{n} \times \hat{n} \times \mathbf{E} = -R\hat{n} \times (\mathbf{H}^+ - \mathbf{H}^-)$$

where $\mathbf{H}^{\pm}$ denotes the fields above and below the surface S_r. As seen from Eq. (2), no special care is required for the evaluation of the integral over S_r, in spite of the discontinuity in the magnetic field. The explicit knowledge of $\mathbf{H}$, however, is required over the surfaces S_0 and S_f (referred to as mesh truncation surfaces) for the construction of a complete discrete system of equations for the solution of $\mathbf{E}$. Specifically, the magnetic field $\mathbf{H}$ over S_0 may be replaced in terms of $\mathbf{E}$ via a boundary integral (BI) or an absorbing boundary condition (ABC), whereas $\mathbf{H}$ on S_f is determined on the basis of the given feeding structure. These truncation schemes will be discussed in later subsections.

2.1. FEM System

In proceeding with the discretization of Eq. (2), it is convenient to reexpress it as

$$F = F_V + F_S \tag{3}$$

where F_V denotes the volume integral contributions and, similarly, F_S accounts for the surface integral terms. To construct a system of equations from Eq. (2), one first needs to subdivide the volume V into a number of small elements such as those shown in Figure 2. Each of these small elements occupies a volume V_e ($e = 1, 2, 3, \ldots, M$), where M denotes the total number of elements filling V. The field in each element is then approximated by a linear expansion (Gong et al., 1994), namely,

$$\mathbf{E}^e = \sum_{j=1}^{m} E_j^e \mathbf{V}_j^e = [\mathbf{V}_e]^T \{E^e\} \tag{4}$$

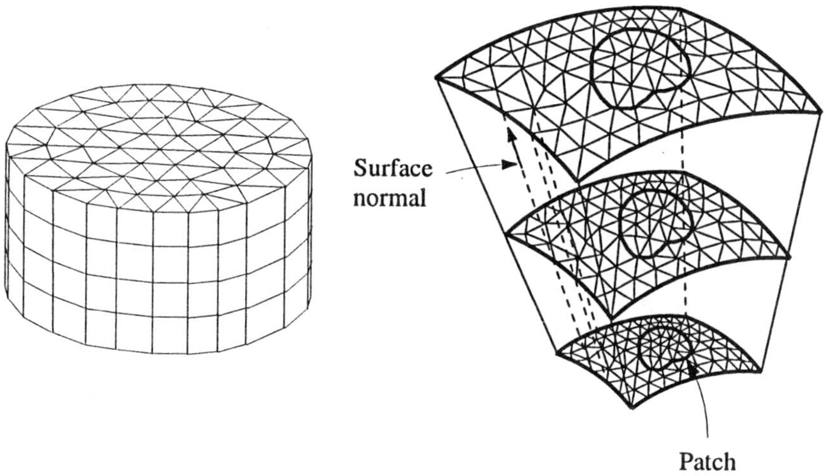

FIGURE 2. Example of tessellation of the cavity and surface.

with

$$[V]_e = [\{V_x\}\{V_y\}\{V_z\}]_e$$

$$\{V_u\} = \begin{pmatrix} V_{u_1} \\ V_{u_2} \\ \vdots \\ V_{u_m} \end{pmatrix}, \quad u = x, y, z$$

$$\{E\}_e = \begin{pmatrix} E_1 \\ E_2 \\ \vdots \\ E_m \end{pmatrix}_e$$

in which $\mathbf{V}_j^e$ are the vector basis functions (Barton and Cendes, 1987; Jin and Volakis, 1991a; Ozdemir and Volakis, 1996), V_{u_i} denote the u ($u = x, y,$ or z) components of the basis functions along the ith edge, m is equal to the number of degrees of freedom in a single element, and the superscript T in Eq. (4) denotes the matrix transpose operation. With this representation, the unknown expansion coefficient E_j^e is simply the field component along the jth edge of the eth element. The edge-based linear expansion (4) is chosen to have zero divergence within each element, as required by Maxwell's equations, and avoids an explicit specification of the field at metallic edges or corners (Webb, 1993); Bossavit and Mayergoyz, 1989; Nedelec, 1980; Peterson and Wilton,

1994). Corresponding node-based field representations, which do not satisfy the zero divergence condition and model the nullspace of the curl operator improperly, yield extraneous solutions unless augmented in some way by adding, for example, a penalty term to Eq. (2) (Rahman and Davies, 1984) or instead by working with potentials (Paulsen and Lynch, 1991) rather than the fields themselves.

The FEM system of equations may be obtained from Eq. (2) via the Rayleigh–Ritz procedure. This amounts to differentiating the functional F_V with respect to each edge field and then setting it to zero. On substituting Eq. (4) ino the volume integrals of Eq. (2), taking the first variation of F_V with respect to **E**, and assembling all elements, one obtains the finite element method (FEM) system

$$\left\{\frac{\partial F_V}{\partial E^e}\right\} = \sum_{e=1}^{M} [A^e]\{E^e\} + \sum_{e=1}^{M} \{K_e\} \tag{5}$$

The sum of the element submatrices $[A^e]$ results in a large sparse matrix and its elements are computed from the volume integral(s) over V in Eq. (2); the column $\{K_e\}$ results from the discretization of the volume integrals over the source volume V_s. They are given by

$$[A]_e = \iiint_{V_e} \{[DV]_e[\mu_r^{-1}][DV]_e^T - k_0^2[V]_e[\varepsilon_r]^T[V]_e^T\}\, dv \tag{6}$$

$$\{K\}_e = \iiint_{V_e} [V]_e \left[jk_0Z_0 \begin{Bmatrix} J_{ix} \\ J_{iy} \\ J_{iz} \end{Bmatrix} + \nabla \times [\mu_r^{-1}] \begin{Bmatrix} M_{ix} \\ M_{iy} \\ M_{iz} \end{Bmatrix} \right] dv \tag{7}$$

$$[DV]_e^T = \begin{bmatrix} \frac{\partial}{\partial y}\{V_z\}^T - \frac{\partial}{\partial z}\{V_y\}^T \\ \frac{\partial}{\partial z}\{V_x\}^T - \frac{\partial}{\partial x}\{V_z\}^T \\ \frac{\partial}{\partial x}\{V_y\}^T - \frac{\partial}{\partial y}\{V_x\}^T \end{bmatrix} \tag{8}$$

With the presence of a lumped load, it is noted that the load may be modeled as a post of length d and cross section s. When the post is very thin and coincident with the kth edge, the contribution from the load to the system can be simplified as $A_{kk} = jk_0Z_0(d^2/Z_L)$. For the short-circuit ($Z_L = 0$) condition, it is straightforward to set the electric field along the post to zero.

We next proceed with the discretization of the surface integral. However, as expected, the task is dependent on the type of mesh truncation being used.

Below we discuss it in connection with three different mesh termination schemes.

2.2. Finite Element–Boundary Integral Method

The hybrid finite element–boundary integral (FE–BI) method was introduced in the early 1970s (Silvester and Hsieh, 1971); McDonald and Wexler, 1972), but its application to electromagnetics did not occur until the 1980s (Jin and Volakis, 1989; Yuan et al., 1990). This method combines the adaptability of the FEM with the rigor of the BI or the method of moments. Because of the sparsity of the FEM matrix, this method is preferable to the volume method of moments formulation (Schaubert et al., 1984), which leads to completely full systems and is not therefore attractive for iterative solution. The overall FE–BI system is partly sparse and partly full, and, typically (for antenna analysis), the partly full BI subsystem is much smaller in size than the entire system. Therefore no significant computational burden is added owing to the introduction of the boundary integral. Moreover, the system can be solved efficiently with only $O(N)$ memory demand and uniform zoning on the mesh termination boundary (Jin et al., 1991; Collins et al., 1990).

In accordance with the finite element–boundary integral method, the magnetic field in the integral over S_0 is represented by an integral equation, which may take the form (Jin et al., 1991; Jin and Volakis, 1991a, b)

$$\mathbf{H} = \mathbf{H}^{g0} + \iint_{S_0} [\hat{z} \times \mathbf{E}(\mathbf{r}')] \cdot \bar{\mathbf{G}}(\mathbf{r}, \mathbf{r}')\, ds' \tag{9}$$

where $\bar{\mathbf{G}}$ is the electric dyadic Green's function of the second kind (Tai, 1994) such that $\hat{n} \times (\nabla \times \bar{\mathbf{G}}) = 0$ is satisfied on a (planar, spherical, or cylindrical) metallic platform. For the antenna problem shown in Figure 1, where the platform is a planar ground plane, $\bar{\mathbf{G}}$ becomes the half-space dyadic Green's function

$$\bar{\mathbf{G}} = jk_0 Y_0 \left(\bar{\mathbf{I}} + \frac{1}{k_0^2} \nabla\nabla \right) \frac{e^{-jk_0|\mathbf{r}-\mathbf{r}'|}}{2\pi|\mathbf{r} - \mathbf{r}'|} \tag{10}$$

with $\mathbf{r}$ and $\mathbf{r}'$ representing the observation and integration points, respectively, and $\bar{\mathbf{I}} = \hat{x}\hat{x} + \hat{y}\hat{y} + \hat{z}\hat{z}$ is the unit dyad. In connection with our problem—that of a cavity recessed in a ground plane—$\mathbf{H}^{g0}$ is equal to the sum of the incident and reflected fields for scattering computations, or equal to zero for the antenna parameter computations. If the platform is not planar, then, as noted earlier, for scattering computations $\mathbf{H}^{g0}$ is equal to the total field including the field scattered by an otherwise metallic or coated surface. That is, if the platform is cylindrical, $\mathbf{H}^{g0}$ is the total field due to a plane wave incident on a smooth metallic or coated cylinder (Kempel and Volakis, 1994; Kempel et al., 1995).

The integral in Eq. (9) can be discretized by introducing the expansion (for the sth surface element)

$$\mathbf{E}(\mathbf{r}) = \sum_{i=1}^{n} E_i^s \mathbf{S}_i^s(\mathbf{r}) = [S]_s^T \{E^s\} \tag{11}$$

with

$$\begin{aligned} [S]_s &= [\{S_x\}\{S_y\}]_s \\ \{S_u\} &= \begin{pmatrix} S_{u_1} \\ S_{u_2} \\ \vdots \\ S_{u_n} \end{pmatrix}, \quad u = x, y \\ \{E\}_s &= \begin{pmatrix} E_{s_1} \\ E_{s_2} \\ \vdots \\ E_{s_n} \end{pmatrix}_s \end{aligned} \tag{12}$$

where $\mathbf{S}_i^s(\mathbf{r})$ are the vector surface basis functions, S_{u_i} are the u $(u = x, y)$ components of the surface vector basis functions along the ith edge, and n denotes the number of edges forming the surface elements comprising S_0. On substituting Eq. (11) into the surface integrals F_S of Eq. (3), taking the first variation of F_S with respect to $\{E\}_s$, and assembling all the surface elements, one obtains

$$\left\{\frac{\partial F_S}{\partial E^s}\right\} = \sum_s \{[B^s]\{E^s\} + \{L_s\}\} \tag{13}$$

where

$$\begin{aligned} [B^s] = \iint_{S_s} \iint_{S_s'} \bigg\{ &-2k_0^2 [S]_s [S]_s^T + 2\left[\frac{\partial}{\partial x}\{S_y\} - \frac{\partial}{\partial y}\{S_x\}\right] \\ &\times \left[\frac{\partial}{\partial x'}\{S_y\}^T - \frac{\partial}{\partial y'}\{S_x\}^T\right]\bigg\} \cdot G_0(\mathbf{r}, \mathbf{r}')\, dS\, dS' \end{aligned} \tag{14}$$

and

$$\{L_s\} = jk_0 Z_0 \iint_{S_s} [S]_s \begin{pmatrix} H_y^{g0} \\ -H_x^{g0} \end{pmatrix} dS \tag{15}$$

Note that in Eq. (14) the elements of the array $[S]_s$ are functions of the observation vector **r**, whereas the elements of $[S]_s^T$ are expressed with respect to the integration point **r**′.

2.3. Finite Element–Absorbing Boundary Condition Method

In accordance with this method, some local boundary condition is enforced on a mathematical surface S_0 or S_{abc} that encloses the region being modeled by the FEM (see Figure 3a). The goal with any absorbing boundary condition (ABC) is to eliminate backward reflections from S_0, and a variety of these boundary conditions have been proposed over the years beginning with those of Bayliss and Turkel (1980) and Engquist and Majda (1977). More recently, other ABCs such as those by Webb and Kanellopoulos (1989) and Chatterjee and Volakis (1993; see also Senior and Volakis, 1995) have been proposed. All ABCs provide an approximate relation between the **E** and **H** fields on the surface S_0, which in most cases is derived by assuming a field expansion in inverse powers of r, where r is the radial distance from the center of S_0. If the ABC annihilates the first $(2m + 1)$ inverse powers of r, it is then referred to as an mth order ABC. The second order ABC derived by Chatterjee and Volakis

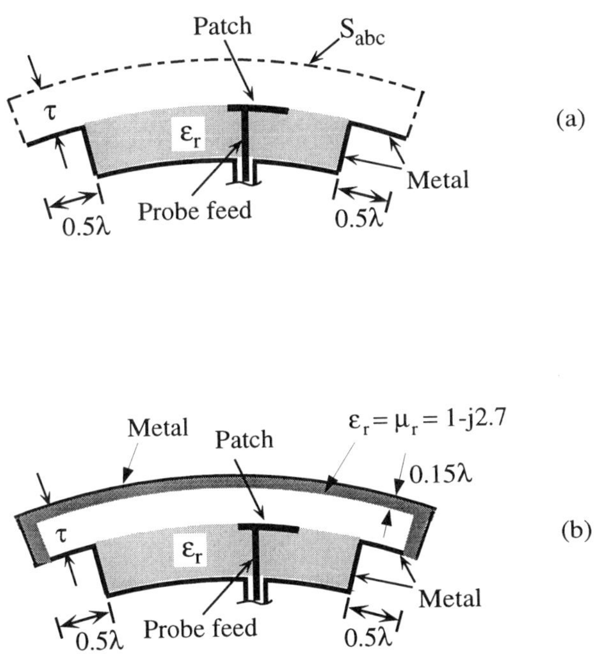

FIGURE 3. Illustration of mesh truncation using ABCs and artificial absorbers.

(1993) is given by

$$jk_0 Z_0(\mathbf{H} \times \hat{n}) = \bar{\bar{\alpha}} \cdot \mathbf{E}_t + \bar{\bar{\beta}} \cdot [\nabla \times \hat{n}(\nabla \times \mathbf{E})_n] + \bar{\bar{\gamma}} \cdot \nabla_t(\nabla \cdot \mathbf{E}_t) \tag{16}$$

where

$$\bar{\bar{\alpha}} = \alpha_1 \hat{t}_1 \hat{t}_1 + \alpha_2 \hat{t}_2 \hat{t}_2$$

$$\bar{\bar{\beta}} = \eta(\hat{t}_1 \hat{t}_1 + \hat{t}_2 \hat{t}_2)$$

$$\bar{\bar{\gamma}} = \frac{\eta}{jk_0}\left(jk_0 + 3\kappa_m - \frac{\kappa_g}{\kappa_m} - 2\kappa_1\right)\hat{t}_1\hat{t}_1 + \frac{\eta}{jk_0}\left(jk_0 + 3\kappa_m - \frac{\kappa_g}{\kappa_m} - 2\kappa_2\right)\hat{t}_2\hat{t}_2$$

$$\alpha_1 = \eta[4\kappa_m^2 - \kappa_g + D(jk_0 - \kappa_1) + \kappa_1^2 + \kappa_m \Delta\kappa]$$

$$\alpha_2 = \eta[4\kappa_m^2 - \kappa_g + D(jk_0 - \kappa_2) + \kappa_2^2 - \kappa_m \Delta\kappa]$$

$$\eta = \frac{1}{D - 2\kappa_m}$$

$$\Delta\kappa = \kappa_1 - \kappa_2$$

$$D = 2jk_0 + 5\kappa_m - \frac{\kappa_g}{\kappa_m}$$

In the above equations, the subscripts n and t denote surface normal and tangential components, $\hat{t}_1, \hat{t}_2$ are the principal surface directions, κ_1, κ_2 are the corresponding curvatures, $\kappa_m = (\kappa_1 + \kappa_2)/2$ is the mean curvature, and $\kappa_g = \kappa_1\kappa_2$ is the Gaussian curvature. For $\kappa_1 = \kappa_2$ (spherical termination boundary), Eq. (16) reduces to the ABC derived by Webb and Kanellopoulos (1989).

Upon substituting Eq. (16) into the surface integral of Eq. (2) and making use of vector differential and integral identities, F_S in Eq. (3) becomes (Chatterjee and Volakis, 1995)

$$F_S = \iint_{S_0} (\alpha_1 E_{t_1}^2 + \alpha_2 E_{t_2}^2)\, dS + \iint_{S_0} \eta(\nabla \times \mathbf{E})_n^2\, dS - \iint_{S_0} (\nabla \cdot \mathbf{E}_t)[\nabla \cdot (\bar{\bar{\gamma}} \cdot \mathbf{E})_t]\, dS \tag{17}$$

Next, expansion (11) is introduced and Eq. (17) is differentiated as in Eq. (13) to obtain the finite element equations. Since the ABC is a local condition, the sparsity of the finite element matrix is preserved. However, for arbitrarily curved boundaries, the matrix will not be symmetric except for planar and spherical boundaries. The system will also be symmetric for cylindrical boundaries only if linear edge-based expansion functions are used. These ABCs have been extensively validated for a number of antenna and scattering configurations (Kempel, 1994; Chatterjee and Volakis, 1995).

2.4. Finite Element–Isotropic Absorber Method

In accordance with this method, the outgoing waves are suppressed by an absorbing dielectric layer placed at some distance from the antenna as shown in Figure 3b. Typically, the layer is backed by metal and the finite element region extends all the way to the metal, at which point the mesh is terminated by setting the tangential component of the electric field to zero. This is equivalent to removing in Eq. (2) the integral over S_0. Obviously, the accuracy of the technique depends on how well the metal-backed absorber suppresses the incoming waves without introducing backward reflections. The plane wave reflection coefficient of the absorber shown in Figure 3b has been minimized over the entire visible angular range. The fact that the dielectric has the same relative permittivity and permeability ensures that there is no reflection from the air–dielectric interface at normal incidence (perfect impedance match), but the performance degrades away from normal with the reflection coefficient reaching unity at grazing. Because the absorber layer is lossy, the entire reflection process is dominated by the air–dielectric interface for all angles of incidence. This limitation led researchers to consider perfectly matched interfaces as discussed next.

2.5. Finite Element–Anisotropic Absorber Method

This mesh termination method is similar to that described in the previous section except that the layer is comprised of an anisotropic lossy dielectric, which has zero reflection coefficient at the air–dielectric interface and can therefore be considered as a perfectly matched interface (PMI) (Sacks et al., 1995). The intrinsic parameters of the PMI layer are

$$\frac{\bar{\bar{\mu}}_r}{a} = \bar{\bar{\varepsilon}}_r = \begin{bmatrix} \alpha - j\beta & 0 & 0 \\ 0 & \alpha - j\beta & 0 \\ 0 & 0 & \dfrac{1}{\alpha - j\beta} \end{bmatrix}$$

where $a = (Z/Z_0)^2$, and Z is the intrinsic impedance of the medium being terminated by the absorber; α, β, and the thickness of the layer are parameters that can be optimized for maximum absorption. For example, Figure 4 shows that as much as a 45-dB reduction can be achieved by selecting $\beta \approx 1$ and that the value of α does not play a major role in this case (Gong and Volakis, 1995a).

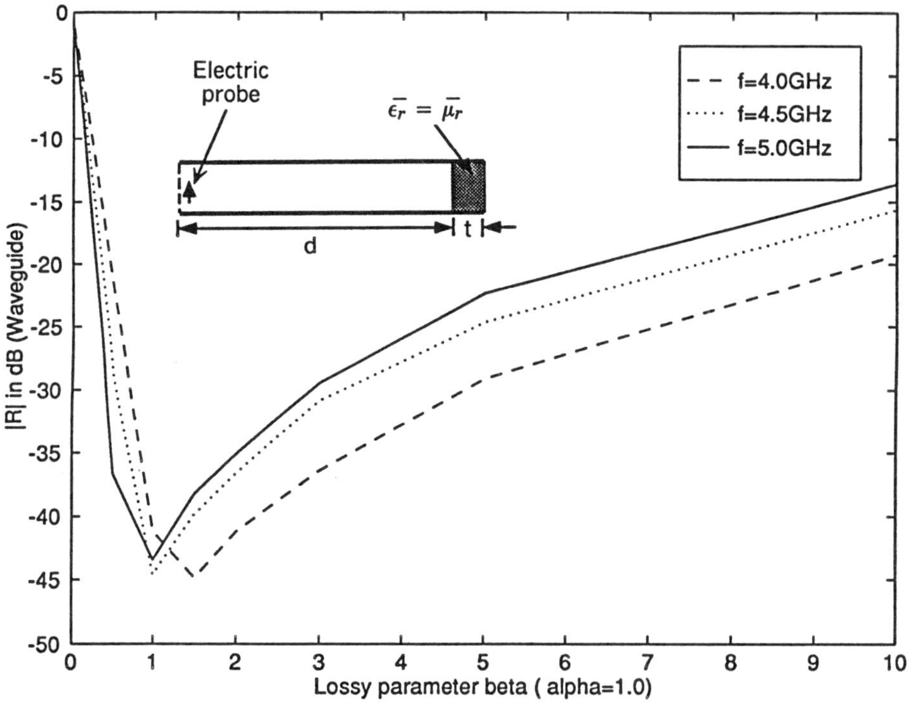

FIGURE 4. Reflection coefficient of the PMI layer in a waveguide as a function of the loss parameter.

2.6. Feed Modeling

For scattering problems where the plane wave incidence is usually considered as the "source," the right-hand-side excitation has been expressed explicitly in Eq. (15) and will not be discussed further. However, for antenna parameter computations, several feeding schemes have been reported (Gong and Volakis, 1995b), associated with various practical feed designs for microstrip antennas. Some of these are discussed below.

Simple Probe Feed. For thin substrates the coaxial cable feed may be simplified as a thin current filament of length l carrying an electric current $I\hat{l}$. Since this filament is located inside the cavity, the first term of the integral in Eq. (7) needs to be considered for this model. Specifically, the ith (global) entry of the excitation vector K_i becomes

$$K_i = jk_0 Z_0 I\hat{l} \cdot \mathbf{V}_i(\mathbf{r}), \quad i = j_1, j_2, \ldots, j_m$$

where $\mathbf{r}$ is the location of the filament, m is the number of (nonmetallic) element edges, and j_m is the global edge numbering index. In general, m such entries are associated with m element edges, and thus the index i goes from j_1 up to j_m. This expression can be further reduced to $K_i = jk_0 Z_0 Il$, provided the ith edge is coincident with the current filament.

Voltage Gap Feed. This excitation is also referred to as a *gap generator* and amounts to specifying *a priori* the electric voltage V across the opening of the coax cable or any other gap. Since $V = \mathbf{E} \cdot \bar{d}$, where $\bar{d}$ is a vector whose magnitude is the gap width, and $\mathbf{E}$ is the electric field across the gap, we have that $E_i = V/(d \cos \theta_i)$, where $\cos \theta_i$ is equal to 1 if the ith edge is parallel to $\bar{d}$. Numerically, this gap voltage model can be realized by first setting the diagonal term A_{ii} equal to unity and the off-diagonal terms A_{ij} $(i \neq j)$ to zero. For the right-hand-side vector, only the entry corresponding to the ith (global) edge across the gap is specified and set equal to the value E_i, whereas all other entries associated with edges not in the gap are set to zero.

Improved Coaxial Cable Feed Model. The simple probe feed of the coaxial cable is accurate only if the substrate is very thin. For thicker substrates, an improved feed model is necessary and this can be achieved by evaluating the functional

$$F_C = -jk_0 Z_0 \iint_{S_f} (\mathbf{E} \times \mathbf{H}) \cdot \hat{z} \, dS \tag{18}$$

over the aperture S_f of the coax cable. Assuming a TEM mode across S_f, the fields within the cable may be expressed as (see Figure 5)

$$\mathbf{E} = \frac{\varepsilon_0}{r} \hat{r}, \quad \mathbf{H} = \frac{h_0}{r} \hat{\phi} \tag{19}$$

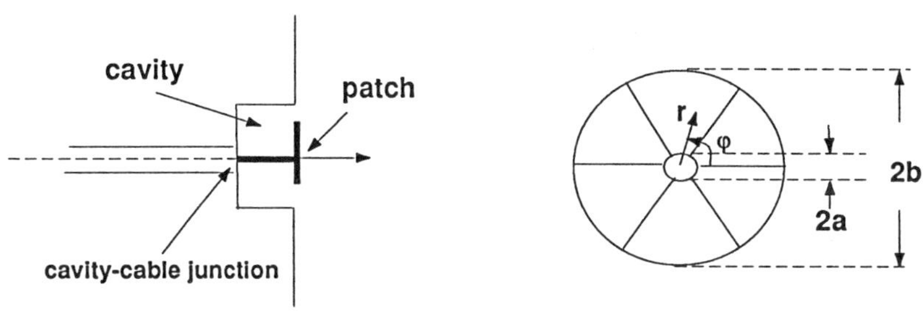

(*a*) (*b*)

FIGURE 5. (a) Side view of a cavity-backed antenna with a coax cable feed. (b) Illustration of the FEM mesh at the cavity–cable junction (the field is set to zero at the center conductor surface).

with

$$h_0 = -\frac{\sqrt{\varepsilon_{rc}}}{Z_0} e_0 + \frac{I_0}{\pi} \tag{20}$$

In these expressions, ε_{rc} is the relative permittivity inside the cable, **E** and **H** are the electric and magnetic fields, respectively, measured at $z = 0$, and I_0 is the center conductor current. Also, (r, ϕ, z) are the polar coordinates of a point in the cable with the center at $r = 0$. We observe that Eq. (20) is the desired constraint at the cable junction in terms of the new quantities h_0 and e_0, which can be used as new unknowns in place of the fields **E** and **H**. It is therefore appropriate to rewrite F_C in terms of these new coefficients and the resultant F_C is given by

$$F_C = -2\pi j k_0 Z_0 e_0 h_0^{\text{src}} \ln\left(\frac{b}{a}\right) \tag{21}$$

where a and b are the radii of the inner and outer cable conductors. The superscripts src stands to indicate that h_0 is treated as a source term in the extremization of the functional F_C.

Before introducing F_C into the system, it is necessary to relate e_0 and h_0 to the constant edge fields associated with the elements in the cavity region that border the cable aperture. Since the actual field has a $1/r$ behavior in the cable, we find that

$$\Delta V = E_i(b - a) = e_0 \ln\frac{b}{a}, \quad i = N_p(p = 1, 2, \ldots, N_C) \tag{22}$$

where ΔV denotes the potential difference between the inner and outer surface of the cable and N_p denotes the global number for the edge across the coax cable. When this condition is used in the functional F_C, it introduces the excitation into the hybrid finite element system without a need to extend the mesh inside the cable or to employ a fictitious current probe. Specifically, F_C becomes

$$F_C = -\frac{\pi}{3} j k_0 Z_0 (b - a) \left\{\frac{I_0}{\pi} - \frac{\sqrt{\varepsilon_{rc}}}{Z_0} \frac{b - a}{\ln(b/a)} E_i\right\}^{\text{src}} \sum_{i=N_p}^{N_C} E_i \tag{23}$$

where instead of using a single edge field, the average field across the cable was computed by the totality of the equal element fields on the cable's aperture, taking advantage of the axisymmetry property of the cable. The factor inside the curly brackets of Eq. (23), with the superscript src, functions as a source in the extremization process. The derivation of $\partial F_C/\partial E_i$ and its incorporation into the system are then straightforward tasks (Gong and Volakis, 1995b).

It is observed that the "constant cavity field" along each mesh edge at the cable junction is just a fictitious field representation and its meaningful physical interpretation is governed by the equipotential constraint (22). Once the hybrid FE–BI system is solved, the input admittance at $z = 0$ is calculated from

$$\begin{aligned} Y_{in} &= \frac{1}{V_0} \oint_{2\pi} \mathbf{H} \cdot \hat{r}\, r\, d\phi \\ &= \frac{2I_0}{e_0 \ln(b/a)} - \frac{1}{Z_C} \end{aligned} \tag{24}$$

where Z_c is the characteristic impedance of the coax cable.

As can be expected, the above feed model assumes the presence of only the dominant (TEM) mode at the cavity–cable junction, an assumption that may not be suitable for certain applications. The model can easily be improved by extending the mesh (say, a distance d) into the cable. The equipotential condition will then be applied at $z = -d$, where all higher order modes vanish.

Aperture-Coupled Microstrip Line Model. As shown in Figure 6, when the microstrip antenna is fed with a microstrip line network underneath the ground plane (cavity's base) via a coupling aperture, special treatment of the feed structure must be considered in the FEM formulation. This is because the microstrip line is usually designed to have a different size and shape as compared to the cavity's geometries. Hence the conventional simulation of treating the entire three-dimensional domain using a single type of elements is not efficient or appropriate for this feed.

Referring to Figure 6, it is appropriate to separate the computational domains because of the small element size required in modeling the guided feed structure. One difficulty encountered when this decomposition is implemented

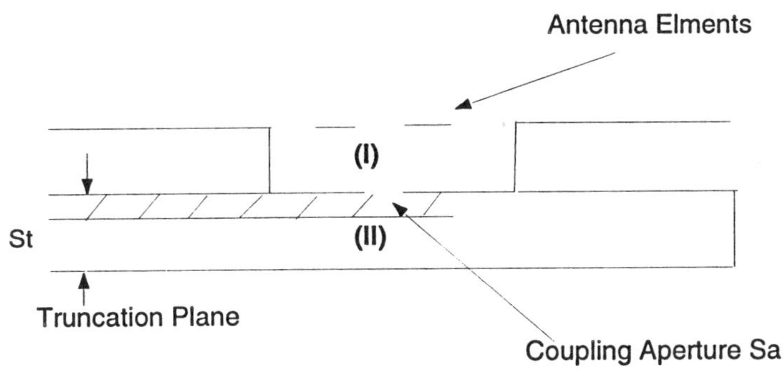

FIGURE 6. Cross section of an aperture-coupled patch antenna, showing the cavity region I and the microstrip line region II for two different FEM computation domains.

is how to model the coupling through the aperture. As an example let us consider a rectangular aperture that has been employed extensively in practice. The cavity fields may be discretized using tetrahedral elements, whereas in the microstrip line region rectangular bricks are the best candidate since the feed structure is rectangular in shape and the substrate is of constant thickness. Although both types of elements employ edge-based field expansions, the meshes across the common area (coupling aperture) are different, and this causes difficulty in enforcing field continuity across the slot aperture. However, since the aperture is very narrow, a "static" field distribution may be assumed at any given frequency. Therefore the *potential* concept may again be applied to relate the fields below and above the aperture. Specifically, the "equipotential" continuity condition is enforced, and to proceed to do so, let us first classify the slot edges as follows:

Tetrahedral Mesh (Cavity Mesh)

- E_j^{b1} $j = 1, 2, 3, \ldots$ vertical edges
- E_j^{b2} $j = 1, 2, 3, \ldots$ diagonal edges

Brick Mesh (Feed Mesh)

- E_j^c $j = 1, 2, 3, \ldots$ vertical edges only

Then the "equipotential" continuity condition requires that

$$E_j^{b1} = \varepsilon_j E_j^c \tag{25}$$

$$E_j^{b2} = \frac{t}{2d}\left(\varepsilon_j E_j^c + \varepsilon_{j+1} E_{j+1}^c\right) \tag{26}$$

in which

$$\varepsilon_j = \begin{cases} +1 \\ -1 \end{cases}$$

whereas t and d are the lengths of the vertical and diagonal edges, respectively. That is, t is simply the width of the narrow rectangular aperture. The coefficient ε_j is equal to ± 1 depending on the sign conventions associated with the meshes above and below the coupling aperture.

The connectivity scheme for entirely different computational domains may be extended by generalizing this concept. It is apparent that this approach makes the FEM implementation straightforward for different geometry/size domains that would be significantly inefficient if only one type of element were used for modeling the structure. In addition, the technique ensures a good system condition since the number of distorted elements in the mesh are minimized.

2.7. Applications

Example 1: RCS of a Rectangular Patch with Lumped and Distributed Loading (FE–BI Method). A traditional approach for increasing the bandwidth of a patch and reducing its in-band radar cross section (RCS) is to place lumped loads between the patch and the ground plane. These lumped loads can readily be modeled in the context of FEM as discussed earlier, but, as shown in Figure 7, they compromise the patch's gain. The results in Figure 7 were generated by an FE–BI code using a single layer of 30×30 brick elements ($0.25\,\text{cm} \times 0.17\,\text{cm}$ in the xy plane) having a height equal to that of the substrate thickness, that is, 0.17558 cm. They clearly demonstrate that the patch's gain is reduced along with its RCS as the four shunt loads decrease in value. More specifically, with a matched load (conjugate of the input impedance) placed at the feed locdation, the RCS reduction (in dB) is twice that of the gain reduction at the

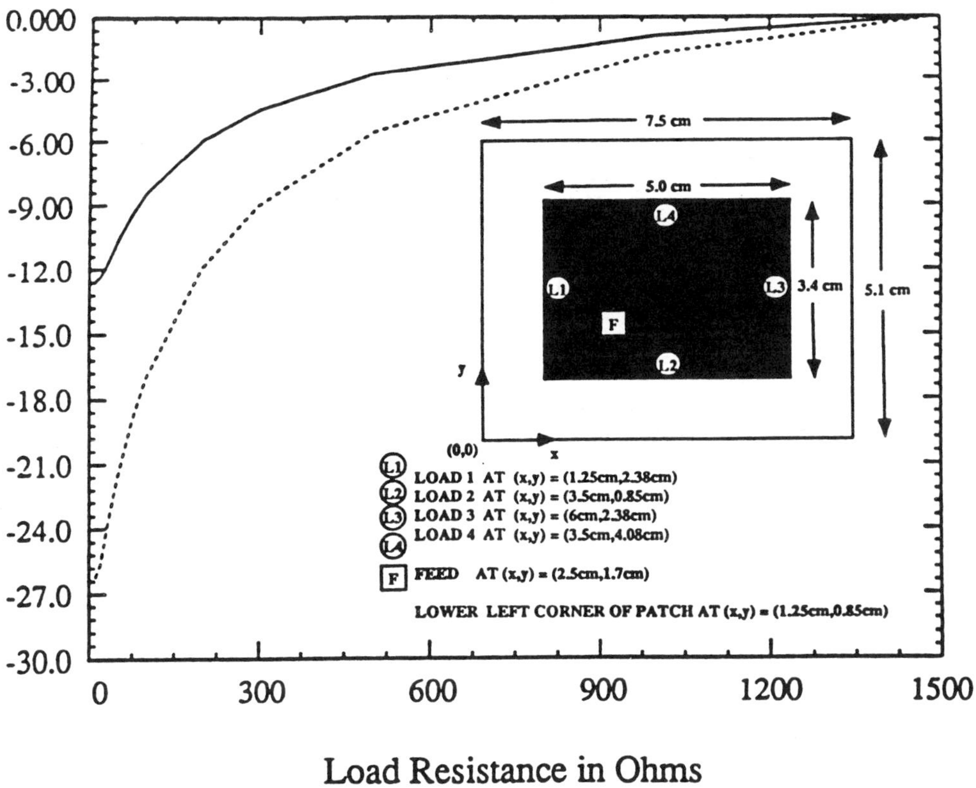

FIGURE 7. Relative normal incidence RCS (dotted line) and relative gains (solid line) reduction (in dB) of the shown rectangular patch at $f = 1.97\,\text{GHz}$ as a function of the four lumped loads placed at the edges of the patch. The cavity is 0.17558 cm deep and is filled with a substrate having a relative dielectric constant of 2.17.

resonance frequency, regardless of the shunt load values, and this is a consequence of the relationship (Jackson, 1990) $\sigma = \lambda^2/\pi G^2$, where G denotes the gain and σ is the RCS. However, FEM modeling can also allow for distributed resistive loading and, if properly selected, distributed loading has the advantage of not appreciably compromising the gain of the patch. A possibility of distributed loading is to place a resistive ribbon (R-skirt) around the periphery of the patch as shown in Figure 8. The ribbon's electrical width increases with frequency and thus its effect is more noticeable at higher frequencies. Indeed, as shown in Figure 8, the 0.15625-cm wide resistive ribbon has a relatively small effect on the RCS (and thus the gain) of the patch at the operating frequency (lowest resonance), but its effect at higher frequency is quite pronounced. Basically, all higher order resonances are suppressed except the lowest, which is only shifted down to 1.9522 GHz from 1.744 GHz. Some reduction in gain and RCS is also observed at the first resonance, but this reduction is one to two orders of magnitude smaller than what is observed at

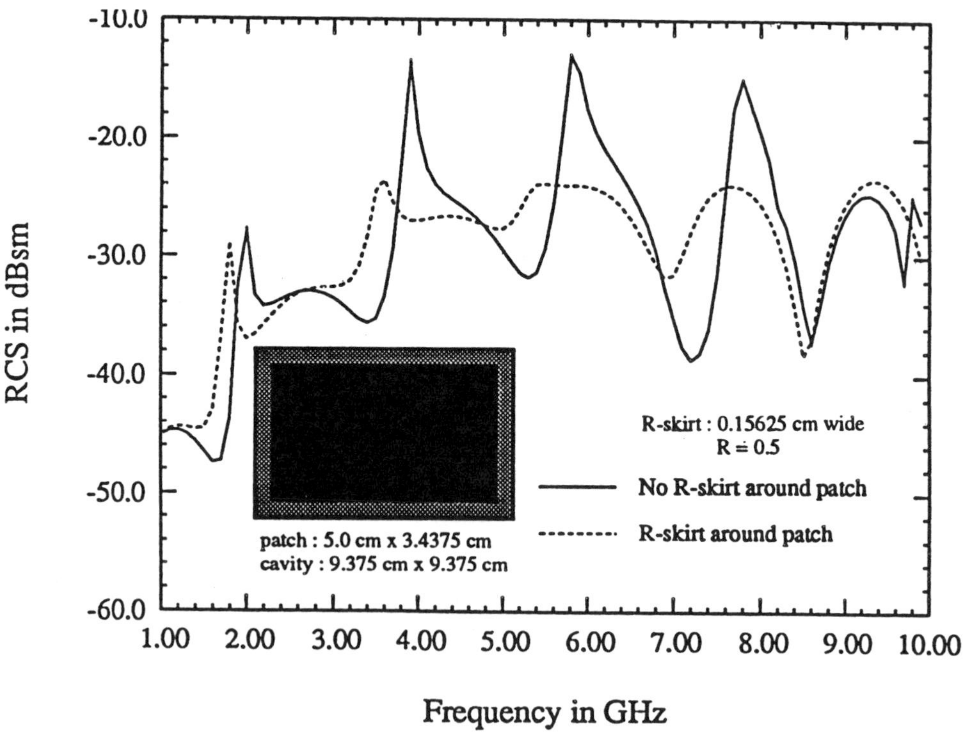

FIGURE 8. Backscatter RCS of the shown patch as a function of frequency. The patch is loaded with a resistive skirt 0.15625 cm wide and of resistivity $R = 0.5Z_0$. The excitation is a plane wave impinging at 70° from normal and in the principal plane, which bisects the longer dimension of the patch.

the higher resonances. Basically, Figures 7 and 8 demonstrate that a combination of lumped and distributed loads can be used to reduce the in-band and out-of-band RCS of the printed antenna (Volakis et al., 1992).

Example 2: Resonance Control of a Rectangular Patch Using Aperture-Backed Cavities and High-Contrast Coatings (FE–BI Method). Of particular interest in antenna design is resonance control. Also, there is a growing interest

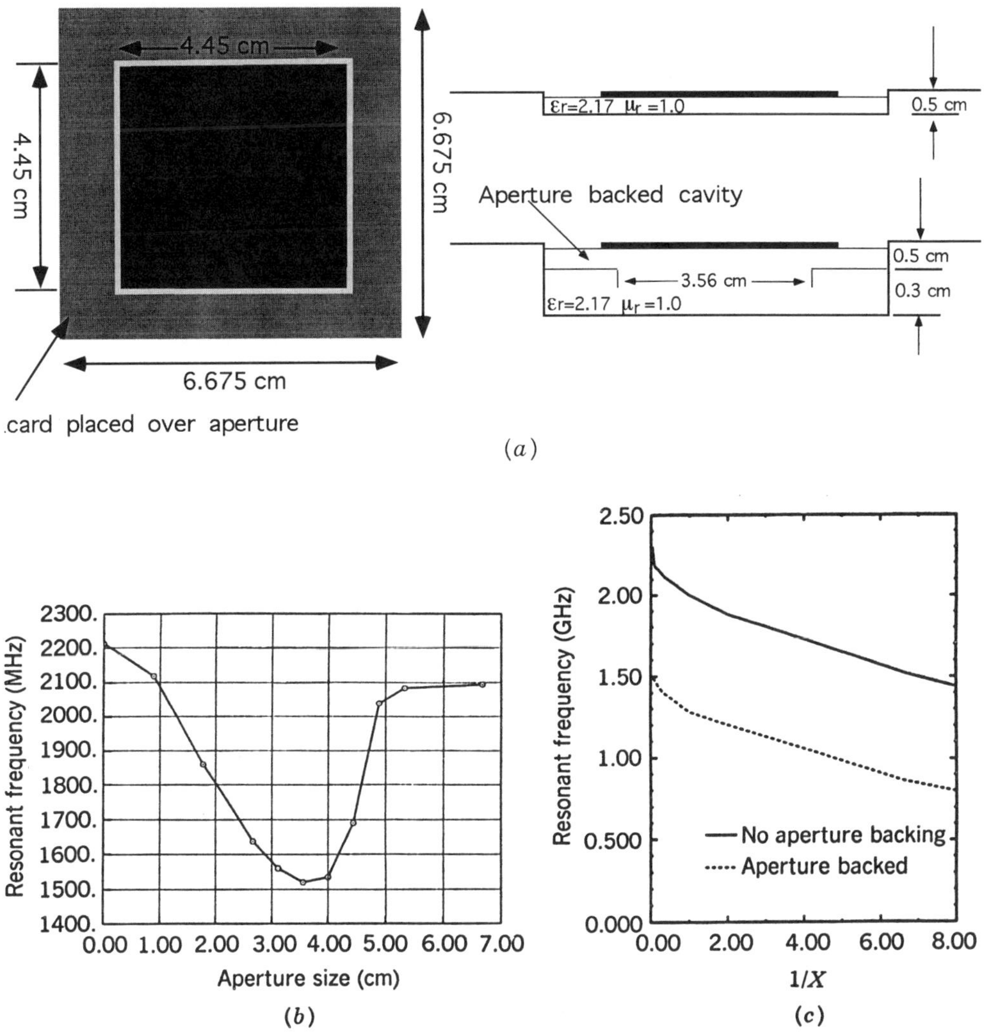

FIGURE 9. Geometry and resonance frequency of the aperture-backed patch with and without substrate: (a) geometry; (b) resonance frequency as a function of the aperture width; (c) resonance frequency as a function of the superstrate's reactance.

in reducing the resonance frequency of a patch antenna without increasing its aperture dimensions. This is accomplished by increasing the substrate thickness and/or using shorting pins between the patch and the ground plane (Kuboyama et al., 1985), by using a high-contrast dielectric sub/superstrate (James and Hall, 1989), or by introducing a second aperture below the patch as shown in Figure 9a. The first of these approaches leads to rather small resonance reductions but the latter two can provide as much as 100% resonance frequency reduction, particularly if used concurrently. The more recent introduction of low-loss, high-contrast dielectric materials (Krupka et al., 1994; Perrota and Hoorfar, 1995) makes the superstrate approach more attractive than before when the performance of the high-contrast dielectrics was compromised by their losses.

The finite element method is particularly suited for the analysis of any cavity configuration and material composition, and it is therefore an ideal method for exploring the performance of different cavity-backed patches. For resonance reduction, we shall examine a 4.45-cm square patch, which resonates at 2.2 GHz. The patch is situated in a rectangular (6.675 cm × 6.675 cm cavity 0.05 cm deep and filled with a dielectric material having $\varepsilon_r = 2.17$. By placing a second cavity 0.3 cm deep below the patch and connected to the first cavity through a rectangular aperture of width w, Figure 9b shows that as much as 30% resonance frequency (down to 1.6 GHz) reduction can be achieved. For a given aperture depth, an optimum aperture width exists for maximum resonance reduction and, typically, this aperture width is only a small fraction of the wavelength less than the patch itself. Basically, as the aperture of the second cavity increases beyond the size of the patch, the field behavior changes since the patch can be considered as simply residing on a thicker substrate.

Having achieved an optimum aperture width of $w = 3.56$ cm for maximum resonance reduction, additional miniaturization can be attained by introducing high-contrast dielectric superstrates. In the context of FEM, the superstrate can either be modeled as a layer inside the recessed cavity or as a reactive card (Senior and Volakis, 1995) placed over the aperture. The latter is actually a material/resistive sheet of resistivity

$$R = -jXZ_0 = \frac{-jZ_0}{k_0\tau(\varepsilon_r - 1)}$$

where τ is the thickness of the superstrate. This simulation is much more convenient and is quite accurate provided τ is a small fraction of the wavelength in the dielectric and this is typically the case. Figure 9c shows that the resonance frequency can be reduced down to 1 GHz when the normalized reactance is $X = 0.2$, corresponding to a thickness of 0.0031 free-space wavelengths when $\varepsilon_r = 256$. When the second aperture is not introduced, the corresponding resonance reduction due to the superstrate alone having $X = 0.2$ is reduced to 1.65 GHz from 2.2 GHz.

Example 3: Input Impedance of a Circular Patch Using Boundary Integral and Artificial Absorbers for Truncating the Finite Element Mesh. As stated earlier, use of absorbing boundary conditions (ABCs) or artificial absorbers (AAs) for truncating finite element meshes leads to fully sparse matrices instead of partly sparse and partly full matrices, as is the case with boundary integral operators. For antenna applications, it is necessary, however, to place the mesh truncation surface some distance from the aperture, as illustrated in Figure 10. This figure displays the cross section of an array of rectangular apertures in a thick ground plane. Both the ABC and the artificial absorber can be used to truncate the mesh, and a question then arises as to which truncation method is more preferred. This issue was examined by considering the number of unknowns and CPU time required for computing the scattering by a $2 \times N_x$ slot array in a thick ground plane (Ozdemir and Volakis, 1994). In the calculations, the incident plane was impinging from above and a local mesh truncation was used only in modeling the lower portion of the mesh. Also, the matrix–vector products for the boundary integral calculation were executed by making use of the FFT since the apertures were planar. The initial study concentrated on establishing an appropriate distance for placing the ABC and artificial absorber on the basis of accuracy, and the conclusion is that sufficient accuracy can be attained by placing the outer mesh boundary at a distance of about 0.3λ from the aperture, as shown in Figure 10a. It has been pointed out that the employed homogeneous artificial absorber is not superior (in terms of absorption) to the second order ABC. However, as shown in Figure 10c, the artificial absorber termination requires less CPU time primarily because of faster solution convergence and reduced unknowns as a result of the metal backing (Ozdemir and Volakis, 1994). The CPU time of the FE–BI implementation is comparable to that of the FE–AA (in spite of the full submatrix) only because the FFT was employed in the iterative solver to perform the matrix-vector product. Thus it may be concluded that the FE–BI implementation is preferred for planar apertures, but for doubly curved surfaces and nonrectangular printed antenna simulations, the FE–ABC and FE–AA methods may be the only alternative.

Having gained an understanding of the accuracy and implementation details related to the FE–AA method, we can proceed with its use for printed antenna analysis. As an example, Figure 11 shows a comparison of the input impedance of a circular patch (1.3 cm in radius) as computed by the FE–BI and FE–AA methods. The patch is placed at the aperture of a circular cavity 0.45 cm deep and filled with a dielectric having $\varepsilon_r = 2.4$. For the FE–BI implementation the mesh was truncated on the aperture and calculations were carried out using tetrahedral elements (Gong et al., 1994). For the FE–AA implementation, the metal backing of the absorber was placed only 0.3λ from the aperture's surface and formed a circular cap of radius 6.07 cm (0.7λ at resonance). About 16 prismatic elements spanned the diameter of the cavity as depicted in the figure and seven elements were used across the height of the computational domain. Of the latter, one prism was used in the substrate, three prisms for the air

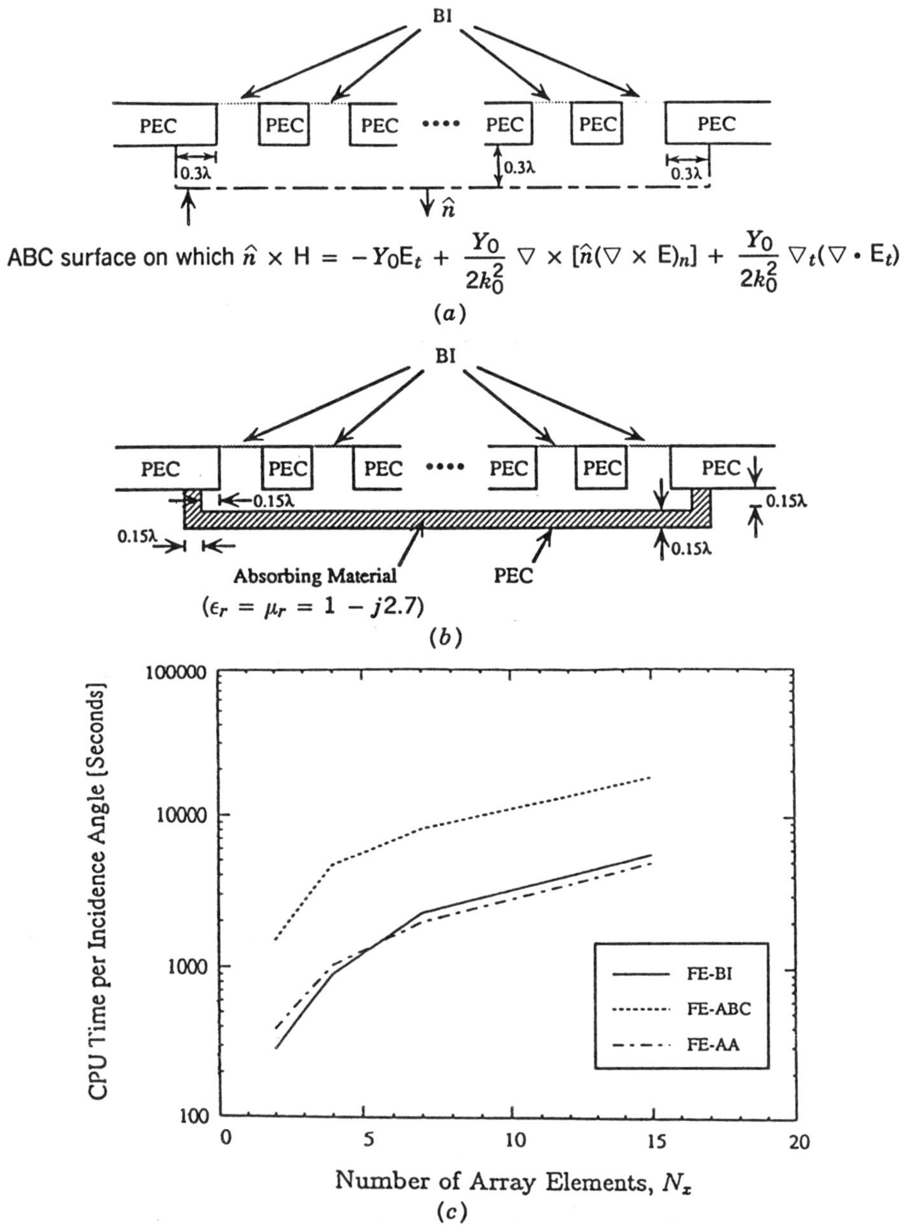

FIGURE 10. Comparison of the CPU time required by each of the finite element formulations to compute the plane wave scattering by the shown $2 \times N_x$ slot array in a thick ground plane. The ground plane thickness is 0.225λ; the slots are separated by 0.225λ along the long dimension and by 0.15λ along the short dimension. Each slot is $1\lambda \times 1\lambda$ wide.

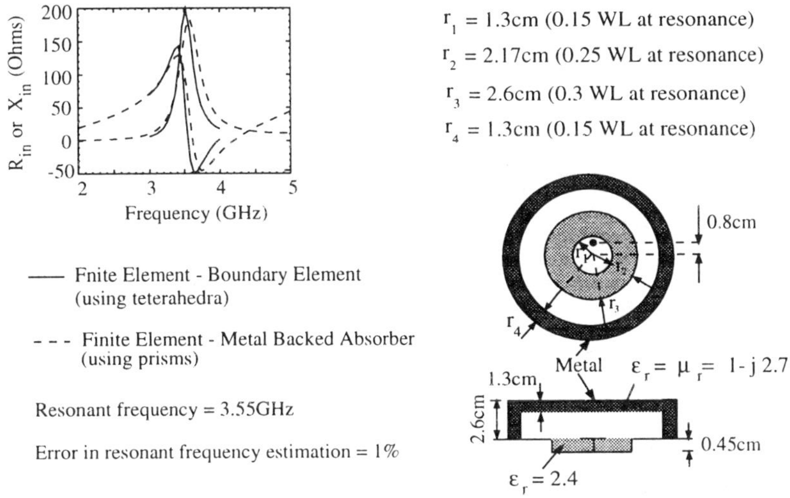

FIGURE 11. Geometry, finite element mesh, and input impedance calculations for a circular patch residing in a circular cavity filled with a dielectric having $\varepsilon_r = 2.4$. The input impedance calculations were carried out using the FE–BI and FE–AA implementations. The first employed tetrahedral elements and the latter used right-angled prisms. The prismatic mesh was generated using a built-in mesh generator and SDRC I-DEAS was employed to create the tetrahedral mesh.

region between the patch and the absorber's surface, and the other three prisms were used to model the absorber thickness. The input impedance calculations were done using a probe feed and, as seen, the approximate FE–AA results are within 1–2% of the FE–BI calculations, which are more rigorous. This level of accuracy is quite acceptable for input impedance calculations and consequently the artificial absorber termination can be used for antenna parameter calculations without a need to extend the mesh far beyond the radiating aperture. Clearly, the advantages of the local mesh truncation schemes are much more pronounced when dealing with printed antennas on curved surfaces. In that case the boundary integral implementation is much more complicated and impractical.

Example 4: Radiation and Scattering of a Rectangular Patch on a Cylindrical Platform (FE–BI and FE–ABC Methods). Patches must often be placed on curved surfaces of airborne platforms and consequently there is interest in studying the radiation and scattering characteristics of patches mounted on these nonplanar surfaces. Figure 12a shows an array of patches mounted on a cylindrical surface and these may be housed in discrete cavities or on continuous (wraparound) substrates. From the point of view of RCS, the discrete array configuration is more attractive because of its low RCS pattern (Kempel and

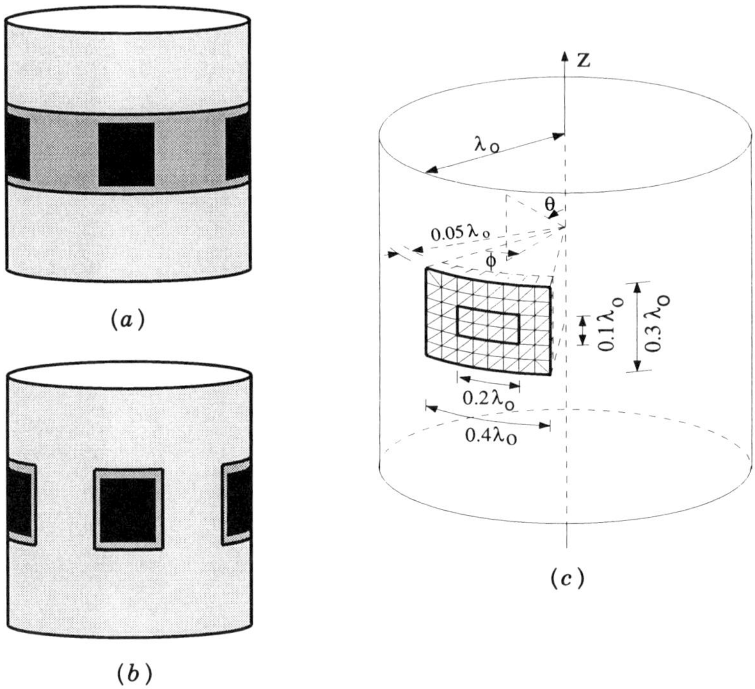

FIGURE 12. Printed antennas on cylindrical platforms: (a) rectangular patches on a continuous dielectric collar; (b) discrete array of cavity backed patches; (c) example of a discretized patch and cavity.

Volakis, 1994). The primary reason for the lower RCS is the elimination of surface wave scattering and it has also been pointed out that the individual cavities are more effective for heat dissipation purposes (Zavosh and Aberle, 1995). A typical mesh comprised of cylindrical shells or prismatic elements is shown in Figure 12c, and the mesh truncation can be either a boundary integral or a conformal absorbing boundary condition. The implementation for the first of these involves the metallic cylinder's Green's function of the second kind (Kempel and Volakis, 1994), which is rather involved. The FE–ABC approach requires conformal ABCs (Chatterjee and Volakis, 1993) or an artificial absorber as discussed earlier and, although approximate, it has the advantage of being applicable to antennas on doubly curved platforms where a Green's function is not available.

It is important to first validate the finite element codes and for this purpose a measurement model was constructed. The model is depicted in Figure 13 and consists of a 0.69 in. × 1.047 in. patch placed on an ogive-cylinder structure. The patch resides on the cylindrical section of the model and atop a 0.062-in. dielectric collar ($\varepsilon_r = 2.33$) recessed in the cylinder's surface as shown. Clearly,

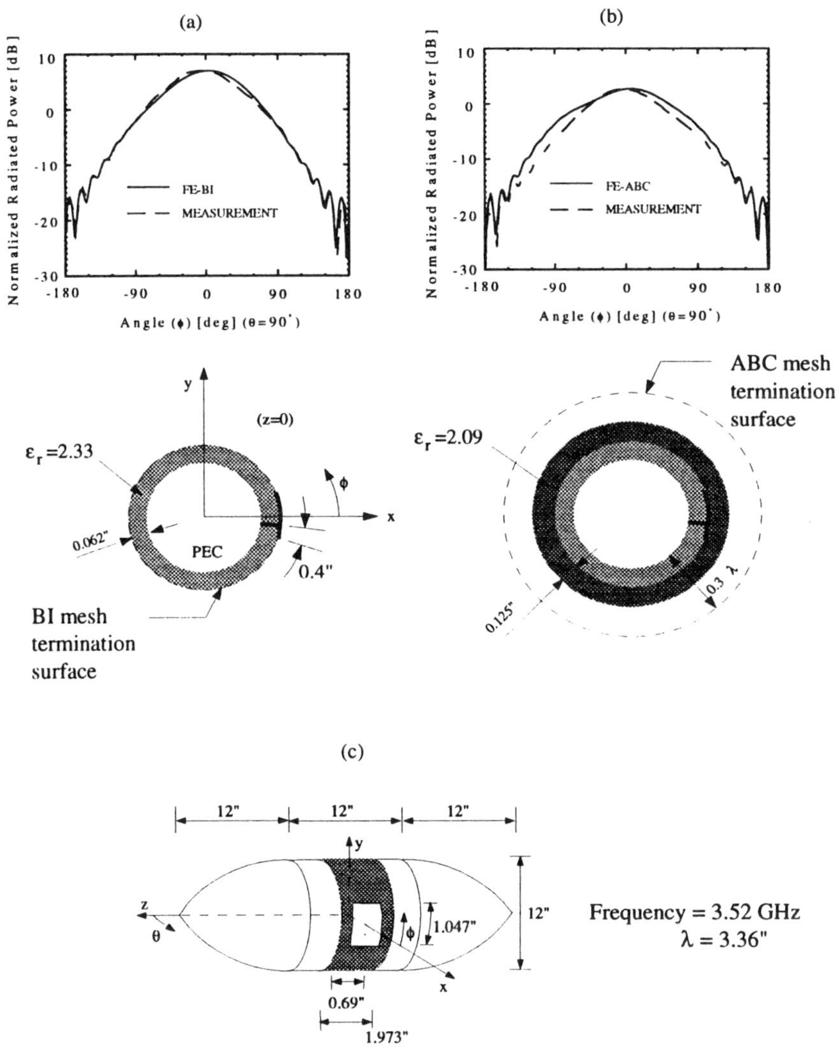

FIGURE 13. Comparison of measured and calculated patterns for a coated and uncoated rectangular patch on a circular cylinder. The calculations for the uncoated patch were carried out by the FE–BI method, whereas those corresponding to the uncoated patch were carried out by the FE–ABC method.

the main reason for terminating the cylinder with the ogival sections is to reduce diffraction contributions, which would otherwise disturb the radiation pattern of the patch. Measurements (in the plane perpendicular to the cylinder) of the patch radiation pattern at 3.56 GHz were obtained by M. Gilreath (NASA Langley Research Center, personal communication, 1994) with and without a 0.125-in. coating of $\varepsilon_r = 2.09$ over its surface. The calculations for the uncoated patch were carried out via the FE–BI and are seen to be in excellent agreement with the measured data. In the case of the coated patch, the FE–ABC method was used since the coated cylinder's Green's function is rather difficult to implement. As before, the ABC was placed only 0.3λ from the cylinder's surface but the agreement with the measured pattern is very good.

Having validated the FEM codes, we now examine the antenna characteristics as a function of the cylinder's radius. Let us, for example, look at the resonance frequency of the rectangular 3.5 cm × 3.5 cm patch. The patch was placed in a 14 cm × 14 cm rectangular cavity recessed in the cylinder's surface, and this cavity is sufficiently large to have no practical effect on the patch's radiation properties (Kempel et al., 1995). Depending on the feed locations, the patch can either be considered as axially or circumferentially polarized. As illustrated in Figures 14a and 14b, the circumferentially polarized patch radiates from the horizontal (curved) edges whereas the axially polarized patch radiates from the vertical edges. Thus curvature changes are expected to have

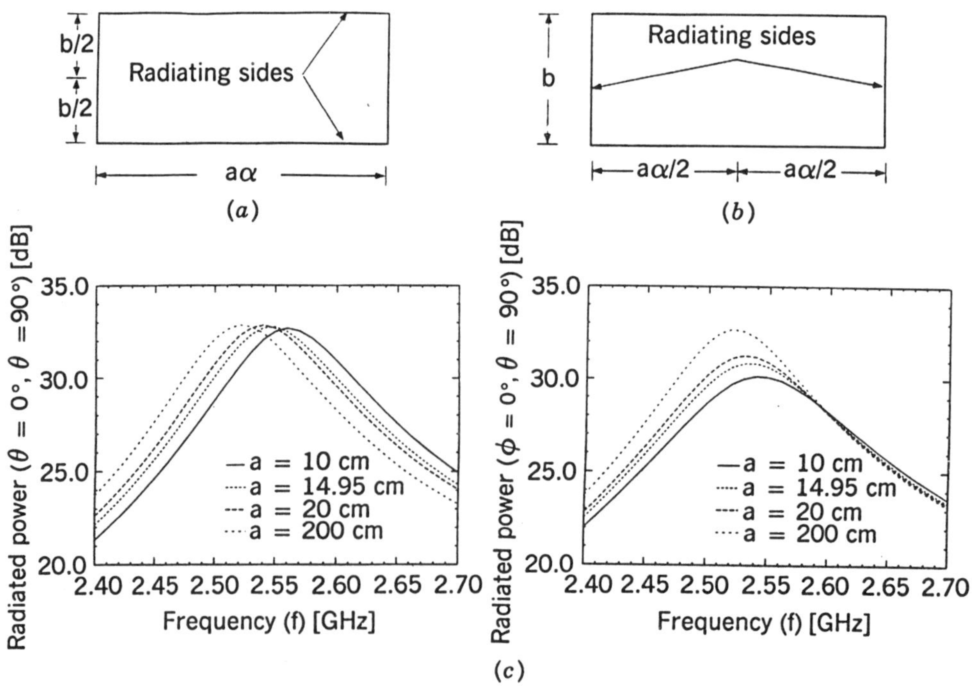

FIGURE 14. Resonance characteristics of rectangular circumferential and axial patches as a function of the cylinder's curvature. The patches are 3.5 cm × 3.5 cm in size and each was placed in a 14 cm × 14 cm × 0.1375 cm cavity filled with a dielectric having $\varepsilon_r = 2.32$ (Kempel et al., 1995).

a more profound effect on the patch's resonance and this is clearly illustrated in the computed results given in Figure 14c. The cylinder's curvature has less effect on the resonance frequency of the axially polarized patches, but the gain decreases with increasing curvature. This is because the vertical radiating edges have increasing angular separation with decreasing radius and the peak radiation of each edge is no longer in the broadside direction.

Example 5: Resonance Frequency of Patch Antenna on a Finite Substrate (FE–BI Method). Wide use of patch antennas for mobile communications (cellular telephones and pagers) necessitates the study of antennas with finite substrates and the finite element method is ideally suited for this purpose. Of particular interest is the prediction of the resonance frequency as a function of the finite substrate width and for this purpose we consider the configuration shown in Figure 15a. The rectangular patch is 2 cm × 3 cm and is placed on a substrate of width w' so that it extends the same amount around the entire patch. For computations, the entire configuration is recessed in a 5.25 cm × 4 cm rectangular metallic cavity that is 0.0787 cm deep. Figure 15b shows the resonance frequency (for $\varepsilon_r = 2.17$, 4.0, and 9.0) as of function of w'/w, where $w = 3$ cm is the patch length. The computations were carried out with a sampling rate of 34 samples per wavelength at the resonance frequency and the resolution of the search routine was 200 MHz. It is seen that the resonance frequency is nearly independent of the substrate width until it becomes comparable in size to the patch itself. When the substrate size is less

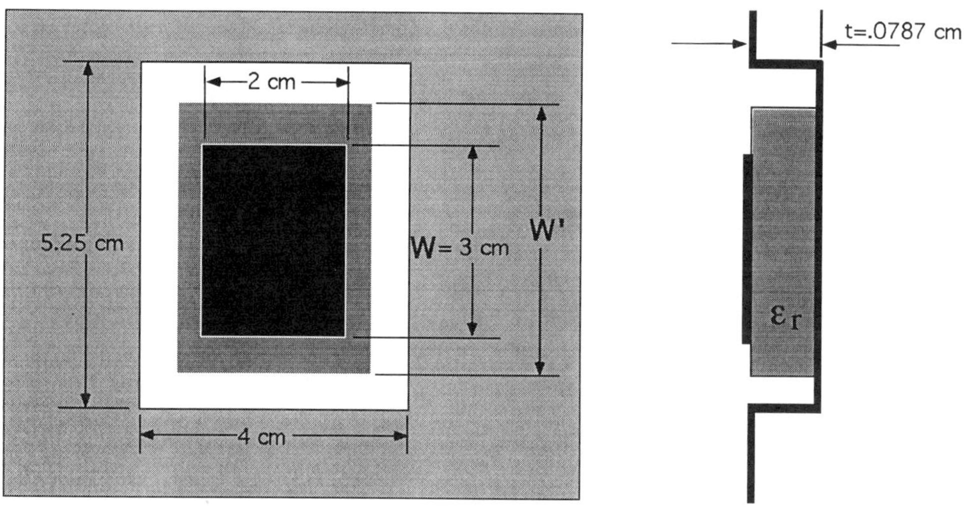

(a)

FIGURE 15. Resonance frequency of a 3 cm × 2 cm patch on a finite substrate 0.0787 cm thick. (*Continues on next page.*)

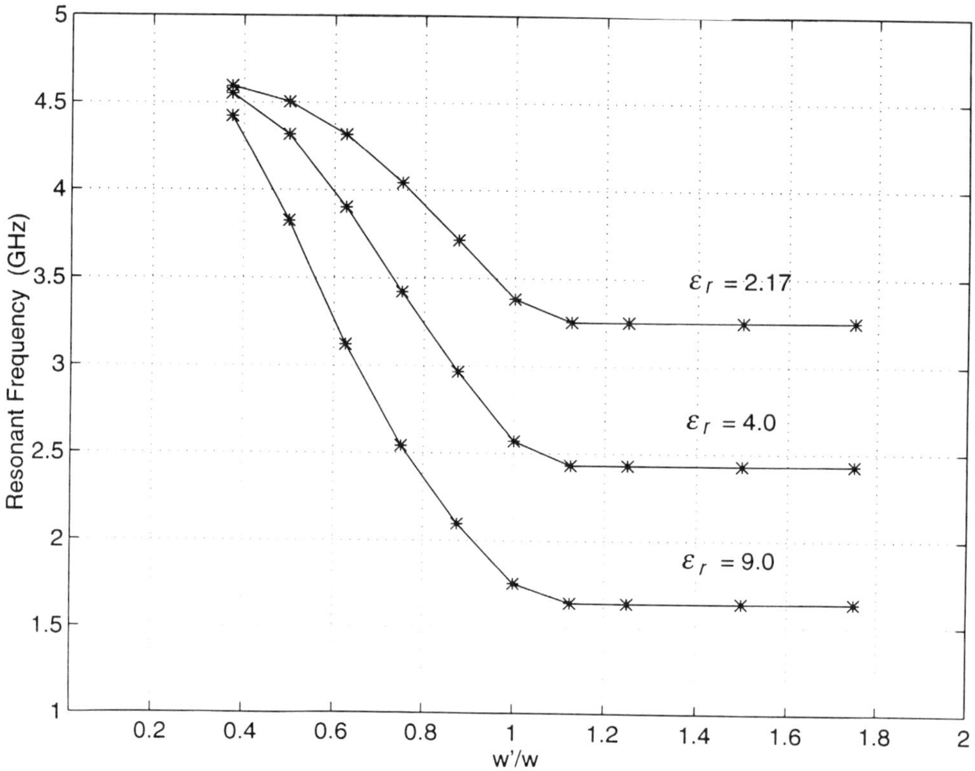

(*b*)

FIGURE 15. (*Continued*).

than that of the patch, the resonance frequency increases almost linearly relative to w'/w and can reach a value as much as three times higher than that of the patch on an infinite substrate.

Example 6: Patch Antenna Radiation on an Aircraft Platform (Hybrid Finite Element–High Frequency Method). Much of the advantage of the FEM comes about from its hybridization with other techniques for the purpose of modeling a diverse set of problems and structures. As stated earlier, the FEM is particularly attractive for modeling structural details and intricate structures and antennas fit in that category. On the other hand, high frequency or ray techniques (Marhefka and Burnside, 1982; Andersh et al., 1994) can handle the scattering by large structures rather efficiently without large computational demands, as would be the case with the FEM method. Thus a hybridization of the finite element and high frequency methods can provide us with an effective approach for pattern computations of antennas on aircraft platforms,

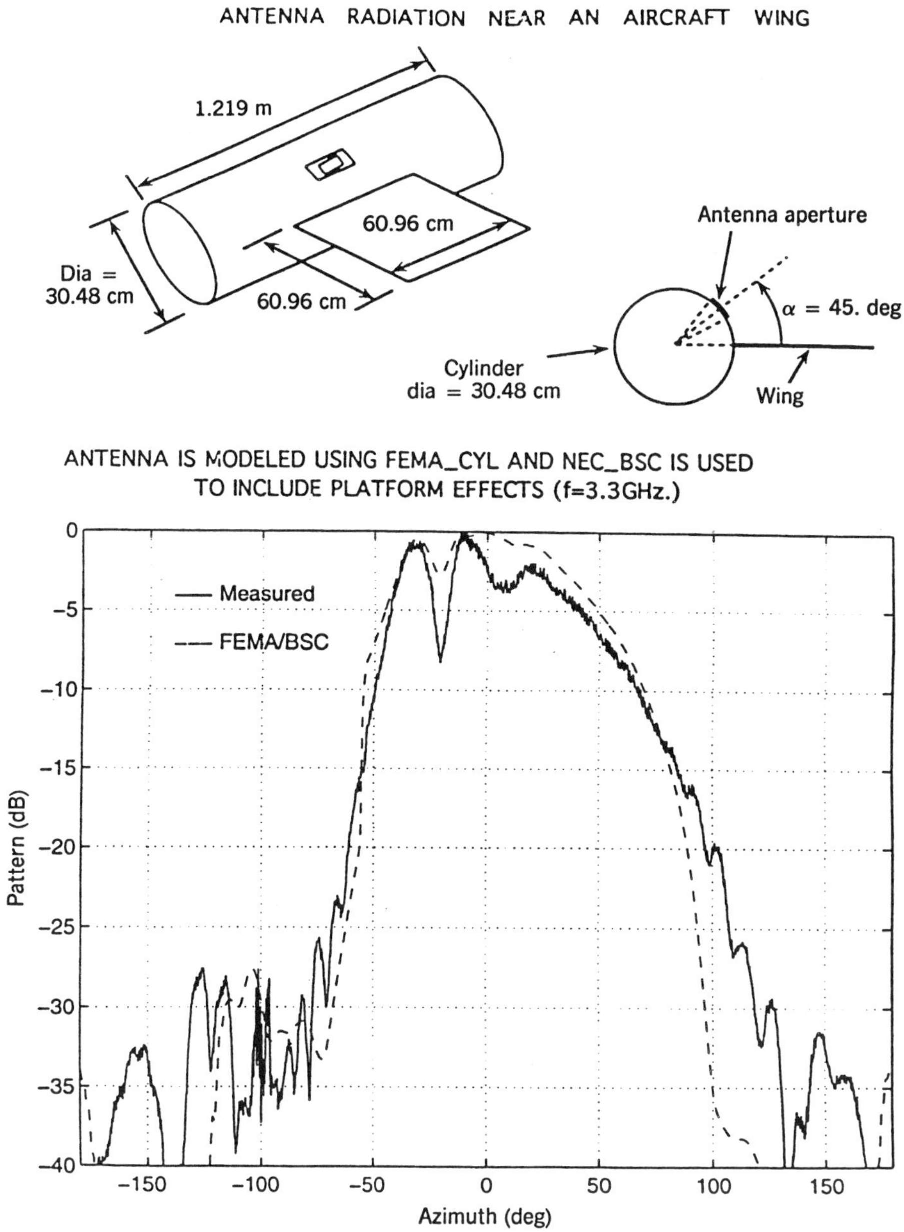

FIGURE 16. Comparison of measured and calculated patterns of the shown patch antenna radiating at 3.3 GHz in the presence of the cylinder–wing structure. The patch dimensions were 3 cm × 2 cm and it was situated in a cavity 6 cm × 5 cm × 0.07974 cm filled with a dielectric having $\varepsilon_r = 2.17$.

including any interference due to the aircdraft substructures. A simple hybridization of the FEM and a high frequency method such as the geometrical theory of diffraction (Marhefka and Burnside, 1982) is to first compute the antenna aperture fields using FEM in the absence of the aircraft substructures such as wings and engines. The aperture fields can then be turned into equivalent magnetic currents, which are subsequently used as sources for the high frequency code. Additional corrections can be considered by iteration or by simply running the FEM analysis code in the presence of both internal and external (due to secondary scattering from the platform) excitations. This hybridization approach was tested by constructing the cylinder + wing model shown in Figure 16 (R. Silva and H. Wang, Naval Warfare Weapons Center, China Lake, CA, personal communication, 1994) and placing a patch antenna on the cylinder's surface as illustrated in the figure. It is seen that the calculations are in very good agreement with the measured data (J. Berrie, Mission Research Corp., Dayton, OH, personal communication, 1994) even after the first iteration. The shown calculations were carried out at 3.3 GHz, the patch was 3 cm × 2 cm with its longer edge parallel to the cylinder's axis. For reference, it should be noted that the pattern in the absence of the wing is symmetric and very similar to that of a patch on an infinite substrate.

REFERENCES

Andersh, D., Hazlett, M., Lee, S. W., Reeves, D., Sullivan, D., and Chu, Y. (1994). XPATCH: A high frequency electromagnetic scattering prediction code and environment for complex three dimensional objects. *IEEE Antennas Propag. Mag.* **APM-36** (1), 65–70.

Antilla, G. E., and Alexopoulos, N. G. (1992). Radiation and scattering from complex 3D curvilinear geometries using a hybrid finite element–integral equation method. *IEEE AP-S Int. Symp. Dig.*, pp. 1758–1761.

Barton, M. L., and Cendes, Z. L. (1987). New vector finite elements for three dimensional magnetic field computation. *J. Appl. Phys.* **61**, 3919–3921.

Bayliss, A., and Turkel, E. (1980). Radiation boundary conditions for wave-like equations. *Commun. Pure Appl. Math.* **33**, 707–725.

Berrie, J. (1994). *Personal Communication*, Mission Research Corp., Dayton, Ohio.

Bossavit, A., and Mayergoyz, I. (1989). Edge-elements for scattering problems. *IEEE Trans. Magn.* **MAG-25** (4), 2816–2821.

Chatterjee, A., and Volakis, J. L. (1993). Conformal absorbing boundary conditions for the vector wave equation. *Microwave Opt. Tech. Lett.* **6**, 886–889; see also correction: **7**, 323–324 (1994).

Chatterjee, A., and Volakis, J. L. (1995). Conformal absorbing boundary conditions for 3D problems: Derivation and applications. *IEEE Trans. Antennas Propag.* **AP-43**, 860–866.

Chatterjee, A., Jin, J.-M., and Volakis, J. L. (1993). A finite element formulation with absorbing boundary conditions for three dimensional scattering. *IEEE Trans. Antennas Propag.* **AP-41**, 221–226.

Collins, J. D., Jin, J.-M., and Volakis, J. L. (1990). A combined finite element-boundary element formulation for solution of two-dimensional problems via CGFFT. *Electromagnetics* **10**, 423–437.

Cwik, T. (1992). Coupling finite element and integral equation solutions using decoupled boundary meshes. *IEEE Trans. Antennas Propag.* **AP-40**, 1496–1504.

Engquist, B., and Majda, A. (1977). Absorbing boundary conditions for the numerical simulation of waves. *Math. Comput.* **31**, 629–651.

Gong, J., and Volakis, J. L. (1995a). Two novel schemes for truncating finite element meshes. *IEEE AP-S Int. Symp. Dig.*, pp. 80–83.

Gong, J., and Volakis, J. L. (1995b). An efficient and accurate model of the coax cable feeding structure for FEM simulations. *IEEE Trans. Antennas Propag.* **AP-43**, 1474–1478.

Gong, J., Volakis, J. L., Woo, A., and Wang, H. (1994). A hybrid finite element boundary integral method for analysis of cavity-backed antennas of arbitrary shape. *IEEE Trans. Antennas Propag.* **AP-42**, 1233–1242.

Jackson, D. R. (1990). The RCS of a rectangular microstrip patch in a substrate-superstrate geometry. *IEEE Trans. Antennas Propag.* **AP-38**, 2–8.

James, J. R., and Hall, P. S. (1989). *Handbook of Microstrip Antennas.* Peter Peregrinus, London.

Jin, J.-M., and Volakis, J. L. (1989). A new technique for characterizing diffraction by inhomogeneously filled slots of arbitrary cross section in a thick conducting plane. *IEEE Electron. Lett.* **25** (17), 1121–1122.

Jin, J.-M., and Volakis, J. L. (1991a). Electromagnetic scattering by and transmission through a three-dimensional slot in a thick conducting plane. *IEEE Trans. Antennas Propag.* **AP-39**, 543–550.

Jin, J.-M., and Volakis, J. L. (1991b). Scattering and radiation from microstrip patch antennas and arrays residing in a cavity. *IEEE Trans. Antennas Propag.* **AP-39**, 1598–1604.

Jin, J.-M., Volakis, J. L., and Collins, J. D. (1991). A finite element-boundary integral method for scattering and radiation by two- and three-dimensional structures. *IEEE Antennas Propag. Mag.* **33** (3), 22–32.

Kempel, L. C. (1994). Radiation and scattering from cylindrically conformal printed antennas. Ph.D. Thesis, University of Michigan, Ann Arbor.

Kempel, L. C., and Volakis, J. L. (1994). Scattering by cavity-backed antennas on a circular cylinder. *IEEE Trans. Antennas Propag.* **AP-42**, 1268–1279.

Kempel, L. C., Volakis, J. L., and Sliva, R. (1995). Radiation by cavity-backed antennas on a circular cylinder. *IEE Proc. Part H: Microwaves, Opt. Antennas.* **142** (3), 233–239.

Krupka, J., Geyer, R., Kuhn, M., and Hinton, J. H. (1994). Dielectric properties of single crystals of Al_2O_3, $LaAlO_3$, $NdGaO_3$, $SrTiO_3$ and MgO at cryogenic temperatures. *IEEE Trans. Microwave Theory Tech.* **MTT-42**, 1886–1890.

Kuboyama, H., Imai, S., Hirasawa, K., and Fujimoto, K. (1985). Post loaded microstrip antenna for pocket size equipment at UHF. *Proc. Int. Symp. Antennas and Propag.*, pp. 433–436.

Lynch, D. R., Paulsen, K. D., and Boyse, W. E. (1993). Synthesis of vector parasites in finite element Maxwell's solutions. *IEEE Trans. Microwave Theory Tech.* **MTT-41**, 1439–1448.

Marhefka, R. J., and Burnside, W. D. (1982). *Numerical Electromagnetic Code — Basic Scattering Code: NEC-BSC*, ElectroScience Lab. Tech. Rep. 712242-14. Ohio State University, Columbus.

McDonald, B. H., and Wexler, A. (1972). Finite element solution of unbounded field problems. *IEEE Trans. Microwave Theory Tech.* **MTT-20**, 841–847.

Nedelec, J. C. (1980). Mixed elements in R^3. *Numer. Math.* **35**, 315–341.

Ozdemir, T., and Volakis, J. L. (1994). A comparative study of an absorbing boundary condition and an artificial absorber for terminating finite element meshes. *Radio Sci.* **29** (5), 1255–1263.

Ozdemir, T., and Volakis, J. L. (1996). Triangular prisms for edge-based vector finite element analysis of conformal antennas. *IEEE Trans. Antenna Propag.* (submitted for publication).

Paulsen, K. D., and Lynch, D. R. (1991). Elimination of vector parasites in finite element Maxwell's solutions. *IEEE Trans. Microwave Theory Tech.* **MTT-39**, 395–404.

Perrotta, A., and Hoorfar, A. (1995). A study of microstrip antennas on very high permittivity substrate and very small ground plane. *Int. IEEE/AP-S URSI Symp., URSI Dig.*, Newport Beach, CA, p. 284.

Peterson, A. F., and Wilton, D. R. (1994). A rationale for the use of mixed-order basis functions within finite element solutions of the vector Helmholtz equation. *Appl. Comput. Electromagn. Soc. Proc.*, Monterey, CA, pp. 1077–1084.

Pozar, D. M. (1992). Microstrip antennas. *Proc. IEEE* **80**, 79–91.

Rahman, B. M. A., and Davies, J. B. (1984). Penalty function improvement of waveguide solutions by finite elements. *IEEE Trans. Microwave Theory Tech.* **MTT-32** (8), 922–928.

Sacks, Z. S., Kingsland, D. M., Lee, R., and Lee, J. F. (1995). A perfectly matched anisotropic absorber for use as an absorbing boundary condition. *IEEE Trans. Antennas Propag.* **AP-43**, 1460–1463.

Schaubert, D. H., Wilton, D. R., and Glisson, A. W. (1984). A tetrahedral modeling method for electromagnetic scattering by arbitrarily shaped inhomogeneous dielectric bodies. *IEEE Trans. Antennas Propag.* **AP-32**, 77–85.

Senior, T. B. A., and Volakis, J. L. (1995). *Approximate Boundary Conditions in Electromagnetics.* IEE Press, London.

Silvester, P., and Hsieh, M. S. (1971). Finite element solution of 2-dimensional exterior field problems. *Proc. Inst. Electr. Eng.* **118**, 1743–1747.

Tai, C. T. (1994). *Dyadic Green's Functions.* IEEE Press, New York.

Volakis, J. L., Alexanian, A., and Jin, J. M. (1992). Broadband RCS reduction of rectangular patches using distributed loading. *IEE Electron. Lett.* **28** (25), 2322–2323.

Volakis, J. L., Chatterjee, A., and Kempel, L. C. (1994a). A review of the finite element method for three dimensional scattering. *J. Opt. Soc. Am.* **11** (4), 1422–1433.

Volakis, J. L., Gong, J., and Alexanian, A. (1994b). Electromagnetic scattering from microstrip patch antennas and spirals residing in a cavity. *Electromagnetics* **14**, 63–85.

Webb, J. P. (1993). Edge elements and what they can do for you. *IEEE Trans. Magn.* **MAG-29**, 1460–1465.

Webb, J. P., and Kanellopoulos, V. N. (1989). Absorbing boundary conditions for finite element solution of the vector wave equation. *Microwave Opt. Tech. Lett.* **2** (10), 370–372.

Yuan, X., Lynch, D. R., and Strohbehn, J. W. (1990). Coupling of finite element and moment methods for electromagnetic scattering from inhomogeneous objects. *IEEE Trans. Antennas Propag.* **AP-38**, 386–393.

Zavosh, F., and Aberle, T. (1995). Single and stacked circular microstrip patch antennas backed by a circular cavity. *IEEE Trans. Antennas Propag.* **AP-43**, 746–750.

CHAPTER FOURTEEN

Numerical Error in Finite Element Solutions of Electromagnetic Boundary Value Problems

ANDREAS C. CANGELLARIS

1. INTRODUCTION

This chapter examines the various types of numerical error associated with the finite element approximation of electromagnetic boundary value problems and their subsequent numerical solution using a finite word length computer.

In addition to the *order of approximation* of the discretization of the governing differential equations, there are several other factors that impact the accuracy of the finite element modeling of electromagnetic wave phenomena. These factors include (1) the fineness with which the wavelength is resolved on the numerical grid; (2) the smoothness of the field quantities; (3) the smoothness of the problem data, including geometry and material parameter variations; (4) the types of boundary conditions associated with the specific problem; and (5) the physical size of the geometry in relation to the wavelength of interest. These parameters impact both the numerical solution accuracy as well as the accuracy of other secondary quantities of practical interest such as characteristic impedance and propagation constant of waveguiding structures, resonant frequencies of resonators, passband and stopband characteristics of

Finite Element Software for Microwave Engineering, Edited by Tatsuo Itoh, Giuseppe Pelosi, and Peter P. Silvester.
ISBN 0-471-12636-5

filters, scattering parameters of waveguide discontinuities, and radiation patterns of antennas. Therefore it is important to examine and, if posssible, quantify the dependence of numerical solution accuracy on the aforementioned problem parameters. Such a quantification leads to the development of guidelines for the proper implementation of the finite element method to the solution of a specific problem.

Such an analysis of the various sources of numerical error in finite element approximations of electromagnetic boundary value problems is attempted in this chapter. Section 2 discusses the fundamentals of numerical dispersion, a discretization-induced property of all discrete approximations of wave phenomena. Through the analysis of simple finite element approximations of the Helmholtz equation on regular one- and two-dimensional grids, several closed-form expressions are derived, which quantify the impact of the order of the finite element approximation on the accuracy with which the wavelength and thus the propagation properties of the wave solution are approximated numerically. In Section 3, the impact of geometry and material characteristics as well as the type of boundary conditions on solution accuracy are examined. For this purpose, appropriate error estimates are introduced and, once again, simple element approximations on regular grids are used to help demonstrate and/or quantify the dependence of numerical error on these problem data. Section 4 concerns itself with the conditioning of the linear system of equations that results from the finite element approximation and examines in detail some of the sources of ill-conditioning. In Section 5, the issue of accuracy of the derivatives of the finite element field solution is examined, and techniques to improve the accuracy with which such derivatives are calculated are reviewed. The presentation concludes with some remarks on how the various results of this chapter impact the way the finite element grid should be constructed for accurate modeling of electromagnetic wave phenomena.

2. NUMERICAL DISPERSION AND ITS IMPACT ON DISCRETIZATION ERROR

For the vector boundary value problems of interest, time-harmonic solutions may proceed directly from the double-curl equation for the elecric field, **E**, which, in a source-free, isotropic, and linear medium with position-dependent magnetic and electric properties, has the form

$$\nabla \times \left(\frac{1}{j\omega\mu} \nabla \times \mathbf{E} \right) + j\omega\hat{\varepsilon}\mathbf{E} = 0 \tag{1}$$

The time dependence $\exp(j\omega t)$ is assumed ($j = \sqrt{-1}$), and the complex permittivity, $\hat{\varepsilon} = \varepsilon - j\sigma/\omega$, is used to account for any conduction and/or dielectric losses in the medium.

For the purposes of finite element solutions, a weak form of Eq. (1) is required. For node-based finite element expansions the unknown vector field is approximated in terms of scalar basis functions, ϕ_i,

$$\mathbf{E} = \sum_i \mathbf{E}_i \phi_i \tag{2}$$

where $\mathbf{E}_i$ denotes the unknown vector field value at node i. The relevant weak form, in the spirit of Galerkin's approximation, is

$$\left\langle \left(\frac{1}{j\omega\mu} \nabla \times \mathbf{E} \right) \times \nabla \phi_i \right\rangle + \langle j\omega\hat{\varepsilon}\mathbf{E}\phi_i \rangle = -\oint \frac{1}{j\omega\mu} \hat{\mathbf{n}} \times (\nabla \times \mathbf{E})\phi_i \, ds \tag{3}$$

where $\langle\ \rangle$ and $\oint$ indicate integration over the domain of interest and its boundary, respectively, while $\hat{\mathbf{n}}$ is the outward unit normal on the boundary. For edge element expansions, vector basis functions, $\mathbf{N}_i$, are used for the expansion of the field,

$$\mathbf{E} = \sum_i E_i \mathbf{N}_i \tag{4}$$

where E_i are the unknown coefficients in the expansion. The relevant weak form is

$$\left\langle \left(\frac{1}{j\omega\mu} \nabla \times \mathbf{E} \right) \cdot (\nabla \times \mathbf{N}_i) \right\rangle + \langle j\omega\hat{\varepsilon}\mathbf{E} \cdot \mathbf{N}_i \rangle = -\oint \frac{1}{j\omega\mu} \hat{\mathbf{n}} \times (\nabla \times \mathbf{E}) \cdot \mathbf{N}_i \, ds \tag{5}$$

For two-dimensional problems, a scalar version of Eq. (3) is readily obtained. More specifically, for a transverse magnetic to z (TM_z) polarization, the fields, $\mathbf{E} = \hat{z}E$, $\mathbf{H} = \hat{\mathbf{x}}H_x + \hat{\mathbf{y}}H_y$, are independent of z and Eq. (3) reduces to

$$\left\langle \left(\frac{1}{j\omega\mu} \nabla_{xy} E \right) \cdot \nabla_{xy} \phi_i \right\rangle + \langle j\omega\hat{\varepsilon}E\phi_i \rangle = \oint \frac{1}{j\omega\mu} (\hat{\mathbf{n}} \cdot \nabla E)\phi_i \, dl \tag{6}$$

where $\nabla_{xy} = \hat{\mathbf{x}}\, \partial/\partial x + \hat{\mathbf{y}}\, \partial/\partial y$. For transverse electric to z (TE_z) polarization of the two-dimensional fields, where $\mathbf{H} = \hat{\mathbf{z}}H$, $\mathbf{E} = \hat{\mathbf{x}}E_x + \hat{\mathbf{y}}E_y$, the weak form is easily found from Eq. (6) by duality. Finally, the one-dimensional version of Eq. (6) in the domain $0 \leqslant x \leqslant d$ is

$$\left\langle \left(-\frac{1}{j\omega\mu} \frac{dE}{dx} \right) \frac{d\phi_i}{dx} \right\rangle - \langle j\omega\hat{\varepsilon}E\phi_i \rangle = \frac{1}{j\omega\mu(0)} \left. \frac{dE}{dx} \right|_{x=0} \phi_i(0) - \frac{1}{j\omega\mu(d)} \left. \frac{dE}{dx} \right|_{x=d} \phi_i(d) \tag{7}$$

The most well-known attribute of Galerkin's method, where the solution is sought in a finite-dimensional subspace of the class of admissible functions for the problem of interest using the same set of functions as trial and test functions, is the symmetry of the resulting stiffness matrix given a symmetric weak formulation. However, another important merit of Galerkin's method is that, if a symmetric weak formulation is used, Galerkin's approximate solution exactly conserves energy in the electromagnetic field despite the fact that it satisfies the vector Helmholtz equation only approximately over the domain of interest. This is easily shown starting from Eq. (5) and using the complex conjugate of the field, $\mathbf{E}^*$, as the test function and Faraday's law to introduce the magnetic field in the boundary integral

$$\left\langle \left(\frac{1}{j\omega\mu}\nabla\times\mathbf{E}\right)\cdot(\nabla\times\mathbf{E}^*)\right\rangle + \langle j\omega\hat{\varepsilon}\mathbf{E}\cdot\mathbf{E}^*\rangle = \oint(\hat{\mathbf{n}}\times\mathbf{H})\cdot\mathbf{E}^*\,ds \tag{8}$$

The complex conjugation of Eq. (8), use of Faraday's law to simplify the integrand in the first term on the left, and use of the constitutive relations $\mathbf{B} = \mu\mathbf{H}$, $\mathbf{D} = \hat{\varepsilon}\mathbf{E}$ result in the following equation:

$$\langle -j\omega\mathbf{B}\cdot\mathbf{H}^*\rangle + \langle j\omega\mathbf{D}^*\cdot\mathbf{E}\rangle = \oint(\mathbf{E}\times\mathbf{H}^*)\cdot\hat{\mathbf{n}}\,ds \tag{9}$$

Clearly, the resulting expression is Poynting's theorem for time-harmonic fields. Thus energy conservation in the electromagnetic field is satisfied exactly by the approximate solution.

The importance of the aforementioned result is rather significant. Considering the various potential sources of error in the development of a numerical solution to a boundary value problem, it is definitely advantageous to be able to work with a weak statement that is consistent with the correct physics of the field we are attempting to calculate. For the electromagnetic field problems of interest, the calculated field quantities will be acceptable only if they satisfy both energy conservation and electric charge conservation. As a matter of fact, the latter has been found to be extremely important in the finite element solution of three-dimensional vector electromagnetic problems. As discussed elsewhere in this book, modifications to the weak statement in Eq. (3) and careful selection of the vector basis functions in Eq. (5) are needed to prevent the contamination of the numerical solution from spurious fields caused by the lack of enforcement of charge conservation in the original weak statement.

2.1. The Fundamentals of Numerical Dispersion

The numerical solutions of wave propagation problems by finite methods are dispersive in general, even though their analytic counterparts are not. This *numerical dispersion* is caused by the discretization of Maxwell's equations effected by the numerical procedure. Consider a finite element approximation using node-based elements on a uniform grid with node spacing h. The approximate solution is expressed in terms of its values at the grid nodes. Of relevance here is the sampling theorem, which essentially states that, under certain conditions, a continuous function can be represented completely by its equally spaced samples in the interval of interest. More specifically, consider a square integrable function $E(x)$ and its Fourier transform $\tilde{E}(k)$ related through the Fourier transform pair

$$\tilde{E}(k) = \int_{-\infty}^{\infty} E(x) \exp(-jkx)\, dx \tag{10a}$$

$$E(x) = \frac{1}{2\pi} \int_{-\infty}^{\infty} \tilde{E}(k) \exp(jkx)\, dk \tag{10b}$$

Let $E(x)$ be band-limited; that is, its Fourier transform vanishes outside a finite interval $|k| > k_b$. If this function is sampled, $E_n = E(x = nh)$, $n = 0, \pm 1, \pm 2, \ldots$, and reconstructed as

$$E_s(x) = \sum_{n=-\infty}^{\infty} E_n S(x - nh)$$

where $S(x)$ is the sinc function

$$S(x) = \frac{\sin(\pi x/h)}{(nx/h)}$$

then $E_s(x)$ and $E(x)$ are identical if

$$h \leqslant \pi/k_b \tag{11}$$

As an example of the application of the sampling theorem to wave solutions, consider the case of a standing wave pattern resulting from the interference between a time-harmonic uniform plane wave of angular frequency ω, propagating in the positive x direction, with a plane wave of the same angular frequency and field amplitude propagating in the negative x direction. A homogeneous and lossless medium is assumed, with permittivity ε_1 and permeability μ_1. The total fields are

$$\mathbf{E}^{t} = \hat{\mathbf{z}}[\exp(-jk_1 x) - \exp(jk_1 x)] \tag{12a}$$

$$\mathbf{H}^{t} = -\hat{\mathbf{y}}\eta_1^{-1}[\exp(-jk_1 x) + \exp(jk_1 x)] \tag{12b}$$

The Fourier transform of the magnetic field is

$$\tilde{H}(k) = -\frac{\pi}{\eta_1}\left[\delta(k + k_1) + \delta(k - k_1)\right] \tag{13}$$

Clearly, the field is a band-limited function of bandwidth k_1. Consequently, according to the sampling theorem, and assuming that sinc functions are used for the field interpolation, the numerical grid should be of size

$$h \leqslant \frac{\pi}{k_1} = \frac{\lambda_1}{2} \tag{14}$$

where λ_1 is the wavelength of the plane wave in the medium (ε_1, μ_1). Conversely, on a uniform grid of element size h only those wave solutions with wavelength $\lambda > 2h$ can be approximated accurately. The wavelength $\lambda_c = 2h$ is sometimes called the *cutoff wavelength* of the grid, and the corresponding wavenumber, $k_c = \pi/h$, the *cutoff wavenumber*.

Since the most commonly used finite element interpolation functions are not entire functions but functions of compact support, a resolution much better than two nodes per wavelength is necessary for the accurate representation of the numerical solution on the finite element grid. More specifically, the lower the degree of the interpolating polynomial the finer the wavelength resolution by the finite element grid needs to be in order to ensure the accuracy of the numerical solution. To quantify this effect, one needs to obtain the numerical dispersion relation for the finite element approximation of Maxwell's equations on a specific grid. This quantification is presented in this and the following two subsections.

The development begins with the simple case of the finite element approximation of the weak statement in Eq. (7) for the one-dimensional Helmholtz equation on a uniform grid of element size h using linear interpolation functions. The grid is of infinite extent and the medium properties μ and $\hat{\varepsilon}$ are constant. It is readily verified that the finite element equation for the nth node in the grid is

$$-\left(1 - \frac{\gamma^2 h^2}{6}\right)\hat{E}_{n-1} + 2\left(1 + \frac{\gamma^2 h^2}{3}\right)\hat{E}_n - \left(1 - \frac{\gamma^2 h^2}{6}\right)\hat{E}_{n+1} = 0 \tag{15}$$

where $\hat{E}_n$ is the unknown (discrete) field value at node n, and $\gamma = \alpha + j\beta$ is the wavenumber satisfying the familiar analytic dispersion relation $\gamma^2 = -\omega^2\mu\hat{\varepsilon}$.

Plane solutions to eq. (15) are sought, which for node n have the form

$$\hat{E}_n = A\exp(-\hat{\gamma}nh) \tag{16}$$

where $\hat{\gamma} = \hat{\alpha} + j\hat{\beta}$ is now the *numerical wavenumber*. The dispersion relation for

$\hat{\gamma}$ is easily found by substituting Eq. (16) in Eq. (15)

$$p^2 - 2\,\frac{1 + \gamma^2 h^2/3}{1 - \gamma^2 h^2/6}\,p + 1 = 0 \tag{17}$$

where p is the ratio of the numerical solution at adjacent nodes

$$p = \exp(-\hat{\gamma} h) \tag{18}$$

From Eq. (17) it is clear that the two roots, p_1, p_2, will be such that $p_1 p_2 = 1$ or $\hat{\gamma}_1 = -\hat{\gamma}_2$, corresponding to forward and backward propagating waves. The solution for p is

$$p = \frac{1 + \gamma^2 h^2/3}{1 - \gamma^2 h^2/6} \pm j\,\sqrt{1 - \left(\frac{1 + \gamma^2 h^2/3}{1 - \gamma^2 h^2/6}\right)^2} \tag{19}$$

For the case of a loss-free medium, $\gamma^2 = -\omega^2 \mu \varepsilon = -4\pi^2 \lambda^{-2}$ and Eq. (19) reduces to

$$p = \frac{1 - (2\pi h/\lambda)^2/3}{1 + (2\pi h/\lambda)^2/6} \pm j\,\sqrt{1 - \left(\frac{1 - (2\pi h/\lambda)^2/3}{1 + (2\pi h/\lambda)^2/6}\right)^2} \tag{20}$$

Considering the term under the radical, it is straightforward to show that

$$0 < 1 - \left(\frac{1 - (2\pi h/\lambda)^2/3}{1 + (2\pi h/\lambda)^2/6}\right)^2 < 1$$

for all wavelengths $\lambda > 2h$, that is, for all wavelengths greater than the cutoff wavelength of the grid. Thus, for the lossless case $|p| = 1$, or equivalently, $\hat{\alpha} = 0$, which means that the numerical solution does not exhibit any numerically induced (i.e., artificial) damping. Furthermore, from Eqs. (18) and (19) one has

$$p + p^{-1} = 2\cos\hat{\beta} h = 2\,\frac{1 - (2\pi h/\lambda)^2/3}{1 + (2\pi h/\lambda)^2/6} \Rightarrow$$

$$\hat{\beta} = h^{-1}\cos^{-1}\left(\frac{1 - (2\pi h/\lambda)^2/3}{1 + (2\pi h/\lambda)^2/6}\right) \tag{21}$$

or

$$\frac{\hat{\beta}}{\beta} = \left(\frac{\hat{\lambda}}{\lambda}\right)^{-1} = \left(\frac{\hat{v}_p}{v_p}\right)^{-1} = \frac{\lambda}{2\pi h}\cos^{-1}\left(\frac{1 - (2\pi h/\lambda)^2/3}{1 + (2\pi h/\lambda)^2/6}\right) \tag{22}$$

where $\hat{\lambda}$ is the numerical wavelength and $\hat{v}_p$ is the numerical phase velocity ($\hat{\beta} = 2\pi/\hat{\lambda} = \omega/\hat{v}_p$).

From the above equation it becomes apparent that, in contrast to the analytic result, where $v_p = 1/\sqrt{\mu\varepsilon}$ is independent of frequency, the numerical discretization results in dispersion. When h is a small fraction of λ such that $(2\pi h/\lambda)^2 \ll 1$ one has

$$\left[1 - \frac{1}{3}\left(\frac{2\pi h}{\lambda}\right)^2\right]\left[1 + \frac{1}{6}\left(\frac{2\pi h}{\lambda}\right)^2\right]^{-1} \simeq 1 - \frac{1}{2}\left(\frac{2\pi h}{\lambda}\right)^2$$

For small angle values $\cos t \simeq 1 - t^2/2$ may be used in Eq. (22) to obtain

$$\frac{\hat{\beta}}{\beta} \simeq \frac{\lambda}{2\pi h}\cos^{-1}\left(\cos\left(\frac{2\pi h}{\lambda}\right)\right) = 1$$

Thus, if the length of the element is a small fraction of the wavelength, the numerical dispersion is minimal. Indeed, for $\lambda/h > 20$, $\hat{\beta}/\beta$ is very close to unity.

For the case of lossy media, the resolution of the penetration depth, $d_p = \sqrt{2/(\omega\mu\sigma)}$, becomes an issue of concern. Since

$$\gamma^2 h^2 = -\omega^2\mu\varepsilon\left(1 - j\frac{\sigma}{\omega\varepsilon}\right)h^2 = -(\omega h\sqrt{\mu\varepsilon})^2 + j(h\sqrt{\omega\mu\sigma})^2 \tag{23}$$

the dimensionless quantities $w = \omega h\sqrt{\mu\varepsilon}$ and $s = h\sqrt{\omega\mu\sigma} = h\sqrt{2}/d_p$ are used as parameters in our discussion. Plots of the normalized propagation constant, $\hat{\beta}/\beta$, and the error in the numerical attenuation constant, $(\hat{\alpha} - \alpha)$ versus w with s as a parameter, lead to the following observations (Lynch et al., 1985). First, $\hat{\beta}/\beta$ is only slightly influenced by the presence of loss and is very near to unity for $w \leqslant 0.1\pi$, or $\lambda_0/h > 20$, where $\lambda_0 = 2\pi/(\omega\sqrt{\mu\varepsilon})$ is the wavelength in the absence of loss. Second, the departure of the numerical attenuation constant from its analytical counterpart is significant as s and w increase; however, it is acceptable for $ws \leqslant 0.02$. It is interesting to recast this final result in terms of the penetration depth and the wavelength λ_0:

$$\left(\frac{d_p}{h}\right)\left(\frac{\lambda_0}{h}\right) \geqslant 100\sqrt{2}\pi \tag{24}$$

From this expression it is clearly seen that for media with losses such that the penetration depth is of the order of the undamped wavelength or larger, the selection of the element size is based entirely on the adequate resolution of the undamped wavelength for negligible error in the approximation of the correct propagation constant and attenuation constant in the medium. However, if significant losses are present, such that the penetration depth becomes smaller than the undamped wavelength, selection of the element size must be based on the adequate resolution of the penetration depth. For example, consider the case of a lossy medium with loss tangent $(\sigma/\omega\varepsilon)$ equal to 100. It is then, $d_p/\lambda_0 = \sqrt{2\omega\varepsilon/\sigma} = 0.1\sqrt{2}$, and thus a choice of $h = \lambda_0/20$ fails to satisfy Eq. (24). On the other hand, the selection $h = d_p/10$ results in $\lambda_0/h = 100/\sqrt{2}$, which implies negligible error in the approximation of the propagation constant, while the left-hand side of Eq. (24) assumes the value of $1000/\sqrt{2} > 100\sqrt{2}\pi$, thus ensuring negligible error in the approximation of the attenuation constant.

At this point, before continuing with the investigation of the dependence of the numerical wavenumber on the order of the finite element interpolation functions as well as the shape of the elements for two- and three-dimensional applications, it is important to point out that the overall accuracy of the finite element approximation is not dictated only by the adequate resolution of the wavelength and the penetration depth. As was mentioned in the introduction and will be discussed in detail later, the electrical size of the domain of simulation, the type of boundary conditions used for grid truncation, and the geometry/material complexity impact the discretization error. Therefore one should not conclude that the order of accuracy with which the wavenumber is approximated is also the order of accuracy of the finite element solution.

2.2. Numerical Dispersion and Numerical Anisotropy

In order to quantify the impact of the degree of the interpolating polynomials and the element type and shape on numerical dispersion in the finite element approximation of the scalar Helmholtz equation, the weak statement of Eq. (6) will be used to develop the relevant discrete equations for nodes on an infinite, uniform two-dimensional grid on the xy plane. The mathematical development will emphasize finite element approximations that use nodal elements. The numerical dispersion performance of edge elements will be reviewed briefly at the end of the section.

For two-dimensional nodal elements, the weak statement of Eq. (6) is used to develop the finite element equations. A lossless medium of constant permittivity ε and constant permeability μ is assumed. It is well known that the plane wave with electric field vector

$$\mathbf{E} = \hat{\mathbf{z}} \exp[-jk(x \cos a + y \sin a)] \tag{25}$$

with $k = \omega\sqrt{\mu\varepsilon}$ and propagating at an angle a with respect to the x axis, is the solution to the Helmholtz equation of interest. The uniform grid of Figure 1 is used for the development of the discrete equations. Its nodes are uniformly distributed at a constant distance h from each other in the x and y directions. For the case of the four-node bilinear quadrilateral element, the fact that the elements are square leads to the same discrete equation for each node in the

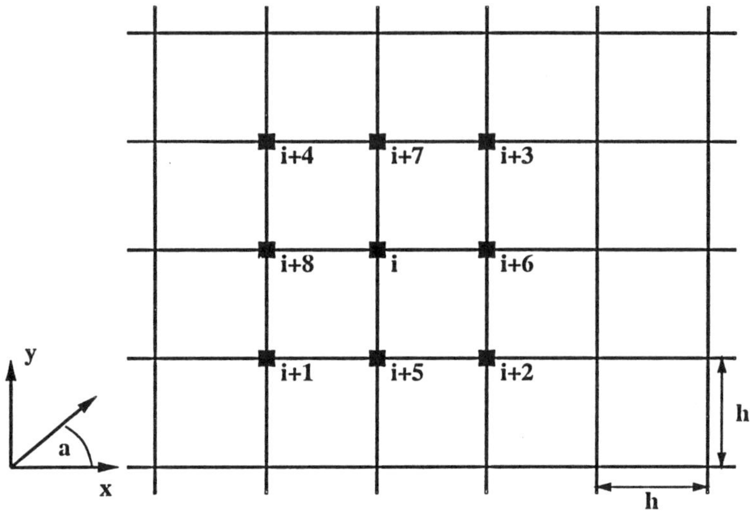

FIGURE 1. Uniform two-dimensional grid for the numerical dispersion analysis of the finite element approximation of the scalar Helmholtz equation. Quadrilateral elements with bilinear interpolating functions are used for the approximation.

grid. For the ith node the equation is

$$\frac{(kh)^2}{36}\left[16\hat{E}_i + \sum_{n=i+1}^{i+4} \hat{E}_n + 4\sum_{n=i+5}^{i+8} \hat{E}_n\right] + \frac{1}{6}\left[2\sum_{n=i+1}^{i+8} \hat{E}_n - 16\hat{E}_i\right] = 0 \quad (26)$$

where E_n is the unknown (discrete) field value at node n. A nonzero plane wave solution to Eq. (26) is sought, of the form as Eq. (25), which for node n with coordinates (x_n, y_n) has the form

$$E_n(x_n, y_n) = \exp[-j\hat{k}(x_n \cos a + y_n \sin a)] \quad (27)$$

Substitution of Eq. (27) into Eq. (26) and the requirement for a nontrivial solution lead to the numerical dispersion equation, which is a nonlinear equation for the numerical wavenumber $\hat{k}$. Clearly, in addition to being dependent on the electrical size of the grid, $\hat{k}$ is also dependent on the direction of propagation of the wave, that is, the angle a. In other words, in addition to numerical dispersion, the finite element solution will exhibit *numerical anisotropy* and thus the numerical phase velocity will be different for different directions of propagation on the grid. It is instructive to compare the dispersion relations we obtain for the bilinear square element for the two special cases of (1) propagation along the axes of the grid ($a = 0°$ and $a = 90°$)

and (2) propagation along the diagonal ($a = 45°$). Since for bilinear elements the interpolation function varies linearly along x and y, the dispersion relation for $a = 0°$ and $a = 90°$ is identical to the one in Eq. (22) (with k, $\hat{k}$ replacing β, $\hat{\beta}$, respectively) obtained for the one-dimensional linear element. For the case $a = 45°$, the dispersion relation becomes

$$\frac{\hat{k}}{k} = \frac{\sqrt{2}}{kh} \cos^{-1}\left[\frac{1 - (kh)^2/6}{1 + (kh)^2/12}\right] \tag{28}$$

A customary way of expressing the error due to numerical dispersion is by means of the phase error per wavelength defined as

$$\Delta\theta_\lambda = (k - \hat{k})\lambda = 2\pi\left(1 - \frac{\hat{k}}{k}\right) \tag{29}$$

In other words, Eq. (29) gives the phase error that the discrete solution accumulates for a propagation distance of one wavelength. The curves "bilinear-($a = 0$)" and "bilinear-($a = 45$)" in Figure 2 depict the dependence of

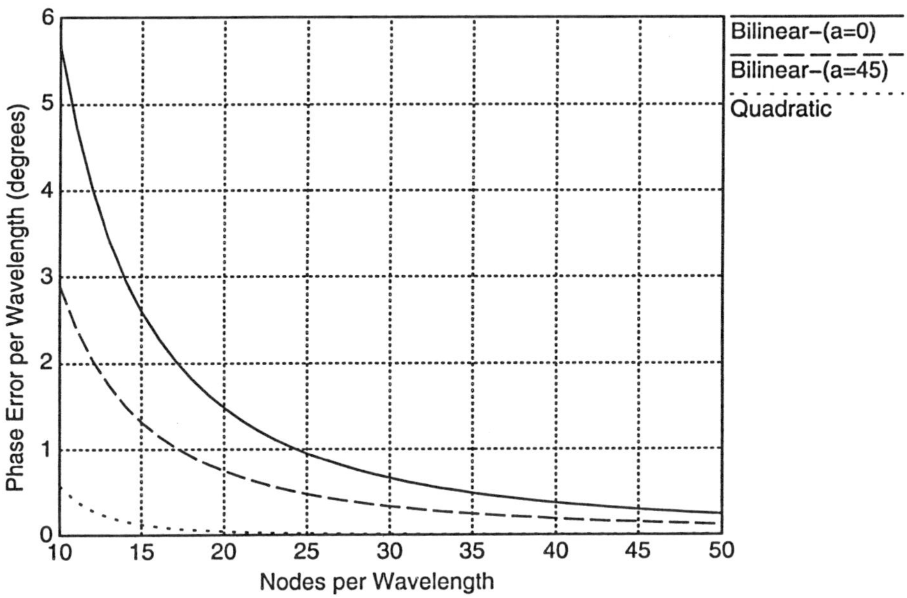

FIGURE 2. Phase error per wavelength versus the number of nodes per wavelength. The numerical anisotropy of the discrete approximation is illustrated by the difference in the numerical dispersion curves for plane wave propagation along the x axis ($a = 0°$) and propagation along the diagonal ($a = 45°$). For quadratic elements, the phase error becomes negligible for resolutions better than 20 nodes per wavelength.

$\Delta\theta_\lambda$ on the number of nodes per wavelength for the aforementioned two cases. It is noted that since the bilinear finite element interpolation functions provide for linear variation along the two axes and bilinear variation along the diagonal, the largest phase error occurs for $a = n90^\circ$, and the smallest phase error occurs for $a = 45^\circ + n90^\circ$, where $n = 0, 1, 2, 3$. The error for any other angle will be bounded by the error for these two sets of angles. From the figure it is clear that, for the case of bilinear elements, a resolution of at least 20 nodes per wavelength is required in order to keep the error due to numerical dispersion and numerical anisotropy at sufficiently low levels.

Instead of increasing the node density, numerical dispersion may also be reduced by increasing the order of the finite element interpolation functions. This intuitively acceptable result may be demonstrated by examining the numerical dispersion relation associated with the quadratic nodal element in two dimensions. Compared to the case of bilinear elements, the dispersion relation for quadratic elements is far more complex since, as depicted in Figure 3, node i is now coupled to 24 other nodes. Furthermore, unlike the bilinear elements, it is now necessary to group the nodes in three groups. Group A consists of nodes $n = i, i+1, \ldots, i+8$; group B consists of nodes $n = i+9$, $i+10, \ldots, i+20$; group C consists of nodes $n = i+21$, $i+22$, $i+23$, $i+24$. It should be noted that h is now the separation distance between the nodes rather than the size of the element. Due to their different relative position in the grid, three different residual equations result, one for each group of nodes. Consequently, complete dispersion analysis requires all three equations to be considered simultaneously. The details of the development of numerical dispersion relations for higher-order elements will not be presented here. The papers

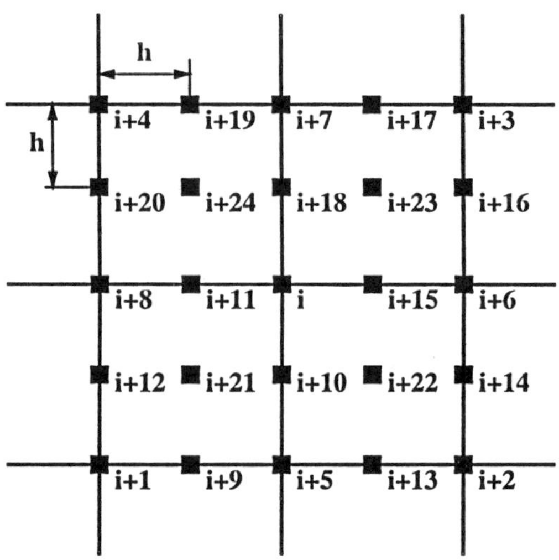

FIGURE 3. Uniform two-dimensional grid with quadratic elements.

by Lee and Cangellaris (1992) and Scott (1994) include rather thorough developments of such numerical dispersion relations.

For the purposes of this presentation, it is sufficient to examine the numerical dispersion relation obtained for plane wave propagation along the x (or y) axis of the quadratic-element grid of Figure 3. The numerical dispersion relation for this case is the same as the one-dimensional dispersion relation for the three-node quadratic element depicted in the one-dimensional grid of Figure 4. The relation is found to be (Lee and Cangellaris, 1992)

$$\frac{\hat{k}}{k} = \frac{1}{2h} \cos^{-1} \left[\frac{3(kh)^4 - 26(kh)^2 + 15}{(kh)^4 + 4(kh)^2 + 15} \right] \tag{30}$$

A simple algebraic analysis reveals that for wavelengths greater than the cutoff wavelength, $\lambda_c = 2h$, the numerical wavenumber is real for all values of kh outside the interval $(\sqrt{2.5}, \sqrt{3})$ or, equivalently, for all values of λ/h outside the interval $(2\pi/\sqrt{3}, 2\pi/\sqrt{2.5})$. For these real values of $\hat{k}$, the numerical solution exhibits no artificial damping. The curve termed "quadratic" in Figure 2 depicts the numerical phase error per wavelength versus the wavelength resolution in the grid. Clearly, the phase error is significantly less for the quadratic element than for the bilinear element. More specifically, a resolution of 15 nodes per wavelength seems sufficient for negligible dispersion when quadratic elements are used for the finite element approximation of wave solutions.

Despite the fact that, at first sight, the aforementioned interval, $(2\pi/\sqrt{3}, 2\pi/\sqrt{2.5})$, in which the numerical wavenumber becomes complex, seems of no practical importance due to the low wavelength resolution associated with it, the work of Scott (1994) makes it worthy of a brief discussion. A simple algebraic manipulation shows that the argument of the inverse cosine function in Eq. (30) is negative and less than -1 for values of λ/h in the aforementioned interval. In view of the fact that, for a real number $q < -1$, it is $\mathscr{R}(\cos^{-1} q) = m\pi$, where m is an odd integer, one concludes that $\mathscr{R}(2\hat{k}h) = m\pi$. In terms of the numerical wavelength, $\hat{\lambda}$, and the size of the element, $\ell = 2h$, this last result may be cast in the form $\ell = \hat{\lambda}/2$. In words, the numerical solution will exhibit artificial damping when the size of the quadratic element is equal to $\hat{\lambda}/2$.

From a more physical point of view, this result can be explained in terms of the nonuniqueness of the solution to the wave equation inside a closed lossless domain. From the uniqueness theorem for electromagnetic fields it is known that in a lossless domain the solution to Maxwell's equation is not unique at the resonant frequencies of the domain (Chew, 1990). For the quadratic element under consideration, the element is the domain, and its length is equal to one-half of the wavelength at the first resonance. This nonuniqueness may be seen directly from the residual equation associated with the internal node

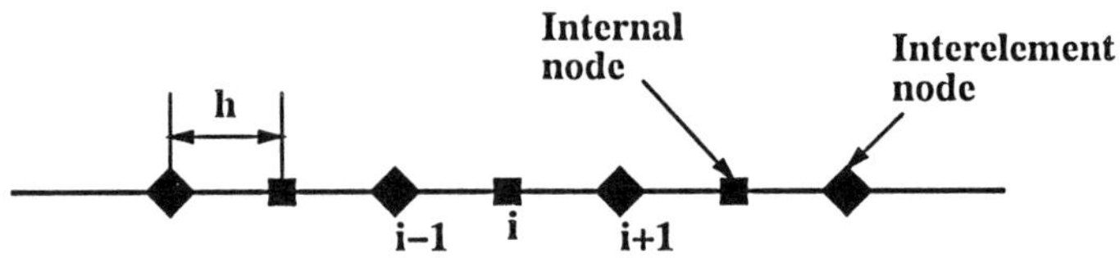

FIGURE 4. Uniform one-dimensional grid with quadratic elements.

in the one-dimensional quadratic element of Figure 4. The equation is

$$[10 + (kh)^2]E_{i-1} + [-20 + 8(kh)^2]E_i + [10 + (kh)^2]E_{i+1} = 0 \tag{31}$$

As expected, the value of the interior node depends only on the values of the end nodes (interelement nodes). Thus the degree of freedom E_i may be eliminated in favor of the degrees of freedom associated with the interelement nodes

$$E_i = \left[\frac{10 + (kh)^2}{20 - 8(kh)^2}\right](E_{i+1} + E_{i-1}) \tag{32}$$

This "interior node condensation" is actually a quite common process in finite element codes for the reduction of the number of unknowns. From Eq. (32) it is now clear that the value of E_i becomes undefined when the denominator is zero. This occurs for $kh = \sqrt{2.5}$, which, from the previous discussion, is equivalent to the case when the element length becomes equal to one-half of the numerical wavelength.

Scott (1994) has generalized this result for nth order Lagrange elements. More specifically, he has shown that for an nth order Lagrange element there exist $n - 1$ intervals in the domain $0 < kh < \pi$, in which the numerical solution exhibits artificial damping. These regions correspond to element lengths $\ell = m\hat{\lambda}/2$, $m = 1, 2, \ldots, n - 1$. The numerical solutions produced by values of kh in these intervals suffer from serious distortion and are totally unreliable. Even though these results have been derived for the simple case of one-dimensional elements, the aforementioned uniqueness argument suggests that they can be extended, at least qualitatively, to two- and three-dimensional elements. The important result is that, when higher-order elements are used in the finite element modeling of wave phenomena, the grid should be constructed in such a way that all elements are of size sufficiently smaller than half a wavelength.

In contrast to the rather thorough investigation of the numerical dispersion characteristics of nodal elements, the numerical dispersion of edge elements has received very little attention. Warren and Scott (1994) investigated first- and second-order, quadrilateral, vector finite elements. They found that the numerical dispersion relations for these simple elements were identical with those obtained for the corresponding nodal elements.

3. IMPACT OF GEOMETRY AND BOUNDARY CONDITIONS ON SOLUTION ACCURACY

While the numerical dispersion studies of the previous section have been useful in quantifying the impact of wavelength resolution on the accuracy with which the physical wavenumber is approximated by the numerical one, it is important to remember that an infinite mesh was assumed for the development of the numerical dispersion relations. However, for realistic problems the mesh must be truncated, and appropriate boundary conditions need to be imposed at the truncation boundaries. Furthermore, the mesh must be constructed properly so that geometric features and material property variations are represented properly. As it turns out, the accuracy of the numerical solution, in addition to being dependent on the number of nodes per wavelength and the order of the interpolating polynomials, is affected also by the electrical size of the computational domain, the type of boundary conditions used for mesh truncation, as well as the geometric and material characteristics. The discussion in this section examines the impact of these parameters on numerical solution accuracy and, whenever possible, gives error estimates for their quantification.

3.1. Error Estimates for Finite Element Approximations of Waves

Let $\mathbf{E}$ be the exact solution and $\tilde{\mathbf{E}}$ the exact solution of the finite element approximation. Clearly, $\tilde{\mathbf{E}}$ would be obtained only if it were possible to carry out all calculations to an infinite number of decimal places. Since, in practice, calculations are carried out with finite precision, the numerical procedures introduce round-off errors, thus resulting in an approximate numerical solution $\hat{\mathbf{E}}$. Consequently, in the finite element solution, the *total error*, $\mathbf{E} - \hat{\mathbf{E}}$, is the sum of the *discretization error*, $\mathbf{E} - \tilde{\mathbf{E}}$, and the *round-off error*, $\tilde{\mathbf{E}} - \hat{\mathbf{E}}$.

Clearly, calculation of the actual error requires knowledge of the exact solution. Thus, initially, there may seem to be little reason to want to compute it. However, even when $\mathbf{E}$ is unknown, it is possible to construct *a priori* error estimates and to determine how the error is affected as the number of elements in the mesh is increased or as the order of the interpolating polynomials is increased.

The *a priori* error estimates that will be discussed next are for the discretization error. Since errors are functions, their magnitude need be measured with respect to some norm. For the purposes of this discussion, the mean-square or L^2-norm will be used only. It is defined by

$$\|\mathbf{e}\|_0 = \left(\int_\Omega |\mathbf{E} - \tilde{\mathbf{E}}|^2 \, d\Omega \right)^{1/2} \tag{33}$$

where Ω is the domain of computation. It is customary to seek *asymptotic estimates* of these error measures in finite element approximations. More specifically, assume that one starts with a uniform finite element grid in which all elements are cubic in shape of equal side-length h. The mesh is then refined by decreasing h, thus increasing the number of elements in the mesh. One wishes to know bounds on the error, measured using Eq. (33), which are given in terms of the mesh size h and are dependent somehow on the data of the problem. Ordinarily, these estimates will be of the form

$$\|\mathbf{e}\|_0 \leqslant Ch^p \tag{34}$$

where C is a constant depending on the data of the problem and p is an integer that depends on the interpolation functions used in the finite element approximation. The exponent p is a measure of the rate of convergence of the method with respect to this particular norm. If p is positive , then $\|\mathbf{e}\|_0$ approaches zero as h tends to zero, and the approximation is said to be convergent to the exact solution with respect to this norm.

A result of great importance in the construction of *a priori* error estimates in finite element approximations is the following. "For any member, $\mathbf{U}_h$, of the class of admissible functions of the finite element approximation it is

$$\|\mathbf{e}\|_0 \leqslant M\|\mathbf{E} - \mathbf{U}_h\|_0 \tag{35}$$

where M is a positive constant." This result says that the magnitude of the discretization error, measured in the L^2-norm, is bounded above by the "distance" (in the L^2-norm sense) of the exact solution $\mathbf{E}$ from any member $\mathbf{U}_h$ in the class of admissible functions. Consequently, the problem of establishing the convergence of the finite element solution is reduced to one of estimating the distance $\|\mathbf{E} - \mathbf{U}_h\|_0$, which is actually a problem of interpolation. This distance depends on the degree of the interpolating polynomials. It is assumed that the exact solution is regular enough so that it is square-integrable and has square-integrable derivatives of order less than or equal to r, where r is an integer >1. With the assumption that the local shape functions are constructed with complete polynomials of degree m, it can be shown that (Strang and Fix, 1973)

$$\|\mathbf{E} - \mathbf{U}_h\|_0 \leqslant C_0 h^{\mu+1}\|\mathbf{E}\|_0 \tag{36}$$

where $\mu = \min(m, r-1)$, and C_0 is a constant independent of $\mathbf{E}$ and h. Use of Eq. (36) in Eq. (35) results in the following estimate for the discretization error:

$$\|\mathbf{e}\|_0 \leqslant M_0 h^{\mu+1}\|\mathbf{E}\|_0 \tag{37}$$

If $\mathbf{E}$ is very smooth (i.e., r is large), which will be the case if the geometry of interest does not involve abrupt material property changes and conducting wedges and corners, then the rate of convergence in the L^2-norm will be $O(h^{m+1})$. For example, with linear, quadratic,... elements, we anticipate $O(h^2)$, $O(h^3)$, ... accuracy asymptotically as h approaches zero. The effect of r is to limit the rate that can be achieved by simply increasing the degree of the local element approximation.

The constant M_0 in Eq. (37) depends on the electromagnetic properties of the media in the domain of interest, the boundary conditions used for the truncation of the finite element mesh, the frequency of operation, and the electrical size of the domain (length of the domain in wavelengths). From the earlier discussion on numerical dispersion, it is clear that the accuracy of the solution depends on the number of nodes per wavelength, and any useful error estimates should include this dependence. Critical to the quantification of the impact of some of the aforementioned parameters on the accuracy of finite element approximations of wave phenomena was the work of Bayliss et al. (1985). They showed that, if the wave solution and the data are smooth enough, and complete polynomials of degree m are used for the finite element interpolation, the error estimate in Eq. (37) becomes

$$\|\mathbf{e}\|_0 \leqslant C_m h^{m+1}(1 + k^{m+2})\|\mathbf{E}\|_0 \tag{38}$$

In the above expression, k is the wavenumber and C_m is a constant that depends on the problem data (geometry/media characteristics and type of boundary conditions) but is independent of h, k. Later work by Peterson and Baca (1991), Lee and Cangellaris (1992), and Cangellaris and Lee (1992) provided further numerical validations of the aforementioned error estimate.

In order to discuss in more detail the computational implications of this last result, it is appropriate to introduce in Eq. (38) the characteristic dimension d of the geometry under consideration. This is justified since the constant C_m depends on the geometry data. Then, the error bound in the L^2-norm is $O((kh)^{m+1}kd)$. This form implies that the number of nodes per wavelength will have to increase with the electrical length of the domain to maintain accuracy. Indeed, let the characteristic length of the domain be N_λ wavelengths, $N_\lambda = d/\lambda$. For linear elements $m = 1$, and the error bounds of Eq. (38) are of $O(k^3h^2d)$ in the L^2-norm. Let $N_h + 1$ be the number of nodes per wavelength, in which case $N_h = \lambda/h$. With these definitions, the error in the L^2-norm can be written as $O(N_h^{-2}N_\lambda)$. Clearly, if λ is decreased while N_h and d are kept fixed, N_λ increases and the error grows with N_λ. Thus, in order to maintain a desirable accuracy in the finite element solution, the number of nodes per wavelength, N_h, needs to increase as the electrical size of the computational domain increases. Alternatively, higher-order elements may be utilized in order to increase the exponent m and thus balance the increase of the error due to the increasing value of N_λ.

In conclusion, the dispersion analysis of the previous section alone is not sufficient to determine the number of nodes per wavelength or the degree of the interpolating polynomials required to achieve a desirable accuracy in finite element simulations of wave phenomena. In addition to numerical dispersion, the electrical size of the domain of interest affects the accuracy of the finite element solution, and its impact was demonstrated and quantified in this subsection. Additional parameters that impact the solution accuracy and need to be accounted for include the smoothness of the data of the problem and the types of boundary conditions involved. These are discussed in some detail in the next two subsections.

3.2. Impact of Data Smoothness on Solution Accuracy

For the purposes of electromagnetic modeling, "data smoothness" refers to abrupt changes in the electromagnetic properties of the media inside the domain of interest, presence of conducting wedges and/or corners, or points where multiple conductor and dielectric interfaces meet. The smoothness of the data impacts the smoothness of the solution and thus its influence on the numerical accuracy is readily apparent from the general expression for the error estimate of Eq. (37) and the discussion pertinent to it.

The quantification of the singular behavior of electromagnetic fields at metallic and/or dielectric wedges and corners has been presented by several authors (Meixner, 1972; Andersen and Solodukhov, 1978; Van Bladel, 1985; De Smedt and Van Bladel, 1986). In order to illustrate the type of field singularities encountered in such problems, consider the simple geometry of a two-dimensional conducting wedge of internal angle α as shown in Figure 5. Even for the case of dynamic problems, the region of field singularity is the immediate neighborhood of the tip of the wedge. In other words, of interest here is only a cylindrical region of radius equal to a small fraction of the wavelength and the axis of the wedge. Thus a static analysis suffices for the description of the singular behavior of the fields. As discussed in the aforementioned references, an electrostatic potential, $\Phi(\rho, \theta)$, for the transverse electric polarization (E_ρ, E_θ, H_z), and a magnetostatic potential, $\Psi(\rho, \theta)$, for the transverse magnetic polarization (H_ρ, H_θ, E_z), are introduced for the analysis. The leading term in

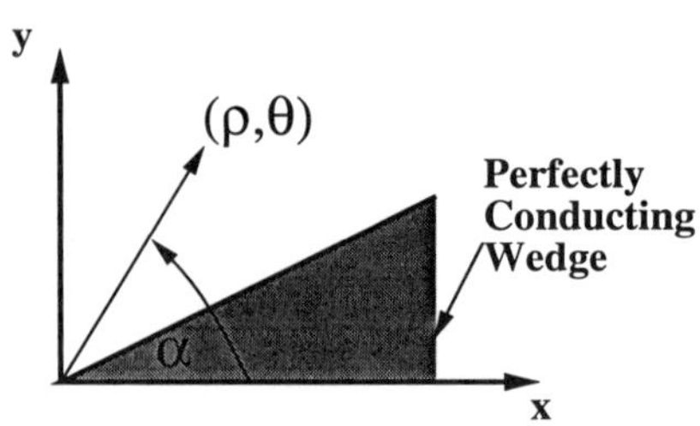

FIGURE 5. Geometry of a two-dimensional, perfectly conducting wedge.

the singularity of the potentials is $A\rho^{\nu} \sin \nu(\theta - \alpha)$, where A is a constant and

$$\nu = \frac{\pi}{2\pi - \alpha} \tag{39}$$

Consequently (concentrating on the leading terms), the corresponding transverse electric field, $\mathbf{E}_{TE}$, and transverse magnetic field, $\mathbf{H}_{TM}$, are given by the expressions

$$\mathbf{E}_{TE} = -A\nu\rho^{\nu-1}[\hat{\rho} \sin \nu(\theta - \alpha) + \hat{\theta} \cos \nu(\theta - \alpha)] \tag{40}$$

$$\mathbf{H}_{TM} = A\nu\rho^{\nu-1}[\hat{\rho} \cos \nu(\theta - \alpha) - \hat{\theta} \sin \nu(\theta - \alpha)] \tag{41}$$

From Eqs. (39)–(41) it is clear that when α is less than π, ν becomes less than 1, and the transverse components of the fields become singular at the tip of the edge. This lack of smoothness of the fields makes the error estimate of Eq. (37) invalid.

While one might argue that this singular behavior of the fields is only local and thus it should be of no consequence on the solution accuracy away from the singularity, the results of numerical experimentation with a variety of geometries that give rise to field singularities point against this conjecture. For example, proper resolution of field singularities is necessary for overall solution accuracy in problems involving coupling through apertures in metallic structures, and for propagation constant and characteristic impedance calculation accuracy in the eigenvalue analysis of planar waveguides such as microstrip, slotlines, and coplanar waveguides. The problem is more difficult in three dimensions. As a matter of fact, modeling of three-dimensional structures with metallic edges and tips has been one of the greatest difficulties of finite element formulations based on nodal elements. One of the advantages of edge elements is that the degrees of freedom are essentially associated with tangential components of the fields. Thus, for the case of the perfectly conducting metallic corner, the physical boundary condition of zero tangential electric field may be imposed simply, without any apparent concern about field singularities. However, one must keep in mind that, unless higher-order vector finite elements are used along with proper mesh refinement, the singular behavior of the fields will not be approximated accurately by a finite element solution based on edge elements.

At this point it is appropriate to review the methods that have been used so far for the accurate finite element modeling of field singularities. These methods fall in two basic classes: direct methods and singular element methods. Both methods have advantages and disadvantages, which are discussed next.

Direct methods involve a straightforward finite element analysis using standard finite elements and appropriate mesh refinement near the region containing the singularity. A procedure for adaptive mesh refinement that will lead to solution of good accuracy has been proposed and demonstrated in two dimensions (Babuska, 1976). The major advantage of this approach is that standard codes can be utilized. However, this is done at the expense of an increase in the number of degrees of freedom used.

Singular element methods are very appealing since rigorous mathematical investigations have shown that, by including enough singular functions in the finite element approximation, it is possible to obtain the same rate of convergence as for smooth problems (Strang and Fix, 1973). The typical approach in the development of such singular elements is to construct interpolating functions that result in the proper singularity appearing either in the expression of the unknown quantity and/or its derivatives. For example, for two-dimensional problems, scalar potentials are used as unknowns in conjunction with the weak statement of Eq. (6). Thus, for such problems, the interpolating functions associated with the singular elements are such that the local derivatives exhibit the correct singularity. Among the various singular elements that have been proposed, those by Akin, for two-dimensional applications (Akin, 1976), and by Stern, for two- and three-dimensional applications (Stern, 1979), appear to be the most advantageous and most convenient to use. Singular elements have also been developed and used successfully in nodal element-based finite element formulations of electromagnetic vector eigenvalue problems (Webb, 1988).

The most obvious difficulty associated with singular elements is the increased complexity in programming for including such elements in standard finite element codes. Furthermore, in complicated geometries, it may be extremely difficult to identify the singularities and utilize the analytic results in the references cited earlier for the construction of special singular elements. As already mentioned, this is especially true in three dimensions.

Clearly, high accuracy in modeling field singularities comes at a price and there seems to be no easy way around it. The practical electromagnetic problems of interest are three-dimensional and often of significant material complexity. In addition to material wedges and corners, highly curved surfaces are present. Proper resolution of large-curvature surfaces by the finite element mesh is necessary for accurate field calculation. Boyse and Seidl (1994) have proposed the following curvature-based sampling criterion: "Finite element interpolation of solutions to Maxwell's equations around curved bodies requires a minimum of four second-order elements per 90° of arc, regardless of wavelength." Furthermore, the enforcement of tangential field and normal flux continuity across material interfaces is critical for acceptable numerical solution accuracy.

Over the past few years, several researchers have been exploring the possibility of combining nodal and edge elements for the development of three-dimensional finite element codes that can handle the aforementioned difficulties and lead to accurate electromagnetic field simulations. The reasoning behind such an approach may be summarized as follows. From a computational point of view, edge elements are much more expensive than nodal elements. Therefore nodal elements should be used over those regions that involve either homogeneous media or media with smooth variations in their electromagnetic properties. Edge elements, on the other hand, should be used at material interfaces as well as in the vicinity of conducting wedges, corners, and, in general, points where several material interfaces meet and thus the definition of a vector normal is not unique. The reason for this is twofold. First, edge elements are designed to account for the continuity of tangential components of the fields across material interfaces. Second, they allow a jump in the normal components of the field strengths such that normal flux continuity conditions across charge-free material interfaces can be enforced. Mur (1988) has shown that the use of such combinations of edge and nodal elements in electric or magnetic field-based finite element formulations yields optimum computational results. In such mixed-element formulations, it is important that the divergence-free character of the fields and the continuity of the flux across interelement boundaries are imposed explicitly in a weighted sense. This is necessary for the elimination of spurious modes and helps increase solution accuracy and improve convergence (Mur, 1991). The combination of mixed nodal and edge elements has also been used for the finite element modeling of transient electromagnetic phenomena (Mur, 1994). Finally, Boyse and Seidl (1994) have used such a combination of edge and nodal elements in conjunction with the finite element approximation of a scalar and vector potential formulation of Maxwell's equations.

3.3. Impact of Boundary Conditions on Solution Accuracy

From the discussion in Section 2 it became clear that the numerical dispersion associated with the discretization of wave phenomena manifests itself as a phase error in the numerical solution. One, then, expects this phase error to translate into amplitude error in the numerical simulation of electromagnetic wave interactions with geometries that involve material discontinuities and other "scattering centers" that give rise to reflections and standing wave patterns. In other words, one expects that the overall error in the finite element solution will be dependent on the types of boundary conditions encountered in the domain of interest.

This dependence was first indicated in the paper by Bayliss et al. (1985). Later work by Cangellaris and Lee (1992) and Lee and Cangellaris (1992) provided further quantification of this dependence of the error on the type of boundary conditions used for grid truncation. In order to discuss their results in some detail, it is worth examining two simple one-dimensional problems for which the finite element solutions can be obtained exactly. The first problem deals with the solution of the scalar Helmholtz equation in the homogeneous domain $0 < x < d$ with constant wavenumber k, subject to boundary conditions of $E(x = 0) = 1$ and $(d/dx + jk)E = 0$ at $x = d$. In other words, a radiation boundary condition is imposed at $x = d$. The weak statement of Eq. (6) is used to develop the finite element approximation on linear elements of length h. Let $M + 1$ be the number of elements in the grid. Thus there are

$M + 2$ nodes in the grid and, since the value of the node at $x = 0$ is fixed, there are $M + 1$ degrees of freedom in the approximation. Cangellaris and Lee (1992) have shown that the theory of difference equations can be used to solve the resulting finite element system. The expression for the solution at node n is given by

$$\hat{E}_n = e^{-j\hat{k}(nh)} \frac{1 - qe^{-j2\hat{k}(d-nh)}}{1 - qe^{-j2\hat{k}d}}, \quad n = 1, 2, \ldots, M + 1 \tag{42}$$

where $\hat{k}$ is given by Eq. (22) (with $\beta, \hat{\beta}$ replaced by $k, \hat{k}$, respectively) and

$$q = \frac{be^{j\hat{k}h} + a/2 - jkh}{be^{-j\hat{k}h} + a/2 - jkh} \tag{43}$$

with a and b given by

$$a = -2\left(1 - \frac{k^2h^2}{3}\right), \quad b = 1 + \frac{k^2h^2}{6} \tag{44}$$

Since the analytic solution for this problem is given by $E(x) = \exp(-jkx)$, the mean-square error can be computed easily.

The second problem replaces the radiation boundary condition at $x = d$ with a (Dirichlet) perfectly conducting wall boundary condition, $E(x = d) = 1$. For this case, the exact solution of the resulting finite element system for the field values at the nodes is

$$\hat{E}_n = \frac{\sin \hat{k}(d - nh)}{\sin \hat{k}d}, \quad n = 1, 2, \ldots, M + 1 \tag{45}$$

This should be compared with the analytic solution $E(x) = \sin k(d - x)/\sin kx$. Figure 6 compares the relative mean-square errors for these two boundary value problems for various nodal densities. In all cases, the error is higher for the Dirichlet boundary condition. Furthermore, the increase in the error with the electrical length of the domain is faster for the Dirichlet boundary condition case. This is because, as mentioned in the beginning of this subsection, the error in the numerical solution for the radiation boundary condition case is essentially phase error, while the error in the Dirichlet boundary condition case is amplitude error.

The aforementioned one-dimensional results were shown to hold true in two dimensions also. The numerical experiments in the paper by Lee and Cangellaris (1992) indicate clearly that higher-degree interpolation functions are needed to control accuracy in simulations of electromagnetic wave interactions in domains of large electrical length. Furthermore, the increase in accuracy was accompanied by savings in CPU time due to the smaller number of degrees of

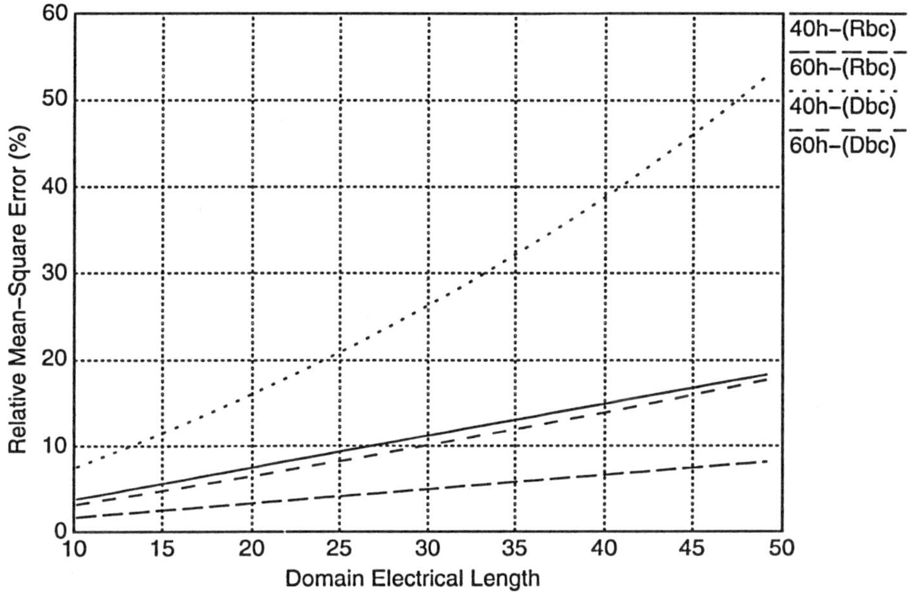

FIGURE 6. Relative mean-square error versus domain electrical length. The finite element interpolation functions are linear. The one-dimensional domain is terminated at its right end either by a radiation boundary condition (Rbc) or a Dirichlet boundary condition (Dbc). For both wavelength resolutions (40 and 60 elements per wavelength) the increase in the error with the domain electrical length is faster for the Dirichlet boundary condition case.

freedom in the higher-order approximation. As a matter of fact, an additional benefit of increasing the degree of the interpolating polynomials is the improvement in the conditioning of the finite element system. This point will become clear in Section 4.

An alternative to local, absorbing boundary conditions for numerical grid truncation, in open-region scattering problems, is the use of global grid truncation schemes that utilize a boundary integral statement, consistent with the physics of the region exterior to the finite element domain, to relate the field and its normal derivative on the truncation boundary. A thorough discussion of a variety of such hybrid, finite element–boundary integral methods for both two- and three-dimensional scattering problems can be found in Jin (1993). A potential source of error in such hybrid formulations is the appearance of "singular" frequencies, that is, frequencies at which the numerical solution exhibits significant deviation from the true solution. This problem occurs only when the truncation boundary is a closed boundary, and it is due to the use of the boundary integral statement. More specifically, these singular frequencies coincide with the resonant frequencies of the cavity formed by the closed boundary and filled with material having the same properties as those of the exterior region. This "interior resonance problem" associated with boundary integral solutions to scattering problems has been studied extensively in the computational electromagnetics literature (see, e.g., Peterson, 1990).

With regard to hybrid finite element–boundary integral formulations, Collins et al. (1992) have proposed a simple modification to the formulation, which eliminates the interior resonance problem. More specifically, the scattered field replaces the total field as the unknown and a small loss is introduced in the medium. From the linearity of Maxwell's equations and the fact that the scattered field is the difference between the total field and the incident field, it is clear that in such a formulation the incident field appears now as a "distributed source" over the finite element domain. Once the scattered field has been computed, the total field is recovered by adding the incident field, which, in general, is known analytically. This process may lead to additional error if the numerically obtained scattered field contains substantial numerical phase error due to numerical dispersion. Clearly, for such formulations, wavelength resolution on the finite element grid should be done as finely as possible.

3.4. Numerical Accuracy in Eigenvalue Problems

The finite element solution of electromagnetic eigenvalue problems has been studied extensively in electromagnetics literature, especially in relation to the analysis of wave propagation in microwave, millimeter-wave, and optical waveguides, and the analysis and design of resonators. Specific examples from the application of the finite element method to such eigenvalue problems are given elsewhere in this book.

The purpose of this discussion is to provide some quantification of the error in the finite element calculation of the eigenvalues and the associated eigenfunctions. For this purpose, let us consider the case of a three-dimensional, perfectly conducting cavity loaded with an inhomogeneous lossless dielectric. For the case of an electric field formulation, the relevant weak statement is given by Eq. (5). Multiplying through by $j\omega$ and imposing the Dirichlet boundary condition $\hat{\mathbf{n}} \times \mathbf{E} = \mathbf{0}$ on the perfectly conducting walls, the following generalized eigenvalue problem is obtained:

$$\mathbf{KE} = \omega^2 \mathbf{ME} \tag{46}$$

where $\mathbf{M}$ and $\mathbf{K}$ are real, symmetric matrices with elements $M_{ij} = \langle \varepsilon \mathbf{N}_i \cdot \mathbf{N}_j \rangle$, and $K_{ij} = \langle (\mu^{-1} \nabla \times \mathbf{N}_i) \cdot (\nabla \times \mathbf{N}_j) \rangle$, respectively.

To quantify the accuracy of the numerical eigenvalues and eigenfunctions, two important results from the error analysis of the aforementioned eigenvalue problem will be used. These results are given below without proof. For further details the reader may consult the material in Chapter 6 in Strang and Fix

(1973). The first result has to do with the error estimate in the finite element calculation of the eigenvalues of the problem in Eq. (46). Let k be the degree of the *complete* polynomials used in the construction of the vector finite element basis function $\mathbf{N}_i$, and h the grid size. Then, for sufficiently smooth data, the approximate eigenvalues, $\hat{s}_n = \hat{\omega}_n^2$, and exact eigenvalues, $s_n = \omega_n^2$, satisfy the inequality

$$0 \leqslant \hat{s}_n - s_n \leqslant Ch^{2k} s_n^{k+1} \tag{47}$$

Thus the numerical eigenvalues are always larger than the analytical ones. Furthermore, the error in the computed eigenvalue varies with the eigenvalue in question, becoming larger for larger eigenvalues. Consequently, the higher eigenvalues will be progressively more difficult to compute to a specific fixed accuracy.

As a simple demonstration of the aforementioned result, consider the case of the finite element prediction of the cutoff frequencies for the transverse magnetic (TM) modes in a perfectly conducting, hollow waveguide of rectangular cross section and dimension a in x and b in y. The waveguide is assumed to be filled with a homogeneous material ε, μ. The relevant eigenvalue equation for this problem is $\nabla_{xy}^2 E_x + \omega^2 \mu\varepsilon E_z = 0$, and the relevant weak statement is that of Eq. (6). If the bilinear elements of Figure 1 are used, the finite element equation at node i is that of Eq. (26) with k^2 replaced with $\hat{\omega}^2\mu\varepsilon$. A numerical dispersion relation for the numerical cutoff frequencies, $\hat{\omega}_{mn}$, can be derived from Eq. (26) by noting that, for the mn eigenmode, the wavenumbers along x and y are of the form $k_x = (m\pi/a)$ and $k_y = (n\pi/b)$, respectively. These two expressions are used in place of $k \cos\alpha$ and $k \sin\alpha$, respectively, in Eq. (27), which, in turn, is used in Eq. (26) to obtain

$$\hat{\omega}_{mn}^2 \mu\varepsilon = \frac{6}{h^2} \frac{4 - \cos\dfrac{m\pi h}{a} - \cos\dfrac{n\pi h}{b} - 2\cos\dfrac{m\pi h}{a}\cos\dfrac{n\pi h}{b}}{4 + 2\cos\dfrac{m\pi h}{a} + 2\cos\dfrac{n\pi h}{b} + \cos\dfrac{m\pi h}{a}\cos\dfrac{n\pi h}{b}} \tag{48}$$

With the assumption that the grid size h is fine enough such that $m\pi/a \ll 1$ and $n\pi/b \ll 1$, the approximation $\cos t \simeq 1 - t^2/2$ may be used in Eq. (48). After some algebraic manipulation, keeping terms up to fourth order only, one obtains

$$\hat{\omega}_{mn}^2 \mu\varepsilon = \left[\left(\frac{m\pi}{a}\right)^2 + \left(\frac{n\pi}{b}\right)^2\right] + \frac{h^2}{6}\left[\left(\frac{m\pi}{a}\right)^2 + \left(\frac{n\pi}{b}\right)^2\right]^2 \tag{49}$$

Equation (49) is consistent with the error estimate of Eq. (46). This is clearly seen once it is realized that bilinear finite elements have complete polynomials of degree $k = 1$ only, and that the term inside the brackets in Eq. (49) is the analytic expression for $\omega_{mn}^2 \mu\varepsilon$.

The second result from the error analysis of the eigenvalue problem of Eq. (46) has to do with the accuracy of the numerical *eigenfunction*. It can be shown that the eigenfunctions are computed with the same accuracy as the corresponding boundary value problem. Thus the error estimates derived earlier are also applicable to the numerical eigenfunctions.

4. CONDITIONING OF THE FINITE ELEMENT MATRIX AND ROUND-OFF ERROR

Finite element approximations to realistic boundary value problems involve extremely large systems of equations. Consequently, loss of significant digits, occurring during the solution process as a result of finite precision arithmetic, becomes an issue of concern. Undoubtedly, some loss of significant digits is expected to occur in the calculation of the stiffness matrix and the forcing vector due to numerical quadrature and round-off. The following result by Wilkinson (1963) is worth mentioning here. Assume that the elements in the $N \times N$ stiffness matrix and the $N \times 1$ forcing vector are all less than 1 in modulus and are known with certainty to only m decimal places; that is, they contain errors less than 0.5×10^{-m}. Assume also that Gaussian elimination with partial pivoting is used for the solution. Then, if the arithmetic is carried out with numbers having more than $m + \log_{10}N$ decimal places, that is, with more than $\log_{10}N$ guarding figures, the solution obtained to m decimal places is the exact solution of a set of N linear equations whose coefficients and constants differ from those of the original equations by less than 0.5×10^{-m}. This result implies that with the proper choice of word length, the rounding errors introduced during the computation will have less effect on the solution than the rounding errors in the elements of the stiffness matrix and the forcing vector. Thus the numerical solution is as accurate as the data warrant.

It is important to note that the above result does not imply that the number of correct significant figures in the numerical solution is the same as in the data. It might be less, but this loss of accuracy is not due to the matrix inversion process but due to the ill-conditioning of the stiffness matrix. If the system of linear equations is nearly singular or, as usually termed, *ill-conditioned*, its solution is extremely sensitive to small changes in the elements of the system matrix and the forcing vector. In some instances, ill-conditioning may be so severe that round-off errors during the solution process will destroy completely all significant digits in the solution, even on a computer with large word length.

There are several sources of ill-conditioning in finite element approximations to boundary value problems in electromagnetics. One of them has to do with the case where solution to Maxwell's equations is sought inside an enclosed, lossless domain, with Dirichlet or Neumann boundary conditions on its boundary, and the frequency of interest is close to one of the resonant frequencies of the "cavity" formed by the enclosed domain. This situation is easily avoided by introducing a slight loss in the media. Ill-conditioning may be caused also by poor discretization in which abrupt variations in mesh size are employed. Similarly, an abrupt change in material properties without a compensating transition in mesh size may also induce ill-conditioning. These sources of ill-conditioning are discussed in more detail in the following subsections.

4.1. The Condition Number

A measure of the degree of ill-conditioning of the finite element system is provided by means of a *condition number*. As shown next, this condition number can be determined from the extreme eigenvalues of the stiffness matrix $\mathbf{K}$. A concept necessary for the mathematical development is that of the norm of a matrix. Given a matrix $\mathbf{K}$, the matrix norm, $\|\mathbf{K}\|$, is defined as

$$\|\mathbf{K}\| = \max_{\mathbf{x}\neq 0} \frac{\|\mathbf{Kx}\|}{\|\mathbf{x}\|} \tag{50}$$

where $\|\mathbf{x}\|$ denotes the usual Euclidean vector norm for the vector $\mathbf{x}$.

Consider the effect of a perturbation $\delta\mathbf{f}$ to the forcing vector $\mathbf{f}$ in the finite element system

$$\mathbf{Kx} = \mathbf{f} \tag{51}$$

A change $\delta\mathbf{x}$ in the solution is expected such that

$$\mathbf{K}(\mathbf{x} + \delta\mathbf{x}) = \mathbf{f} + \delta\mathbf{f} \tag{52}$$

Subtraction of Eq. (51) from Eq. (52) results in

$$\mathbf{K}\,\delta\mathbf{x} = \delta\mathbf{f}$$

or

$$\delta\mathbf{x} = \mathbf{K}^{-1}\,\delta\mathbf{f} \tag{53}$$

Taking norms on both sides of Eq. (53) leads to

$$\|\delta\mathbf{x}\| = \|\mathbf{K}^{-1}\,\delta\mathbf{f}\| \leqslant \|\mathbf{K}^{-1}\|\,\|\delta\mathbf{f}\| \tag{54}$$

Thus a bound has been obtained for the size of the perturbation in the solution. Actually, a bound can be obtained for the relative perturbation $\|\delta\mathbf{x}\|/\|\mathbf{x}\|$, noting that Eq. (50) implies

$$\|\mathbf{K}\|\ \|\mathbf{x}\| \geqslant \|\mathbf{f}\| \tag{55}$$

Combining Eqs. (54) and (55) yields

$$\frac{\|\delta\mathbf{x}\|}{\|\mathbf{x}\|} \leqslant \|\mathbf{K}^{-1}\|\ \|\mathbf{K}\|\ \frac{\|\delta\mathbf{f}\|}{\|\mathbf{f}\|} \tag{56}$$

In Eq. (56) the relative change in the solution vector is related to the relative change in the forcing vector through the factor $\|\mathbf{K}^{-1}\|\ \|\mathbf{K}\|$. This factor is called the condition number,

$$C_{\mathbf{K}} = \|\mathbf{K}^{-1}\|\ \|\mathbf{K}\| \tag{57}$$

Oftentimes the number $\log_{10} C_{\mathbf{K}}$ is used as an estimate of the number of lost digits.

An upper bound on the condition number can be obtained in terms of the maximum and minimum eigenvalues of $\mathbf{K}$ as follows. Taking norms on both sides of the eigenvalue equation $\mathbf{K}\mathbf{x}_e = s\mathbf{x}_e$ results in

$$\|\mathbf{K}\mathbf{x}_e\| = \|s\mathbf{x}_e\| \leqslant |s|\ \|\mathbf{x}_e\| \tag{58}$$

Equation (58) in conjunction with Eq. (50) yields

$$\|\mathbf{K}\| \leqslant |s_{\max}| \tag{59}$$

Now the fact that the eigenvalues of $\mathbf{K}^{-1}$ are the reciprocals of the eigenvalues of $\mathbf{K}$ may be used to obtain from Eq. (58)

$$\|\mathbf{K}^{-1}\| \leqslant |s_{\min}|^{-1} \tag{60}$$

The results in Eqs. (59) and (60) are then used in Eq. (57) to obtain the following expression for the condition number:

$$C_{\mathbf{K}} \leqslant \frac{|s_{\max}|}{|s_{\min}|} \tag{61}$$

This final result will be used in the following subsection to quantify the impact of material characteristics and grid uniformity on the conditioning of the stiffness matrix for finite element approximations of electromagnetic boundary value problems.

4.2. Impact of Material Characteristics and Grid Uniformity on the Condition Number of the Stiffness Matrix

In this subsection we examine the impact of abrupt variations in material characteristics and finite element grid size on the condition number of the matrix resulting from the finite element approximation of the one-dimensional Laplace and Helmholtz equations. The reason for dealing with one-dimensional problems is that they provide for a rather straightforward estimation of the extreme eigenvalues of the stiffness matrix. However, the results from the one-dimensional studies can and will be extended to higher dimensions.

Consider first Laplace's equation over a finite interval along the x axis

$$\frac{d}{dx}\left(\varepsilon(x)\,\frac{d\Phi}{dx}\right) = 0, \quad 0 < x < d \tag{62}$$

with Dirichlet boundary conditions at the two end points of the interval. A weak statement of the form of Eq. (7) is of relevance here. The finite element grid used for the discretization consists of M elements, all of which are of equal length h_1 except for element 2, which is of length h_2. In addition, the permittivity has a constant value ε_1 everywhere except for element 2, where it assumes a different value ε_2. For a linear element basis the resulting stiffness matrix after the application of the boundary conditions is of the form

$$\mathbf{K} = \begin{bmatrix} p_1 & -p_1 & 0 & 0 & 0 & 0 & \cdots & 0 & 0 \\ -p_1 & p_1 + p_2 & -p_2 & 0 & 0 & 0 & \cdots & 0 & 0 \\ 0 & -p_2 & 2p_1 & -p_1 & 0 & 0 & \cdots & 0 & 0 \\ 0 & 0 & -p_1 & 2p_1 & -p_1 & 0 & \cdots & 0 & 0 \\ \vdots & \vdots & \vdots & \vdots & \vdots & \vdots & \ddots & \vdots & \vdots \\ 0 & 0 & 0 & 0 & 0 & 0 & \cdots & -p_1 & p_1 \end{bmatrix} \tag{63}$$

where $p_1 = \varepsilon_1/h_1$ and $p_2 = \varepsilon_2/h_2$. $\mathbf{K}$ is an $(M-1) \times (M-1)$ matrix. Even though, for a given value of M, the eigenvalues of $\mathbf{K}$ can be found in a precise manner, this necessitates extra computation. An attractive alternative is to use the Gershgorin circle theorem. This theorem states that the eigenvalues s_i of a matrix $\mathbf{K}$ of dimension $N \times N$ with elements K_{ij} lie in the union of the Gershgorin disks G_i defined by the relations

$$|s - K_{ii}| \leqslant \sum_{j=1}^{N} |K_{ij}|, \quad i = 1, 2, \ldots, N \tag{64}$$

on the complex s-plane. Note that the disks are centered at points K_{ii} in the complex s-plane and have radius equal to the sum of the magnitudes of the off-diagonal elements for $i = 1, 2, \ldots, N$. Obviously, Gershgorin's theorem only provides bounds on the eigenvalues. Nevertheless, such bounds are in most cases sufficient for a qualitative study of the effects of variations in material characteristics and grid size on the conditioning of the matrix.

Application of Eq. (64) to the second and fourth row of Eq. (63) yields the following inequalities, respectively:

$$0 \leqslant s \leqslant 2\left(\frac{\varepsilon_1}{h_1} + \frac{\varepsilon_2}{h_2}\right) = 2\,\frac{\varepsilon_1}{h_1}\left(1 + \frac{\varepsilon_2}{\varepsilon_1}\frac{h_1}{h_2}\right) \tag{65a}$$

$$0 \leqslant s \leqslant 2\,\frac{\varepsilon_1}{h_1} \tag{65b}$$

Let us consider first the case $\varepsilon_1 = \varepsilon_2$. Then from Eq. (65a) it becomes apparent that if $h_2 \ll h_1$ the bound for the extreme eigenvalue increases proportionally to the ratio h_1/h_2 and, qualitatively, this indicates a deterioration in the condition of the finite element matrix. From a more intuitive point of view, this deterioration in the conditioning of the matrix is expected since two nodes in the grid tend to coalesce, as h_2 becomes much smaller than h_1, and thus the matrix becomes singular. A similar behavior is observed in two dimensions as elements become slender; for example, triangles degenerate toward line segments or quadrilateral elements degenerate toward triangles.

The same effect is obtained for a homogeneous grid, $h_1 = h_2$, used to discretize a domain over which the dielectric permittivity exhibits abrupt large changes $\varepsilon_2 \gg \varepsilon_1$. As a matter of fact, Eq. (65a) suggests that abrupt increases in the value of the dielectric permittivity must be accompanied by proportional decreases in the grid size in order to preserve the good conditioning of the finite element system.

A similar procedure can be used in conjunction with the Helmholtz equation

$$\frac{d}{dx}\left(\mu^{-1}\frac{dE}{dx}\right) + \omega^2 \varepsilon E = 0, \quad 0 < x < d \tag{66}$$

with Dirichlet boundary conditions at $x = 0$ and $x = d$. Once again, the finite element grid used for the discretization consists of M elements, all of which are of equal length h_1 except for element 2, which is of length h_2. It is assumed that μ is constant throughout the domain, while ε assumes a constant value ε_1 everywhere except element 2, where it assumes a different value ε_2. For a linear element basis the resulting stiffness matrix after the application of the boundary conditions is of the form

$$\mathbf{K} = \begin{bmatrix} a_1 & b_1 & 0 & 0 & 0 & 0 & \cdots & 0 & 0 \\ b_1 & a_1 + a_2 & b_2 & 0 & 0 & 0 & \cdots & 0 & 0 \\ 0 & b_2 & 2a_1 & b_1 & 0 & 0 & \cdots & 0 & 0 \\ 0 & 0 & b_1 & 2a_1 & b_1 & 0 & \cdots & 0 & 0 \\ \vdots & \vdots & \vdots & \vdots & \vdots & \vdots & \ddots & \vdots & \vdots \\ 0 & 0 & 0 & 0 & 0 & 0 & \cdots & b_1 & a_1 \end{bmatrix} \tag{67}$$

where

$$a_i = 1 - \frac{\omega^2 \varepsilon_i \mu h_i^2}{3}, \quad b_i = -\left(1 + \frac{\omega^2 \varepsilon_i \mu h_i^2}{6}\right), \quad i = 1, 2 \tag{68}$$

Application of the Gershgorin circle theorem to rows 2 and 4 yields the following inequalities, respectively:

$$-\frac{1}{2}\left(\frac{2\pi h_1}{\lambda_1}\right)^2\left(1 + \frac{\varepsilon_2}{\varepsilon_1}\frac{h_2}{h_1}\right) \leqslant s \leqslant 2\left(1 + \frac{h_1}{h_2}\right) - \frac{1}{6}\left(\frac{2\pi h_1}{\lambda_1}\right)^2\left(1 + \frac{\varepsilon_2}{\varepsilon_1}\frac{h_2}{h_1}\right) \tag{69a}$$

$$\left(\frac{2\pi h_1}{\lambda_1}\right)^2 \leqslant s \leqslant 4 - \frac{1}{3}\left(\frac{2\pi h_1}{\lambda_1}\right)^2 \tag{69b}$$

where λ_1 is the wavelength in medium (μ, ε_1), and the fact that $\omega^2 \mu \varepsilon_1 = (2\pi/\lambda_1)^2$ has been used.

The case $\varepsilon_1 = \varepsilon_2$ is considered first. From Eq. (69a) it is apparent that, if $h_2 \ll h_1$, the bound for the extreme eigenvalue increases almost proportionally to the ratio h_1/h_2 [because of the term $2(1 + h_1/h_2)$ on the right-hand side of the inequality]. Of course, this widening of the spectrum indicates a deterioration in the conditioning of the finite element system.

For the case of a homogeneous grid, $h_1 = h_2 = h$, and a medium with an abrupt large change in ε, $\varepsilon_2 \gg \varepsilon_1$, Eq. (69a) indicates that the widening of the spectrum can be controlled by selecting a fine enough discretization, $h \ll \lambda_1$. From the discussion on numerical dispersion it must be clear that the choice of h is controlled by the minimum wavelength in the domain of interest. More specifically, since $\varepsilon_2 \gg \varepsilon_1$, h must be chosen such that $h/\lambda_2 \leqslant 0.05$ for sufficiently low numerical dispersion. Consequently, the inequality $h \ll \lambda_1$ is easily satisfied and the good conditioning of the finite element system is maintained.

Finally, it might be of interest to point out that, in addition to controlling numerical dispersion and maintaining the good conditioning of the finite element matrix, good wavelength resolution prevents the occurrence of (artificial) numerical reflections caused by changes in grid fineness. This can be demonstrated by examining the numerical reflection coefficient associated with the interface between two regions of a one-dimensional grid, one of element size h_1 and the other of element size h_2. Let ε_1, μ_1 and ε_2, μ_2 be the electromagnetic properties in the two regions. Assuming plane wave propagation and linear finite elements, the numerical solution for the wave at node n in grid h_1 will be of the form $E_n = \exp[-j\hat{k}_1(nh_1)]$, where the numerical wavenumber $\hat{k}_1$ is obtained from Eq. (22). From this, the corresponding magnetic field at position $(n + 1/2)h$ is computed as follows:

$$\hat{H}_{n+1/2} = -\frac{1}{j\omega\mu_1}\frac{e^{-j\hat{k}_1[(n+1)h_1]} - e^{-j\hat{k}_1(nh_1)}}{h_1}$$

Finally, the numerical wave impedance, $\hat{\eta}_1$, is computed as the ratio of the numerical electric and magnetic fields at position $(n + \frac{1}{2})h$. After some algebraic manipulation one obtains

$$\hat{\eta}_1 = \eta\,\frac{\hat{k}_1 h_1}{2}\,\frac{\cos\left(\dfrac{\hat{k}_1 h_1}{2}\right)}{\sin\left(\dfrac{\hat{k}_1 h_1}{2}\right)} \tag{70}$$

where $\eta = \sqrt{\mu_1/\varepsilon_1}$. Clearly, $\hat{\eta}_1 \to \eta$ as $\hat{k}_1 h_1 \to 0$. A similar expression holds for the numerical wave impedance, $\hat{\eta}_2$, of grid h_2. A numerical reflection coefficient for the interface between the two grids may be defined as follows:

$$\hat{R} = \frac{\hat{\eta}_2 - \hat{\eta}_1}{\hat{\eta}_2 + \hat{\eta}_1} \tag{71}$$

Clearly, if h_1 and h_2 are chosen such that the wavelengths on the two grids are well-resolved ($k_1 h_1 \ll 1, k_2 h_2 \ll 1$), the difference between $\hat{R}$ and the analytic reflection coefficient is negligible. Figure 7 depicts the numerical reflection coefficient versus the number of elements per wavelength in grid h_2 for the case $\varepsilon_1 = \varepsilon_2 = \varepsilon$, $\mu_1 = \mu_2 = \mu$, and $h_1/\lambda = 0.04$. The analytic result for the reflection coefficient is zero. Note that the numerical reflection coefficient becomes negligible for values of h_2 such that $h_2/\lambda < 0.05$.

4.3. Round-off, Mesh Refinement, and Higher-Degree Elements

The final subject to be discussed in this section is the competition between mesh refinement and round-off error. From the error estimates of Section 3 it is clear that for a uniform mesh, well-behaved data, and computations with exact arithmetic, the finite element solution should approach the analytic

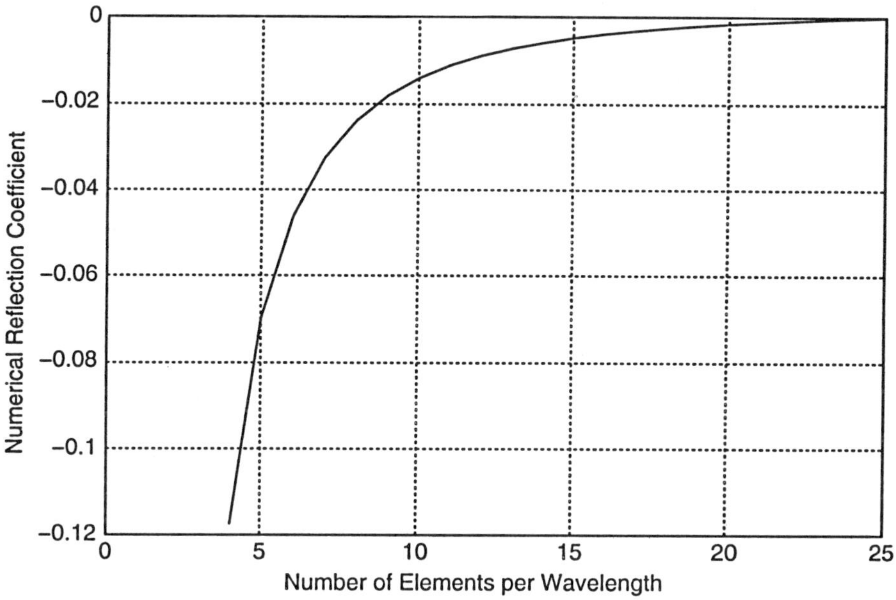

FIGURE 7. Numerical reflection coefficient for the interface between two one-dimensional, linear finite element grids with different wavelength resolutions. The material parameters are the same in both grids. Wavelength resolution in grid 1 is fixed at 25 elements per wavelength. Wavelength resolution in grid 2 is varied from 4 to 25 elements per wavelength.

solution as h is refined, at a rate dictated by the degree of the interpolating polynomials. However, the results in the previous section indicate that, as the mesh is refined, the eigenvalue spectrum of the stiffness matrix widens; hence the system becomes progressively more ill-conditioned. Consider, for example, Eq. (69b). As $h_1 \to 0$, $s_{\text{max}} \to 4$, while $s_{\text{min}} \to 0$. This progressive ill-conditioning of the stiffness matrix as h is refined results in an increase in the round-off error, which counteracts the decrease in approximation error. At some point, approximation error and round-off become equal, and any further mesh refinement will actually result in deterioration in solution accuracy.

Extensive mathematical and numerical studies have shown that the round-off error does not depend strongly on the degree of the interpolating polynomials (Strang and Fix, 1973). It depends mainly on h and, for the second-order problems of interest in electromagnetics, the dependence is of the form $C_K = O(h^{-2})$. Consequently, the way to maintain accuracy and postpone the deleterious effect of round-off is to increase the order of the trial functions. The numerical experiments in Scott (1994) support this conjecture in a very illustrative manner.

5. ACCURACY OF DERIVATIVES OF FINITE ELEMENT SOLUTIONS

Derivatives of finite element approximations to electromagnetic fields are essential for the calculation of several quantities of practical interest. For example, for the quasi-transverse electric and magnetic characterization of planar waveguides (microstrip line, stripline, coplanar waveguide, etc.), the finite element solution for the electrostatic potential needs to be differentiated in order to calculate either the energy stored in the system or the charge density on the conductors. As another example, numerical differentiation of the near fields of a scatterer or an antenna, calculated using the finite element method, is needed for the subsequent calculation of the far fields to provide radar cross-section or antenna gain results.

The difficulty associated with the numerical differentiation of finite element solutions is rather obvious. First-order derivatives, obtained from a finite element solution calculated with $O(h^2)$ accuracy, exhibit $O(h)$ accuracy only. Furthermore, finite element solutions based on error-minimization processes are usually contaminated with a small oscillatory error term that may further worsen the accuracy of a numerically calculated derivative.

For the case of the two-dimensional Laplace and Poisson equations, relevant to the quasi-transverse electric and magnetic analysis of planar waveguides, several elegant schemes have been proposed for the accurate calculation of higher-order derivatives of finite element solutions (Silvester, 1991; Silvester and Omeragić, 1993; Franz and Kasper, 1995). The method by Franz and Kasper is based on the observation that the solution to Laplace's equation on a circular disk of radius α may be cast (in polar coordinates ρ, θ) in the form

$$\Phi(\rho, \theta) = \frac{a_0}{2} + \sum_{n=1}^{\infty} \rho^n(a_n \cos n\theta + b_n \cos n\theta) \tag{72}$$

where the origin is assumed to coincide with the center of the disk. If Φ is known on the circumference of the disk, the coefficients in the aforementioned expansion are calculated from the integrals

$$a_n = \frac{1}{\pi\alpha^n} \int_0^{2\pi} \Phi(\alpha, \theta) \cos n\theta \, \mathrm{d}\theta \tag{73a}$$

$$b_n = \frac{1}{\pi\alpha^n} \int_0^{2\pi} \Phi(\alpha, \theta) \sin n\theta \, d\theta \tag{73b}$$

For the purposes of a discrete solution, Φ is calculated at a set of N equidistant points on the circumference of the disk, $\Phi_i = \Phi(\alpha, i2\pi/N)$, $i = 1, 2, \ldots, N$, and the integrals in Eq. (73) are calculated efficiently by means of discrete fast transforms. Calculation of the first-order derivatives of Φ at the center of the disk is now straightforward in view of the fact that

$$\left.\frac{\partial \Phi}{\partial x}\right|_{\rho=0} = a_1 \tag{74a}$$

$$\left.\frac{\partial \Phi}{\partial y}\right|_{\rho=0} = b_1 \tag{74b}$$

As far as scattering and radiation problems are concerned, differentiation of the numerical solution is usually needed for subsequent far-field calculations. Consider, for example, the case of electromagnetic scattering by a three-dimensional target in free space. The finite element solution, based on the weak statement of Eq. (5), is generated within a closed volume V that contains the scatterer. Let S be any closed surface inside V that surrounds the scatterer. Then the electric field at any point $\mathbf{r}$ exterior to the closed surface S can be calculated from the Stratton–Chu formula

$$\begin{aligned}\mathbf{E}(\mathbf{r}) = -\oint_S [&G(\mathbf{r}, \mathbf{r}')(\hat{\mathbf{n}}' \times \nabla' \times \mathbf{E}(\mathbf{r}')) + (\hat{\mathbf{n}}' \times \mathbf{E}(\mathbf{r}')) \times \nabla' G(\mathbf{r}, \mathbf{r}') \\ &+ (\hat{\mathbf{n}}' \cdot \mathbf{E}(\mathbf{r}'))\nabla' G(\mathbf{r}, \mathbf{r}')] \, dS'\end{aligned} \tag{75}$$

where ∇' denotes differentiation with respect to the coordinates of the "source" point $\mathbf{r}'$ on the surface S, $\hat{\mathbf{n}}'$ is the unit normal on S pointing into the volume enclosed by S, and $G(\mathbf{r}, \mathbf{r}')$ is the solution to the scalar Helmholtz equation due to a point source of strength -1 at $\mathbf{r} = \mathbf{r}'$ in the volume exterior to S.

The impact of the loss of accuracy in the numerical differentiation of $\mathbf{E}$ on the subsequent calculation of the far field using the expression of Eq. (75) has not been addressed in detail. The difficulty here is expected to be more severe than the Laplace equation case, due to the oscillatory nature of the electromagnetic field quantities. More specifically, phase accuracy in the quantities that appear under the integral in Eq. (75) is essential for the accurate calculation of the far field. Some limited results have been presented by Baucke et al. (1993) for two-dimensional scattering problems involving scatterers of rather small electrical size. To date, the most detailed study of this issue has been given by Meyer (1994). Meyer applied a p-adaptive finite element/boundary element method to two-dimensional scattering and radiation problems. His studies showed that the p-adaptive method did not produce improved convergence rates for radiation cross-section values when the energy norm in the finite element region was used as the criterion for the adaptation. He concluded that a mesh consisting of only third-order elements was required in order to obtain highly accurate radar cross-section results.

Meyer's results are consistent with the earlier discussion on numerical dispersion. Apparently, use of the third-order elements allowed for practically negligible numerical dispersion in the finite element solution, which, in turn, led to highly accurate calculation of first-order derivatives through numerical differentiation. However, one should be able, in principle, to apply methods such as those mentioned earlier for Laplace's equation, to calculate highly accurate derivatives of the finite element solution.

6. CONCLUDING REMARKS

This chapter concerned itself with the various sources of error that arise in the numerical simulation of electromagnetic wave phenomena using finite element techniques. A general conclusion that may be drawn from the aforementioned presentation is that the development of an appropriate finite element grid is critical for solution accuracy of dynamic problems.

First and foremost, the grid must provide for adequate wavelength resolution in order to minimize numerical dispersion and thus control phase error. The number of degrees of freedom per wavelength, required to achieve a fixed accuracy, is dependent on the degree of the interpolating polynomials, the electrical size of the domain, and the boundary conditions used for grid truncation. Use of higher-order elements reduces the dimensionality of the approximation, improves the conditioning of the finite element system, and leads to increased accuracy in the numerical differentiation of the finite element solution for postprocessing purposes.

Furthermore, problem geometry and material properties impact the way the finite element grid is constructed. Electromagnetic fields exhibit singular behavior at material wedges and corners (especially metallic), which cannot be resolved properly by the polynomial interpolating functions used for the finite element approximation. Thus grid refinement or use of special elements are required in the vicinity of such geometric features to improve solution accuracy. In addition, abrupt changes in the electric permittivity and the magnetic permeability of materials need to be accompanied by appropriate grid size changes in order to both prevent the occurrence of (nonphysical) numerical reflections and maintain the conditioning of the finite element matrix. Finally, the grid generation process should be such that element degeneracy (i.e., triangles degenerating toward line segments, quadrilaterals approaching triangles, or tetrahedrons degenerating to quadrilaterals) never occurs. Such elements lead to ill-conditioning and may eventually render the finite element solution useless.

REFERENCES

Akin, J. E. (1976). The generation of elements with singularities. *Int. J. Numer. Methods Eng.* **10**, 1249–1259.

Andersen, J. B., and Solodukhov, V. V. (1978). Field behavior near a dielectric wedge. *IEEE Trans. Antennas Propag.* **AP-26**, 598–602.

Babuska, I. (1976). The self-adaptive approach in finite element methods. In *The Mathematics of Finite Elements and Applications II, MAFELAP 1975* (J. R. Whiteman, ed.), pp. 125–142. Academic Press, London.

Baucke, R. C., D'Angelo, J., and Crabtree, G. D. (1993). Far-field transforms for finite element frequency-domain solutions. *IEEE Trans. Antennas Propag.* **AP-41**, 511–514.

Bayliss, A., Goldstein, C. I., and Turkel, E. (1985). On accuracy conditions for the numerical computation of waves. *J. Comput. Phys.* **59**, 396–404.

Boyse, W. E., and Seidl, A. A. (1994). A hybrid finite element method for 3-D scattering using nodal and edge elements. *IEEE Trans. Antennas Propag.* **AP-42**, 1436–1442.

Cangellaris, A. C., and Lee, R. (1992). On the accuracy of numerical wave simulations based on finite methods. *J. Electromagn. Waves Appl.* **6**, 1635–1653.

Chew, W. C. (1990). *Waves and Fields in Inhomogeneous Media.* Van Nostrand-Reinhold, New York.

Collins, J. D., Jin, J.-M., and Volakis, J. L. (1992). Eliminating interior resonances in finite element-boundary integral methods for scattering. *IEEE Trans. Antennas Propag.* **AP-40**, 1583–1585.

De Smedt, R., and Van Bladel, J. (1986). Field singularities at the tip of a metallic cone of arbitrary cross section. *IEEE Trans. Antennas Propag.* **AP-34**, 856–870.

Franz, J., and Kasper, M. (1995). Superconvergent finite element solutions of Laplace and Poisson equation. *Conf. Rec. Conf. Comput. Electromagn. Fields, 10th, Berlin, 1995*, pp. 64–65.

Jin. J.-M. (1993). *The Finite Element Method in Electromagnetics.* Wiley, New York.

Lee, R., and Cangellaris, A. C. (1992). A study of discretization error in the finite element approximation of wave solutions. *IEEE Trans. Antennas Propag.* **AP-40**, 542–549.

Lynch, D. R., Paulsen, K. D., and Strohbehn, J. W. (1985). Finite element solution of Maxwell's equations for hyperthermia treatment planning. *J. Comput. Phys.* **58**, 246–269.

Meixner, J. (1972). The behavior of electromagnetic fields at edges. *IEEE Trans. Antennas Propag.* **AP-20**, 442–446.

Meyer, F. J. C. (1994). The two-dimensional finite element/boundary element method in electromagnetics: formulation, applications, error estimates and mesh adaptive procedures. Ph.D. Dissertation, University of Stellenbosch.

Mur, G. (1988). Optimum choice of finite elements for computing three-dimensional electromagnetic fields in inhomogeneous media. *IEEE Trans. Magn.* **MAG-24**, 330–333.

Mur, G. (1991). Finite-element modeling of three-dimensional time harmonic electromagnetic fields in inhomogeneous media. *Radio Sci.* **26**, 275–280.

Mur, G. (1994). The finite-element modeling of three-dimensional electromagnetic fields using edge and nodal elements. *IEEE Trans. Antennas Propag.* **AP-41**, 948–953.

Peterson, A. F. (1990). The "interior resonance" problem associated with surface integral equations of electromagnetics: Numerical consequences and a survey of remedies. *Electromagnetics* **10**, 293–312.

Peterson, A. F., and Baca, R. J. (1991). Error in the finite element discretization of the scalar Helmholtz equation over electrically large regions. *IEEE Microwave Guided Wave Lett.* **1**, 219–222.

Scott, W. R. (1994). Errors due to spatial discretization and numerical precision in the finite-element method. *IEEE Trans. Antennas Propag.* **AP-42**, 1565–1570.

Silvester, P. P. (1991). Differentiation of finite element approximations to harmonic functions. *IEEE Trans. Magn.* **MAG-27**, 3774–3779.

Silvester, P. P., and Omeragić, D. (1993). Differentiation of finite element solutions of Poisson equation. *IEEE Trans. Magn.* **MAG-29**, 1993–1996.

Stern, M. (1979). Families of consistent conforming elements with singular derivative fields. *Int. J. Numer. Methods Eng.* **14**, 409–421.

Strang, G., and Fix, G. J. (1973). *An Analysis of the Finite Element Method.* Prentice-Hall, Englewood Cliffs, NJ.

Van Bladel, J. (1985). Field singularities at metal-dielectric wedges. *IEEE Trans. Antennas Propag.* **AP-33**, 450–455.

Warren, G. S., and Scott, W. R. (1994). An investigation of numerical dispersion in the vector finite element method using quadrilateral elements. *IEEE Trans. Antennas Propag.* **AP-42**, 1502–1508.

Webb, J. P. (1988). Finite element analysis of dispersion in waveguides with sharp metal edges. *IEEE Trans. Microwave Theory Techn.* **MTT-36**, 1819–1824.

Wilkinson, J. H. (1963). *Rounding Errors in Algebraic Processes.* Prentice-Hall, Englewood Cliffs, NJ.

CHAPTER FIFTEEN

Parallel Hybrid Iterative/Direct Solution Methods

ROBERT D. FERRARO

1. INTRODUCTION

In solving finite element problems, the differential equation is ultimately reduced to a set of linear equations that must be solved by some method. Various different approaches can be used to solve this linear equation set. In some instances, an explicit method is used to advance the differential equation in time and thus the solution of the linear equation set is accomplished automatically in the process. In most cases, however, a matrix of coefficients must actually be formulated and some numerical method employed to obtain one or more solutions to this equation set.

There are generally two broad categories of methods for solving linear equations: direct methods and iterative methods. Direct methods typically involve the factorization of a matrix into some combination of upper and lower triangular matrices, which may easily be used in backsolving with the right-hand side to obtain the solution. Once the factorization is accomplished, little additional effort is required to solve problems that differ from the original only in the right-hand side (which usually represents the boundary conditions or driving function). Direct methods, depending on the sparsity of the linear system, involve operation counts that scale approximately as order nb^2, where n is the number of degrees of freedom, and b is the average bandwidth of the sparse matrix. Although various numbering techniques may be used to

Finite Element Software for Microwave Engineering, Edited by Tatsuo Itoh, Giuseppe Pelosi, and Peter P. Silvester.
ISBN 0-471-12636-5

minimize the bandwidth of sparse matrices, the fact that the problem stems from a finite element formulation implies that general geometric considerations will couple the bandwidth to the number of degrees of freedom and the dimensionality of the problem, so that b is on the order of the square root of n for two-dimensional problems, and the two-thirds power of n in three-dimensional problems. General discussions of direct methods may be found in many places, e.g., Duff et al. (1986) or Press et al. (1986).

Iterative methods, on the other hand, typically involve sparse matrix–vector multiplies and the construction of orthogonal vectors to decide on new search directions at each iteration step (for examples, see Jacobs, 1981). The sparse matrix–vector multiply can be computed in order nf operations, where n is the number of degrees of freedom, and f is the average number of nonzero coefficients in each equation. For finite element formulations, f is a number that depends on the type and degree of finite elements employed and is independent of the size of the problem. Obtaining a solution to the linear equations set involves repeated matrix–vector multiply operations until some convergence criterion is satisfied. This typically involves something less than order n operations; thus the iterative methods have somewhat of an advantage in terms of operation count compared to direct methods when dealing with very large systems.

More importantly, in current parallel supercomputing environments, locality of memory access is the most important consideration in the design of any method for solving linear systems. In every parallel supercomputer available, the programmer must pay a price for accessing memory that is not local to a processor. The largest scalable parallel computers today (such as the IBM SP-2, Intel Paragon, Cray T3D, SGI POWER CHALLENGEarray, and Convex Exemplar) have processors with local memory interconnected through some fast network hierarchy. Though some distributed memory systems, like the Cray T3D and Convex Examplar SPP, allow the user a shared memory programming model, efficiency considerations require the user to manage the locality of data access, since referencing data that is not in the processor's local memory is always a significant cost operation. Thus in designing parallel algorithms, special attention must always be given to arranging data in memory to minimize the number of remote memory accesses across the interconnection network, or at least organizing them into block references that can be fetched efficiently. This usually leads to algorithms that operate on localized blocks of data rather than striding uniformly or randomly through memory. Parallel algorithms for direct and iterative methods have been published in many places in the last few years (Fox et al., 1988, 1994; Barrett et al., 1993; Cook and Sadecki, 1993; Heath and Raghavan, 1994; Kumar et al., 1994). Parallel computer manufacturers usually provide some type of parallel factorization and parallel iterative method in library form, which may be employed in solving finite element problems. Although direct methods are typically more stable than iterative methods, for large problem sizes, their operation counts and parallel scaling are such that they are not feasible for use in solving very large problems. As tens of millions of finite elements become used in electromagnetics problems, it will become imperative to apply an iterative method to obtain any solution whatsoever.

A wide variety of iterative methods have been employed to solve finite element problems. The most popular among these are the Krylov subspace orthogonalization methods, on which conjugate gradient and its variations are based. Explicit orthogonalization methods such as GMRES are also very popular and robust but require substantially higher operation counts to achieve convergence. This is due to the fact that the latter method explicitly constructs orthogonal search directions based on some number of previously saved iterations. This process becomes more expensive with every new iteration and thus is typically carried only for a limited number of steps before some restart algorithm is used. An excellent discussion of some of the theory and all of the issues associated with iterative methods may be found in Barrett et al. (1993, and references contained therein), and will not be repeated here.

In this chapter, a novel combination of direct and iterative methods will be considered, which can be employed successfully in solving finite element problems. This combination, referred to here as hybrid methods, originated from the observation that a parallel decomposition of a finite element mesh resulted in some number of subproblems, which could be viewed as independent finite element problems and which could be solved in parallel without communication. The bandwidth of each of these problems was substantially smaller than the bandwidth of the total linear system, and thus direct methods could be employed efficiently on these subproblems when it was not feasible to do so on the global problem.

In parallelizing a finite element method, consideration must be given not only to the parallel linear equations solution method but also to the parallel assembly of the coefficient matrix of the linear system, and the boundary conditions or excitation terms that constitute the right-hand side of the linear system (see, for example, Ferraro et al., 1990; Lyzenga et al., 1988). Since finite elements are geometric objects with geometric connectivity and spatial relationships, the sparsity pattern of the coefficient matrix mimics the spatial pattern and connectivity of the finite elements from which the matrix is constructed. The parallel partitioning of the finite element mesh (such as described in Ding and Ferraro, 1995) may be used directly to construct a decomposed sparse coefficient matrix for use by iterative methods, since the matrix–vector multiplies can be done in parallel without regard to the order in which the operations are carried out.

This chapter will briefly review the issues in constructing efficient parallel conjugate gradient algorithms and preconditioners and then will present the hybrid method as it is used to solve finite element problems. The basic conjugate gradient algorithm for positive definite symmetric systems will be used to illustrate the hybrid algorithm, but the hybrid method can be applied in a straightforward fashion to any sparse linear equation set by substituting the iterative and direct methods appropriate to the particular matrix type. A discussion of the method's parallel efficiency and scaling will follow.

2. PARALLEL CONJUGATE GRADIENT ALGORITHMS

Conjugate gradient algorithms involve two major computational operations: a matrix–vector product and vector–vector inner product. Parallel architectures impose certain constraints on the organization of the data involved in these operations in order to maximize efficiency. Today's scalable architectures consist of processors with local memory connected together via some high-speed communication network, which imposes a penalty on data access for data not located in the processor's local memory. The magnitude of this penalty varies from architecture to architecture but is always a consideration. Thus all efficient implementations of conjugate gradient algorithms on parallel computers pay close attention to the decomposition of the entries of the matrix and vectors involved in the operation. The most efficient decomposition depends on the sparsity of the matrix and the characteristics of the problem from which the matrix is derived, but, as a general rule, elements of the matrix are uniquely decomposed on the processors while elements of vectors may be duplicated to avoid communication.

Equally important to parallel efficiency is the distribution of work among processors. Each processor should have approximately equal amounts of work to perform at all times in order to achieve the highest efficiency. Thus, besides data decomposition, load balance is a critical consideration in any parallel algorithm. For conjugate gradient algorithms, the majority of work is in the matrix–vector product and since that work is directly proportional to the number of nonzero entries in the matrix, a load-balanced implementation corresponds to a uniform distribution of the nonzero matrix elements across processors. Attention should also be paid to the work involved in computing the vector–vector inner products, but this work is usually small compared to the matrix–vector multiply.

The basic conjugate gradient algorithm for linear systems that results in symmetric positive definite matrices is given in Figure 1. This algorithm is intended to solve the linear system

$$\mathbf{Ax} = \mathbf{b} \tag{1}$$

Here, and in Figure 1, $\mathbf{A}$ is a symmetric, positive-definite matrix corresponding to the coefficients of the linear system and $\mathbf{b}$ is the right-hand-side vector. In the conjugate gradient algorithm of Figure 1, $\mathbf{x}^{(i)}$ is the solution at each iteration step, $\mathbf{r}^{(i)}$ is the residual vector at each iteration step, and α and β are scaling coefficients used to determine the new conjugate search directions $\mathbf{p}^{(i)}$ and $\mathbf{q}^{(i)}$. This algorithm involves one matrix–vector product, two vector inner products, and three vector scale & add operations (SAXPYs) in the inner loop.

$$\mathbf{p}^{(0)} = \mathbf{r}^{(0)} = \mathbf{b} - \mathbf{A}\mathbf{x}^{(0)}$$
$$\rho_0 = \mathbf{r}^{(0)} \cdot \mathbf{r}^{(0)}$$
while $\|\mathbf{r}^{(i)}\|$ *not converged*
$$\mathbf{q}^{(i)} = \mathbf{A}\mathbf{p}^{(i)}$$
$$\alpha_i = \frac{\rho_{i-1}}{\mathbf{p}^{(i)} \cdot \mathbf{q}^{(i)}}$$
$$\mathbf{x}^{(i)} = \mathbf{x}^{(i-1)} + \alpha_i \mathbf{p}^{(i)}$$
$$\mathbf{r}^{(i)} = \mathbf{r}^{(i-1)} - \alpha_i \mathbf{p}^{(i)}$$
$$\rho_i = \mathbf{r}^{(i)} \cdot \mathbf{r}^{(i)}$$
$$\beta_i = \frac{\rho_i}{\rho_{i-1}}$$
$$\mathbf{p}^{(i)} = \mathbf{r}^{(i)} + \beta_i \mathbf{p}^{(i-1)}$$
end while

FIGURE 1. The conjugate gradient algorithm for symmetric, positive-definite linear systems.

The theory behind conjugate gradient algorithms can be found in many places and will not be discussed here. Instead, this section focuses on the issues of designing an efficient parallel implementation for conjugate gradient algorithms and their variations.

Parallelism in conjugate gradient algorithms essentially comes from parallelism in the matrix–vector multiply, the inner products, and the SAXPYs. The remainder of the operations involved in the conjugate gradient algorithm in Figure 1 are trivial compared to these. In the main loop of the algorithm the sequence of operations to be performed is fixed by the algorithm itself. However, certain operations can proceed in parallel since there are no dependencies. In particular, the update of the residual vector and the solution vector do not depend on each other and could be performed in parallel. Neither of these parallel operations can proceed before the matrix–vector product is completed and the scaling coefficient α is computed from its result. Likewise, the matrix–vector product of the next iteration may not proceed until the residual vector has been updated. Thus there are two points at which all processors must synchronize before they may proceed further with the algorithm. It is important that the workload remain balanced between synchronization points so that all processors participate in all of the computation that needs to take place and that none go idle during that period. Therefore the parallel implementation of the matrix–vector product should be tailored to the details of the matrix itself so that it achieves its maximum parallel efficiency and the work for the vector inner products and SAXPYs should be equally divided among processors. Note that this requirement does not necessarily impose a rigid constraint on how the matrix and vectors are decomposed in parallel. Any decomposition that achieves these goals will allow a parallel conjugate gradient algorithm to perform well.

In exact arithmetic, the conjugate gradient algorithm applied to a set of n linear equations converges to the solution uniformly (in some norm) and in n iterations. The residual error is generally reduced at each iteration so that the solution comes closer and closer to the exact solution with each iteration step. Thus when only finite precision is required of the answer, the conjugate gradient may arrive at a sufficiently good solution much earlier than the n iterations required for exact convergence. This means that, in practice, conjugate gradient algorithms converge in less than order n^2 operations. When problem sizes are large, conjugate gradient is very attractive compared to direct factorization methods. (The reader is reminded that the uniform convergence property of the conjugate gradient algorithm is applicable only to positive definite symmetric linear systems. The variants of conjugate gradient for other types of linear systems do not, in general, have such convergence properties. However, the reduction of the residual error to some stopping criteria generally produces a uniform error solution, so that less than order n^2 operations are the norm for conjugate gradient and all of its variations.)

The basic conjugate gradient algorithm, however, is almost never used because large linear systems often have poor condition numbers. Finite precision arithmetic in combination with poor conditioning causes convergence to be slow or can prevent convergence altogether. Methods have been developed to improve the conditioning of the problem and speed the rate at which the conjugate gradient iterations achieve the required precision. These methods, known as preconditioners, essentially transform the linear equation set into a new set of equations, which are better suited to solution by conjugate gradient methods. Written formally, a preconditioner $\mathbf{M} = \mathbf{M}_1\mathbf{M}_r$ takes the linear system in Eq. (1) and transforms it by

$$(\mathbf{M}_1^{-1}\mathbf{A}\mathbf{M}_r^{-1})(\mathbf{M}_r\mathbf{x}) = (\mathbf{M}_1^{-1}\mathbf{b}) \tag{2}$$

into

$$\mathbf{A}'\mathbf{x}' = \mathbf{b}' \tag{3}$$

Writing $\mathbf{M}$ as the product of a left preconditioner $\mathbf{M}_1$ and a right preconditioner $\mathbf{M}_r$ allows the properties of Eq. (1) to be preserved in Eq. (3), provided that $\mathbf{M}_1$, $\mathbf{M}_r$, and thus $\mathbf{M}$ are appropriately chosen. For example, if $\mathbf{A}$ in Eq. (1) is symmetric and positive definite, and $\mathbf{M}_1 = \mathbf{M}_r^{\mathrm{T}}$, then $\mathbf{A}'$ will also be symmetric and positive definite.

An effective preconditioner essentially transforms the matrix into another matrix, which is closer to the identity than the original. The ideal preconditioner is in fact the matrix inverse itself! However, the cost of computing the inverse is the cost of solving the problem in the first place. A detailed discussion

$\mathbf{r}^{(0)} = \mathbf{b} - \mathbf{A}\mathbf{x}^{(0)}$
solve $\mathbf{M}\mathbf{z}^{(0)} = \mathbf{r}^{(0)}$
$\rho_0 = \mathbf{r}^{(0)} \cdot \mathbf{z}^{(0)}$
$\mathbf{p}^{(0)} = \mathbf{z}^{(0)}$
while $\|\mathbf{r}^{(i)}\|$ *not converged*
  $\mathbf{q}^{(i)} = \mathbf{A}\mathbf{p}^{(i)}$
  $\alpha_i = \dfrac{\rho_{i-1}}{\mathbf{p}^{(i)} \cdot \mathbf{q}^{(i)}}$
  $\mathbf{x}^{(i)} = \mathbf{x}^{(i-1)} + \alpha_i \mathbf{p}^{(i)}$
  $\mathbf{r}^{(i)} = \mathbf{r}^{(i-1)} - \alpha_i \mathbf{p}^{(i)}$
  solve $\mathbf{M}\mathbf{z}^{(i)} = \mathbf{r}^{(i)}$
  $\rho_i = \mathbf{r}^{(i)} \cdot \mathbf{z}^{(i)}$
  $\beta_i = \dfrac{\rho_i}{\rho_{i-1}}$
  $\mathbf{p}^{(i)} = \mathbf{z}^{(i)} + \beta_i \mathbf{p}^{(i-1)}$
end while

FIGURE 2. The preconditioned conjugate gradient algorithm for symmetric, positive-definite linear systems.

of preconditioning is also beyond the scope of this chapter. Preconditioning is an art rather than a science, and the reader is referred again to Barrett et al. (1993) for a variety of methods in common use.

In Figure 2, the conjugate gradient algorithm has been modified to include the application of a symmetric positive-definite preconditioner, which is represented by solving the linear system $\mathbf{Mz} = \mathbf{r}$. The preconditioner is almost never applied directly to the matrix, because that would destroy the sparsity characteristics of the matrix, and thus the advantages of sparse vector–matrix multiply. Rather, the algorithm is modified to include a linear system solve at each iteration. The preconditioner has presumably been chosen so that the linear system solve can be accomplished quickly! However, the application of a preconditioner to a decomposed linear system on a parallel processor proves to be a serious complication to the efficiency of parallel conjugate gradient. The linear equation solve that represents the application of the preconditioner can be as complicated to implement in parallel as any of the direct factorization methods themselves or it may require a high setup cost to construct in the first place. Unless the preconditioner chosen has data decomposition characteristics that match the matrix decomposition characteristics, the efficiency of the conjugate gradient multiply can be ruined by the inefficiency of applying the preconditioner. An inefficient preconditioner or one that is expensive to construct in a parallel environment can more than cancel the benefit of the reduced iteration count that it was meant to provide. So particular care must be taken in choosing a preconditioner in a parallel implementation.

The basic conjugate gradient algorithm discussed above is useful only for positive-definite linear systems, but there are a variety of extensions and variations of this algorithm, which can be applied to linear systems with other problems. Biconjugate gradient, for example, can be used for complex systems, quasi-minimum residual (QMR) method for complex symmetric systems, and conjugate gradient squared and biconjugate gradient stabilized for complex indefinite systems. Each of these variations involves a different method of computing the search directions, but the basic operations remain the same: matrix–vector products, vector inner products, SAXPYs, and some preconditioning scheme that is applied at each iteration step. Thus parallel efficiency for any of these methods will be achieved by efficient parallel implementation of these core operations. Ignoring the preconditioner, the parallel scaling for the conjugate gradient algorithm is the same as the scaling for parallel matrix multiply. Very high efficiency can be achieved by careful attention to minimizing the communications involved in these operations.

The introduction of a preconditioner can destroy this scaling property. Thus even though the number of iterations needed to achieve the solution is substantially reduced by the use of a preconditioner, the total execution time may actually rise if the parallel efficiency for the application of the preconditioner itself is poor. In some situations, it is in fact better to use a very simple preconditioner like diagonal scaling (Jacobi preconditioning), which has excellent parallelization properties, than it is to use a sophisticated preconditioner like incomplete Cholesky factorization. Even though the iteration count is higher with diagonal scaling, the execution time to solution may be as good as or better than the more sophisticated method.

3. HYBRID DIRECT–CONJUGATE GRADIENT ALGORITHMS

The implementation of finite element methods for parallel computers provides an opportunity to employ a unique method of combining direct and iterative linear equation solvers, which, when taken together, enjoy better parallel scaling and convergence than either alone. The typical implementation of a finite element method uses a domain decomposition that splits the finite element mesh into compact submeshes so that each processor has (approximately) an equal portion of the entire problem. From the individual processor's viewpoint, it has a complete finite element problem. Thus one can employ a direct method to solve the finite element subproblem without regard to the meshes that are being processed on other processors. Consider the two-dimensional finite element mesh shown in Figure 3. In this example, the mesh has been partitioned among four processors, such that elements in the mesh reside in one and only one processor. The boundaries between partitions lie

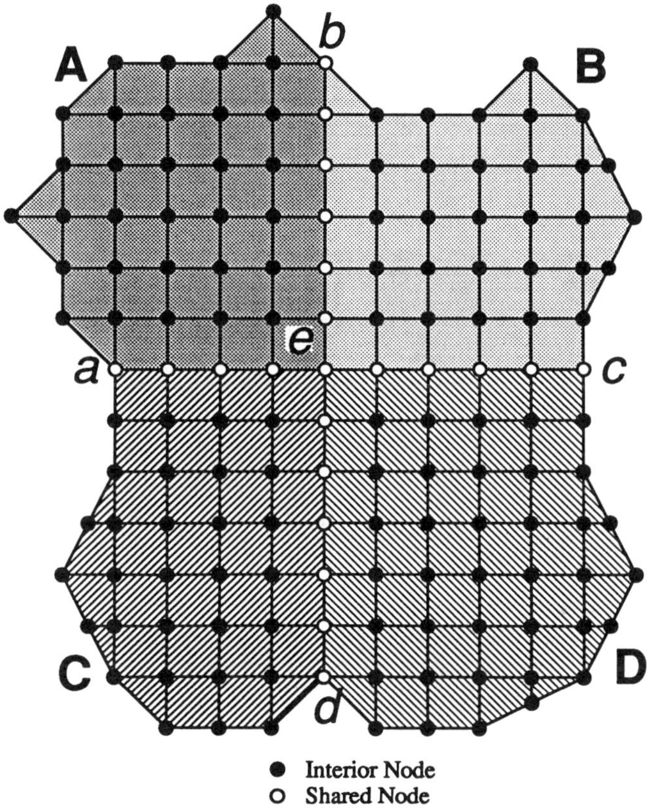

FIGURE 3. A simple finite element mesh partitioned among four processors. Elements are uniquely assigned to partitions A, B, C, and D. Nodes on partition boundaries *a*, *b*, *c*, *d*, and the center node *e* are assigned to multiple partitions.

along element edges or faces, so that nodal points (and degrees of freedom corresponding to those points) serve as a dividing line between processors. Taken individually, each processor has a partition that is a finite element problem unto itself with boundary conditions that are dependent on the results obtained in other processors. In Figure 3 the nodal points interior to a partition are represented by filled circles and nodal points lying on partition boundaries are denoted by open circles. Were it not for the fact that the values on the partition boundaries are coupled to results in neighboring processors, a standard sequential finite elements method could be used without modification to solve the interior problem.

The elements in this simple example belong uniquely to a single partition. In a standard sequential finite element formulation, each element contributes additively to the global *stiffness* matrix **K** and *force* vector **f** (see Hughes, 1987) as follows:

$$\mathbf{K} = \sum_{\mathrm{e}} \boldsymbol{K}^{(\mathrm{e})} \tag{4a}$$

$$\mathbf{f} = \sum_{\mathrm{e}} \boldsymbol{f}^{(\mathrm{e})} \tag{4b}$$

Here $\boldsymbol{K}^{(\mathrm{e})}$ and $\boldsymbol{f}^{(\mathrm{e})}$ are the contributions of a single element to the global stiffness matrix and global force vector, respectively, and the summation is over all elements in the mesh. In a partitioned mesh such as shown in Figure 3, the contributions of all the elements may be computed in an *embarrassingly parallel* fashion (i.e., without interprocessor communication), since each partition is assigned to its own processor. The stiffness matrix $\mathbf{K}^{(p)}$, which corresponds to a partition contribution, is computed on its processor using Eq. (4a) for the elements contained in the partition, and the global stiffness matrix may be recovered (if necessary) by summing over processors p:

$$\mathbf{K} = \sum_{p} \mathbf{K}^{(p)} \tag{5}$$

The force vector $\mathbf{f}$ is computed similarly using Eq. (4b) on each processor, but the results must be summed across all processors p so that the values at the shared nodes correctly include contributions from elements in different partitions.

For iterative methods that require only matrix–vector products and inner products, it is not necessary to recover the global stiffness matrix. These methods only require the result of the matrix–vector product, which can be computed in parallel directly from the individual partition stiffness matrices $\mathbf{K}^{(p)}$:

$$\mathbf{Kx} = \sum_{p} (\mathbf{K}^{(p)}\mathbf{x}^{(p)}) \tag{6}$$

Here $\mathbf{x}^{(p)}$ is the portion of a global vector that corresponds to the unknowns in partition p. The vector $\mathbf{x}^{(p)}$ will contain entries for shared nodes on partition boundaries, which are duplicated in more than one processor. In partition A in Figure 4, for example, $\mathbf{x}^{(\mathrm{A})}$ would consist of the unknowns associated with the interior nodes of the partition and the shared nodes labeled a, b, and e, but no others. The matrix–vector multiply of Eq. (6) will automatically produce the correct results in parallel for interior nodes. The summation of Eq. (6) must be performed only for the shared nodes; that is, the shared nodes have contributions from elements in multiple partitions, which must be summed together to obtain the correct result.

This observation leads to the possibility of using standard sequential direct factorization methods to remove the degrees of freedom interior to the

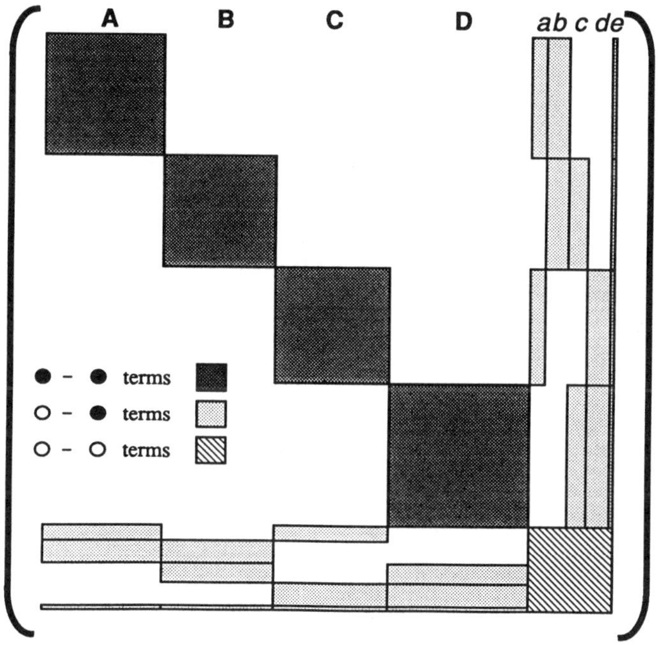

FIGURE 4. The structure of the matrix of coefficients for the linear system that results from the finite element problem of Figure 3. Unshaded areas are zero values.

partitions entirely from the problem, leaving only the shared nodal points on partition boundaries to be solved in parallel. This idea can be represented formally by writing out the linear equation set that corresponds to the finite element problem in the following manner. First, the degrees of freedom attached to partition interior nodal points are numbered in order by partition, followed by the degrees of freedom that are attached to the partition boundary nodes. For edge element problems, degrees of freedom are associated with edges instead of nodes, but the same numbering methodology applies. The global matrix that corresponds to the partitioned finite element problem can now be seen to consist of a set of matrix blocks that correspond to the coupling of interior points of each partition, the coupling terms between the interior points and the boundary points in each partition, and finally the entries corresponding to coupling among the boundary points themselves as represented pictorially in Figure 4 for the partitioned mesh shown in Figure 3. Returning to the usual linear algebra notation of Eq. (1), this linear sysem can be written as

$$\begin{bmatrix} \mathbf{A}_{ii} & \mathbf{A}_{is} \\ \mathbf{A}_{si} & \mathbf{A}_{ss} \end{bmatrix} \begin{bmatrix} \mathbf{x}_i \\ \mathbf{x}_s \end{bmatrix} = \begin{bmatrix} \mathbf{b}_i \\ \mathbf{b}_s \end{bmatrix} \tag{7}$$

where $\mathbf{A}_{ii}$ are the blocks that result from pairs of interior nodes, $\mathbf{A}_{is}$ are the blocks that result from interior node/shared node pairs, and $\mathbf{A}_{ss}$ are the blocks that result from pairs of shared nodes. Note that the large blocks of partition interior nodes in this matrix are coupled to each other only through shared nodes and thus may be formally removed from the problem by simple algebra. The upper equation in Eq. (7) may be solved for $\mathbf{x}_i$ to obtain

$$\mathbf{x}_i = \mathbf{A}_{ii}^{-1}(\mathbf{b}_i - \mathbf{A}_{is}\mathbf{x}_s) \tag{8}$$

Here it is understood that $\mathbf{A}_{ii}^{-1}$ is a shorthand notation for a factorization of $\mathbf{A}_{ii}$. Introducing this expression for $\mathbf{x}_i$ into the lower equation of Eq. (7) results in a reduced linear equation set consisting only of shared partition boundary points:

$$(\mathbf{A}_{ss} - \mathbf{A}_{si}\mathbf{A}_{ii}^{-1}\mathbf{A}_{is})\mathbf{x}_s = (\mathbf{b}_s - \mathbf{A}_{si}\mathbf{A}_{ii}^{-1}\mathbf{b}_i) \tag{9}$$

or

$$\mathbf{A}_r\mathbf{x}_s = \mathbf{b}_r \tag{10}$$

where

$$\mathbf{A}_r = (\mathbf{A}_{ss} - \mathbf{A}_{si}\mathbf{A}_{ii}^{-1}\mathbf{A}_{is}) \tag{11a}$$

$$\mathbf{b}_r = (\mathbf{b}_s - \mathbf{A}_{si}\mathbf{A}_{ii}^{-1}\mathbf{b}_i) \tag{11b}$$

The process of eliminating the interior points couples every shared point on every partitioned boundary with every other shared point.

It is a consequence of Eq. (5), which is a property of the finite element method, that Eq. (11a) partitions completely and independently among processors. Each of the matrix blocks in Eq. (11a) turns out to be partitioned among processors so that

$$\mathbf{A}_r^{(p)} = (\mathbf{A}_{ss}^{(p)} - \mathbf{A}_{si}^{(p)}(\mathbf{A}_{ii}^{(p)})^{-1}\mathbf{A}_{is}^{(p)}) \tag{12}$$

and

$$\mathbf{A}_r = \sum_p \mathbf{A}_r^{(p)} \tag{13}$$

Each of these reduced matrices $\mathbf{A}_r^{(p)}$ are dense matrices. It should be noted that Eq. (8) may also be computed independently in each partition for that partition's interior nodes and thus may be done in parallel without communication. The reduced matrices could be recombined across all processors and redistributed for a parallel factorization step, but the decomposition method allows these matrices to be used directly by an iterative scheme (e.g., a conjugate gradient method) to complete the solution of Eq. (10) for the shared nodes. Note that the solution obtained for the reduced equations is exactly the desired values for the degrees of freedom on the shared nodes. To obtain the solution values on the interior nodes is a simple matter of applying Eq. (8) to the solution for the shared nodes. The factorizations allow for forward and backward substitution in parallel on each processor without communication. Additionally, for simple changes in global boundary conditions, the factorizations of the interior node matrix blocks may be retained so that multiple right-hand sides may be done successively in an efficient manner.

This method produces reduced matrices whose characteristics are the same as those of the global matrix from which they were derived; that is, if the global matrix is symmetric, the reduced matrix is also symmetric; if the global matrix is Hermitian, then the reduced matrix is Hermitian. Thus any of the standard iterative methods can be used to solve the reduced equation set since the matrix–vector multiplies are the fundamental operations and each conjugate gradient method differs only in how the results are combined to form new search directions. Additionally, for poorly conditioned problems, the application of a direct factorization method with pivoting can improve the conditioning of the system and allow the conjugate gradient algorithm to converge on the reduced equations when it would not have converged on the global system.

In constructing the software to implement this method, the essential considerations are the numbering of the unknowns locally within each processor, so that the matrix structure naturally falls into the form shown in Figure 4. The algorithm for this hybrid method is presented in Figure 5. The segregation into the interior and shared node blocks allows sequential algo-

For each partition in parallel:

Factor $\mathbf{A}_{ii}^{(p)}$

Compute $\mathbf{A}_{r}^{(p)} = (\mathbf{A}_{ss}^{(p)} - \mathbf{A}_{si}^{(p)}(\mathbf{A}_{ii}^{(p)})^{-1}\mathbf{A}_{is}^{(p)})$

Compute $\mathbf{b}_{r}^{(p)} = \mathbf{A}_{si}^{(p)}(\mathbf{A}_{ii}^{(p)})^{-1}\mathbf{b}_{i}^{(p)}$

Compute and distribute $\mathbf{b}_r = \mathbf{b}_s - \sum_p \mathbf{b}_r^{(p)}$

Solve $(\sum_p \mathbf{A}_r^{(p)})\mathbf{x}_s = \mathbf{b}_r$ *using a parallel iterative solver*

For each partition in parallel:

Compute $\mathbf{x}_{i}^{(p)} = (\mathbf{A}_{ii}^{(p)})^{-1}(\mathbf{b}_{i}^{(p)} - \mathbf{A}_{is}^{(p)}\mathbf{x}_s)$

FIGURE 5. The parallel hybrid algorithm. The factor and iterative solve steps must be tailored to match the properties of the underlying linear system. For symmetric systems, some additional algebraic recombinations are possible.

rithms for the computation of $\mathbf{A}_{ii}^{-1}$ (factorization) to be applied unchanged. Many are available; see, for example, Press et al. (1986). The application of Eq. (12) may be done using standard library routines like the BLAS of Lawson et al. (1979) or, if special storage methods are used, can be written simply as matrix–vector multiplies. The final reduced matrix, which is essentially dense, may be used with a BLAS SEGMV routine, which is typically available on all parallel platforms and optimized for its architecture. Thus the only communication required is a global summation of vectors at the end of each call to SEGMV.

Since the iterative part of this method is applied only to the unknowns that reside on partition boundaries, this method has scaling properties that are better than an iterative method applied to the original problem. The operation count for conjugate gradient on the global finite element problem scales as order n^2, where n is the number of degrees of freedom. In this method, the number of unknowns on the partition boundary goes as $n^{1/2}$ for two-dimensional problems or $n^{2/3}$ for three-dimensional problems, so that the iterative portion of the solver converges in either order n operations in two dimensions or order $n^{4/3}$ in three-dimensions. The remainder of the operations involved in the method scale as the number of interior points. As problem size grows with the number of processors, this workload is a constant independent of problem size. Thus if we compare the implementation of a conjugate gradient method with diagonal preconditioning to this hybrid method, we see that the hybrid method outperforms the conjugate gradient method after only a small number of processors.

In Figure 6 the scaling behavior of each of these methods is plotted versus number of processors for a series of scaled problems with 1600 grid points per processor. That is, a base problem consisting of 1600 unknowns was solved on a single processor, and scaled problems consisting of $1600p$ unknowns were solved for $p = 2, 4, 8, 16, \ldots$. The data shown are the ratio of execution time for solution of each of these problems normalized to the execution time for the 1600-grid problem on a single processor. The biconjugate gradient method scales as p, where p is the number of processors, while the hybrid method applied to the same problem scales approximately as the square root of p. It is clear from the graph that for large problem sizes on large numbers of processors, the hybrid method is the clear winner.

Although the results presented here are for a Cholesky factorization of the interior degrees of freedom followed by a conjugate gradient solution of the shared degrees of freedom, it is expected that this scaling behavior will carry over to any appropriate combination of direct and iterative method. The direct method is a constant cost, while the iterative method operates on a problem size that reduced from the original by the surface-to-volume ratio (since the shared nodes are the surfaces of a partitioned volume). Although this method is most easily understood in terms of and applies to parallel finite element problems, it relies only on the assumption that the linear equations set being solved represents some local coupling among unknowns (like that typically

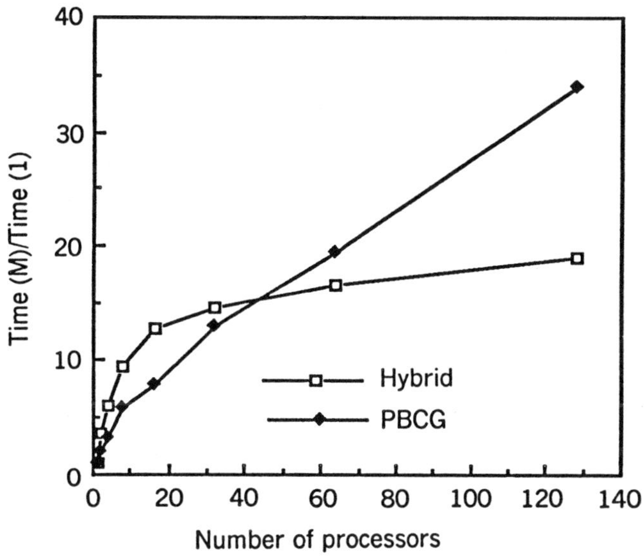

FIGURE 6. Scaling characteristics of the hybrid method compared to a diagonally preconditioned conjugate gradient method for the same problems.

arising from the solution of partial differential equations via finite difference, finite element, or finite volume methods) and that a geometric compact partitioning of the unknowns is possible.

ACKNOWLEDGMENTS

Many people have contributed to the development of this idea (over several years and projects) into a working software package. In particular, the author wishes to acknowledge G. A. Lyzenga, as the originator of the idea a number of years ago at the Jet Propulsion Laboratory, and to thank H. Ding for his contributions in extending the software to finite difference problems and in measuring the method's scaling properties.

This work was performed at the Jet Propulsion Laboratory, California Institute of Technology, under a contract with the National Aeronautics and Space Administration. Reference herein to any specific commercial product, process, or service by trade name, trademark, manufacturer, or otherwise does not constitute or imply its endorsement by the United States Government or the Jet Propulsion Laboratory, California Institute of Technology.

This research was performed in part using the Intel Touchstone Delta and Paragon Systems operated by Caltech on behalf of the Concurrent Supercomputing Consortium. Access to this facility was provided by the Jet Propulsion Laboratory, and NASA Office of Aeronautics.

REFERENCES

Barrett, R., Berry, M., Chan, T., Demmel, J., Donato, J., Dongarra, J., Eijkhout, V., Pozo, R., Romine, C., and van der Vorst, H. (1993). *Templates for the Solution of Linear Systems: Building Blocks for Iterative Methods.* SIAM, Philadelphia.

Cook, R., and Sadecki, J. (1993). Sparse matrix vector product on distributed memory MIMD architectures. *Proc. SIAM Conf. Parallel Process. Sci. Comput., 6th, 1993*, p. 429.

Ding, H. Q., and Ferraro, R. D. (1995). Slices: A scalable partitioner for finite element meshes in *Proc. SIAM Conf. Parallel Process. Comput., 7th, 1995.*

Duff, I. S., Erisman, A. M., and Reid, J. K. (1986). *Direct Methods for Sparse Matrices.* Oxford Univ. Press, London.

Ferraro, R. D., Cwik, T., Jacobi, N., Liewer, P. C., Lockhart, T. G., Lyzenga, G. A., Parker, J., and Patterson, J. E. (1990). Parallel finite elements applied to the electromagnetic scattering problem. In *Proceedings of the 5th Distributed Memory Computing Conference* (D. W. Walker and Q. F. Stout, eds.), p. 417. IEEE Computer Society Press, Los Alamitos, CA.

Fox, G. C., Johnson, M. A., Lyzenga, G. A., Otto, S. W., Salmon, J. K., and Walker, D. W. (1988). *Solving Problems on Concurrent Processors*, Vol. 1, Chapter 7. Prentice-Hall, Englewood Cliffs, NJ.

Fox, G. C., Williams, R. D., and Messina, P. C. (1994). *Parallel Computing Works!* Morgan Kaufmann, San Francisco.

Heath, M. T., and Raghavan, P. (1994). Performance of a fully parallel sparse solver. In *Proceedings of Scalable High Performance Computing Conference,* p. 334. IEEE Computer Society Press, Los Alamitos, CA.

Hughes, T. J. R. (1987). *The Finite Element Method.* Prentice-Hall, Englewood Cliffs, NJ.

Jacobs, D. A. H. (1981). The exploitation of sparsity of iterative methods. In *Sparse Matrices and their Uses* (I. S. Duff, ed.). Academic Press, London.

Kumar, V., Grama, A., Gupta, A., and Karypis, G. (1994). *Introduction to Parallel Computing*, Chapter 11. Benjamin/Cummings, Redwood City, CA.

Lawson, C., Hanson, R., Kincaid, D., and Kroch, F. (1979). Basic linear algebra subprograms for FORTRAN usage. *ACM Trans. Math. Software* **5**, 308–325.

Lyzenga, G. A., Raefsky, A., and Nour-Omid, B. (1988). *Implement Finite Element Software on the Hypercube,* Proc. 3rd Hypercube Conf. p. 1755. ACM Press, New York.

Press, W. H., Flannery, B. P., Teukolsky, S. A., and Vetterling, W. T. (1986). *Numerical Recipes*, pp. 31–37. Cambridge Univ. Press, New York.

CHAPTER SIXTEEN

Self-Adaptive Procedures for Waveguiding Structure Analysis

MAGDALENA SALAZAR-PALMA and
LUIS-EMILIO GARCÍA-CASTILLO

1. INTRODUCTION

The finite element method is a powerful and flexible tool that may be applied to solve a given electromagnetic problem in a wide variety of arbitrarily shaped structures and configurations. Its versatility is one of its greatest advantages.

The accuracy of the method greatly depends on a proper discretization of the domain of definition of the problem, namely, on the number and distribution of the elements in which the domain is subdivided and, as a consequence, on the nodes where the degrees of freedom are defined. In general, the method is used in an iterative way applying it to a sequence of meshes, with an increasing number of degrees of freedom, until a convergent solution is obtained. By mesh it is understood the domain discretization in the general sense (elements and nodes) given above.

Once a finite element code is available, the user's most cumbersome task is to create an adequate mesh. Thus automatic procedures are highly desired. This chapter addresses the issue of self-adaptive mesh techniques that provide numerical results to a prespecified degree of accuracy. Those procedures may briefly be described as follows. An initial coarse mesh is generated from the definition of the geometry of the problem. The finite element method is applied to it and the error of the computed solution is estimated both locally (in each

Finite Element Software for Microwave Engineering, Edited by Tatsuo Itoh, Giuseppe Pelosi, and Peter P. Silvester.
ISBN 0-471-12636-5

element) and globally. Those estimates are used to decide if the mesh should be refined and, if that is the case, which regions should be enriched. A refined mesh is then created. The procedure will continue until a stopping criterion is met. This technique provides a powerful tool to automatically generate meshes with local errors of the same order of magnitude. The achievement of these types of meshes, called equilibrated or optimal, ensures a high rate of convergence of the finite element method, minimizing the overall computational time required to obtain the final result.

Due to editorial limitations, the self-adaptive mesh procedure is described only in the context of the analysis of waveguiding structures that are invariant with respect to the direction of wave propagation (called the longitudinal axis), so that the electromagnetic problem may be defined as a two-dimensional one. Namely, Maxwell's equations must be solved in the source-free region defined by the cross section of the structure employing either a quasistatic or a dynamic analysis.

The outline of the chapter is as follows. Some results regarding the rate of convergence of the finite element method are summarized in Section 2. In Section 3 a concrete self-adaptive scheme for the quasistatic analysis of transmission line structures is presented. The procedure has been validated by applying it to various structures. In Section 4 the extension of the previous procedure to the dynamic analysis of waveguiding structures is described briefly together with some validation results. Section 5 summarizes the features of self-adaptive procedures.

2. SOME RESULTS REGARDING THE RATE OF CONVERGENCE OF THE FINITE ELEMENT METHOD

This discussion is limited to two-dimensional deterministic second-order boundary value elliptic problems, discretized by means of Lagrange finite elements.

Let the definition of the problem be given by

$$\mathscr{L}u = f_\Omega, \quad \text{in } \Omega; \qquad \zeta u = f_\Gamma, \quad \text{in } \Gamma \tag{1}$$

where $\mathscr{L}$ and ζ are differential operators, Ω is the domain of definition of the problem, Γ is its boundary, f_Ω and f_Γ are given functions, and u is the scalar unknown. It is assumed that the exact solution, u, belongs to the Hilbert space $H^{\nu+1}(\Omega)$, with ν a real positive number.

Let the domain be discretized by means of a regular and quasi-uniform mesh, such that the following expressions hold:

$$\sigma_i \leqslant \frac{h_i}{\rho_i} < \sigma; \quad \frac{h}{h_i} < \tau; \quad \text{for } i = 1,\dots,N_e \tag{2}$$

where h_i is the diameter of the ith element; ρ_i is its roundness; N_e is the number of elements of the mesh; $h = \max(h_i)$, $i = 1,\ldots,N_e$; and σ and τ are finite real numbers, independent of h. The diameter of an element is the maximum euclidean distance between any two vertices. The roundness is the maximum diameter of the circumferences that might be circumscribed by the element. The first expression defines the aspect ratio, σ_i, of the ith element and the concept of a regular mesh as that in which the aspect ratio of any element is limited by σ. The second expression defines a quasi-uniform mesh.

Let $\tilde{u}$ be the approximate solution obtained by the application of the finite element method employing Lagrange elements of p order. The error incurred is defined as $e = u - \tilde{u}$. A measure of the error might be given by the norm in the Hilbert space of order m, $H^m(\Omega)$, defined as

$$\|e\|_{H^m(\Omega)} = \|e\|_{m,\Omega} = \left[\sum_{\alpha=0}^{m} \int_\Omega |D^\alpha e|^2 \, d\Omega\right]^{1/2}; \quad D^\alpha e = \frac{\partial^\alpha e}{\partial^{\alpha_1} x \partial^{\alpha_2} y} \tag{3}$$

where (x, y) are the cartesian coordinates; α_1 and α_2 are non-negative integer numbers and $\alpha = \alpha_1 + \alpha_2$. The error may also be measured by the energy norm according to

$$\|e\|_E = \langle e, \mathscr{L}e\rangle^{1/2} \tag{4}$$

where $\langle \cdot, \cdot \rangle$ is the inner product,

$$\langle v, w\rangle = \int_\Omega v^* w \, d\Omega \tag{5}$$

and v^* is the complex conjugate of v.

With the above-mentioned assumptions, the following expression holds:

$$\|e\|_{m,\Omega} \leqslant C_1 h^{k+1-m} |u|_{k+1,\Omega}, \quad k = \min(p, v), \quad 0 \leqslant m \leqslant k + 1 \tag{6}$$

where C_1 is a constant independent of h and u, but dependent on the type and order, p, of the elements employed and

$$|u|_{H^m(\Omega)} = |u|_{m,\Omega} = \left[\sum_{\alpha=m} \int_\Omega |D^\alpha u|^2 \, d\Omega\right]^{1/2} \tag{7}$$

is the seminorm of u in the Hilbert space of order m.

Expression (6) is obtained by maximizing the interpolation error incurred in the ith element, Ω_i, to the complete discretized domain, $\Omega = \cup \Omega_i$. The interpolation error between a function, u, and its interpolant, Πu, is defined as

$u - \Pi u$, and for u belonging to $H^{\nu+1}(\Omega_i)$, with $\nu \geqslant 0$, its measure is given by

$$\|u - \Pi u\|_{m,\Omega_i} \leqslant C_2 \frac{h_i^{k+1}}{\rho_i^m} |u|_{k+1,\Omega_i}, \quad k = \min(p, \nu), \quad 0 \leqslant m \leqslant k + 1 \tag{8}$$

with the same meaning for C_2. The process of maximizing Eq. (8) to obtain Eq. (6) assumes that the interpolation error is the dominant one, so that all other sources of error (geometrical and physical discretization, numerical quadrature, round-off and truncation errors, etc.) have been properly handled.

For the case of elliptic problems the energy norm turns out to be a measure that is more convenient than expression (6). It is given by

$$\|e\|_E \leqslant C_3 h^k, \quad k = \min(p, \nu) \tag{9}$$

where C_3 is a constant independent of h, but now dependent on u and the type and order, p, of the elements employed.

From Eq. (9) it may be concluded that if the solution of the problem, u, is smooth enough, then $\nu > p$, giving $k = p$, so that successive uniform subdivisions of the mesh will give an error, which energy norm will tend to zero at the rate of h^p.

If, instead, the solution is not smooth enough, then $\nu < p$, giving $k = \nu$, so that the rate of convergence with successive uniform refinements is only h^ν. The use of higher-order elements will not affect the rate of convergence, only the accuracy.

As stated by Babuska (1977), among others, there are several reasons for a function not to be sufficiently smooth: nonsmoothness of the boundaries (corners, wedges, edges, etc.), abrupt changes in the values of the physical constants characterizing different regions, discontinuities in the boundary conditions, singular sources, and so on. In these cases an important conclusion may be derived from expression (8). Namely, in a quasi-uniform mesh the interpolation error is higher in those elements located in regions where the function is not smooth enough. Thus a better approach should be to generate a mesh with smaller elements in those regions than in the rest, trying to obtain a nonuniform equilibrated mesh with local errors of the same order of magnitude.

It also seems to be more convenient to write any measure of the error in terms of the total number of nodes, N, instead of using the parameter related to the size of the mesh, h, that might not be meaningful if nonuniform meshes are employed.

In doing so, expression (9), for quasi-uniform meshes becomes

$$\|e\|_E \leqslant C_4 N^{-(1/2)\min(p,\nu)} \tag{10}$$

where C_4 is a constant depending on u and p.

If nonuniform equilibrated meshes are used, the rate of convergence is

$$\|e\|_E \leqslant C_5 N^{-p/2} \tag{11}$$

where C_5 is a constant depending on u and p. Equation (11) turns out to be independent of the smoothness of the function.

Expressions (10) and (11) are the basis for the so-called h version of the finite element method, in which the number of degrees of freedom is increased, subdividing the elements of the original mesh either uniformly or selectively. In particular, self-adaptive techniques using h refinements are based on Eq. (11), which states that if a nonuniform subdivision strategy of the mesh is selected so that successive meshes tend to be of the equilibrated type, then the maximum rate of convergence of the h version will be achieved as if quasi-uniform meshes were used in a smooth enough problem, regardless of the actual smoothness of the solution.

If instead of subdividing the elements the increment on the number of nodes is obtained by increasing the order of the elements of the original mesh (p version), the following expression holds:

$$\|e\|_E \leqslant C_6 N^{-\beta} \tag{12}$$

where C_6 is a constant depending on the type of elements employed, β and u; β is a positive real number that depends on the quality of the mesh and the smoothness of the solution. For a quasi-uniform mesh and a smooth function the rate of convergence is equal to or higher than that of the h version; with the same type of mesh, if the solution is nonsmooth and corners are located in the vertices of the elements, $\beta = v$, giving a rate of convergence double than that corresponding to a quasi-uniform h version. In the case of a nonuniform mesh the rate of convergence may be higher, depending on its quality and on the type of p refinement employed, uniform or nonuniform. Among the disadvantages of the p version is the increment on the bandwidth of the matrices of the global system of equations of the discretized problem. Both the p and the h versions are limited by numerical stability and memory and storage requirements.

If the enrichment of the mesh is obtained by both, subdividing the elements and increasing the order (h–p version), the rate of convergence turns out to be

$$\|e\|_E \leqslant C_7 e^{-\alpha N^{\beta}} \tag{13}$$

where C_7 is a constant depending on the type of elements, α, β, and u; α and β are real positive numbers that depend on the smoothness of the solution and the quality of the mesh; in general, $\beta \geqslant \frac{1}{3}$. The h–p version is more difficult to implement.

The previous expressions give *a priori* error estimates. They are due to Raviart and Thomas (1983), Babuska and Suri (1987), and Babuska and Guo (1988), among others. See also the references given by these authors.

In most electromagnetic and waveguiding problems the exact solution is not sufficiently smooth because any of the possible nonsmoothness sources mentioned previously may be present, as well as proximity between conductors, or among dielectric corners and conductor wedges, and so on. Thus automatic nonuniform h, p, or h–p refinement procedures generating optimal meshes should be most convenient in order to obtain the maximum rates of convergence. A measure of the global and local error is needed. The *a priori* error estimates make use of unknown constants dependent on the exact solution of the problem; thus, in general, they can not be employed. An alternate approach is to utilize *a posteriori* error estimates that may be computed as a postprocess once an approximate solution, $\tilde{u}$, is obtained. This is the way in which self-adaptive methods, as described in Section 1, are conceived.

It must be emphasized that a self-adaptive procedure will automatically take into account any source of nonsufficient smoothness of the solution, no matter if it is originated by sharp corners or by any other cause and regardless of the actual measure of its local or global smoothness. What is only required is a measure of the error incurred and a mesh refinement strategy able to produce equilibrated meshes. As a consequence, these techniques will provide meshes well adapted to the electromagnetic field behavior.

3. A SELF-ADAPTIVE SCHEME FOR THE QUASISTATIC ANALYSIS OF TRANSMISSION LINES

Let consider a waveguiding structure with $N_c + 1$ perfect conductors ($N_c \geqslant 1$), immersed in an inhomogeneous anisotropic medium consisting of N_d dielectric regions with relative dielectric permittivity and permeability given by tensors that may be represented by 3×3 real and symmetric matrices, $\boldsymbol{\varepsilon}_\mathbf{r}(x, y)$ and $\boldsymbol{\mu}_\mathbf{r}(x, y)$, respectively, that may be decomposed in their transverse part, 2×2 real symmetric matrices, $\boldsymbol{\varepsilon}_{\mathbf{rt}}(x, y)$ and $\boldsymbol{\mu}_{\mathbf{rt}}(x, y)$, respectively, and their longitudinal coefficient, $\varepsilon_{rzz}(x, y)$ and $\mu_{rzz}(x, y)$, respectively; thus coefficients xz, yz, zx, zy are assumed to be zero. The cross section of the structure configures the domain of the problem, Ω, whose boundary is given by $\Gamma = \Gamma_c \cup \Gamma_m$, being $\Gamma_c = \cup \Gamma_{ci}$; Γ_{ci} is the contour of the ith conductor in the cross section of the structure (electric wall); and Γ_m are the symmetry walls (or magnetic walls).

As is well known, N_c quasi-TEM fundamental modes without cutoff frequency may propagate in such a structure. The reader is addressed to the works by Wei et al. (1984), Venkataraman et al. (1985), Horno et al. (1990), and Gentili and Salazar-Palma (1995), where, among others, the theory related to the characterization of the quasi-TEM modes or transmission line modes is given. See also the references given by these authors. The coefficients of the primary parameter matrices **C**, or per unit length electrostatic induction matrix (also called capacitance matrix), **L**, or per unit length inductance matrix, that characterize the transmission line, may be obtained by solving a number of electrostatic problems, each of them with a given permittivity tensor characterizing the propagating medium and given boundary conditions.

The following subsections describe the main issues related to the design of a self-adaptive mesh technique to solve any of those electrostatic problems: namely, its formulation and discretization, the definition and computation of the error estimates, or the mesh refinement strategy.

3.1. Problem Formulation and Discretization Employing Lagrange Elements

The primal formulation of the quasistatic analysis of a transmission line configuration is as follows:

"Given $\varepsilon_{rt}(x, y)$ in Ω, find the electrostatic scalar potential, $\phi(x, y)$, with $\phi = V_i$ in Γ_{ci}, for $i = 0, \ldots, N_c$, such that

$$\mathbf{\nabla_t} \cdot \boldsymbol{\varepsilon}_{\mathbf{rt}}(x, y) \mathbf{\nabla_t} \phi(x, y) = 0, \quad \text{in } \Omega \tag{14}$$

$$\mathbf{a_n} \cdot \boldsymbol{\varepsilon}_{\mathbf{rt}}(x, y) \mathbf{\nabla_t} \phi(x, y) = 0, \quad \text{in } \Gamma_m\text{''} \tag{15}$$

The dot stands for the scalar product of two vectors; $\mathbf{\nabla_t}$ means the two-dimensional gradient operator; $\mathbf{a_n}$ stands for the unit normal vector; and V_i is the voltage corresponding to the ith conductor (with $V_0 = 0$ for the reference conductor, $i = 0$).

A weak formulation may be obtained minimizing the functional:

$$W_n(\phi) = \frac{1}{2} \int_\Omega \mathbf{\nabla_t} \phi \cdot \boldsymbol{\varepsilon}_{\mathbf{rt}} \mathbf{\nabla_t} \phi \, dS \tag{16}$$

with $\phi = V_i$ in Γ_{ci}, for $i = 0, \ldots, N_c$, that gives the normalized electrostatic energy, $W_n = \varepsilon_0^{-1} W$, where ε_0 is the vacuum permittivity and W the electrostatic energy.

Triangular Lagrange elements of first and second order are used; in the last case curved elements are utilized for the analysis of structures with curved contours. All computations are performed in the master element. Real elements are related to the master element by transformations using first-order (straight elements) or second-order (curved ones) Lagrange basis functions (see Zienkiewicz and Taylor, 1989).

The discretization of the continuous problem is done following the standard finite element procedures through the Rayleigh–Ritz method. In each element a set of equations involving the so-called local stiffness matrix and the local vector of degrees of freedom is obtained. After assembling the equations of all elements and imposing the electric wall boundary conditions, a global system of equations is achieved with the unknowns being the values of the potential at each node. The global stiffness matrix is real, positive definite, symmetric, and sparse. Solvers specially designed for that type of problem are used. The Cholesky method employs a skyline storage procedure; for the conjugate gradient method only the nonzero coefficients are stored. Once the discretized problem is solved, the approximate value of the potential, $\tilde{\phi}$, at any point of a given element, may be obtained according to

$$\tilde{\phi}(x, y) = \sum_{i=1}^{n} \psi_i(x, y)\, d_i \tag{17}$$

where n is the number of nodes of the element employed, ψ_i is the ith basis function, and d_i is the ith degree of freedom. The approximate solution, $\tilde{\phi}$, belongs to $H^1(\Omega)$, being continuous across Ω, with continuous derivatives in each element and discontinuous derivatives across element edges; namely, the derivative in the normal direction will be discontinuous. This behavior introduces an error in the computed solution because the exact solution should have continuous derivatives within each dielectric region; on the other hand, it allows them to be discontinuous across dielectric interfaces, where what should be continuous is the normal component of the displacement vector, $\boldsymbol{\varepsilon}_{\mathrm{rt}}\boldsymbol{\nabla}_{\mathrm{t}}\phi$, that is the secondary variable of the problem.

3.2. *A Posteriori* Error Estimates

Different *a posteriori* error estimates have been proposed, based on principles like nodal perturbation, complementary or dual formulations (which require the problem to be solved twice), averaging or recovery techniques of the primary or secondary variable, complete or incomplete residuals, and so on. Several measures for the error have been also suggested. Adamiak (1992) does a brief review and gives references to electromagnetic applications; additional works in this field are those by Hoole et al. (1989), Drago et al. (1992), Daigang and Kexun (1994), Chellamuthu and Ida (1994), and Golias et al. (1994). See also the references given by these authors.

The objective of the problem under consideration is to obtain an approximate value of the electrostatic energy, from which the desired values of the **C** or **L** matrix coefficients will be derived. Thus the method that seems more suited and economical is that based on complete residuals employing an energy norm.

Kelly et al. (1983) demonstrate that for second-order deterministic elliptic boundary value problems, an estimate, ϵ_G, of the exact global error measured in the energy norm, $\|e\|_E$, is given by

$$\epsilon_G^2 = \sum_{i=1}^{N_e} \epsilon_i^2 \tag{18}$$

where ϵ_i is the error estimate of the ith element or local error estimate, given by

$$\epsilon_i^2 = \frac{h_i^2}{24\varepsilon_i p}\int_{\Omega_i} r_s^2\, d\Omega + \frac{h_i}{24\varepsilon_i p}\sum_{\substack{\Gamma_k\subset\Gamma_{\Omega_i}\\ \Gamma_k\not\subset\Gamma_D}}\int_{\Gamma_k} r_e^2\, d\Gamma \tag{19}$$

with h_i being the diameter of the ith element, Ω_i; ε_i, the minimum eigenvalue of the matrix, $\boldsymbol{\varepsilon}_{\mathrm{rt}}$, of the ith element; p, the order of the element employed; r_s, the surface residual; Γ_k, the kth edge of the ith element; Γ_{Ω_i}, the contour of the ith element; Γ_D, the boundaries with Dirichlet type conditions, $\Gamma_D = \Gamma_c$; and r_e, the singular residual in the element edges. The surface residual, r_s, is given by

$$r_s = -\boldsymbol{\nabla}_{\mathrm{t}}\cdot\boldsymbol{\varepsilon}_{\mathrm{rt}}\boldsymbol{\nabla}_{\mathrm{t}}\tilde{\phi}, \quad \text{in } \Omega_i,\ i = 1,\ldots,N_e \tag{20}$$

and refers to the primary variable; the edge residual, r_e, is given by

$$r_e = \mathbf{a}_{\mathrm{n}}\cdot\boldsymbol{\varepsilon}_{\mathrm{rt}}\boldsymbol{\nabla}_{\mathrm{t}}\tilde{\phi}, \quad \text{for } \Gamma_k\subset\Gamma_N; \qquad r_e = [\![\mathbf{a}_{\mathrm{n}}\cdot\boldsymbol{\varepsilon}_{\mathrm{rt}}\boldsymbol{\nabla}_{\mathrm{t}}\tilde{\phi}]\!]_k, \quad \text{for } \Gamma_k\not\subset\Gamma \tag{21}$$

where Γ_N stands for the boundaries of the domain with a Neumann type condition, namely, the magnetic walls, Γ_m; and $[\![\cdot]\!]_k$ means the jump across the kth edge of the magnitude between double brackets; r_e allude to the secondary variable.

Expression (19) has been demonstrated for isotropic structures employing rectangular Lagrange elements with h_i being the longest edge of the element and ε_i being the value of the ith element physical constant. It has been generalized to anisotropic material and triangular elements.

Once the finite element method has been applied to a given mesh, the local error estimates are obtained through a postprocess in each element according to Eq. (19), which may be written in terms of a sum of products of matrices by the local vector of degrees of freedom. The coefficients of those matrices may be calculated at the same time as those of the local stiffness matrix, because they only depend on the coordinates of the real element. Thus they may be precomputed and stored. The postprocess is reduced to identifying the local vector of degrees of freedom of each element and performing the corresponding multiplications. In order to compute the jump across edges it is also required to identify the elements sharing a given internal edge as well as edges lying in boundaries with Neumann conditions.

Kelly et al. (1983) state that the error estimate is correct if the effectivity index

$$i_e = \frac{\epsilon_G}{\|e\|_E} \tag{22}$$

is close to unity; $\|e\|_E$ is the exact value of the energy norm of the error.

3.3. Mesh Refinement Strategy

The indication of whether a given mesh should be refined or not is obtained by comparing the global error estimate, ϵ_G, with the energy norm of the computed solution, $\|\tilde{\phi}\|_E = [2\tilde{W}_n(\tilde{\phi})]^{1/2}$, where $\tilde{W}_n$ is calculated from Eq. (16), substituting $\tilde{\phi}$ for ϕ. The refinement process takes place whenever the following expression holds:

$$\frac{\epsilon_G^2}{\|\tilde{\phi}\|_E^2} \geqslant r_g \tag{23}$$

where r_g is the global refinement parameter, selected by the user, that gives the prespecified degree of accuracy; typical values are from 10^{-3} (0.1%) to 10^{-4} (0.01%).

Once the decision to enrich a mesh is taken, the next issue is to determine which elements should be refined. In order to answer this question a characteristic local error estimate, ϵ_c, may be defined as being the maximum local error estimate, $\epsilon_M = \max(\epsilon_i)$, $i = 1,\ldots,N_e$, or the optimal local error estimate, ϵ_o:

$$\epsilon_o = \left(\frac{\sum_{i=1}^{N_e} \epsilon_i^2}{N_e} \right)^{1/2} \tag{24}$$

that is, the value of the local error estimate in each element of an optimal mesh.

An element is refined whenever the following expression holds:

$$\epsilon_i^2 \geqslant r_l \epsilon_c^2 \tag{25}$$

where r_l is a local refinement parameter selected by the user, between 0 and 1.

Several enrichment procedures are available; h, p, and h–p refinement algorithms have already been mentioned; r refinement procedures consist of a relocation of the nodes of the previous mesh; r–h techniques are also possible. Another approach is the generation of a completely new mesh, whose element sizes are based on the local error estimates; this method requires a sophisticated mesh generator and is not easy to implement in existing codes. Most of the work done in electromagnetics makes use of h procedures, although the other techniques have also been implemented; see, for example, the works mentioned in Section 3.2 and those by Golias and Tsiboukis (1991), Fernandez et al. (1993), Giannacopoulos and McFee (1994), and Tang et al. (1994). See also the references given by these authors.

A selective h refinement procedure has been chosen. Elements fulfilling condition (25), called primary elements, will be subdivided into four elements. The algorithm selected to subdivide them must ensure that degenerated elements (i.e., elements with very small angles) will not be created. It should also guarantee a smooth transition between regions with elements of large and small sizes, in order to obtain an adequate condition number of the stiffness matrix. In order to obtain a conforming mesh (i.e., a mesh in which nodes placed on edges or vertices shared by more than one element are defined as nodes of all those elements), depending on the algorithm employed, it may be necessary to extend the refinement to nonprimary elements (called secondary elements) contiguous to the primary ones, which will be subdivided into two, three, or, eventually, four elements. The subdivision algorithm must also ensure no degeneracies in this process. The refinement should not be extended far away from the primary elements.

Several element subdivision algorithms are available. The most simple one generates an internal node in each primary element and connects it to its vertices; thus the refinement is not extended to other elements; the use of reshaping procedures (Delaunay or similar) tries to avoid degeneracies. Other methods giving better transition performances have been tested. A method fulfilling all the requirements mentioned in the previous paragraph has been selected. It is based on the work of Rivara (1989). If curved contours are not present, a primary element is bisected by its longest edge and the new node is connected to the midpoints of the other two edges, as shown in Figure 1a. Figure 1b shows the subdivision method employed when there are curved contours, in order to avoid degeneracies. Secondary elements are subdivided into two or three elements following the rule of bisection by the longest edge; Figure 1c corresponds to the case in which a new node has appeared in edge AC, which is not the longest edge; Figure 1d refers to the case in which a new node has appeared in the longest edge AB; eventually, a secondary element should be divided into four elements following Figure 1a or 1b.

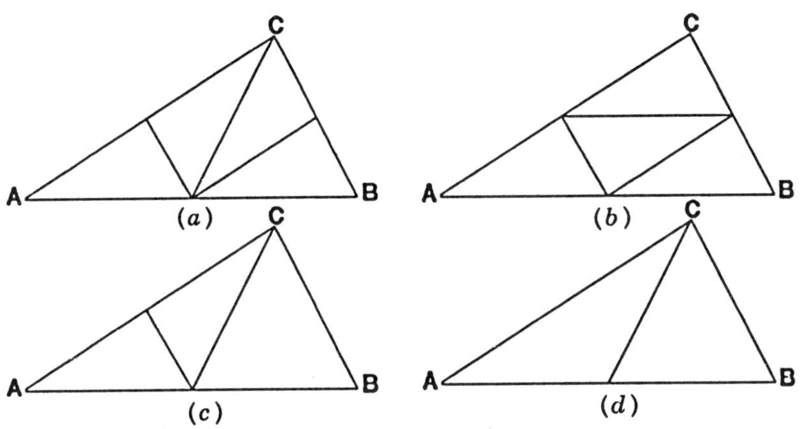

FIGURE 1. Element subdivision: (a, b) primary elements and (c, d) secondary elements.

3.4. Self-Adaptive Mesh Algorithm

Figure 2 shows the flow diagram of the previous self-adaptive procedure implemented by Salazar-Palma and Hernández-Gil (1989).

Two mesh generators have been employed to produce the initial discretization. The first one is based on subdivisions of rectangular shapes into triangles. The second one employs the Delaunay algorithm.

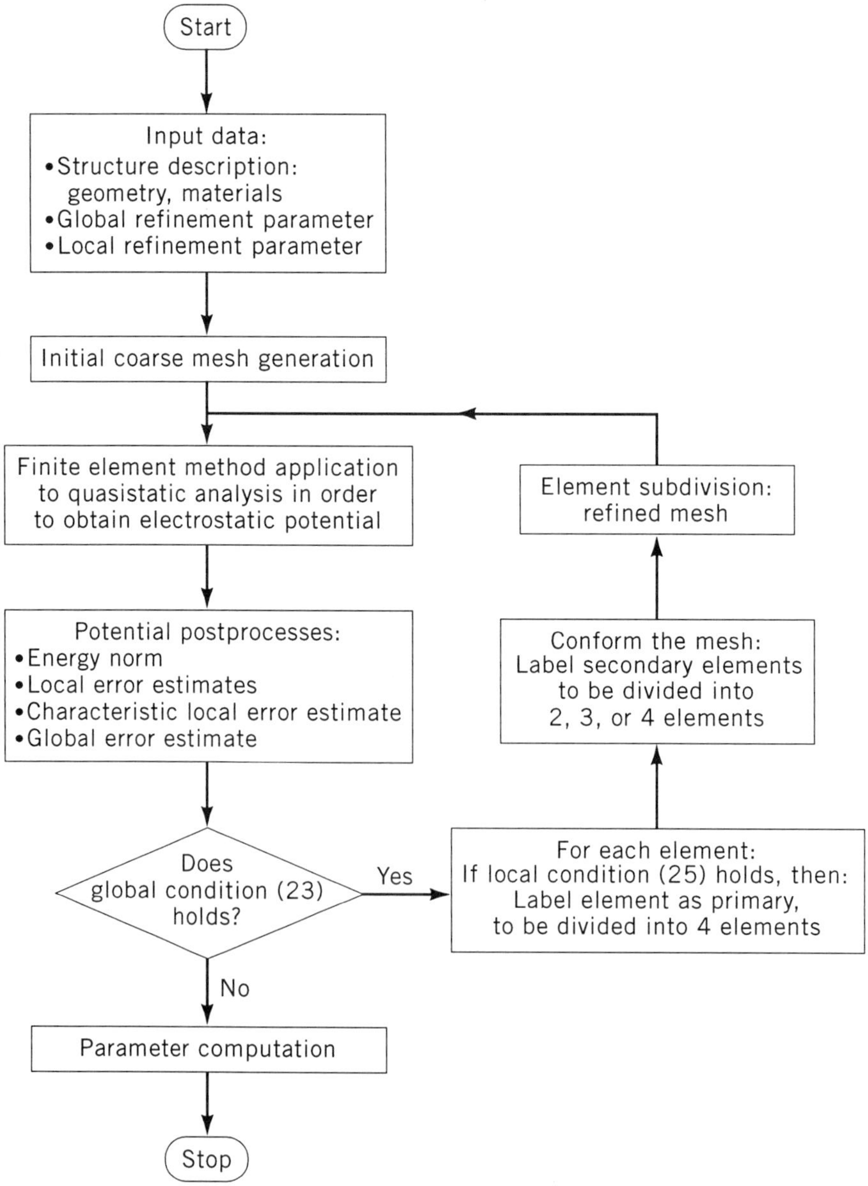

FIGURE 2. Flow diagram of the self-adaptive mesh procedure.

At each mesh step, element and node renumbering techniques are used to minimize the storage requirements. The solution corresponding to the first mesh is obtained through the Cholesky method. For the rest of the meshes a conjugate gradient procedure with Cholesky incomplete factorization preconditioning is utilized. The iterative procedure is started employing the solution of the previous mesh from which the value of the potential in the new nodes has been interpolated. Thus the method converges in a few iterations.

The stopping criterion of the self-adaptive process may be given by Eq. (23), by the relative difference between the last two computed values of the normalized energy (or the desired parameter), or by a maximum number of meshes.

3.5. Procedure Validation

The method has been validated by applying it to structures having analytical solutions, as well as to additional transmission lines analyzed by other authors and methods. The following examples have been chosen.

Square Coaxial Structure. Figure 3a shows the cross section of the structure. The exact value of the normalized capacitance, $C_n = C/\varepsilon_0 = 2W_n$, determined by Terakado (1976) using a conformal transformation is 10.234. The potential has an asymptotic behavior near a corner that may be estimated as $r^{2/3}$, where r is the distance from the corner to the point where it is computed. The electric field will be singular at corners, with an estimated asymptotic behavior given by $r^{-1/3}$. The value of v may be estimated as $\frac{2}{3}$; thus, if quasi-uniform refinements are used, the rate of convergence should be given by $N^{-1/3}$ no matter what the order of the element employed. If an adaptive procedure is utilized, a rate of $N^{-p/2}$ should be obtained.

Figure 3b shows the initial coarse mesh from which both the quasi-uniform refinement process and the self-adaptive process take place. The quasi-uniform refinements are achieved by successive subdivisions of each element into four elements. Figures 3b–f show the evolution of the mesh in the self-adaptive algorithm with $r_l = 0.5$ and second-order elements.

Figure 4 compares results for the rate of convergence of both types of analysis for first- and second-order elements. The vertical axis represents, in log scale, the energy norm of the relative error calculated as

$$(e_r)_E = \left| \frac{\tilde{W}_n - W_n}{W_n} \right|^{1/2} \tag{26}$$

where $\tilde{W}_n$ is the computed value and W_n the exact one. The horizontal axis refers, in log scale, to the number of nodes, N. The inset in Figure 4 gives the theoretical slopes of the curves when uniform refinements are used ($-\frac{1}{3}$) and

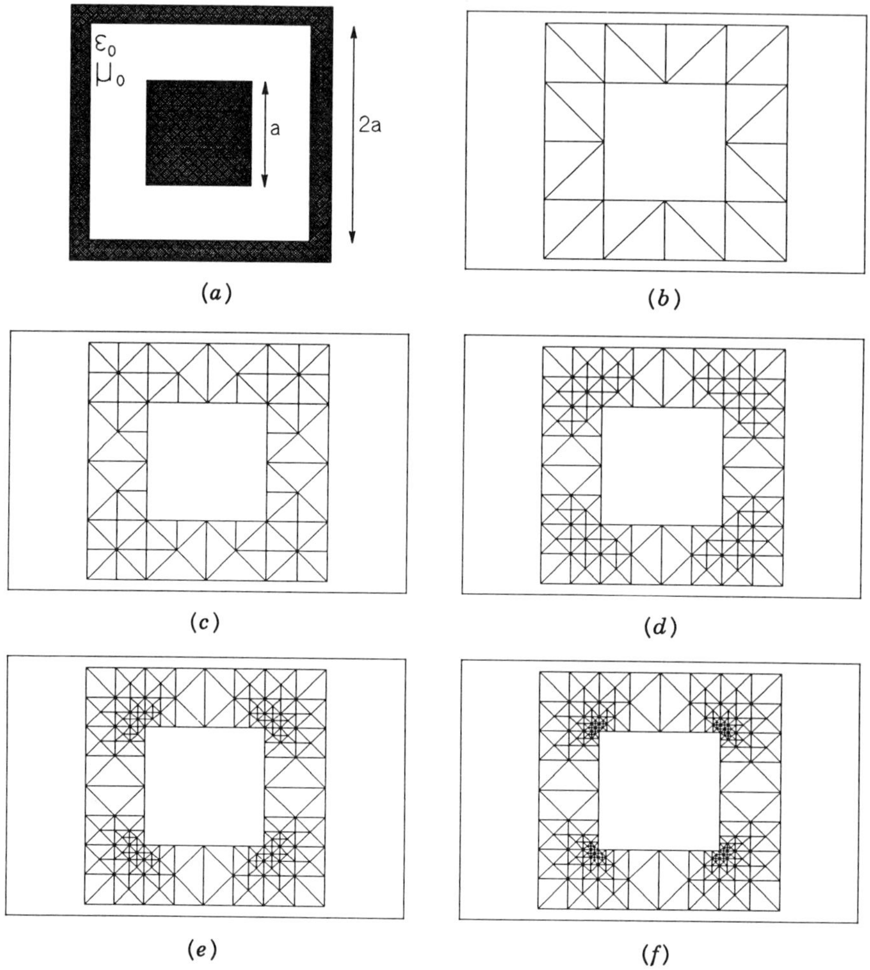

FIGURE 3. Square coaxial structure: (a) configuration and (b–f) mesh evolution for the self-adaptive procedure with $r_l = 0.5$ and second-order elements.

when the adaptive procedure is employed ($-\frac{1}{2}$ for first-order elements and -1 for second-order elements). It can be observed that with uniform refinements the rate of convergence is very close to the theoretical one for both first- and second-order elements, while, with the adaptive procedure, it immediately reaches the maximum negative values, $-\frac{1}{2}$ and -1, respectively. This result allows one to conclude that both the error estimate and the refinement algorithm have adequately been chosen. A further confirmation of those conclusions is given by the correct behavior of the effectivity index, as Figure 5 shows.

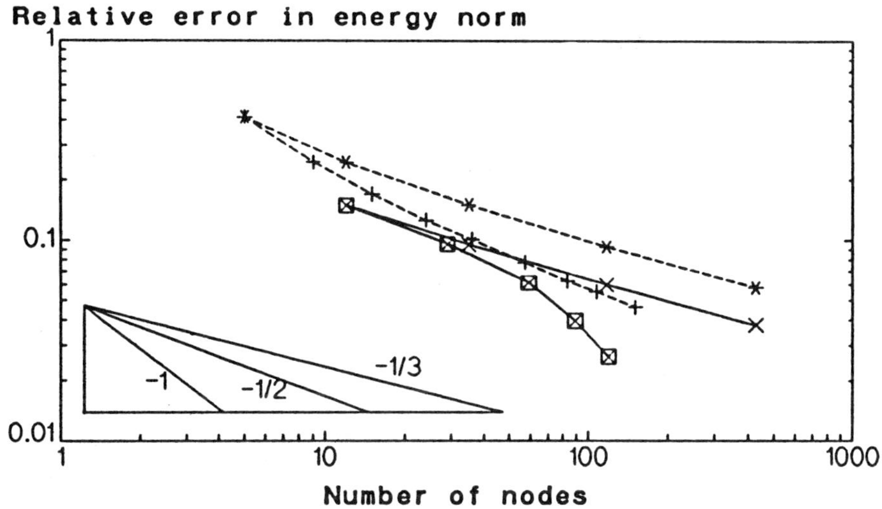

FIGURE 4. Rate of convergence for the analysis of one-eighth of the square coaxial structure. Symbols: -- * --, uniform refinement, first-order elements; -- + --, adaptive refinement, first-order elements; — × —, uniform refinement, second-order elements; — ⊠ —, adaptive refinement, second-order elements.

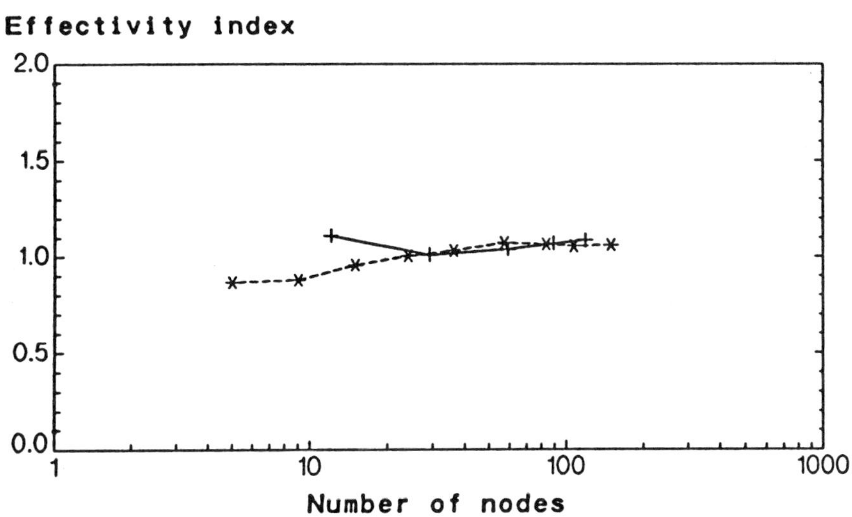

FIGURE 5. Effectivity index of the global error estimate for the self-adaptive analysis of the square coaxial structure. Symbols: -- * --, first-order elements; — + —, second-order elements.

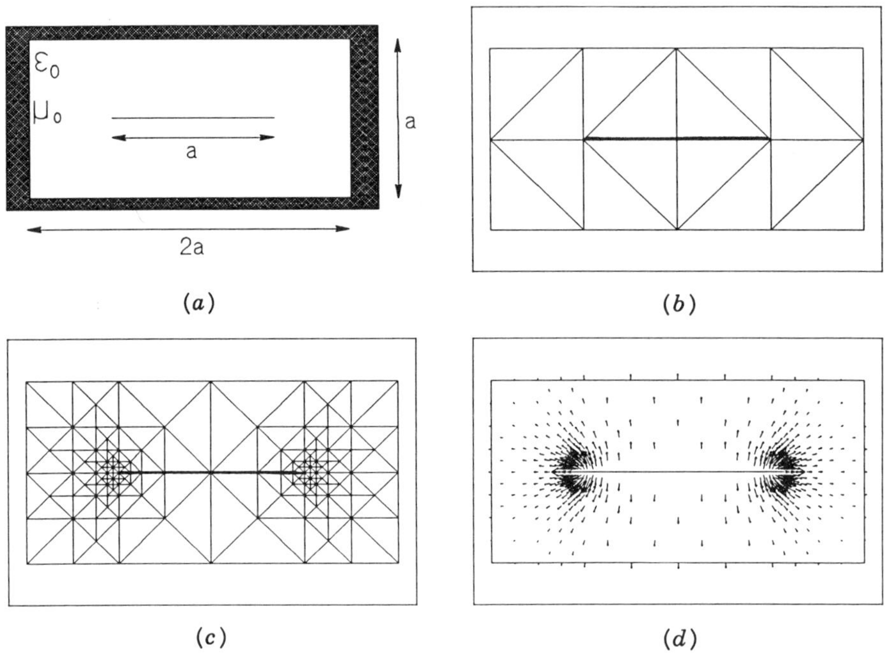

FIGURE 6. Symmetric stripline: (a) structure; (b) initial coarse mesh; (c) fifth mesh, with second-order elements and $r_l = 0.5$; and (d) electric field.

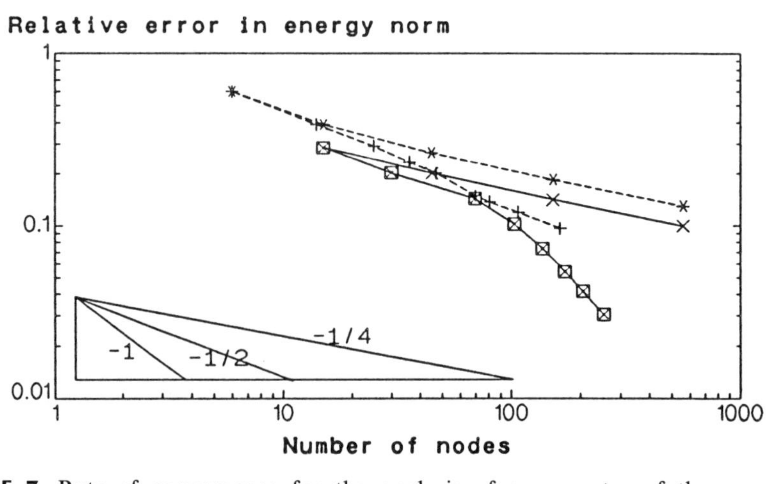

FIGURE 7. Rate of convergence for the analysis of one-quarter of the symmetric stripline. Symbols: $-- * --$, uniform refinement, first-order elements; $-- + --$, adaptive refinement, first-order elements; $— \times —$, uniform refinement, second-order elements; — ⊠ —, adaptive refinement, second-order elements.

Shielded Symmetric Stripline. Figure 6a shows the structure. The exact result for the normalized capacitance is given by Tippet and Chang (1976), using conformal transformation, as 5.87688. The potential, ϕ, has an estimated behavior near the edges of the strip given by $r^{1/2}$. The electric field is singular at the edges, behaving as $r^{-1/2}$, which is the highest possible order for a singularity. Figure 6b shows the initial mesh; the fifth step in the adaptive procedure, employing second-order elements is given in Figure 6c; Figure 6d shows the electric field; it may be observed that the mesh is well adapted to it. Figure 7 compares the rates of convergence as in the preceding example. Results confirm the conclusions of the analysis of the previous structure.

Coaxial Line Partially Filled with Dielectric. Figure 8a shows the structure configuration. It has been chosen in order to study the high-permittivity dielectric ($\varepsilon_r = 25$) corner behavior. One-eighth of the structure was discretized using second-order straight and curved elements. Figure 8b shows the initial mesh. Figure 8c displays a mesh step of the adaptive procedure; the complete structure is shown. The electric field configuration can be seen in Figure 8d. The mesh is refined near dielectric corners, where the field is singular, and in regions of high field intensity. Figure 9 compares results with those obtained

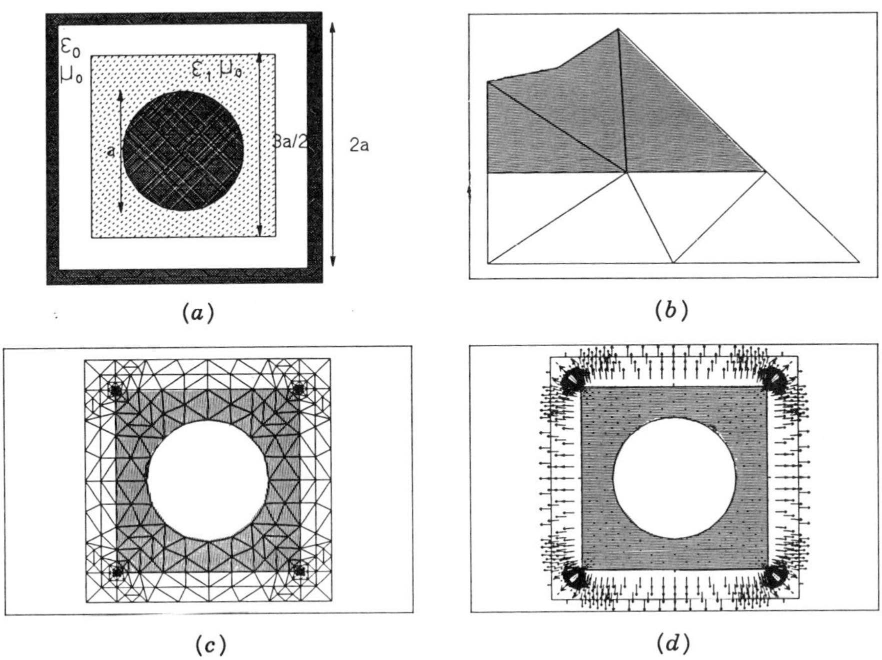

FIGURE 8. Coaxial line partially filled with dielectric: (a) structure; (b) initial mesh of one-eighth; (c) final adapted mesh (complete structure); and (d) electric field.

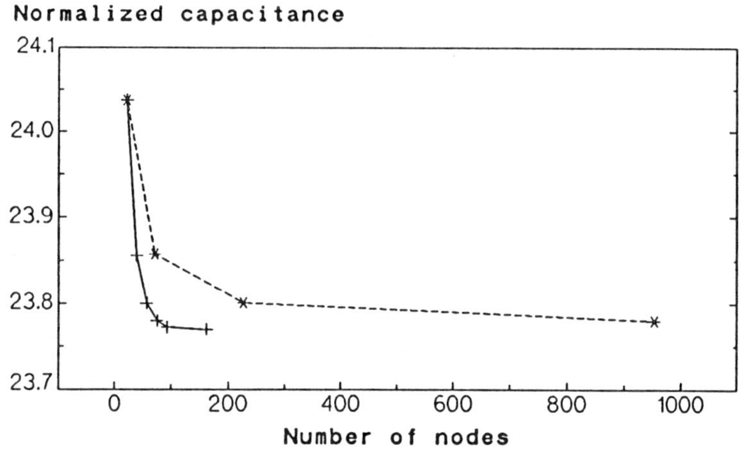

FIGURE 9. Results for the analysis of the coaxial line with dielectric corner, using second-order elements. Symbols: — + —, self-adaptive procedure; – – * – –, uniform refinement technique.

through a uniform refinement procedure. As a comparison for the overall computational time, it may be mentioned that a value of $C_n = 23.77993$ was achieved with 75 nodes and 9.66 seconds (in a PC 486, 50 MHz, 20 Mb of RAM) with the adaptive algorithm, while 953 nodes and 29.33 seconds were required to obtain $C_n = 23.7798$ with uniform refinements.

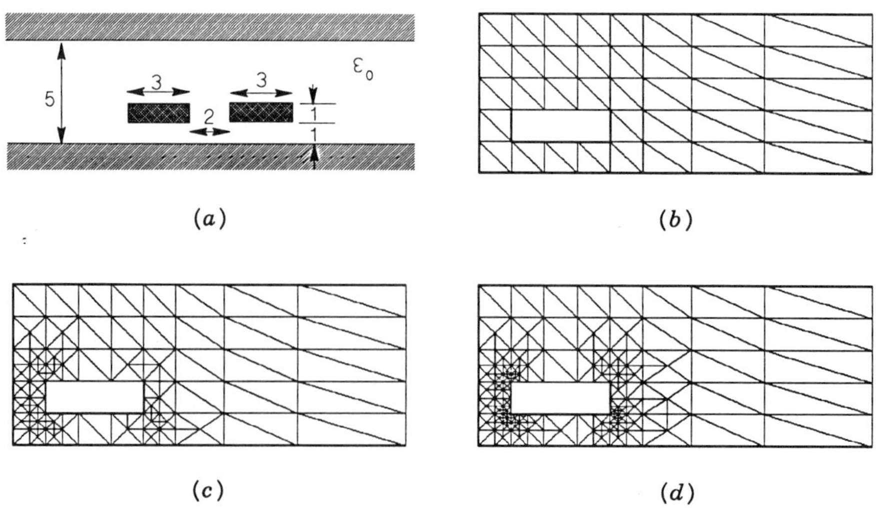

FIGURE 10. Two rectangular conductors between ground planes: (a) structure; (b) initial mesh; and (c, d) mesh steps.

TABLE 1. Results for the Capacitance Parameters of Two Rectangular Conductors Between Ground Planes

Capacitance (pF/m)	This Procedure	Wei et al. (1984)	Weeks (1970)
$C_{11} = C_{22}$	63.41	62.33	63.07
$C_{12} = C_{21}$	−5.899	−5.931	−5.724

Two Rectangular Conductors Between Infinite Ground Planes. The symmetric structure under study is shown in Figure 10a. One-half of it has been analyzed using an artificial lateral boundary sufficiently far away. Figures 10b–d give the initial mesh and the third and sixth mesh steps for the analysis of the odd mode of the coupler. It can be seen how the refinement of the mesh in the lower left corner of the active conductor is higher than in the other corners; the procedure automatically takes into account its proximity to two electric walls (the lower ground plane and the electric symmetry wall), while the other corners are near to one or none of them. The mesh is also more

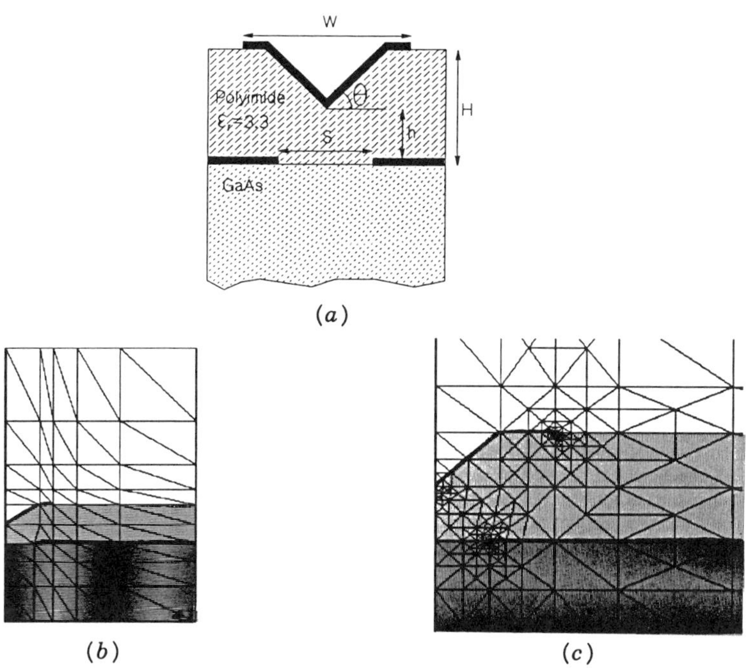

FIGURE 11. Valley microstrip: (a) structure; (b) initial mesh, $\theta = 35°$, $H = 10\,\mu$m, $h = 5\,\mu$m; and (c) zoom over the valley region in a refined mesh.

TABLE 2. Results for the Analysis of the Valley Microstrip

S (μm)	W (μm)	$Z_c(\Omega)$ This Method	$Z_c(\Omega)$ Hasegawa et al. (1992)
0	14.28	50.02	51
	24	39.9	41
	30	35.06	36
12	20	47.96	48.5
	26	41.61	42
	34	35.12	35.2
24	24	49.24	50.8
	34	39.14	40.8
	40	34.67	36

refined in the upper left corner than in the lower right one because although both of them are equally close to an electric wall, a second corner is closer to the former than to the latter. Table 1 compares results for the even and odd capacitance with those by Wei et al. (1984), employing an integral formulation and a point matching method of moments, and those by Weeks (1970), using a least-square method. Agreement is better than 3%.

Valley Microstrip. The valley microstrip has been proposed recently as an alternative to conventional microstrip in order to reduce losses. The structure is shown in Figure 11a. It is a good example of a geometry with some complexity. Figure 11b, c displays the initial mesh of half of the structure and a zoom over a given step in the self-adaptive process. A different refinement in each corner that depends not only on its interior angle and the surrounding medium but on the proximity to other corners, conductors, and dielectric interfaces leads to a well adapted mesh. Table 2 compares results for the characteristic impedance with those obtained by Hasegawa et al. (1992), also using a finite element method. Agreement is within 3%.

4. A SELF-ADAPTIVE SCHEME FOR THE DYNAMIC ANALYSIS OF WAVEGUIDING STRUCTURES

The self-adaptive procedure of the previous section may be extended to the dynamic analysis of waveguiding structures with one or more conductors and any number of dielectric regions. They may also be pure dielectric structures involving two or more dielectrics. The permittivity and permeability are given by matrices having the properties mentioned in Section 3 plus the existence of their inverses.

4.1. Problem Formulation and Discretization by Means of Lagrange Edge Elements

The problem may be formulated in terms of the electric, $\mathbf{E}e^{-j\beta z}$, or the magnetic, $\mathbf{H}e^{-j\beta z}$, field. Starting from Maxwell's equations, the well-known double-curl equations are obtained. With the previous assumptions, their operator turns out to be self-adjoint and positive definite. The selected unknown, $\mathbf{E}$ or $\mathbf{H}$, called $\mathbf{V}$ in the following, is decomposed into its transverse and longitudinal components, $\mathbf{V} = \mathbf{V}_t + jV_z\mathbf{a}_z$, where $\mathbf{a}_z$ is the unit vector in the longitudinal direction. The secondary variable, $\mathbf{H}$ or $\mathbf{E}$, is called $\mathbf{U}$. The permittivity and permeability tensors are also decomposed into their transverse and longitudinal matrices. Using the normalizations given in Table 3 the problem is defined as:

"Given $\mathbf{f}_{rt}$, f_{rzz}, $\mathbf{g}_{rt}$, g_{rzz}, and β_a, find s^2, $\mathbf{V}_{tn}(x, y)$, and $V_{zn}(x, y)$, with $\mathbf{a}_n \times \mathbf{V}_{tn} = 0$ and $V_{zn} = 0$ in Γ_D, such that

$$\nabla_t \times f_{rzz}^{-1}\nabla_t \times \mathbf{V}_{tn} + \beta_a\mathbf{a}_z \times \mathbf{f}_{rt}^{-1}\nabla_t \times V_{zn}\mathbf{a}_z - \beta_a^2\mathbf{a}_z \times \mathbf{f}_{rt}^{-1}\mathbf{a}_z \times \mathbf{V}_{tn} - s^2\mathbf{g}_{rt}\mathbf{V}_{tn} = 0 \quad (27)$$

$$\nabla_t \times \mathbf{f}_{rt}^{-1}\nabla_t \times V_{zn}\mathbf{a}_z - \beta_a\nabla_t \times \mathbf{f}_{rt}^{-1}\mathbf{a}_z \times \mathbf{V}_{tn} - s^2 g_{rzz}V_{zn}\mathbf{a}_z = 0 \quad (28)$$

TABLE 3. Variables for the Dynamic Formulation

$\mathbf{V}(x, y)$	$\mathbf{E}(x, y)$	$\mathbf{H}(x, y)$
$\mathbf{f} = f_0\mathbf{f}_r$	$\boldsymbol{\mu} = \mu_0\boldsymbol{\mu}_r$	$\boldsymbol{\varepsilon} = \varepsilon_0\boldsymbol{\varepsilon}_r$
$\mathbf{g} = g_0\mathbf{g}_r$	$\boldsymbol{\varepsilon} = \varepsilon_0\boldsymbol{\varepsilon}_r$	$\boldsymbol{\mu} = \mu_0\boldsymbol{\mu}_r$
$\mathbf{U}$	$\mathbf{H}$	$\mathbf{E}$
Γ_D	Electric walls	Magnetic walls
Γ_N	Magnetic walls	Electric walls
sign	$+1$	-1

$$\mathbf{V}_n = g_0^{1/2}\mathbf{V}, \quad \mathbf{U}_n = f_0^{1/2}\mathbf{U}$$

$$\mathbf{U}_{tn} = \frac{\text{sign}}{s}\left[\beta_a\mathbf{f}_{rt}^{-1}\mathbf{a}_z \times \mathbf{V}_{tn} - \mathbf{f}_{rt}^{-1}\nabla_t \times V_{zn}\mathbf{a}_z\right]$$

$$U_{zn}\mathbf{a}_z = \frac{\text{sign}}{s} f_{rzz}^{-1}\nabla_t \times \mathbf{V}_{tn}$$

in Ω; and

$$\mathbf{a}_\mathbf{n} \times (\mathbf{f}_\mathbf{rt}^{-1}\nabla_\mathbf{t} \times V_{zn}\mathbf{a}_\mathbf{z} - \beta_a \mathbf{f}_\mathbf{rt}^{-1}\mathbf{a}_\mathbf{z} \times \mathbf{V}_\mathbf{tn}) = 0, \quad \text{in } \Gamma_N \tag{29}$$

$$\mathbf{a}_\mathbf{n} \times (f_{rzz}^{-1}\nabla_\mathbf{t} \times \mathbf{V}_\mathbf{tn}) = 0, \quad \text{in } \Gamma_N\text{''} \tag{30}$$

where $\mathbf{f}_\mathbf{rt}$, f_{rzz}, $\mathbf{g}_\mathbf{rt}$, g_{rzz}, $\mathbf{V}_\mathbf{tn}$, V_{zn}, Γ_D, and Γ_N are given according to Table 3; $\beta_a = a\beta$ is the normalized propagation constant, with a being a normalization geometric factor and β being the propagation constant of the mode under study; $s = ak_0$ is the normalized frequency, with k_0 being the vacuum wavenumber; finally, $\times$ means the vectorial product between two vectors. An eigenvalue problem is considered, in which s^2 is the eigenvalue, the couple $\mathbf{V}_\mathbf{tn}$, V_{zn} are the eigenvector, and β_a represent input data. A weak formulation of the problem may be achieved by minimizing the real variational principle from which the above expressions are derived.

The triangular Lagrange-edge element of first order is employed. Table 4 shows its master element and its basis functions, where (λ, ξ, η) are the baricentric coordinates, with $\lambda = 1 - \xi - \eta$; and $\mathbf{a}_\mathbf{x}$ and $\mathbf{a}_\mathbf{y}$ are the unit vectors in the x and y directions, respectively. The longitudinal component, V_{zn}, is discretized using a Lagrange type approximation, while for the transverse one, $\mathbf{V}_\mathbf{tn}$, an edge $H(rot)$ one is used (see Nédélec, 1980), according to

$$\tilde{V}_{zn} = \sum_{i=1}^{3} \psi_i d_{zi}, \quad \tilde{\mathbf{V}}_\mathbf{tn} = \sum_{i=4}^{6} \mathbf{N}_\mathbf{i} d_{ti} \tag{31}$$

where d_{zi} stands for the ith degree of freedom of the Lagrange type, defined as the value of V_{zn} at the ith node, for $i = 1, 2, 3$, while d_{ti} is the ith degree of freedom of the edge $H(rot)$ type, defined as

$$d_{ti} = \text{signe}(j)L(j)\mathbf{a}_\mathbf{r}(j) \cdot \mathbf{V}_\mathbf{tn}(x_i, y_i), \quad j = i - 3 \tag{32}$$

for $i = 4, 5, 6$, where (x_i, y_i) are the ith node coordinates; signe(j) is $+1$ or -1,

TABLE 4. Lagrange Edge Master Element

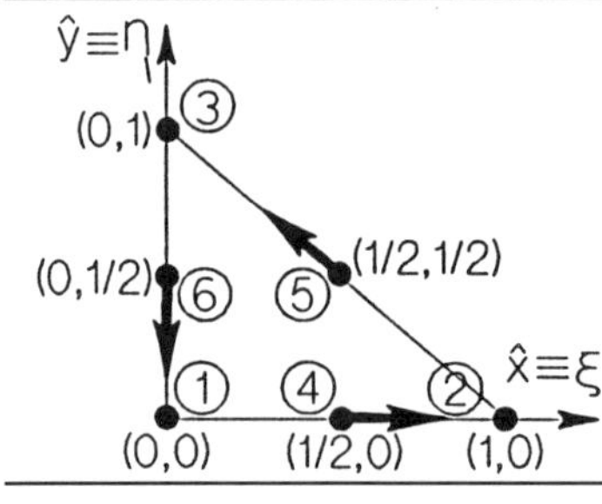

$\psi_1 = \lambda$
$\psi_2 = \xi$
$\psi_3 = \eta$
$\mathbf{N}_4 = (1 - \eta)\mathbf{a}_\mathbf{x} + \xi\mathbf{a}_\mathbf{y}$
$\mathbf{N}_5 = (-\eta)\mathbf{a}_\mathbf{x} + \xi\mathbf{a}_\mathbf{y}$
$\mathbf{N}_6 = (-\eta)\mathbf{a}_\mathbf{x} + (\xi - 1)\mathbf{a}_\mathbf{y}$

depending on the sign of the jth edge; $L(j)$ is the length of the jth edge, and $\mathbf{a}_r(j)$ is the unit vector tangential to the jth edge at the ith node.

The transformations between the master element and the real ones follow the standard procedure. Vectors are transformed according to Nédélec (1980):

$$\mathbf{v} = \mathbf{J}^{-1}\hat{\mathbf{v}} \tag{33}$$

where $\hat{\mathbf{v}}$ stands for a vector in the master element; $\mathbf{v}$ is for the transformed one in the real element and $\mathbf{J}$ is the geometric transformation Jacobean matrix.

Following the finite element standard procedures, an eigenvalue problem results involving a sparse, real, symmetric, positive-semidefinite global stiffness matrix and a positive-definite global mass matrix. It is solved using a subspace inverse iteration method specially designed for problems with those type of matrices and skyline storage. With the elements employed the nullspace of the curl–curl operator, as demonstrated by Bermúdez de Castro and Gómez Pedreira (1992), is correctly modeled, obtaining zero eigenvalue solutions that will not pollute the spectrum (the so-called spurious modes); see also Peterson (1994). The described method has been implemented by García-Castillo and Salazar-Palma (1992).

4.2. Extension of the Self-Adaptive Procedure

Following the work by Babuska and Rheinboldt (1978), it is seen that in an elliptic positive-definite eigenvalue problem, once a solution $\tilde{s}^2$, $\tilde{\mathbf{V}}_{\mathrm{tn}}$, $\tilde{V}_{zn}$ is obtained, an indicator of the error incurred in the ith element is given by

$$\epsilon_i^2 = \frac{h_i^2}{24e_i p}\int_{\Omega_i} \mathbf{r}_{\mathrm{s}}\cdot\mathbf{r}_{\mathrm{s}}\, d\Omega + \frac{h_i}{24e_i p}\sum_{\substack{\Gamma_k\subset\Gamma_{\Omega_i}\\ \Gamma_k\not\subset\Gamma_D}}\int_{\Gamma_k} \mathbf{r}_{\mathrm{e}}^*\cdot\mathbf{r}_{\mathrm{e}}\, d\Gamma \tag{34}$$

where e_i is the minimum eigenvalue of $\mathbf{f}_{\mathrm{r}}^{-1}$, and $\mathbf{r}_{\mathrm{s}}$ and $\mathbf{r}_{\mathrm{e}}$ are the surface residual vector and edge residual vector, respectively, given by

$$\mathbf{r}_{\mathrm{s}} = \mathbf{r}_{\mathrm{st}} + j r_{sz}\mathbf{a}_{\mathrm{z}} \tag{35}$$

and

$$\mathbf{r}_{\mathrm{e}} = j\mathbf{a}_{\mathrm{n}}\times\tilde{\mathbf{u}}_{\mathrm{tn}} + \mathbf{a}_{\mathrm{n}}\times\tilde{u}_{zn}\mathbf{a}_{\mathrm{z}}, \quad \text{for } \Gamma_k\subset\Gamma_N \tag{36}$$

$$\mathbf{r}_{\mathrm{e}} = [\![j\mathbf{a}_{\mathrm{n}}\times\tilde{\mathbf{u}}_{\mathrm{tn}} + \mathbf{a}_{\mathrm{n}}\times\tilde{u}_{zn}\mathbf{a}_{\mathrm{z}}]\!]_k, \quad \text{for } \Gamma_k\not\subset\Gamma \tag{37}$$

where $\mathbf{r}_{\mathrm{st}}$ and r_{sz} are given by the left-hand term of Eqs. (27) and (28), respectively, when the approximate solutions $\tilde{s}^2$ and Eq. (31) are substituted in them; $\tilde{\mathbf{u}}_{\mathrm{tn}}$ and $\tilde{u}_{zn}\mathbf{a}_{\mathrm{z}}$ are given by the expressions between parentheses of the left-hand term of Eqs. (29) and (30), respectively, that turn out to be proportional to the secondary variables, $\tilde{\mathbf{U}}_{\mathrm{tn}}$ and $\tilde{U}_{zn}\mathbf{a}_{\mathrm{z}}$, respectively (see Table 3).

Once the local error indicators are computed, the self-adaptive mesh algorithm proceeds in a similar way to that explained for the quasistatic case. It must be pointed out that the self-adaptive process should take place for each mode because their field configurations are different and from Eqs. (35), (27), and (28) it is also evident that the surface residual is a function of the corresponding eigenvalue. This self-adaptive scheme for dynamic analysis has been implemented by Salazar-Palma et al. (1994).

4.3. Procedure Validation

A number of structures have been analyzed confirming the efficiency of the method. The following cases have been selected.

Rectangular Waveguide Half-Filled with Dielectric. The structure consists of a $2a \times a$ rectangular waveguide filled in its left half with an isotropic dielectric with $\varepsilon_r = 2.25$. LSE and LSM modes may propagate; they may be analyzed with a transverse resonance method giving exact solutions for s^2 for different values of β_a, as shown by Collin (1960). This structure has been used by many authors to validate their codes and to check the appearance of spurious modes. The analysis with the finite element method described in Section 4.1 has given accurate results, confirming also what was previously mentioned regarding spurious modes. Figure 12 shows the initial mesh, a mesh step for the analysis

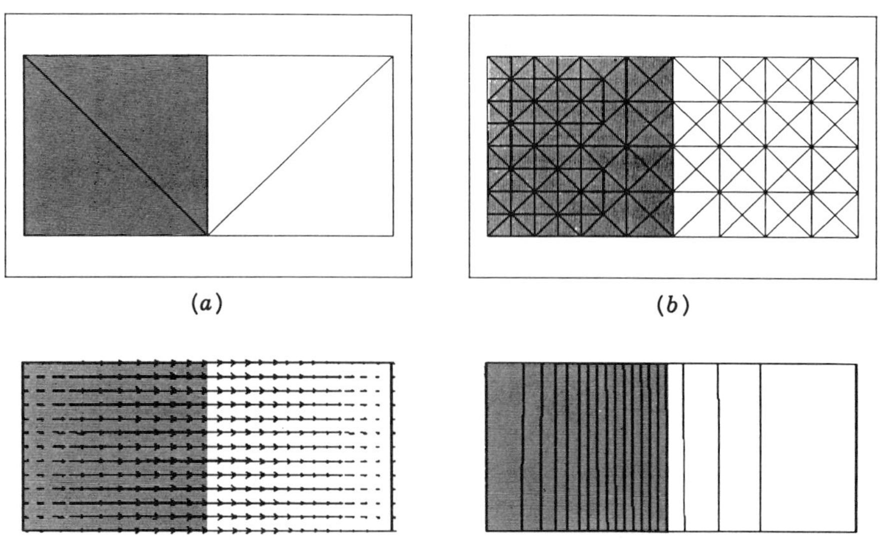

FIGURE 12. Rectangular waveguide half-filled with dielectric: (a) initial mesh; (b) mesh step for the analysis of the first mode; (c) $\mathbf{H}_t$; and (d) H_z.

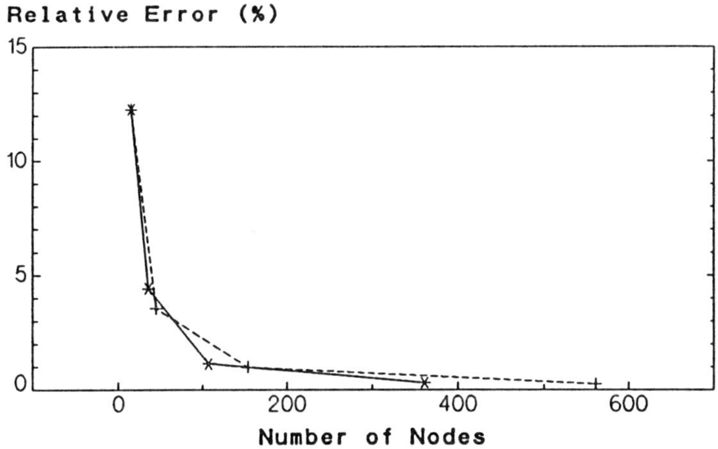

FIGURE 13. Rectangular waveguide half-filled with dielectric—first mode, relative error in s^2 ($\beta_a = 1$) versus the number of nodes. Symbols: -- + --, uniform refinement process; —∗—, self-adaptive procedure.

of the first LSE mode and its magnetic field configuration. The **H** field formulation has been used. Both components of the field have a smooth behavior. It may be seen that the mesh is refined in the region where the magnetic field intensity is greater. For the **H** formulation electric walls are boundaries with Neumann conditions not exactly satisfied. The adaptive procedure shows a greater refinement near the electric wall of the left side, indicating a corresponding greater error. Figure 13 demonstrates that convergence may be accelerated by using the self-adaptive procedure.

L-Shaped Waveguide. Figure 14a, b shows the structure under study and the initial mesh. This structure has been selected in order to analyze the behavior of the different modes at the corner and to show how the refinement of the mesh depends on the field configuration of the mode to be analyzed. Figure 15a, b gives the fourth mesh for the analysis of the first TE mode and its electric field configuration. The **E** formulation has been used, with $\beta_a = 0$. The field becomes singular at the corner and the mesh is refined accordingly. Table 5 gives results and a comparison with those obtained by Swaminathan et al. (1990), using a surface integral formulation. An agreement within 0.2% is obtained.

Figure 16a, b shows the fourth mesh for the analysis of the second TE mode and the electric field configuration, using the **E** formulation. The diagonal of the structure bisecting the corner behaves as an electric wall and, as stated by Bach Andersen and Solodukhov (1978), the electric field is not singular at the wedge. The mesh is refined in the areas of higher field intensity (at both sides of the diagonal).

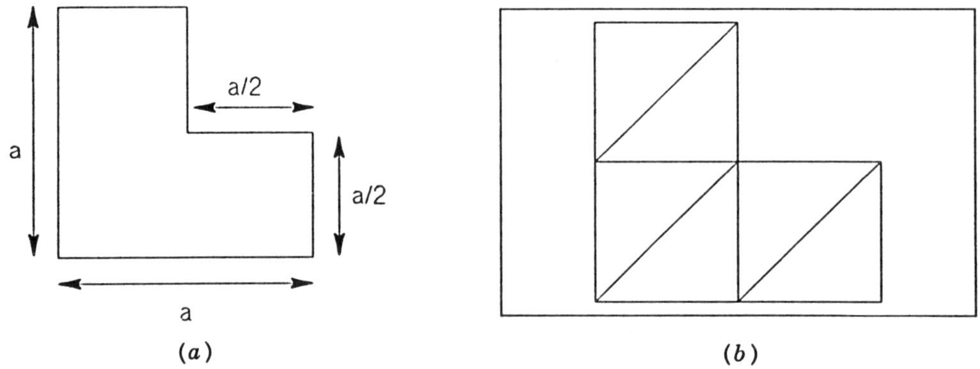

FIGURE 14. L-shaped waveguide: (a) structure and (b) initial mesh.

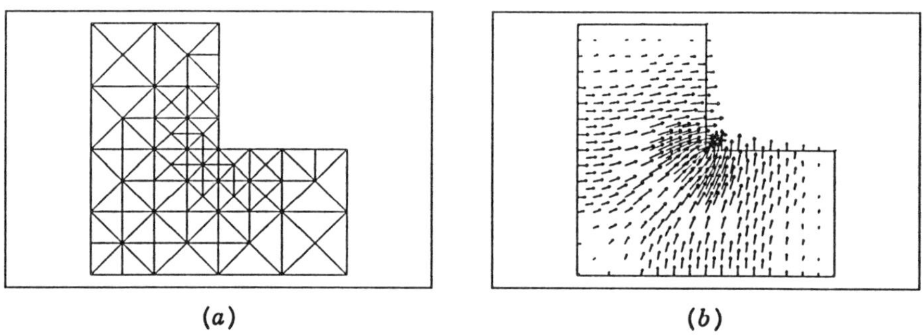

FIGURE 15. L-shaped waveguide, first TE mode: (a) fourth mesh and (b) electric field.

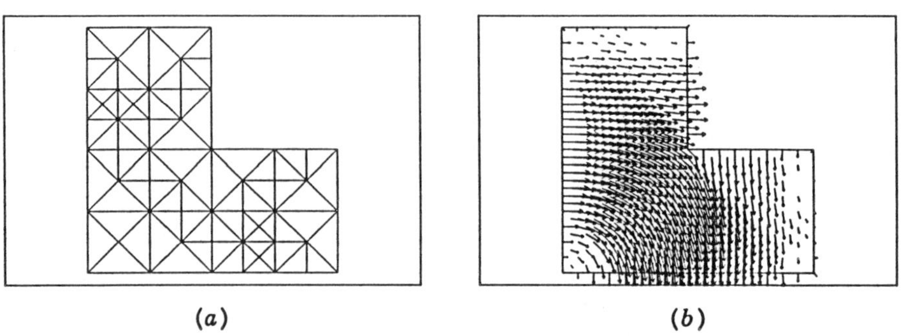

FIGURE 16. L-shaped waveguide, second TE mode: (a) fourth mesh and (b) electric field.

TABLE 5. L-Shaped Waveguide: Results for the Square of the Cutoff Wavenumber, k_c^2, of the First TE Mode, with $a = 1.27$

Mesh	Number of Nodes	k_c^2
1	21	2.681323
2	41	3.303096
3	145	3.516722
4	227	3.574351
5	351	3.585899
Swaminathan et al. (1990)		3.578528

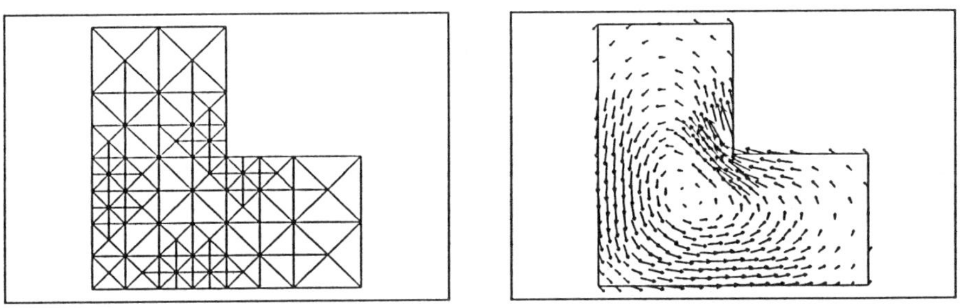

(a) (b)

FIGURE 17. L-shaped waveguide, first TM mode: (a) fourth mesh and (b) magnetic field.

TABLE 6. L-Shaped Waveguide: Results for the Square of the Cutoff Wavenumber, k_c^2, of the First TM Mode, with $a = 1.27$

Mesh	Number of Nodes	k_c^2
1	21	29.75980
2	65	25.73466
3	209	24.50415
4	605	24.44148
5	835	23.46113
Swaminathan et al. (1990)		23.694503

Figure 17a, b gives the fourth mesh for the analysis of the first TM mode and the magnetic field configuration. The **H** formulation has been used, with $\beta_a = 0$. It is again observed that the mesh is refined in accordance with field behavior. Note the greater refinement in areas of higher field intensity near the conductors that are now boundaries with Neumann conditions. Table 6 gives results and compares them with those obtained by the authors previously mentioned.

Shielded Microstrip. The structure under study is shown in the inset of Figure 20; dimensions are $b = 10$ mm, $d = 0.635$ mm, $h = 6.35$ mm, $w = 3$ mm, $t = 0.3$ mm, and $\varepsilon_r = 9.8$. Three cases for its cross section have been considered (rectangular, $\theta = 90°$; trapezoidal, $\theta = 45°$; and $\theta = 135°$). One-half of the structure has been analyzed in order to study the quasi-TEM fundamental

(*a*) (*b*)

FIGURE 18. Shielded microstrip, $\theta = 90°$: (a) initial mesh and (b) zoom over the last mesh (fifth one, 501 nodes).

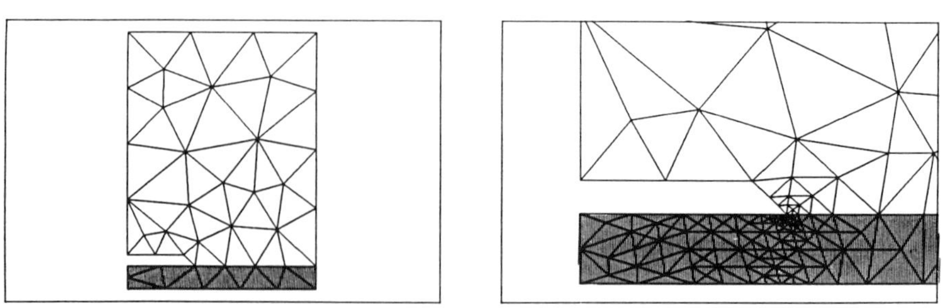

(*a*) (*b*)

FIGURE 19. Shielded microstrip, $\theta = 135°$: (a) initial mesh and (b) zoom over the last mesh (sixth one, 537 nodes).

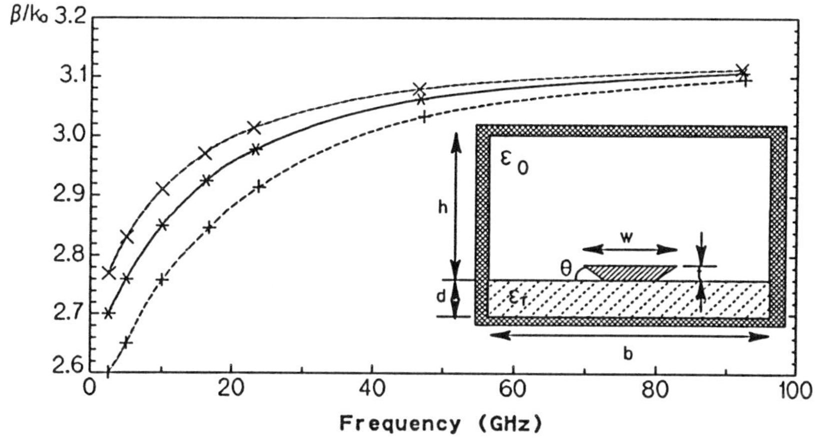

FIGURE 20. Shielded microstrip, dispersion of β/k_0. Symbols: – –, $\theta = 45°$; —, $\theta = 90°$; · – ·, $\theta = 135°$. Lines: Alam et al. (1994) and Olyslager et al. (1993) Markers: this method.

mode. The **H** formulation has been employed. The transverse magnetic field will be singular at the corners. Figure 18a, b refers to the case of a rectangular cross section. Figure 19a, b corresponds to the trapezoidal case with $\theta = 135°$. Both figures show the initial mesh and a zoom over the last mesh generated by the self-adaptive algorithm with a prespecified accuracy of 0.01%. In accordance with field behavior, the refinement in the lower corner of the microstrip is higher than in the upper one in both cases. The analysis of the trapezoidal cross section requires one more step than the rectangular one in accordance with the higher order of the singularity of the lower corner of the strip. Figures 20 and 21 give results for the dispersion of the normalized propagation constant and characteristic impedance and compare them with those obtained by Alam et al. (1994) (using also a finite element method but with a 12-node Lagrange edge element and a higher total number of nodes) and Olyslager et al. (1993) (employing a spectral domain method of moments). Agreement is excellent.

Anisotropic Coplanar Waveguide. The structure is shown in the inset of Figure 22, which gives results for the dispersion of $\varepsilon_{\text{eff}}\mu_{\text{eff}} = (\beta/k_0)^2$ of the fundamental quasi-TEM mode, for four different substrates including the case of dielectric and magnetic biaxial anisotropy (diagonal matrices). In Figure 22 those matrices are given by $(\xi_{rxx}, \xi_{ryy}, \xi_{rzz})$, where ξ stands for either the permittivity or the permeability. Dimensions are $c_1/a = 0.4$, $c_2/a = 0.1$, $c_3/a = 0.5$, $d_1 = d_2 = d_3 = d_4 = d_5 = b/5$; the shielding waveguide is the WR28: $a = 7.112$ mm, $b = 3.556$ mm. Excellent agreement is shown between results obtained with the finite element method and those reported by

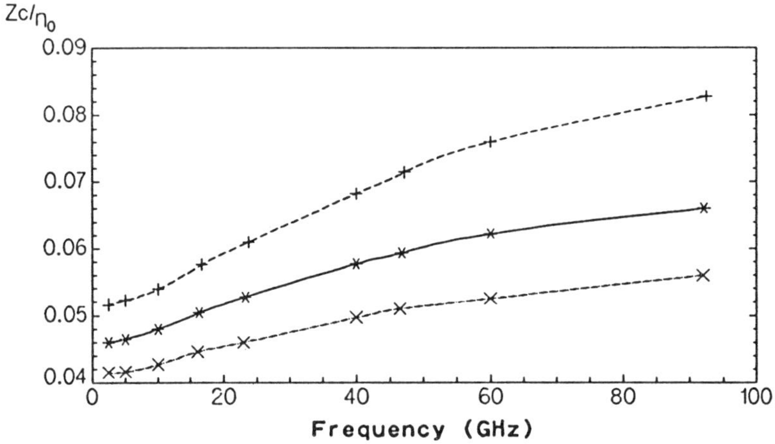

FIGURE 21. Shielded microstrip, dispersion of the normalized characteristic impedance, Z_c/η_0, where $\eta_0 = (\mu_0/\varepsilon_0)^{1/2}$. Symbols: – –, $\theta = 45°$; —, $\theta = 90°$; · – ·, $\theta = 135°$. Lines: Alam et al. (1994) and Olyslager et al. (1993). Markers: this method.

Mazé-Merceur et al. (1993) using a spectral domain procedure. Figure 23a shows the initial mesh employed for the self-adaptive technique. Figure 23b, c refers to the analysis of the case with dielectric and magnetic anisotropy using the **E** formulation. Figure 23d, e corresponds to the analysis employing the **H** formulation. In both cases the mesh is well adapted to the field configuration.

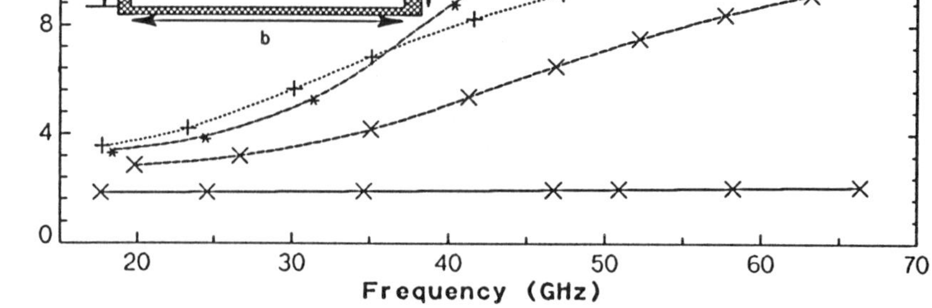

FIGURE 22. Coplanar waveguide, dispersion of $\varepsilon_{\text{eff}}\mu_{\text{eff}}$ of the fundamental mode. Symbols: · – ·, $\varepsilon_r = (3, 5, 3)$, $\mu_r = (5, 1, 1)$; ···, $\varepsilon_r = 3$, $\mu_r = 5$; – –, $\varepsilon_r = 3$, $\mu_r = (5, 1, 1)$; —, $\varepsilon_r = 3$, $\mu_r = 1$. Lines: Mazé-Merceur et al. (1993) Markers: this method.

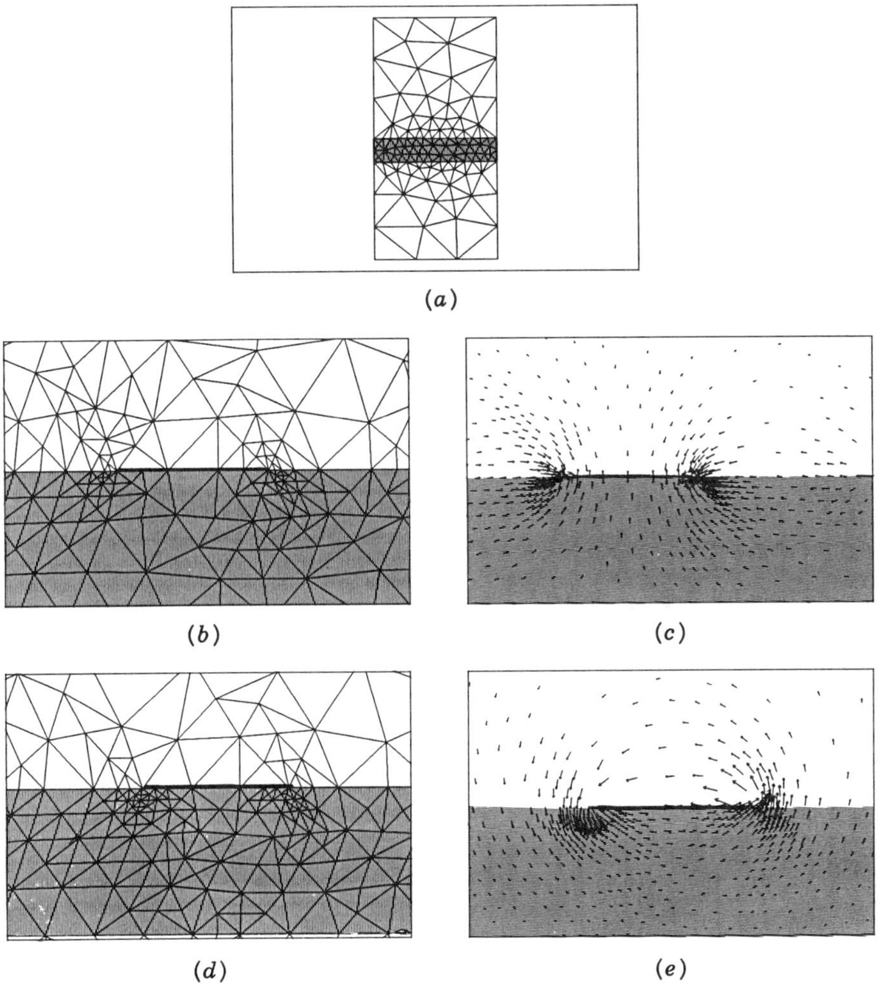

FIGURE 23. Anisotropic coplanar waveguide. (a) Initial mesh. Zoom over the strip region for (b) fourth mesh, **E** formulation; (c) $\mathbf{E}_t$; (d) fourth mesh, **H** formulation; and (e) $\mathbf{H}_t$.

5. CONCLUDING REMARKS

A self-adaptive mesh procedure for the finite element analysis of waveguiding structures has been outlined in Sections 3 and 4. It provides a flexible, powerful, and efficient numerical tool able to:

- automatically generate optimal meshes, with local errors of the same order of magnitude and well adapted to the field configuration, starting from coarse discretizations of the domain;

- give the maximum rate of convergence regardless of the smoothness of the exact solution of the problem;
- produce results with a prespecified degree of accuracy;
- optimize the overall computational time in the application of the finite element method.

Although a concrete technique for two-dimensional problems has been described, its features are common to any self-adaptive mesh algorithm. These procedures turn out to be essential tools in obtaining user-friendly software codes for the analysis of one-, two-, and three-dimensional electromagnetic problems.

REFERENCES

Adamiak, K. (1992). Local error indicator in finite element analysis of Laplacian fields based on the Green integration formula. *Int. J. Numer. Methods Eng.* **33**, 1625–1642.

Alam, M. S., Hirayama, K., Hayashi, Y., and Koshiba, M. (1994). Analysis of shielded microstrip lines with arbitrary metallization cross section using a vector finite element method. *IEEE Trans. Microwave Theory Techn.* **MTT-42**, 2112–2117.

Babuska, I. (1977). Singularity problems in the finite element method. In *Formulations and Computational Algorithms in Finite Element Analysis* (K. J. Bathe, J. T. Oden, and W. Wurdenlich, eds.), pp. 748–792. MIT Press, Cambridge, MA.

Babuska, I., and Guo, B. Q. (1988). The *h–p* version of the finite element method for domains with curved boundaries. *SIAM J. Numer. Anal.* **25**, 837–861.

Babuska, I., and Rheinboldt, W. C. (1978). A posteriori error estimates for the finite element method. *Int. J. Numer. Methods Eng.* **12**, 1597–1615.

Babuska, I., and Suri, M. (1987). The *h–p* version of the finite element method with quasiuniform meshes. *RAIRO Mathe. Modell. Numer. Anal.* **21**(2), 199–238.

Bach Andersen, J., and Solodukhov, V. V. (1978). Field behavior near a dielectric wedge. *IEEE Trans. Antennas Propag.* **AP-26**, 598–602.

Bermúdez de Castro, A., and Gómez Pedreira, M. D. (1992). Mathematical analysis of a finite element method without spurious solutions for computation of dielectric waveguides. *Numer. Mathe.* **61**, 39–57.

Chellamuthu, K. C., and Ida, N. (1994). *A posteriori* element by element local error estimation technique and 2D & 3D adaptive finite element mesh refinement. *IEEE Trans. Magn.* **MAG-30**, 3527–3530.

Collin, R. E. (1960). *Field Theory of Guided Waves.* McGraw-Hill, New York.

Daigang, W., and Kexun, J. (1994). *p*-version adaptive computation of FEM. *IEEE Trans. Magn.* **MAG-30**, 3515–3518.

Drago, G., Molfino, P., Nervi, M., and Repetto, M. (1992). A "Local field error problems" approach for error estimation in finite elements analysis. *IEEE Trans. Magn.* **MAG-28**, 1743–1746.

Fernandez, F. A., Yong, Y. C., and Ettinger, R. D. (1993). A simple adaptive mesh generator for 2-D finite element calculations. *IEEE Trans. Magn.* **MAG-29**, 1882–1885.

García-Castillo, L. E., and Salazar-Palma, M. (1992). A non-standard finite element method for the dynamic analysis of microwave waveguiding and transmission line structures. *Conf. Proc.—Eur. Microwave Conf., 22nd, Proc.* pp. 1012–1017.

Gentili, G. G., and Salazar-Palma, M. (1995). The definition and computation of modal characteristic impedance in quasi-TEM coupled transmission lines. *IEEE Trans. Microwave Theory Tech.* **MTT-45**, 338–343.

Giannacopoulos, D., and McFee, S. (1994). Towards optimal h–p adaptation near singularities in finite element electromagnetics. *IEEE Trans. Magn.* **MAG-30**, 3523–3526.

Golias, N. A., and Tsiboukis, T. D. (1991). Adaptive refinement in 2-D finite element applications. *Int. J. Numer. Modell: Electron. Networks, Devices Fields* **4**, 81–95.

Golias, N. A., Tsiboukis, T. D., and Bossavit, A. (1994). Constitutive inconsistency: Rigorous solution of Maxwell equations based on a dual approach. *IEEE Trans. Magn.* **MAG-30**, 3586–3589.

Hasegawa, T., Banba, S., and Ogawa, H. (1992). A branchline hybrid using valley microstrip lines. *IEEE Microwave Guided Wave Lett.* **2**, 76–78.

Hoole, S. R. H., Jayakumaran, S., Ananadaraj, A. W., and Hoole, P. R. (1989). Relevant, purpose based error criteria for adaptive finite element mesh generation. *J. Electromagn. Waves Appli.* **3**, 167–177.

Horno, M., Mesa, F. L., Medina, F., and Marqués, R. (1990). Quasi-TEM analysis of multilayered, multiconductor coplanar structures with dielectric and magnetic anisotropy including substrate losses. *IEEE Trans. Microwave Theory Techn.* **MTT-38**, 1059–1068.

Kelly, D. W., Gago, J. P. de S. R., Zienkiewicz, O. C., and Babuska, I. (1983). *A posteriori* error analysis and adaptive processes in the finite element method. Part I. Error analysis. *Int. J. Numer. Methods Eng.* **19**, 1593–1619.

Mazé-Merceur, G., Tedjini, S., and Bonnefoy, J. L. (1993). Analysis of a CPW on electric and magnetic biaxial substrate. *IEEE Trans. Microwave Theory Techn.* **MTT-41**, 457–461.

Nédélec, J. C. (1980). Mixed finite elements in R^3. *Num. Math.* **35**, 315–341.

Olyslager, F., De Zutter, D., and Blomme, K. (1993). Rigorous analysis of propagation characteristics of general lossless and lossy multiconductor transmission lines in multilayered media. *IEEE Trans. Microwave Theory Tech.* **MTT-41**, 79–88.

Peterson, A. F. (1994). Vector finite element formulation for scattering from two-dimensional heterogeneous bodies. *IEEE Trans. Antennas Propag.* **AP-42**, 357–365.

Raviart, P. A., and Thomas, J. M. (1983). *Introduction à l'analyse numérique des equations aux dérivées partielles.* Masson, Paris.

Rivara, M. C. (1989). Selective refinement/derefinement algorithms for sequences of nested triangulations. *Int. J. Numer. Methods Eng.* **28**, 2889–2906.

Salazar-Palma, M., and Hernández-Gil, J. F. (1989). Self adaptive mesh scheme for the finite element analysis of anisotropic multiconductor transmission lines. *IEEE MTT-S Int. Symp. Dig.*, pp. 507–510.

Salazar-Palma, M., García-Castillo, L. E., and Gentili, G. G. (1994). A software package for accurate computation of frequency dependent propagation and circuital parameters of inhomogeneous anisotropic arbitrary shaped multiconductor transmission lines. *Cont. Proc.—Eur. Microwave Conf., 24th*, Vol. 2, pp. 1709–1714.

Swaminathan, M., Arvas, E., Sarkar, T. K., and Djordjevic, A. R. (1990). Computation of cutoff wavenumbers of TE and TM modes in waveguides of arbitrary cross sections using a surface integral formulation. *IEEE Trans. Microwave Theory Tech.* **MTT-38**, 154–159, see comments, pp. 1761–1762.

Tang, Y. Q., Liang, Y. P., Wu, H. R., Meng, X. L., Xu, S. C., and Fan, G. X. (1994). An adaptive finite element computation of h–p version for magnetic field problems. *IEEE Trans. Magn.* **MAG-30**, 3519–3522.

Terakado, R. (1976). The characteristic impedance of rectangular coaxial line with ratio 2:1 of outer-to-inner conductor side length. *IEEE Trans. Microwave Theory Tech.* **MTT-24**, 124–125.

Tippet, J. C., and Chang, D. C. (1976). A new approximation for the capacitance of a rectangular-coaxial-strip transmission line. *IEEE Trans. Microwave Theory Tech.* **MTT-24**, 602–604.

Venkataraman, J., Rao, S. M., Djordjevic, A. R., Sarkar, T. K., and Naiheng, Y. (1985). Analysis of arbitrarily oriented microstrip transmission lines in arbitrarily shaped dielectric media over a finite ground plane. *IEEE Trans. Microwave Theory Tech.* **MTT-33**, pp. 952–959.

Weeks, W. T. (1970). Calculation of coefficients of capacitance of multiconductor transmission lines in the presence of a dielectric interface. *IEEE Trans. Microwave Theory Tech.* **MTT-18**, 35–43.

Wei, C., Harrington, R. F., Mautz, J. R., and Sarkar, T. K. (1984). Multiconductor transmission lines in multilayered dielectric media. *IEEE Trans. Microwave Theory Tech.* **MTT-32**, 439–450.

Zienkiewicz, O. C., and Taylor, R. L. (1989). *Finite Element Method.* McGraw-Hill, New York.

CHAPTER SEVENTEEN

Software Considerations

BILL TROWBRIDGE

1. INTRODUCTION: REQUIREMENTS FOR MICROWAVE ENGINEERING

That no single method will suffice for solving all the field problems associated with microwave engineering appears to be the common experience—a view supported by the very real industrial need for the various techniques described in the earlier chapters of this book. Thus frequency domain methods based on differential equations are well suited to solving inhomogeneous nonlinear problems but are less suited to modeling open boundary problems. Whereas an integral method, like the method of moments, copes naturally with the far-field region, it becomes increasingly hard to implement when dealing with inhomogeneous problems. Nevertheless, while it is abundantly clear in practice that particular problems are most effectively solved by special methods, nearly all the techniques described in the literature are claimed to be general methods and the reasons for specialization are often attributed to limitations in computing resource. On the one hand, to discretize the open space to a sufficiently far distance when solving a problem using a differential formulation will clearly exceed, sooner or later, the computer memory available, on the other hand, the corresponding computing power required to solve a fully populated matrix arising in an integral formulation will rapidly become prohibitive.

Therefore, to be able to deal adequately with the wide range of problems arising in microwave engineering, it is neccssary that the engineer has access to codes utilizing a range of methods. These may be used specifically or in

Finite Element Software for Microwave Engineering, Edited by Tatsuo Itoh, Giuseppe Pelosi, and Peter P. Silvester.
ISBN 0-471-12636-5

TABLE 1. Numerical Methods for Microwave Engineering

Field Solvers	Class	Interior	Exterior
Finite difference–time domain (FD–TD)	Differential time domain; no matrix inversion	Well suited for orthogonal geometry	Requires radiation-absorbing boundary conditions
Transmission-line modeling (TLM)	Differential time domain; no matrix inversion; circuit modeling	Well suited for orthogonal geometry, readily couples to circuits	Requires radiation-absorbing boundary conditions
Method of moments (MM)	Integral frequency domain; dense matrix	Well suited to homogeneous linear materials	Well suited
Finite element (FE)	Differential normal frequency domain; sparse matrix	Well suited to inhomogeneous nonlinear materials	Requires radiation-absorbing boundary conditions

combination as the problem dictates. For example, the finite element method can be used for analyzing an open-ended dielectrically loaded waveguide with a specified waveshape at one end provided the code has facilities for applying an absorbing boundary condition to model the open end. Alternatively, a hybrid finite element–moment method code could be used to deal with the open boundary directly. In either approach, if the problem has fast transients or time-dependent parameters, then the use of a time domain method is probably indicated; but if the steady-state response is required at only a few frequencies, then a frequency domain method is likely to be preferred.

In Table 1 are shown the basic methods with some indications of their strengths and weaknesses. The term "finite element" is often used to describe all methods (differential, integral) that discretize the spatial continuum into subregions, either meshes or elements, but an alternative meaning also remains widely understood where the term is reserved for differential methods discretized into elements.

From the software structure point of view, despite the diversity in both application and method, there are, fortunately, many common features. In broad terms a design for a software package should address the specification of the device's geometric and material properties, allow for a flexible interface to the selected solution technique, and provide appropriate and relevant procedures for the analysis and processing of results. These issues commonly known as pre- and postprocessing will be dealt with in more detail in Section 3.

There are three main categories of computer software for electromagnetic analysis and synthesis:

1. *Problem-specific* codes that solve a particular problem.
2. *Device-oriented* programs that are customized to investigate a single class of device but allow a limited variation in geometry and materials.
3. *General-purpose* packages that claim to solve for a whole range of problems with arbitrary geometric and material specification.

In the next section these main categories are discussed in more detail by considering the role and requirements of users. Some newer developments in software structures in the area of optimization are outlined in Section 4; some opinions on the impact of advanced hardware systems, from supercomputers down to personal computers, will be given in Section 5. Finally, marketing trends will be discussed in Section 6.

2. USERS AND CODES

2.1. Taxonomy of Code Developers

Electromagnetics applications are widespread in industry, national research laboratories, and academia and these institutions, over the past twenty years of so, have been responsible for the production of vast amounts of computer code, mostly now obsolete because it is a rapidly changing subject. Nevertheless, this activity continues unabated, although the impact of commercial systems has meant that code production has become better focused. The main categories of software are as follows.

Research. This category is concerned with *ad hoc* computer programs for solving problems that arise during research and development. This software can normally only be used by the researchers themselves and has little value outside the particular application that prompted the development in the first place. Occasionally, original and far-reaching ideas can emerge though, and it must be acknowledged that some of the more important developments in the past have started this way. Unfortunately, the priorities of the research militate against the easy transfer of the routines to other applications; for instance, the problem geometry is often "hard wired" and the use of standard procedures in databases, numerical methods, and graphics will not in general be employed. The positive aspect of all this is clear though — the user usually gets what he/she wants!

Shareware. Sometimes the developer of research software can offer routines to the community by the "shareware" route. Thus specialized research algorithms can be placed in subject-based special interest groups on the Internet or a bulletin board so that workers sharing similar interests can exchange software tools. For such routines to be of use, the originator has to supply at least some documentation and maybe even some support. For the subject area under consideration here, the most successful programs on offer are either utilities of general use or modules that are designed to interface with widely used existing packages. The former case includes topics like the coding of analytic solutions, conformal mapping routines and postprocessing "add-ons" like animation software; the latter includes function-specific processors (modelers, meshers, solvers, etc.) written to conform to standard databases or company-specific databases to which commercial software can interface; for example, a three-dimensional geometric modeler that is STEP (**St**andard for the **E**xchange of **P**roduct data) compliant [1].

Commercial. Software that is available commercially can be subdivided into two classes: first, application-specific or packaged software that is designed to solve a subset of the field equations; and second, a set of software tools that can be used to customize solutions or indeed partially build a package, for example, the NAG library [2]. The distinction, however, is blurred since users in the first class may well use the second; moreover, for the tools to be useful a degree of packaging is often required. Both classes have to be thoroughly documented and supported and almost certainly their users need training. An extreme example of the first class would be a complete system to solve Maxwell's equations for any realistic geometry and materials. This ideal is hardly possible with current technology and compromises have to be made. For a recent survey of the functionality of commercial codes see Molinari (1990).

2.2. User Expectations

Users of the above resources fall into two main divisions. On the one hand, there are the software developers, who are experts in both electromagnetic theory and numerical methods and are involved in the research itself of new computational tools for analysis and synthesis of devices. This group may well use commercial software for components that are adequate for their needs; for example, if the main purpose of the research is to develop an improved method for determining modes in a microwave cavity then a standard package for data preparation and analysis of results would be sufficient. On the other hand, there are the users of commercial software who are nonexperts in the numerical technology and may have a narrow knowledge of the underlying algorithms. Engineers of this type are to be found in industry and they often require field solutions in order to develop the design and evaluation of engineering devices. Consequently, the expectations of the two groups are very different.

3. ELEMENTS OF THE CLASSIC PACKAGE FOR ELECTROMAGNETIC ANALYSIS

3.1. Electromagnetic Computing Environment

A simplified outline diagram of a software environment, appropriate for the analysis of electromagnetic fields, is shown in Figure 1. The simple heuristic procedure shown in the figure is typical of most current packages in which the user develops a design by selecting parameters that are expected, from experience and engineering principles, to generate solutions that satisfy the design goals. To be compatible with the state of the art the preprocessor should provide a geometric modeler that allows for parametric data input; for example, to specify a circular waveguide the user should only need to specify the standard parameters of radius and length. The solution processor must also allow automatic mesh generation and furthermore incorporate an adaptive "solver" to compute fields or potentials to a specified accuracy. For microwave analysis the solution processor must be sufficiently versatile to be able to solve for either time domain or frequency-driven problems and to compute eigenvalues and eigenvectors. Also, depending on the type of basic method used (i.e., differential or integral), the solution algorithms must be able to deal with the far-field problem, for example, absorbing boundary conditions in the case of

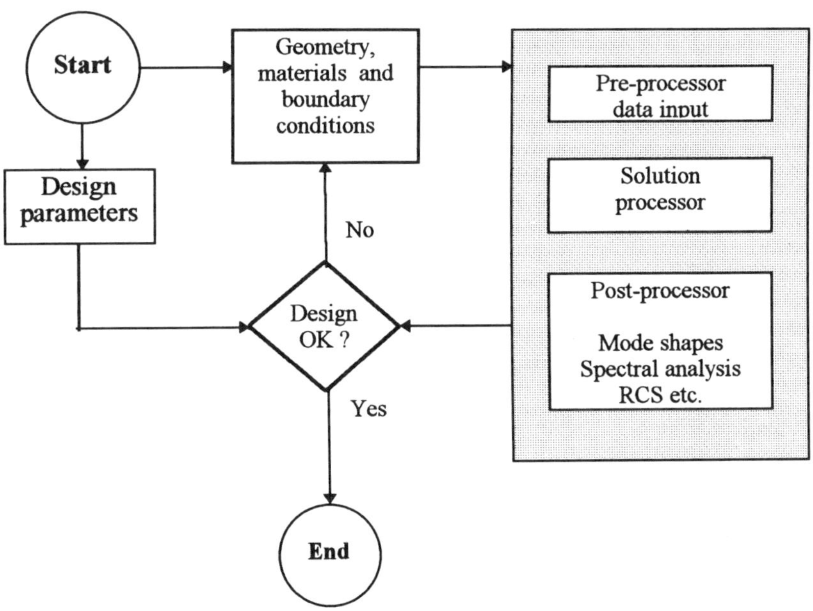

FIGURE 1. Simple heuristic environment for electromagnetic field computation.

the former. The components for the software environment are now discussed in more detail.

3.2. Graphical User Interface

The primary purpose of the software environment is to allow the user to prescribe all the necessary data for a problem. These data typically include the geometries of the structures to be analyzed, their electromagnetic material properties, and all the initial values and boundary data to ensure that the problem is properly defined. The software should provide error checking and editing facilities, utilizing both graphical and keyboard interaction. The usual method of achieving this is for the user to control all the code's functions by means of a program called a graphical user interface (GUI). A typical GUI in use for three-dimensional microwave cavity analysis is shown in Figure 2, in which the user selects items from a menu to initiate commands for the tasks required to define a model for analysis.

A schematic diagram for a typical integration of a GUI, database, and solution processes is shown in Figure 3. In this system each module has an instance of the GUI appropriate to its functionality; this the geometric modeler

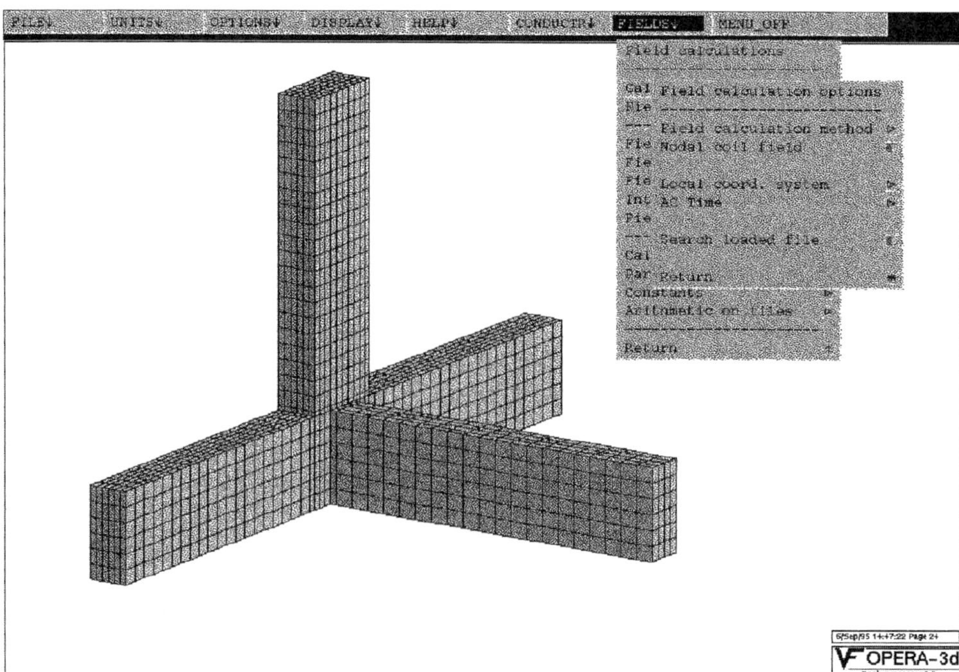

FIGURE 2. A typical GUI for inputting data to define a microwave cavity.

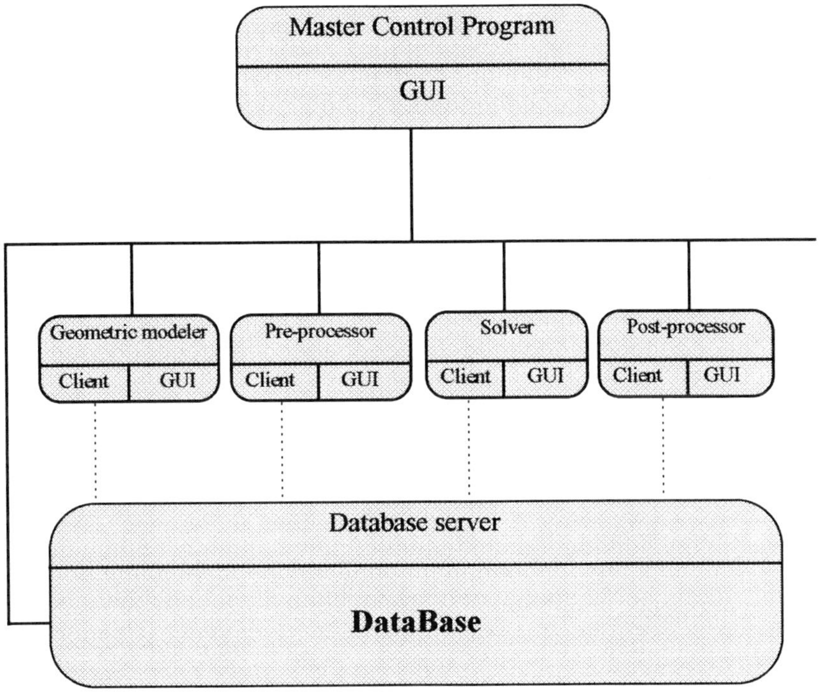

FIGURE 3. Extendible, flexible electromagnetic analysis environment.

will have menu- or icon-driven commands for entering points, lines and surfaces to construct geometric primitives, and the preprocessor itself to provide further commands for assigning material attributes and boundary conditions. The mesh generation phase may also be initiated within the preprocessor or be part of the solution processor in the case of mesh adaptation under *a posteriori* error analysis control (see Chapter 14). Each instance of the GUI is initiated by an initialization file containing the layout and command structure data for each process. The whole system is under the control of a Master Control Program and all applications may be run as concurrent processes. In the system illustrated in Figure 3 the communication between processes and the database is carried out by means of UNIX domain sockets with client–server protocols [3]. Figure 2 also shows an automatically generated mesh initiated by the mesh command but the free-space mesh is not shown.

3.3. Database

Underlying the entire software system there must be a database. Ideally, the user is well protected from the details of the database but is very much dependent on its quality. The database design should be flexible, extendible, and portable. For example, the database must provide all the entities required for storing the data needed for defining the geometry, materials, and boundary conditions; it must allow modifications and the merging of model data as well as including any additional data arising from "second thoughts" and any revisions in design concepts. Finally, the system should provide importing and exporting facilities from and to other software packages. In order for the database to work effectively and be fully portable, there are several important aspects to consider.

1. A common data model should be established. This will allow different applications to access the same sort of data from the database without knowing which application generated it in the first place.
2. The database should be capable of supporting the complete data model.
3. The database should have a well-defined interface that is usable by all of the applications.

Fortunately, the international community has produced several data models that are relevant to electromagnetic analysis; see the ISO STEP standard [1]. Although this work was originally initiated by needs of computational mechanics generally and is deficient in some areas appropriate to electromagnetics, recent work has addressed these special requirements [3]. For example, data models are being defined for many of the necessary data types to perform an electromagnetic analysis:

- Mesh topology
- Connections to geometry
- Physical properties of regions
- Boundary conditions
- Information to control analysis

The STEP standard is gaining respect and developers of commercial software are beginning to ensure that their future releases are STEP compliant. In the meantime they are using a variety of methods based on older attempts at standardization, the so-called industry standards—for example IGES, DFX, and I-DEAS—which are not universal.

3.4. Geometric Modeler

The first preprocessing tool required is the geometric modeler by which means the user specifies the geometry of the device. Although this task can be achieved by specifying primitive shapes starting from points, lines, and surfaces in a hierarchical manner and then subsequently using these primitives to generate a library of parameterized macros, the end user would normally start by utilizing the macros themselves and build the complete model for a device by

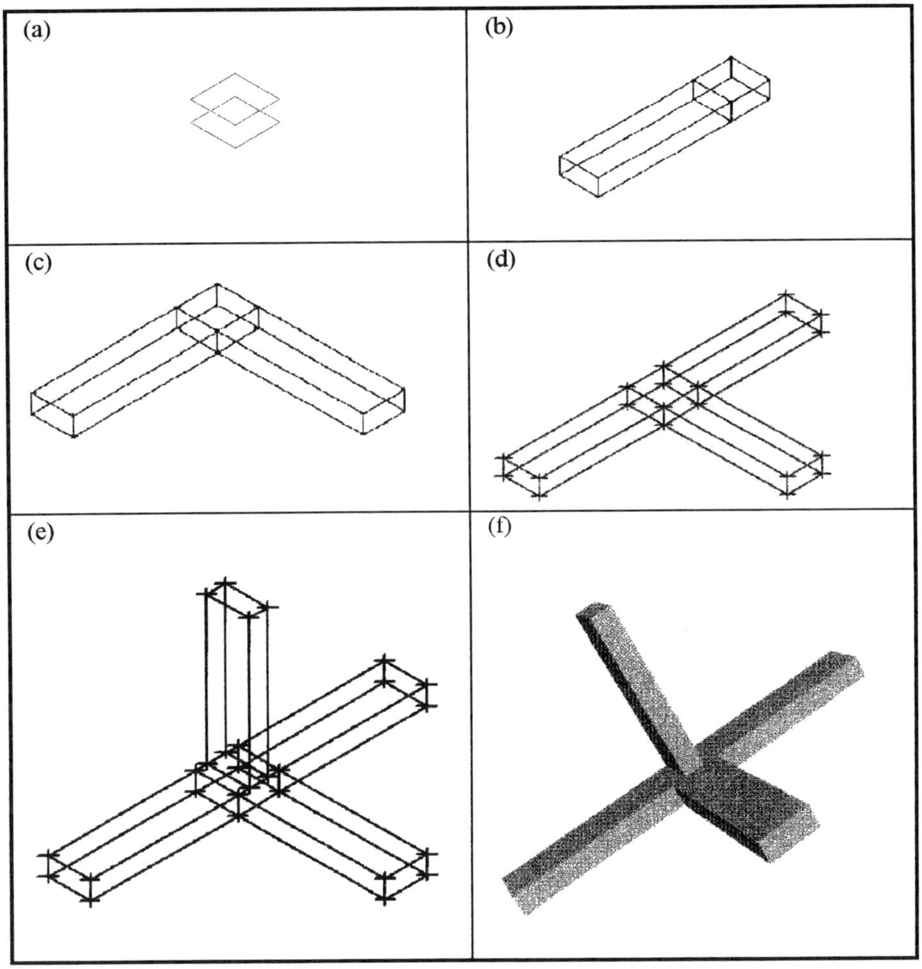

FIGURE 4. Modeling the Magic-T W 16 waveguide: (a) core formed by extrusion of a rectangle; (b) arm formed by extrusion of rectangle; (c–e) other arms also formed by extrusion and replication; and (f) the completed solid model.

using boolean and extrusion operations. In Figure 4, an example of the use of a geometric modeler to construct the magic-T waveguide structure is shown. The complete model consists of four orthogonal limbs and can be constructed in several ways. One approach, which is illustrated in the figure, begins by defining the central core, a parallelepiped, and then extruding faces to form the limbs. In a more realistic case, the finite thickness of the guide walls may have to be taken into account and so extrusions of annular objects would be required. An alternative method would be to construct the object from simpler primitives, for example, single hollow rectangular structures, and then use boolean operations of intersection and union.

There is an extensive literature on all aspects of geometric modeling, particularly in the field of structural design, and since the concepts are fundamental to all areas, the book edited by Kodliyalam and Saxena (1994) is a good example as it relates the special requirements of finite element analysis. There are also a number of commercially available systems [4, 5] that embody some of the facilities outlined above.

3.5. Mesh Generator

This topic has been dealt with in Chapter 14. However, from the software point of view other issues involving the integration into the overall system environment have to be addressed. The various solution methods listed in Table 1 intrinsically have differing discretization requirements. Broadly speaking, methods based on integral equations require meshes in the active regions only, for example, conductors and dielectrics, but methods based on differential equations require meshes in the free space as well. For discretizing active regions, fully automatic mesh generators are available and most commercial codes offer these facilities; see Chapter 14. In recent years a number of techniques for automatic meshing have been discussed in the literature, in particular, the works by Schroeder and Shephard (1990). George and Hermeline (1992), and George and Seveno (1994) are interesting and useful. In three dimensions, the problem of meshing convex primitives is relatively straightforward, particularly if the basic elements are constrained to be tetrahedra, but for reentrant regions there are challenging problems, particularly in generating the internal elements. The work of Joe (1991, 1994) addresses some of the difficulties and has been influential. This author supplies a public domain package (GEOMPACK) available from the Internet, which can be used as the basis for particular problems.

In cases where meshing is required for both the active regions and free space, as for the classic finite element method, then the degree of meshing difficulty is far more dependent on context. Thus if the device has a complicated geometric structure with several regions of differing material attributes, conductivity, permittivity, and so on, and when the union of the active components is performed with the free-space global region, it may be very hard to ensure a properly posed mesh that takes into account both the gradations needed to represent the wavelengths involved and the geometry. In this latter case it is often necessary for the meshing to be performed automatically in stages on subregions with the user controlling the final mesh assembly.

Consequently, in the integrated environment illustrated in Figure 3 the preprocessor must have a range of tools to construct the mesh, depending on the solution process to be used, the complexity of the active regions, and the wavelengths involved. All these tools should communicate with the database from which the geometric data defined by the modeler can be imported and to which the final mesh data, topology, and geometry can be exported for subsequent use by the solvers.

3.6. Solution Processor

The algorithms for the solution process have been widely discussed in this book. From the software construction point of view, it is important that the implementations of these algorithms conform to the database for the overall analysis system and if a wider usage is envisaged then the database adopted should be based on a standard data model of the type discussed earlier.

At the heart of the solver is the algebra processor that solves the system of equations. Apart from the time domain–finite difference method and TLM (see Table 1), implicit solutions are needed and solutions of large systems of equations have to be carried out. Some of the well-known methods for this have been dealt with elsewhere in this book (see Chapter 14), and different methods are employed to deal with particular problems. For example, in electromagnetics applications using standard finite elements, the preconditioned conjugate gradient method with an operation count $\sim n \log n$ has been widely employed (see Silvester and Ferrari, 1990, p. 326), But in methods based on integral equations the system matrix is normally fully populated and the Gaussian elimination method, which has an operation count $\sim n^3$, is often used.

The use of parallel and concurrent processing is becoming increasingly important (see Chapters 9 and 15), and the choice of solution algorithm should take this into account. It is clear, apart from the source and coefficient calculation, that not all the algorithms based on differential methods are intrinsically parallel and special techniques are needed, whereas the integral method, involving fewer unknowns, can be parallelized easily; see, for example, Bryant, Roberts et al. (1990) on the use of multiprocessors to solve field problems using the boundary element method. However, because of the high costs and lack of software tools, industry has been slow to adopt parallel hardware.

3.7. Postprocessing

The postprocessing element of the design system must provide an interactive capability for examining the numerical solutions graphically, but this function is not only a post "viewer" but, more significantly, a processor that provides a library of macros to allow the user to perform computations based on the field solutions. In Figure 5 the propagation of a TE mode wave is displayed by showing contours of the modulus of the **E** field. In the example shown, the input signal of 10 GHz is defined at the end of the arm in the z direction and is propagated in the x-directed arms only. The example also shows, as expected with the "magic T," that there is negligible propagation in the y direction. Figure 6 shows the TE (10 GHz) wave propagation out into free space from

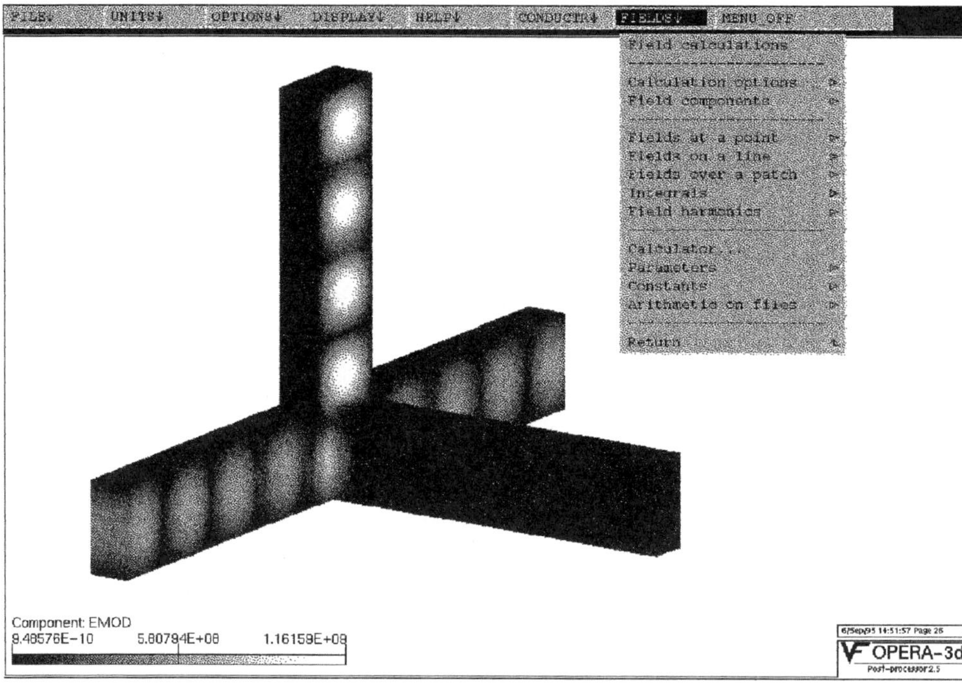

FIGURE 5. Post processing of the WG 16 (Magic-T) standard guide, showing E-field wave modes. The arms are aligned to a right-handed cartesian axis system.

the end of a WG 16 standard waveguide arm. An absorbing boundary condition has been applied to a surface at a distance of 1.5 wavelengths.

Examining the field map solution, as illustrated in figure 6, is necessary but by no means sufficient, because postprocessing tasks in general are very dependent on the context of the problem solved and on the methods used. The system should be able to integrate the solution space in various ways to compute global quantities, like the characteristic impedance. Often radiation patterns, in the form of directed polar diagrams, are needed. This latter procedure requires processing the field variables over defined lines, for example, circles, and the processor must have available to it the full armory of numerical mathematical techniques like interpolation, integration, efficient searching, and graphical editing. For time-domain solutions, both the fields at a location have to be plotted as well as contour maps at a given time. For many applications, integral transforms are important so macros for Fourier and Wavelet operations are also needed. Not all derived tasks can be performed in the postprocessor itself and other quantities like the S matrix or eigenvalues and vectors will be more properly carried out at the solver stage, but special postprocessing routines will be needed to optimally display these

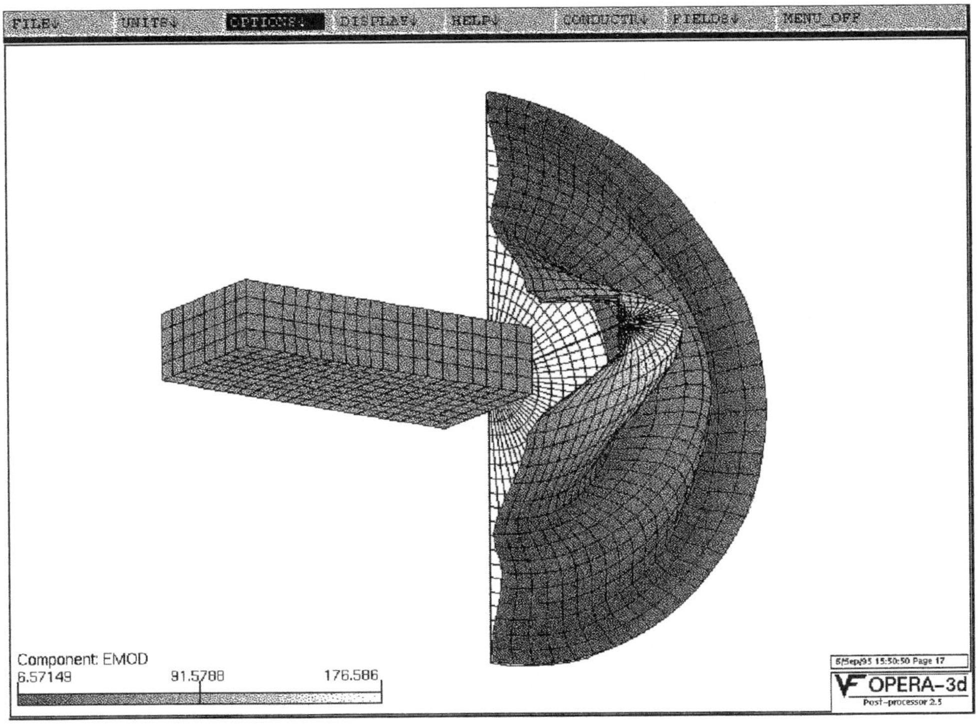

FIGURE 6. Radiation from an open waveguide.

quantities. The details of all these concepts will be found in the other chapters of this book.

3.8. Availability

The question remains of how the user proceeds: should one design one's own environment or acquire a commercial package? The creators of the system reported in reference [3] opted for a new architecture following the above principles because there was no proprietary system having all of these features available in the marketplace. However, various subsystems exist that are more or less portable although even making minor changes to the appearance of a GUI, say—and there are many GUIs on the market—requires expert knowledge.

In the postprocessing area the user will always find new ways of processing and displaying results, so it is important that the system allows sufficient flexibility for this. The code developer will study the market and must have the essential resources to succeed; but the ordinary user will need to study very carefully what is offered.

4. OPTIMIZATION AND FUTURE DEVELOPMENTS

Electromagnetic products permeate modern life and it is taken for granted that the designer has made the best possible use of the electromagnetic fields in the device to provide the best performance at least cost. Unfortunately, the discovery of the best choice of size, shape, and power characteristics for the components in the device, even using the best of today's computer simulations, is very time consuming and costly. It is therefore likely to be incomplete. There are significant delays in bringing improved products to market. Opportunities for even better products are being missed. To accelerate the production of electromagnetic products by the designer a simulation environment is needed that includes the following key features:

1. Full three-dimensional variational geometry; changes in dimensions automatically incorporated into the model.
2. Automatic three-dimensional meshing and mesh adaption to secure accurate field calculation.
3. Optimization based on objectives set in terms that directly affect the quality and competitiveness of the product, supported by a choice of optimization strategies and optimization methods to meet different needs.
4. Efficient execution on a range of computing equipment from clusters of workstations to large-scale shared or distributed memory systems.
5. Powerful postprocessing to ensure the designer remains in control of the process.

The goal of the environment is to automate the synthesis of optimal electromagnetic devices from performance specifications. The current design procedure is to use analysis techniques to predict the performance of a device, once it has been fully specified, then to use the same analysis methods heuristically to refine the parameters to try and achieve a better solution (see Figure 1). Although a number of excellent computer codes are available for analysis of electromagnetic devices, the computing resources needed to explore the full design space is prohibitive. The use of high-performance computing does, of course, accelerate the process, but the enormous number of cases needed rapidly leads to a "combinatorial explosion." For example, even if there are just two states of n parameters there will be 2^n trials. The human designer needs to be able to call on the services of systematic optimization software to accelerate, manage, and control the process, and even though this has been done with significant success in a number of special cases, there is a lack of skilled designers able to create customized systems. The result is electromagnetic products that are not as effective, or as competitive, as they might be.

The need is evident. The solution is still a matter for research and development but the algorithms for three-dimensional modeling of complex geometry have now progressed to a usable state. The concepts of variational geometry can be added so that the shape of geometric entities can be controlled by engineering parameters. Thus objective functions expressed in product quality terms (field shape, energy distributions, costs, etc.) can be evaluated as required by the automatic optimizer with data under the control of the geometric modeler. Also, from work on three-dimensional error estimation, it is now feasible to develop automatic three-dimensional computational meshes and to refine these adaptively to meet accuracy requirements. Because these basic building blocks are becoming available, a practical design optimization system can now, at last, be built with a high probability of success.

The data flow for such a system is shown schematically in Figure 7. A key objective of this software environment is that the designer should be able to vary dimensions and other features of the product and have the consequences of the changes evaluated. Since the many changes that could be made can easily generate a combinatorial explosion in the number of cases to be considered, it is essential that a systematic approach be adopted. For example, a generic parameterized variational geometry input facility, which will auto-

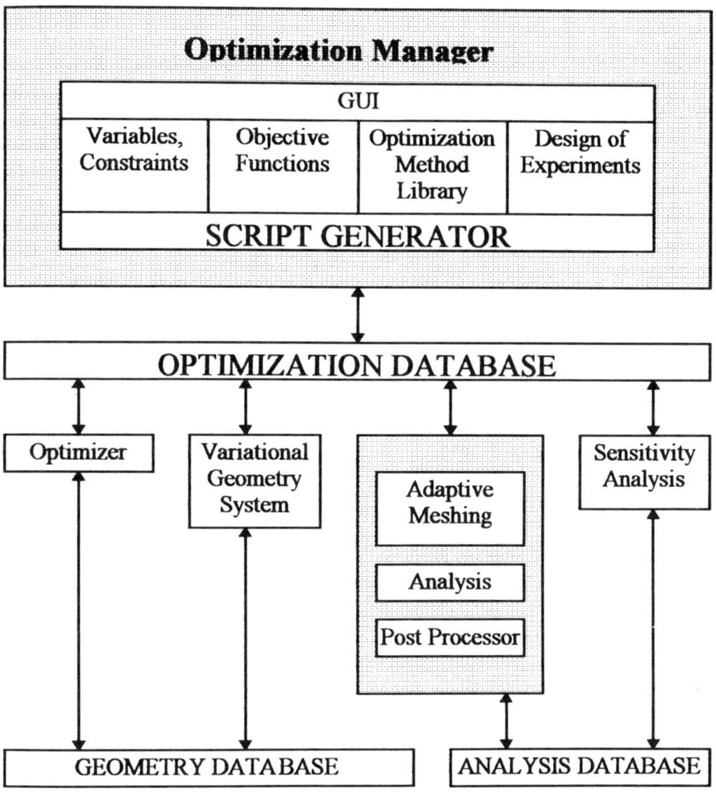

FIGURE 7. Optimizing environment for electromagnetic devices.

matically make intelligent changes to ensure the integrity of the product in consequence of individual changes made to product dimensions and so on, whether by the user or the optimizer, is essential. The system also requires robust automatic meshing that responds to changes in geometry. This can be achieved by generating a three-dimensional mesh that is adaptively refined to ensure that three-dimensional numerical error estimates are within acceptable limits.

It is also important to offer the user convenient means for the construction of a strategy that embodies appropriate optimization methods. Thus an *optimization manager* is required, which, using the results of previous analyses and information about the sensitivity of the objective function to changes in parameters (held in an *optimization database*), enables the user to control the optimization process. Both individual manual changes to parameters or systematic changes over limited domains may be requested. These changes can be implemented using the variational geometry modeler along with the adaptive meshing system. The optimization manager can be script-driven. These scripts can be automatically produced, based on requirements specified by the user (which parameters are allowed to be varied, how the objective function is to be evaluated in the postprocessor, which optimization process should be used, etc.). Since the environment aims to be general purpose and flexible, leaving the user the precise definition of optimization constraints and requirements in each specific case, it will be necessary to implement an *optimization method library* containing standard methods such as gradient search, simulated annealing, and genetic algorithms; further methods can be added, as required, by the user or supplier. For a recent survey of trends on optimization in electromagnetics applications the proceedings of the Compumag & Computational Electromagnetics (CEFC) are recommended [6, 7].

It is essential that the system is a coherent tool for the end-user, well integrated with the design environment. It is important that the data structures used to transmit product details and intermediate results between components of the system be carefully designed and that the end-user has a convenient interface to the system facilities. In this context the use of a data exchange standard like STEP, discussed earlier, would be appropriate.

5. HARDWARE ISSUES

The impact of advanced hardware, that is, supercomputers on the one hand, and the ever increasing power of personal computers on the other, with all types of computer in between, will have a major effect on the design of software systems. There is a clear division between the low-cost personal system and the large-scale costly system affordable by large organizations. It is more often the ratio of computer processing power to hardware cost that is important, and in this respect the modern workstations are competitive with supercomputers. Furthermore, it is not always the central processor unit (CPU) power that is critical, as the system matrices obtained by the solution algorithm methods do not usually fit in the main memory for large problems. In this case, the solution time is constrained much more by the overall system parameters than by the CPU power. The term workstation formerly reserved for UNIX-based systems, with a relative high entry cost ($10,000), has in recent years included personal computers under DOS or Windows operating syatems, with a very low entry cost ($\sim$$1000), which is opening up new possibilities. Nevertheless, the use of high-performance computing will be essential in the advanced environment described in the last section.

A rapid turnaround or interactive working is essential to retain the designer's active involvement in the process, to cut the time to market, and to improve the quality of the product by more comprehensive search of possibilities. The aim should be to fully employ the computational facilities available to the designer whether this be clustered (or even, in extreme cases, single) workstations, a more closely coupled parallel computer, or a system using distributed or shared memory. Obviously, the rate at which a given design can be developed is dependent on the power of the computing facilities available. The objective should be to achieve the best possible cost/performance ratio. Fortunately, many of the algorithmic steps in field computation are "inherently parallel" and the techniques of portable parallel programming with predictable performance will assist in this objective being achieved.

6. MARKETING ISSUES

It is now accepted that computer modeling for both the conceptual research and design stages of advanced devices is essential and often the only way to proceed. With most components it is no longer cost effective to base designs on analytical means followed by costly and inflexible experimentation and prototyping. What started out some years ago as a free exchange of software between developers has now become a business with real costs involved. It is no longer acceptable for developers to regard software only as a cultural extension to their normal work in research. Even good software, in a rapidly changing world, has a very limited "half-life" and the efforts of well-trained professional in the area are needed to keep in step. There is no essential difference between hardware and software in this respect; both require development, maintenance, and support.

With the enormous expansion in computer literacy among professionals of every type, user expectations are escalating rapidly. Software packages like MATHCAD[8] allow a sophisticated set of tools for most of the standard mathematical processes and with the introduction of hypertext "notebooks" a large corpus of foundation knowledge is instantly available; see, for example, the text by Edminster (1993). This type of software is now standard for the everyday tasks of word processing, spreadsheet analysis, and company and personal data. All these tools are instantly available at desktop computers and the user will soon expect a similar environment for solving problems in the more advanced areas under discussion. Thus the designer will demand an environment for applications that provides an intuitive command structure, robust communication between modules (including nonengineering modules), and access to engineering knowledge information. This may seem unrealistic but already most of these components are viable with only integration lacking. The type of GUI, database, and standards similar to the considerations mentioned in Section 3 should play a central role in achieving the integration. Figure 8 attempts to show the idea of an integrated system allowing the user to carry out all aspects of the design.

As the Internet continues to expand, will users demand access, distribution, and support of software through facilities like "worldwide web"? It can be imagined that future applications of finite elements may be carried out using this global environment. The growth of special interest groups with technical forums is already having an impact and software, often of doubtful quality it must be said, is being broadcast far and wide. As always, it is up to the consumer to be on guard and exercise caution—caveat emptor on the Internet as well as at the supermarket. For users who have Internet access it is apparently all too easy but someone has to pay and organizations will have to limit unbridled use of the system for downloading and casual and unfocused

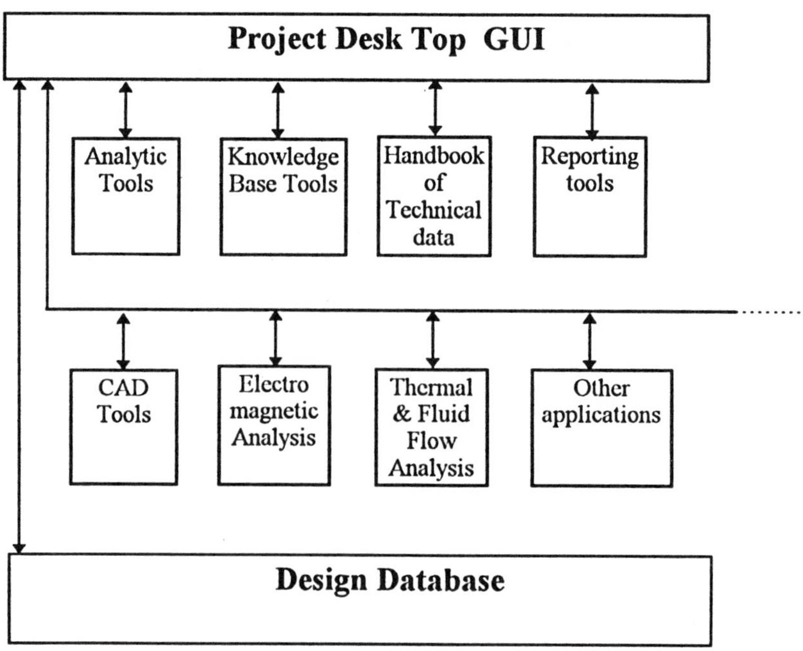

FIGURE 8. Software tools for the future.

browsing. Hopefully, order will prevail at some point to regularize the worst excesses. However, software vendors in the mass market areas of data processing are now issuing upgrades and bug fixes, and in time this service will be made available in the engineering applications under discussion in this book. Some of these issues are likely to have a profound effect on the applications software market.

Finally, some general comments on software packages may be helpful. First, the existence of a claimed feature in a code does not necessarily mean that it is easily accessible or that it will perform optimally under all conditions. It is very important to benchmark a code for the problem before acquiring it; in this context the set of canonical problems defined by the two benchmarking groups TEAM & ACES [9, 10] could be relevant. Second, the fact that a feature is not specifically mentioned does not necessarily mean that it is not provided, or that similar capabilities cannot be obtained by other means. Third, it is often the case that facilities like "ease of use" and integrity and consistency in the command language structure cannot be easily described. Thus the perceived quality of a code becomes a matter of personal taste. Last, the most important characteristics of electromagnetic codes are robustness, efficiency, and accuracy. Not only are these qualities likely to vary between different codes but also for a given code operating under different conditions of use.

This chapter has tried to give a summary overview of the main issues involved in the selection and architecture of components for a software environment for electromagnetic analysis and design. The details of the subsystems have not been addressed and the interested reader should consult the literature for more information. In conclusion, it can be said that the ease of use and robustness of EM codes are generally high, but an insight into the physical phenomena involved and a sufficient preparation in using the package both by training and close study of the documentation are paramount. Is the user always in the best position to know what he/she needs? Do the code builders know? What one will need in the future depends on many factors apart from one's own personal circumstances and aspirations. As with most human endeavors, the most significant advances have come to fruition because developers, in the present case code builders, have listened to the "happy few."

REFERENCES

Bryant, C. F., Roberts, M. H., and Trowbridge, C. W. (1990). Implementing a boundary integral method on a transputer system. *IEEE Trans. Magnetics* **26** (2), 819–822.

Edminster, J. A. (1993). *Theory and Problems of Electromagnetics.* Schaum's Outline Ser., McGraw-Hill, New York.

George, P. L., and Hermeline, F. (1992). Delaunay's mesh of a convex polyhedron in dimension d. Application to arbitrary polyhedra. *IJNME* **33**, 975–995.

George, P. L., and Seveno, E. (1994). The advancing-front mesh generation method revisited. *IJNME* **37**, 3605–3619.

Joe, B. (1991). GEOMPACK—a software package for the generation of meshes using geometric algorithms. *Adv. Eng. Software* **13**, 325–331.

Joe, B. (1994). Tetrahedral mesh generation in polyhedral regions based on convex polyhedron decomposition's. *IJNME* **37**, 693–713.

Kodliyalam, S., and Saxena, M. (1994). *Geometry and Optimization Techniques for Structural Design.* Computational Mechanics Publications, Southampton.

Molinari, G. (1990). Classification of software available and criteria of approach. In *Industrial Application of Electromagnetic Computer Codes* (Y. R. Crutzen, G. Mrlinari, and G. Rubinacci, eds.), pp. 81–100. Kluwer Academic Publishers, Dordrecht, The Netherlands.

Schroeder, W. J., and Shephard, M. S. (1990). A combined Octree/Delaunay method fully automatic 3-D mesh generator. *IJNME* **29**, 37–55.

Silvester, P. P., and Ferrari, R. L. (1990). *Finite Elements for Electrical Engineers.* Cambridge Univ. Press, Cambridge, UK.

NOTES

1. DEVA STEP Database. For information contact: J. Van Maanen, Rutherford Appleton Laboratory, Chilton, Didcot, OX11 0QX, UK. See also EXPRESS Language Reference Manual, ISO TC184/SC4/WG5-n14, April 1991.
2. NAG Ltd. Computer Software, Unit 1, Jordan Hi., Oxford, UK.
3. MIDAS. Electromagnetics Integrated Design System. EC-ESPRIT Project 7294. For information contact: Emanuele Picco, Ansaldo Ricerche CPI, Corso F. M. Perrone 118, Genova 16161, Italy.
4. ACIS. Spatial Technology Inc., 2425 55th Street, Bldg A., Boulder, CO 80301-5704, USA.
5. CASCADE. MATRA Datavision, 536 Route de las Seds, 13127 Vitrolles, France.
6. 9th Conference on the Computation of Electromagnetic Fields (COMPUMAG '93) (1993), *IEEE Transactions on Magnetics*, Vol. 30, MAG-5, pp. 3403–3467.
7. Sixth Biennial IEEE Conference on Electromagnetic Field Computation (CEFC '94) (1994), *IEEE Transactions on Magnetics*, Vol. 31, MAG-3, pp. 1928–2020.
8. MATHCAD. Mathsoft Inc., 101 Main Street, Cambridge, MA 02142, USA.
9. *Applied Computational Electromagnetics Society Journal*, Vol. 8, No. 2, pp. 203–232, 1993.
10. Testing of Electromagnetic Methods (TEAM), COMPEL, Vol. 9, Suppl. A, 1990.

CHAPTER EIGHTEEN

Profiles of Current Finite Element Software

PETER P. SILVESTER

1. INTRODUCTION

Most chapters in this book survey existing finite element methods, advance the art by describing recent developments not yet incorporated in commercial software, or show how such software can be put to use in solving microwave field problems in ways far from obvious. This final chapter addresses the software itself. It briefly describes the field of finite element software and the wider area of engineering software, attempting to extrapolate trends from the historical development and present state of the field. It then seeks to illustrate the trends and history, by describing a sampling of commercial offerings. It must be kept in mind, of course, that the world of software evolves continuously and the world of computing devices evolves even faster, so the details of any given program are likely to become obsolescent even as the ink dries. Nevertheless, statements like "this program runs under the Portholes 3.1 operating system on Moon workstations" are useful even if no longer accurate, for they hint at the probable direction of software evolution.

2. SOFTWARE LONGEVITY

The first, and perhaps most remarkable, feature of engineering software, shared with scientific software and indeed with many other large-scale products of the

Finite Element Software for Microwave Engineering, Edited by Tatsuo Itoh, Giuseppe Pelosi, and Peter P. Silvester.
ISBN 0-471-12636-5

programmer's art, is its longevity. Typical engineering programs endure for decades, not mere years, and the lifetime of a successful major program is probably comparable to the professional lifetime of an engineer. For example, the NASTRAN structural analysis program, initially developed for building the space vehicles that put the first man on the moon, is now over thirty years old. The SPICE circuit analysis program has begun its third decade of life and shows no sign of faltering. The same can be said for the MAGNET package used to analyze magnetic fields, for SUPERFISH used for analyzing electron devices, for PDE2D, and for many others. Of course, longevity is not unique to engineering software; the UNIX operating system is in its third decade and appears to be enjoying vigorous growth.

The long lifetimes of successful engineering programs contrast sharply with the lifetimes ordinarily expected of computer hardware. Measured in terms of the price of a given device or of a given operation—say, the price of a byte of memory, or the cost of performing a floating-point multiplication—computer power has grown at a remarkably steady pace over the past fifty years or so. The growth rate has remained at an almost constant 36% annually. Occasional hiccups or anomalies have sometimes been caused by market-constraining influences or sudden demand fluctuations, but these have generally been compensated by subsequent sharp movements in the opposite direction. At 36% per annum, a computer becomes obsolete in its fourth year of life. Although the machine may continue to operate satisfactorily much longer, it is not cost effective to keep it running longer; replacing it is cheaper than continuing to pay its cost of maintenance, which is usually reckoned at 5–10% of the original capital cost. Hence the lesson to be learned by software makers and software users is that *a good engineering program will be moved from one computer to another at least five or six times* in its life. It must be sufficiently robust, and sufficiently flexible in its initial construction, to survive this near-continuous modification.

Finite element programs of commercial quality require at least 10^5–10^7 lines of code, the exact number depending on the programming language, the operating system, and other variables. A good programmer or software engineer is generally reckoned to average at most 10^1–10^2 properly designed and fully documented lines per day, the lower productivity being valid for the larger programs. Design, coding, and testing of a medium-sized finite element program therefore entails a cost in the neighborhood of ten or fifteen programmer-years. This work can be accomplished most efficiently by a working group of about three or four people; with larger numbers, the overhead costs of communication and personnel training grow rapidly. The time taken to produce a good engineering program is thus about three or four years, roughly the lifetime of one generation of computer hardware! The important point to note is that *a good engineering program will be built and tested on an obsolescent computer*. Its maintanance and modification therefore begin even before the first copy is released for field service. Redesign and rebuilding are therefore inherent in the product; indeed, they must be provided for in the original design.

It goes almost without saying that successive versions of a program incorporate new features and exploit new hardware capabilities. After three or four major revisions, there may be little of the original code left, almost everything having been altered or updated. Such gradual replacement or growth is generally termed *software evolution*, to distinguish it from the more mundane process of error correction.

3. SOFTWARE CREATION AND EVOLUTION

Creation of commercial engineering software is usually best done by a small, tightly knit working team, preferably with distinct and complementary skills. Tight-deadline projects like the original creation of NASTRAN are the exception that proves the rule, by producing software at relatively high cost. Finite element programs are high-technology products; they are mathematically complex and involve complicated data structures, so the software designer must not only be knowledgeable in computer science but must have a good grasp of current numerical analysis and, in the case of microwave programs, a strong knowledge of electromagnetics. As the program nears completion, only the designer best knows the reasons why individual design decisions were taken; and only the designer fully appreciates the strengths and weaknesses of design details. No one is better qualified to undertake the redesign and program modification that must invariably begin even as the first finished copy is readied for shipment to the first customer. This, combined with the pride of workmanship of a small team that has labored together for years toward a common goal, often causes the same software team to continue working on the program. The work may continue for years or even decades, the program becoming closely identified with the people and vice versa.

This longevity of both the software and its supporting technical personnel is of immense value to the prospective user or purchaser of finite element software. It permits the user to forecast the evolution of programs, at least in their scientific content, because a software team demonstrably very good at building high-order finite element software is unlikely to make a sudden switch to finite differences, TLM methods, or some other totally new technique. The examples of commercial products shown later illustrate this point: most of the smaller and medium-sized companies specialize in one kind of software or another, oriented around the specific skills of their best people. The larger companies do likewise, except where they act solely as marketing agents for programs written and maintained by an independent smaller company. In other words, the finite element software industry is mainly technology-driven, not market-driven. It sells what it knows how to produce; it does not divert its efforts to satisfy customer needs — not unless these are within easy reach of the skills that already exist in-house.

Because individual skills are so important in software creation and evolution, forecasting the future evolution of a product is often not difficult. It may suffice to know who the principal designers are, or indeed to assess what fits their talents and experience by reading the technical papers they publish about their products. Outsiders may well predict long-term future directions better than the company's own marketing staff, who are (or should be!) well attuned to customer needs but usually know much less about software technology or finite element mathematics. *Who wrote it and who maintains it*? is clearly the key question. Here the smaller companies can give a convincing answer: *I did and I do*. This may explain why many well-known finite element software companies are rather small by general corporate standards, ranging from a small handful of people to perhaps a few dozen.

4. PACKAGE DESIGN

Most commercial finite element software currently available is spoken of as *packages* rather than *programs*, to indicate that it consists of a multiplicity of independent or semi-independent programs designed to work together. (The word *system* is losing its meaning through overuse; one firm that squirts oil on automobile undercarriages advertises its product as a *corrosion inhibition system*.) The component programs are usually

1. Preprocessor.
2. Solver.
3. Postprocessor.
4. Utilities.

In well-built interactive software, the relative amount of code in each might assume proportions like 35:15:45:5, though of course these figures will vary widely.

The solver is what most people would consider "the finite element program". It generates element descriptions; assembles these into global finite element matrices; attaches such boundary conditions, constraints, or source functions as may be appropriate; and then solves the equations. This is by far the most computation-intensive part of the package, especially in three-dimensional analysis, where computing times of hours or even days are not uncommon. It requires little if any user input, so it is often structured as a pure batch process: it reads its input from files, does its work, then deposits the results in files.

The preprocessor is sometimes referred to, albeit inaccurately, as "the data input module". It provides facilities, usually graphical or at least graphics-assisted, for describing the device to be analyzed, the excitations to be impressed upon it, and the boundary conditions or other constraints to be imposed. Interfaces are frequently provided for importing data from drafting programs or from geometric modelers. The preprocessor may also handle *solver directives*, user instructions to the solver. These may include data management; for example, they may specify the file formats or file names to be used by the solver. They may also include parameters that delimit the numerical procedures, for example, the number of eigenvalues to be computed or the closure error permitted in an iterative solution. These preprocessor actions may clearly be partitioned into two parts: the electromagnetic problem description on the one hand, and solver directives on the other. Some designers separate these into two independent preprocessor programs, thereby permitting the user to repeat an analysis without needing to respecify the problem.

Despite their large size, postprocessors are a relatively ill-developed part of finite element software. They are often large because they consist of many problem-specific or use-specific routines, with few truly general-purpose ingredients. For example, some postprocessors will provide several special-purpose plotting options, such as cartesion components of the electric field and the magnetic field, but will not enable the user to plot components of other vectors (e.g., magnetic vector potential, Hertz potential, or Poynting vector) nor to plot components with respect to an arbitrary coordinate system. Similarly, many postprocessors generate useful algebraic quantities like scattering matrices or impedance values, but do not allow the user to compute any quantities not preselected by the designer. An ideal postprocessor might well provide a general-purpose plotting package and a general-purpose vector calculus package, allowing the user to link together the pieces so as to calculate anything inherent in the stored data. However, such truly general-purpose postprocessors are rare. They can be difficult to sell, not only because they lack instant-answer buttons ("just press this button and the scattering parameters pop out"), but perhaps also because a startling number of microwave engineers have had little contact with field theory since graduation and take fright at having to deal with "all those curls and divergences."

Partly because individual designers feel very much identified with their own products, there is a regrettable lack of standardization in finite element software (indeed, in *any* software) for electromagnetics. Every designer seems to build data structures and files a little differently, so the advantages of software modularity are largely lost. In principle, it should be possible to use the best available solver program (from company X) in conjunction with the postprocessor best suited to the task (but from supplier Y). In practice, that rarely works because the file structures are different.

5. PRACTICAL ASPECTS

The use of almost any electromagnetics software, in particular finite element software, requires training. To start, users require basic training to run a package at all; they must acquire a conceptual model of what goes on inside the software and how to control these processes by the available commands.

This, however, in only the easy part, and anyway the package vendor can do much to help it along. A much larger amount of time is likely to be spent in fitting the new package into the local working procedures; here the package vendor may be able to give some help but the main burden must fall on the user. Quite often a new user begins by attempting to replicate on the computer an established manual design procedure, only to discover weeks or months later that the manual procedure was invented so as to avoid some intricate calculation that is easy to perform in finite element software. Streamlining the procedure may require altering other parts of the workflow, or it may require a greater understanding of field theory than local personnel feel confident about. The result is that a substantial amount of both money and working time is invested learning how to use the new tool. The amounts are hard to assess, not least because accounting systems differ; some people feel that at least 1–2 times the price of the software package should be budgeted for training.

The relatively high cost of training makes comparison shopping between packages a bit difficult. It is true that the various commercial packages are coming to resemble each other as they evolve, so that learning to use several is not proportionately more costly than learning one; nevertheless, there is an extra cost involved. Many commercial packages are available either for purchase or on lease, the vendor hoping that a three-month lease will convince the prospective customer to pay the full purchase price. So far as such commercial issues affect software design at all, this approach is to be regretted. On a short-term lease, it is hard not to follow the ancient exhortation to "sell the sizzle, not the steak." The result may be a reduced emphasis on, for example, general-purpose postprocessors that are more rewarding in the long run but fail to provide much sizzle in the short run.

6. SOFTWARE DESCRIPTIONS

The remainder of this chapter comprises a selection of software descriptions, in a more or less standardized format. In many respects, they should really be referred to as company descriptions, not software descriptions; but in view of the frequently close identification of designers with products, company descriptions and software descriptions are in some sense equivalent anyway.

As of late 1995, about 30–40 companies were active in the microwave finite element software market. The descriptions that follow, ranging as they do from one-man operations to international giants, are intended to give an illustrative overview. They are clearly not intended to be exhaustive. It will be noted that they include software which their makers do not identify as based on finite elements. In some cases, there is simply a difference in nomenclature. In others, the software currently on offer is quite close in spirit and mathematical skills to finite elements, suggesting that the same design and production teams are fully capable of entering the finite element marketplace at a future date.

6.1. CEDRAT S.A.

Corporate Profile. CEDRAT, founded in 1978, specializes in CAD software for electrical engineering and electromagnetism. Its aim is to answer to all needs in this area either through the software it offers, including training courses and support services, or through consultancy services ranging from simple computations to prototypes.

Applications include all kinds of rotating machines (DC, synchronous, induction machines), transformers, electromagnetic, piezoelectric, and magnetostrictive actuators and sensors, magnetic recording heads, magnetic bearings, and high voltage devices. Analysis includes magnetic separation, nondestructive testing, and induction heating.

Key Software. FLUX2D is a finite element software package that analyses the electric, magnetic, and thermal behaviour of either two-dimensional or axisymmetric devices in their static, transient, or dynamic states. It is able to take into account external electric circuits, the rotor motion of rotating machines, and coupled thermal problems.

FLUX3D is a finite element program that analyses the electric, magnetic, thermal, and microwave behavior of three-dimensional structures in their static or dynamic states.

PHI3D is a boundary element package that analyses the electric and magnetic behavior of three-dimensional structures in their static or dynamic states.

ATILA is a finite element software package that analyses the elastic, piezoelectric, and magnetostrictive behavior of three-dimensional structures in their static, transient, or dynamic states. ATILA takes into account electromechanical and fluid–structure couplings.

All these programs include parametrized geometric preprocessors with automatic mesh generation, specialized boundary conditions to take into account infinitely extending space as well as all kinds of symmetry. Each program also includes a materials database and a postprocessor. They are available on PCs as well as on most workstation computers. Most end-users are industrial companies and university research laboratories. Limited versions are available to answer to teaching needs.

Other software, for example, the Cedrat-Circuit package for power electronic circuits, is also available from CEDRAT.

Consultancy Services. Consultancy services performed by CEDRAT Recherche represent more than 50% of the total activity. CEDRAT Recherche offers computation, optimization, design, measurement, and prototyping according to the needs of the customer.

Future Directions. Future development at CEDRAT is expected to continue in the present direction. The microwave application that exists in a specialized version of FLUX3D currently will be implemented in the standard version. The current program version is restricted to closed structures such as waveguides and resonators. It computes the resonant modes of the given geometric structure and finds their frequencies as well as the Q of each. It also gives curves of the scattering parameters as functions of frequency for multiport structures and provides displays of any user-defined quantities that can be derived from the computed values. All capabilities of FLUX2D will be included in FLUX3D: coupling to external circuits, transients, and motional effects. A new mesh generator able to combine many different kinds of elements will be included in the industrial version of FLUX3D by 1996 or 1997. The preprocessor is independent of FLUX3D, so that it can be used with other software.

Contacts

CEDRAT S. A.
10 Chemin du Pre Carre
ZIRST 4301
38943 MEYLAN Cedex, France
Telephone: +33 76 90 50 45
Fax: +33 76 90 16 09
E-mail: cedrat@cedrat-grenoble.fr

Magsoft Corporation
1223 Peoples Avenue
Troy, NY 12180, USA
Telephone: +1 518-271-1352
Fax: +1 518-276-6380
E-mail: philippe@gander.epe.rpi.edu

Inquire for distributors in Austria, Brazil, Canada, Korea, Spain, Switzerland, Taiwan, and other countries.

6.2. CST—Computer Simulation Technology

Corporate Profile. CST—Computer Simulation Technology distributes the program package MAFIA. CST was founded in 1992 in Darmstadt, Germany, as CST Gesellschaft für Computer-Simulationstechnik mbH. The company's purpose is the development, application, and distribution of simulation software.

Since its foundation, this company has undergone a steady and successful growth, with equal shares of customers in Europe, North America, and Japan. Its key product is the program package MAFIA, which has been developed by the MAFIA group since the early 1980s at DESY (Hamburg, Germany), KFA (Jülich, Germany), and the Los Alamos National Laboratory. In the beginning

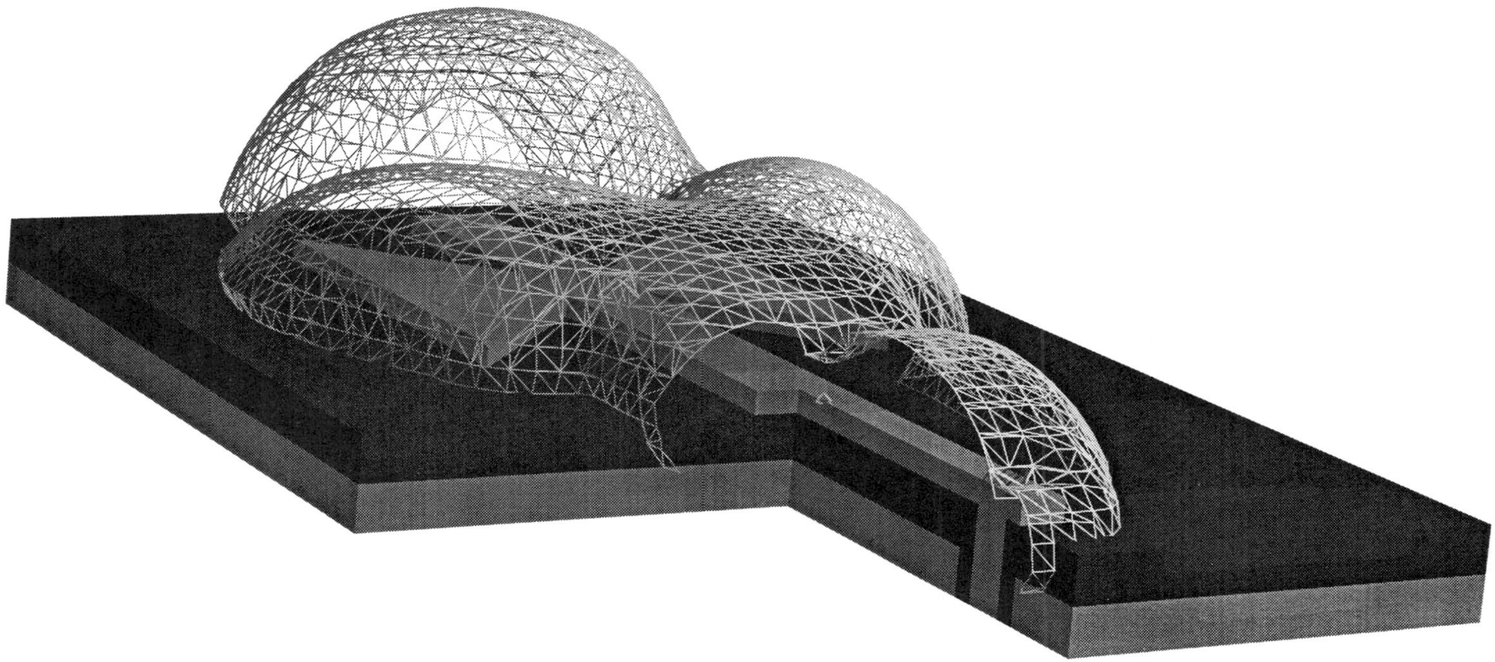

FIGURE 1. Screen display of three-dimensional microwave structure for field analysis (CST): geometric structure.

the group focused on the development of components of accelerators for high-energy physics, but recent applications have been much more widespread.

Key Products and Services. MAFIA is a general-purpose program for the computation of electromagnetic fields. It is based on the Finite Integration Technique by Thomas Weiland, which may be interpreted as a finite volume technique for solving Maxwell's equations. The contour and surface integrals used in Maxwell's equations are carried out on a discrete spatial grid consisting of nodes, edges, surfaces, and volumes, thus preserving the analytical properties of the differential operators (curl grad $\varphi = 0$, div curl $\mathbf{A} = 0$). The material equations are approximated to the second order. As a consequence, the implicit fulfillment of the continuity equations is guaranteed.

The main capabilities of MAFIA include magnetostatic and electrostatic field analysis, low- and high-frequency eddy currents, and the computation of eigenmodes in cavities and waveguides. It can accommodate high-frequency fields with general time dependence as well as particle–field interactions and combined electromagnetic–thermal problems.

Due to MAFIA's general-purpose nature its applications are very widespread. Among other things, MAFIA can be used for microwave component design (scattering parameters, wave analysis, etc.) and for particle accelerator structures (cavities, electron guns, beam dynamics, klystron development). Other applications have included antenna design and electromagnetic compatibility studies.

The program package is divided into a preprocessor, a postprocessor, and a number of solver modules. It allows parametric input of structures and supports optimization. The package is available for rental or purchase. CST also offers consulting services and will perform calculations on behalf of clients. MAFIA runs on all modern workstations, at present including various HP, IBM, Sun, and DEC machines.

Future Directions. A steady growth of the present capabilities will be continued in future years. The main topics are:

- Coupled problems including velocity terms and thermal effects
- Nonlinear problems in time and frequency domains
- Acoustics
- Automatic optimization
- Combination with other methods
- Use of general irregular grids

The main aim is to develop MAFIA to be a general tool that anybody can use, even engineers not trained in electromagnetics or field computation.

Contacts

CST — Computer Simulation Technology
Lauteschlägerstraße 38
D-64289 Darmstadt, Germany
Telephone: +49/(0)6151/717057
Fax: +49/(0)6151/718057
E-mail: r137@temf00.temf.e-technik.th-darmstadt.de

6.3. Field Precision

Company Profile. Field Precision provides stock and custom software for finite element field solutions on IBM-standard personal computers. Versions of the low-priced EMP package for electrostatic and magnetostatic design have been marketed since 1987. The package is in use at research institutes and universities worldwide. Other packages address the design of charged particle guns and optics systems, Hall-effect probes, and electromagnetic resonators. Field Precision has also developed custom software for national laboratories and industries in the areas of high-power-density hydrodynamics and the response of biological systems to electromagnetic fields.

Key Products and Services. Field Precision's key product is EMP (Version 3.1), a versatile two-dimensional finite element code for electrostatic and magnetostatic solutions. The interfaces, documentation, and on-line help have been refined over several years so that the programs can be learned quickly and applied by independent users. The programs are sold through a one-time purchase price with unlimited technical support. Features include an AutoCAD interface for boundary input and an interactive graphics program for solution analysis. Runs typically require 10–30 seconds on a Pentium computer. EMP handles nonlinear materials, anisotropic materials, and permanent magnets. The programs are 32-bit applications that can accommodate up to 60,000 mesh points. They can run in the background under Windows as DOS tasks.

Other stock software products include the following:

E3. A three-dimensional electrostatic field solution package that accommodates up to 1,200,000 mesh points on a personal computer with 16 MB of RAM. E3 handles linear dielectrics and defined regions of space charge. Program advantages include high speed, easy input of geometries, and full graphics and hardcopy support.

TRAK 2.0. A complete charged particle optics and gun design package for cylindrical or slot guns. The programs can handle relativistic particles, self-consistent space charge, and beam-generated magnetic fields. TRAK automatically creates particle distributions to simulate Child-law emission or self-consistent field emission.

SUPERFISH 3.1. A program to find electromagnetic modes in cylindrical or linear resonant structures with the option of dielectric or ferrite inclusions.

PAC 1.0. Near-field electromagnetic field calculations in imperfect dielectrics, with applications to biological systems, electroplating and plasma processing.

NELSON 1.0. Magnetic field calculations with self-consistent eddy currents in homogeneous, nonlinear media.

Future Directions. In 1996, Field Precision will introduce new products in two areas: broad-distribution low-cost software for personal computers and specialized software for advanced simulations. The flagship product will be the TRIOPS series of two-dimensional finite element programs for IBM-standard personal computes. Packages include an autonomous CAD utility for boundary definition and a universal mesh generator. Several standardized solution modules will be available.

- Electrostatics with nonlinear and anisotropic materials.
- Magnetostatics with nonlinear materials and permanent magnets.
- Near-field electromagnetics with resistive/dielectric materials.
- Magnetic fields with eddy currents.
- Charged particle guns and optics systems.
- Time-dependent thermal transport in solids with the option for temperature-dependent conductivity.
- Hall-effect probes of arbitrary geometry.

Each module includes an optimized field solver and dedicated interactive analysis program with graphics and hardcopy support. The target market is universities, laboratories, and small companies. The price for the basic static field solution package will be under $800 (U.S.).

Also, in 1996 two specialized simulation packages for high-performance personal computers will be completed. The first solves coupled problems of thermal diffusion and electromagnetic propagation in nonlinear media for applications to biological systems. The second is a two-dimensional hydrodynamics code with a unique finite element formulation. The program includes support for the Los Alamos National Laboratories SESAME equation-of-state tables.

Contacts

Field Precision
Attn: Dr. Stanley Humphries, Jr., Technical Director
PO Box 13593
Albuquerque, NM 87111, USA
Telephone: 505-296-6689
Fax: 505-294-0222

6.4. HP EEsof Division of Hewlett-Packard Corporation

Corporate Profile. The HP EEsof Division of Hewlett-Packard produces a broad range of system, circuit, and electromagnetic simulation software products to support designers in high-speed, RF, microwave, and millimeter-wave applications. HP EEsof has research and development facilities in Santa Rosa (CA), Westlake Village (CA), and Gent (Belgium).

Key Products and Services. HP EEsof's finite element EM simulation product is The High Frequency Structure Simulator (HFSS), which first shipped in 1990. HFSS is now in its fourth major release.

HFSS has a structured process flow that guides users through the processes of geometry creation, solution setup, and post processing. Solid models can be imported via several different file formats including DXF, UNV, and STL. Meshing and adaptive mesh refinement processes are completely automatic.

HFSS integrates two-dimensional and three-dimensional tangential vector finite elements to model electromagnetic problems in arbitrarily shaped structures. Modal electric field excitations for arbitrary transmission-line cross sections are computed to user-specified accuracy through an adaptive triangular mesh refinement process. These two-dimensional field solutions are then automatically mapped as boundary conditions on the three-dimensional structure using transfinite element methods. Circuit parameters are determined automatically to user-specified accuracy as part of an adaptive tetrahedral mesh refinement process. HFSS can model open-region situations using absorbing boundary conditions. Voltage and current sources are available in addition to modal part excitations. HFSS includes an integrated solid modeler, circuit and antenna parameter presentations, and field solution visualization capabilities.

HFSS is used to analyze three-dimensional structures in commercial and government applications including multilayer circuit design (HSD, RFIC, RFPCB), machined component design, and packaging problems at microwave and millimeter-wave frequencies. HFSS is also used to support education and academic research activities at universities worldwide.

HFSS is available on HP 9000/700, SUN, and IBM workstation platforms. The user interface (including both solid modeling and postprocessing capabilities) can be purchased separately from the finite element engines, as required by the customer's desired configuration.

Future Directions. HFSS will continue to evolve to address important applications in major market segments, both commercial- and military-focused applications.

Contacts. Inquiries about HFSS (Model HP 85180A) and other HP EEsof products can be obtained by calling (800) 452-4844.

6.5. Integrated Engineering Software

Corporate Profile. Integrated Engineering Software develops electromagnetic design and analysis 2D/3D computer-aided engineering software based on its unique boundary element method.

The boundary element method is a variation of the moment method, merging both the moment method and the finite element method. The finite element method solves the partial differential equation governing the field, and the unknown solved for is the potential. The boundary element method solves the integral equation formulation of the problem, and the unknown that is solved for is the source of the field. Integral operators, usually bounded and often completely continuous, admit a wider selection of trial functions than differential operators. Integration is a smoothing operation and the results obtained are smooth, while differentiation introduces discontinuities.

Founded in 1984, Integrated's corporate head office is located in Winnipeg, Manitoba, allowing easy access to large markets in the United States and Europe. Sales outside North America are also supported by European and Asian representatives.

Key Products and Services. Integrated's boundary element software packages include MAGNETO (2D/RS) and AMPERES (3D), magnetostatic field solvers; ELECTRO (2D/RS) and COULOMB (3D), electrostatic field solvers; OERSTED (2D/RS) and FARADAY (3D), time-harmonic field solvers; and KELVIN (2D/RS), a heat transfer analysis solver. Originally developed for DOS-based PC systems, Integrated's software is also available on Sun Microsystems, HP, Silicon Graphics, IBM RS/6000, DEC, and Windows-based PC systems.

MAGNETO (2D/RS) is a magnetostatic field solver for applications such as AC/DC motors, solenoids, transformers, recording heads, magnets, and charging fixtures. MAGNETO caters to linear, nonlinear, and permanent magnet materials. MAGNETO calculates magnetic field components and distributions, potentials, forces, torques, and mutual and self-inductances.

AMPERES (3D) is a magnetostatic field solver for applications such as magnetizing fixtures; permanent magnet assemblies; circuit breakers; recording heads and reclosures; electric (AC/DC) motors; magnetic shielding; and cyclotrons, solenoids, and transformers. AMPERES calculates magnetic field components and distributions, potentials, forces, torques, and mutual and self-inductances.

OERSTED (2D/RS time-harmonic) models electric (AC/DC) motors, crack or fault detection systems, busbars, charging fixtures, induction heating/motors, cables, magnets, recording heads, coils, and transformers. OERSTED calculates induced (eddy) currents, magnetic and elastic fields, magnetic vector potentials, skin and proximity effects, ohmic loss, stored energy, forces, torques, resistance, and inductance.

FARADAY (3D) is an eddy current solver for applications such as induction heating, crack or fault detection (nondestructive testing), recording heads, transformers, coils, and cables. FARADAY calculates eddy current distribution, magnetic field distribution, power loss and stored energy, force/torque, and resistance and inductance.

ELECTRO (2D/RS) is an electrostatic field solver for applications such as insulators, bushings, grounding electrodes, transmission lines, telecommunication cables, and microstrip and integrated circuits. ELECTRO calculates voltage and electric field components, electric field distributions, potentials, mutual and self-capacitances, inductances, and impedance and propagation constants.

COULOMB (3D electrostatic) models insulators, bushings, and grounding electrodes; voltage gradients around high voltage shields; power transmission lines; telecommunication cables; microstrip and integrated circuits; and parts and assemblies subject to electric fields. COULOMB calculates voltage and electric field components, electrical field distributors, potentials, self and mutual capacitances, inductances, and particle trajectories.

KELVIN (2D/RS) is a heat transfer analysis solver for applications such as electromagnetic devices, electronic packaging, heat exchangers, appliances, automotive components, aerospace components, and aerodynamic heating and cooling. KELVIN calculates temperature, temperature gradients, and heat flow distributions. In addition, KELVIN models heat sources/sinks as well as convection and radiation boundary conditions.

Future Directions. Integrated plans on continued enhancements and additions to its current lineup of electromagnetic and thermal analysis design tools. Since customer suggestions and requirements are directly incorporated into development plans, it is Integrated's customers who will be determining the direction of product development in the coming years.

Contacts. Information regarding Integrated Engineering Software and its products may be obtained at the following addresses in Canada, Japan, and Switzerland:

World Headquarters
Integrated Engineering Software
46-1313 Border Place
Winnipeg, Manitoba R3H 0X4
Telephone: (204)632-5636
Fax: (204)633-7780
Modem: (204)633-7819
E-mail: info@integrated.mb.ca

In Asia
SEA Corporation
Tawaramachi City Building 7F
10-13, Kotobuki 2 chome, Taito-ku
Tokyo 111 Japan
Telephone: 81-3-3847-0711
Fax: 81-3-3847-0616

In Europe
Asper Energy Systems
Plattenstrasse 98
Meilen, CH-8706 Switzerland
Telephone: 011-41-1-923-2664
Fax: 011-41-1-923-6917

6.6. Gerrit Mur

Developer Profile. The FEMAX package is developed and maintained by Dr. G. Mur, together with a varying number of his doctoral students, in the Faculty of Electrical Engineering of the Delft University of Technology. The FEMAX package is the result of long-term fundamental research into the development of finite element codes dedicated to electromagnetics, seeking to achieve maximum accuracy of the results in relation to the computational effort required.

Key Products and Services. The FEMAX finite element package consists of the FEMAX3 package for computing three-dimensional time-harmonic electromagnetic fields and the FEMAXT package for computing three-dimensional fields having an arbitrary time dependence. By using a transient, the latter package can also be used for computing time-harmonic fields and is as such competitive with the time-harmonic package, depending on the problem at hand.

The main advantage of the FEMAX packages, as compared with most commercially available software, is that the electric and/or magnetic field strength is computed directly and not through the intermediate use of (vector) potentials. Because of this, electric or magnetic fields computed employing this method are much more accurate with the same computational effort (storage requirements and computation time). The efficiency of the packages is further optimized by a balanced and consistent use of both edge- and nodal-based elements, combining the best properties of both worlds. The user can adjust the criteria for deciding which type of element must be used according to individual accuracy requirements.

These packages can compute electromagnetic fields in arbitrarily inhomogeneous (lossy) configurations that may contain anisotropic media. These fields may be excited by arbitrary source distributions and/or boundary conditions. Various types of boundary conditions are available, including absorbing boundary conditions for modeling radiation into the unbounded free space surrounding the domain of computation. Fields incident from the surroundings can also be taken into account in the absorbing boundary conditions.

The FEMAX package uses the general-purpose SEPRAN finite-element package for a number of general finite element operations; it operates on both UNIX and VAX/VMS operating systems.

Future Directions. The main goal is the further optimization and refinement of the methods already available. A package for computing three-dimensional static (or stationary) electric and magnetic fields is expected to be ready in 1996. In addition to this, there is a plan to expand the pre- and postprocessing options of all FEMAX packages. In this respect, existing pre- and postprocessing codes may be used.

Contacts. Further information about the theoretical background of the FEMAX packages, about their capabilities, and about the conditions for using them may be obtained directly from:

Dr. Gerrit Mur
Laboratory of Electromagnetic Research
Faculty of Electrical Engineering
Delft University of Technology
P.O. Box 5031
2600 GA Delft, The Netherlands
Telephone: +31 15 2786294
Fax: +31 15 2786194
E-mail: g.mur@et.tudelft.nl

6.7. PROFI ENGINEERING SYSTEMS GmbH

Corporate Profile. PROFI ENGINEERING SYSTEMS GmbH was founded in 1984 as an engineering office for numerical field calculations and changed in 1988 to a software and systems house. The heart of its activity is development, marketing, and support of PROFI, a program for numerical field calculation. Its first version was created during the university education of the company's founder in the late 1970s. As of 1995 the company had more than 30 employees, of which about 15 were active in software development, with an annual turnover of about DM 15M.

Product Description. PROFI, also known as PSP, has been used for static and transient magnetic field analysis, including moving parts, harmonic and transient eddy currents; thermal or electrostatic fields; and calculation of current fields. Two-dimensional problems are formulated in terms of the magnetic vector potential; three-dimensional problems use the magnetic scalar potential.

The field problem is discretized by the finite difference method and solved using iterative methods. The program contains a file of user-modifiable data such as the nonlinear material characteristics of the materials. PSP also contains an optimization module. Its mesh generator supports parametric data entry. PSP runs on most well-known UNIX workstations and under VMS.

Outlook. Since 1994 PROFI ENGINEERING SYSTEMS GmbH has been involved in a government-sponsored project to develop a program module to solve transient three-dimensional problems. This addition to PSP is expected to be ready in mid-1996.

Contacts. Further information may be obtained by writing:

PROFI ENGINEERING SYSTEMS GmbH
Herrn Martin Dotterweich
Otto-Röhm-Straße 26
64293 Darmstadt, Germany
Telephone: +49 6151-899195
Fax: +49 6151-895726

6.8. Granville Sewell

PDE2D. PDE2D is a general-purpose partial differential equation solver that handles nonlinear systems of steady-state, time-dependent, and eigenvalue equations in general two-dimensional regions and in three-dimensional boxes. It has been used to solve a wide range of engineering and scientific problems, including many microwave applications. A list of published articles in which microwave problems are solved using PDE2D (or its predecessors) can be obtained from the author at the address below.

Program History. Granville Sewell has been developing PDE2D since 1974; earlier versions, called TWODEPEP and PDE/PROTRAN, were sold by IMSL, Inc. from 1980 to 1991. In 1991, Granville Sewell obtained the rights to PDE/PROTRAN, replaced the high-level PROTRAN language interface with an iteractive interface, and began marketing the new product under the name of PDE2D. Three-dimensional capabilities were added in 1994.

Mathematical Method. Two-dimensional problems are solved using a Galerkin finite element method, with up to fourth-order isoparametric triangular elements, and three-dimensional problems are solved using a collocation finite element method, with tri-cubic Hermite basis functions. The two-dimensional triangular grid can be refined and graded automatically. For problems in two

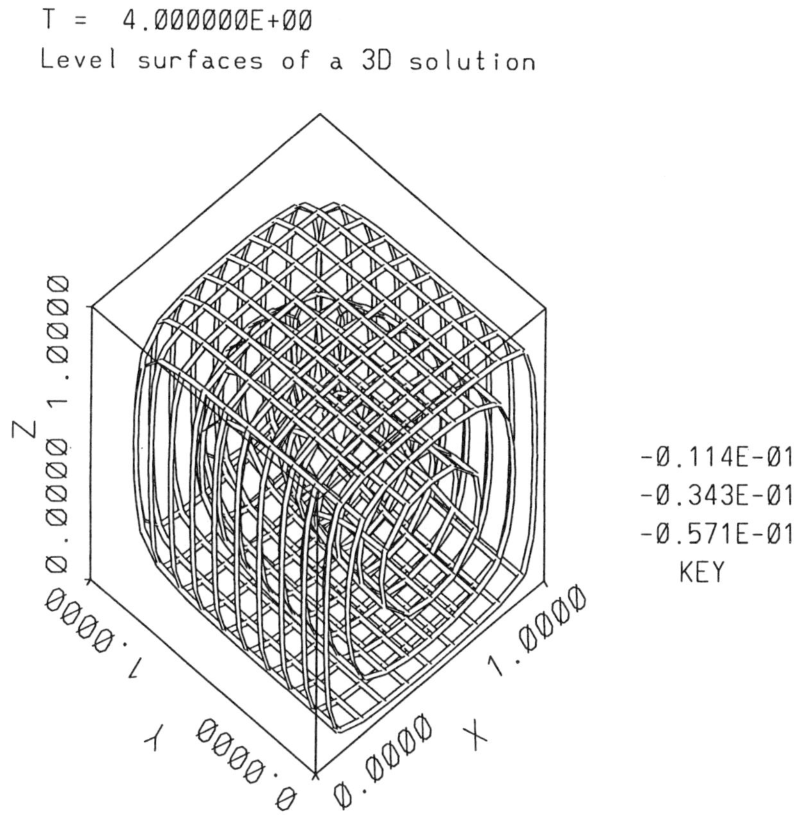

FIGURE 2. Three-dimensional field display (PDE2D): level curves of field solution, drawn suitably for graphic plotting or screen display.

dimensions, the linear systems are solved using a band solver, a frontal method (out-of-core band solver), a biconjugate gradient iterative method, or a minimal degree sparse system solver. Three-dimensional linear systems are solved using a frontal method. Both two- and three-dimensional eigenvalue problems are solved using the inverse power method. Both two- and three-dimensional time-dependent problems have adaptive time step control. Extensive graphical output abilities are provided, including a unique contour surface plotting algorithm for representing three-dimensional solutions.

Ease of use of generality have always been primary design goals for PDE2D, and the interactive interface allows users to construct their PDE2D programs by simply answering a series of questions, without ever consulting a user's manual. In fact, there is no user's manual, as all documentation, including examples, is on-line.

Contact. Further information about PDE2D can be obtained from:

Granville Sewell
Mathematics Department
University of Texas at El Paso
El Paso, TX 79968, USA
Telephone: +1 915-747-6845
Fax: +1 915-747-6502
E-mail: sewell@math.utep.edu

6.9. Sonnet Software, Inc.

Corporate Profile. Sonnet Software focuses closely on the analysis of planar circuits and concentrates on applications in the general microwave area.

Key Software. The Sonnet program uses a Galerkin method of moments on a surface mesh for electromagnetic analysis of three-dimensional planar circuits. The third dimension is represented by vias, since the circuit is embedded in

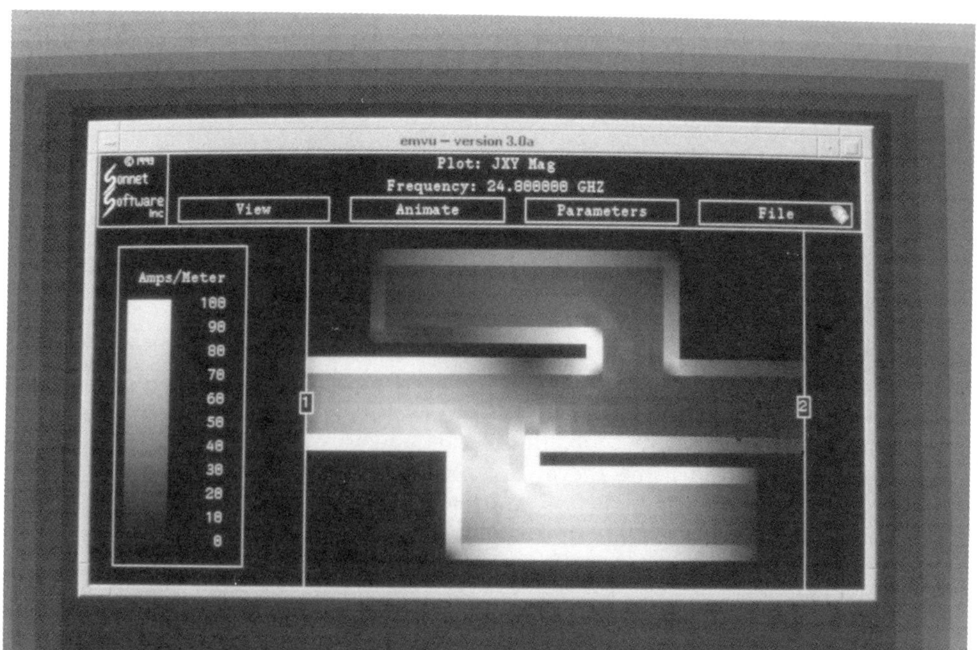

FIGURE 3. Screen display of quasi-two-dimensional microwave structure (Sonnet): field layout. Third dimension is represented by vias.

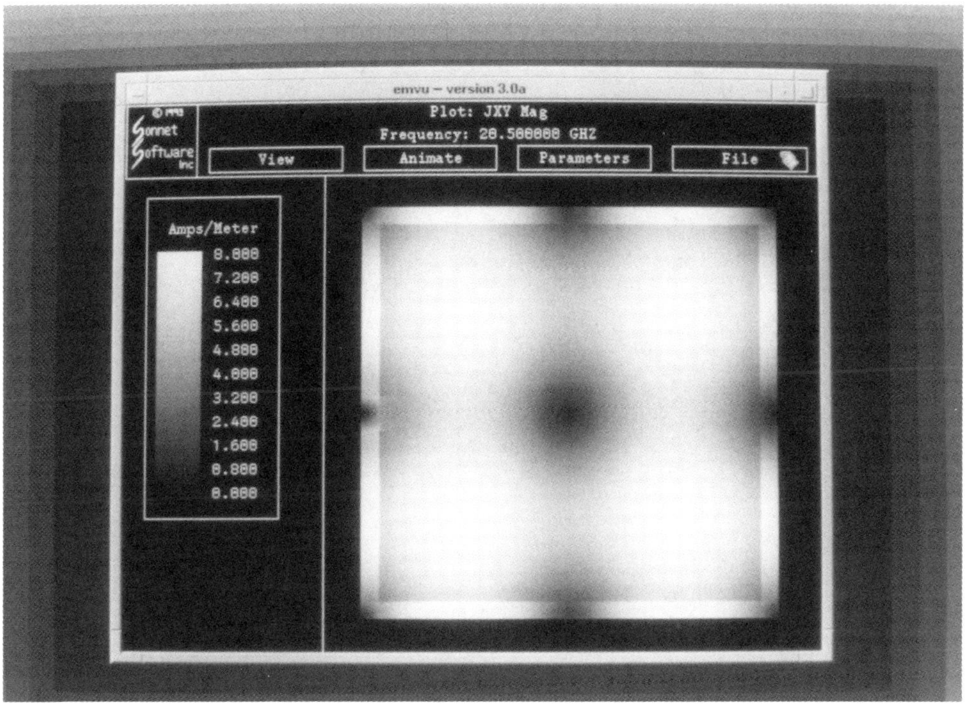

FIGURE 4. Screen display of microwave structure (Sonnet). This form of display is better suited to rapid screen presentation than to printer plots.

layered dielectric; hence it is called planar. Because it uses surface meshing, Sonnet is particularly well suited to planar circuits. On the other hand, Sonnet is inappropriate for arbitrary three-dimensional structures. While its primary application is in microwave and wireless circuit design, applications also include high-speed digital circuits and electromagnetic compatibility problems.

The Sonnet software is modularly constructed; the geometric modeler, electromagnetic field analyzer, and postprocessors are separate programs. Consequently, this package interfaces with other well-known engineering software, including HP/EEsof Series IV, Cadence Analog Artist, OSA automatic optimization software, GDSII Stream, and SPICE, and it recognizes AutoCAD *.DXF files. Input is the circuit geometry either from one of the above sources or input via the Sonnet graphical editor. Output from the field analyzer program takes the form of S-parameters, Y-parameters, Z-parameters, or SPICE netlists, or its output may be directed to graphic postprocessors that plot current distributions or antenna patterns. These may be viewed in full color and animation (sweeping frequency) may be used. The software is available on the IBM-PC and most UNIX workstations.

Contacts

Sonnet Software, Inc.
135 Old Cove Road
Suite 203
Liverpool, NY 13090, USA
Telephone: +1 315 453-3096
Fax: +1 315 451-1694
E-mail: sales@sonnetusa.com
WWW: http://www.sonnetusa.com

Index